FLORA OF RUSSIA

THE EUROPEAN PART AND BORDERING REGIONS

VOLUME VII

MAGNOLIOPHYTA (=ANGIOSPERMAE)
Magnoliopsida (=Dicotyledones)

Editor-in-Chief
N. N. Tzvelev

Translated from Russian

A.A. BALKEMA / ROTTERDAM / BROOKFIELD / 2002

Translation of:
Flora Evropeiskoi Chasti SSSR, tom VII
Nauka Publishers, St. Petersburg, 1994

Translator & General Editor: Dr. V.S. Kothekar

Published by: A.A. Balkema, a member of Swets and Zeitlinger
        Publishers
        www.balkema.nl and www.szp.swets.nl

ISBN (Set)   90 5410 750 2
ISBN (Vol. 7)   90 5410 757 X

Printed in India

4   UDC 582.683.2(470.1/6)

The seventh volume of *Flora of Russia: The European Part and Bordering Regions* contains descriptions of 552 species from 117 genera in the family Asteraceae (family 155 in the list of families, excluding the subfamily Cichorioideae). It includes all the wild as well as major introduced species growing in the European part of Russia and bordering regions, their characteristics, taxonomic status, general description including distribution in the regions of the *Flora*, habitat conditions and chromosome numbers. Complete and sequential typification of all taxonomic categories is the special feature of the volume.

33 illustrations, 1 map.

Editor-in-Chief and Volume Editor
N.N. Tzvelev

Secretary, Editorial Board
D.V. Geltman

Editorial Board
**S.K. Czerepanov, T.V. Egorova,
D.V. Geltman, I.V. Sokolova, N.N. Tzvelev**

Contributors
**S.K. Czerepanov, O.V. Czerneva, D.V. Geltman,
I.A. Gubanov, S.S. Ikonnikov, G.Yu. Konechnaya,
L.I. Krupkina, T.G. Leonova, E.V. Mordak,
V.V. Protopopova, N.N. Tzvelev**

---

*Pagination of the original Russian text—General Editor.

Foreword

The seventh volume of *Flora of Russia: The European Part and Bordering Regions*[1] is devoted to the largest and economically important family of the flora of the USSR—the family of composites (Asteraceae or Compositae), excluding the subfamily Cichorioideae which is included in the next, eighth volume of the *Flora* (already published in 1989). The work on compilation of this volume was entrusted to a group of authors, mostly staff members of the V.L. Komarov Botanical Institute of the Academy of Sciences of the USSR. The following is their individual contribution: D.V. Geltman— genus *Carduus;* S.S. Ikonnikov—genera *Antennaria* and *Leontopodium*; G.Yu. Konechnaya—tribes Senecioneae and Eupatorieae, genus *Saussurea*; L.I. Krupkina—genera *Karelinia, Filago, Logfia,* and *Bombycilaena;* T.G. Leonova—genus *Artemisia;* E.V. Mordak— genus *Serratula;* N.N. Tzvelev—tribes Anthemideae (excluding the genus *Artemisia*) and Astereae, genera *Omalotheca, Filaginella, Gnaphalium, Helichrysum, Xerochrysum, Adenocaulon, Crisium, Picnomon, Lamyra, Carthamus,* and *Cnicus;* S.K. Czerepanov—genera *Centaurea, Acroptilon, Phalacrachena, Hyalea,Crupina,* and *Chartolepis;* O.V. Czerneva—genera *Echinops, Carlina, Xeranthemum, Arctium, Cousinia,* and *Jurinea*. Moreover, I.A. Gubanov (Moscow State University) treated genera *Inula, Pulicaria, Carpesium Pallenis,* and *Telekia*; V.V. Protopopova (Institute of Botany, Academy of Sciences of the Ukrainian SSR)— tribe Heliantheae. The description of the family and its tribes as also the key to the identification of genera is provided by N.N. Tzvelev. In all, the volume includes 552 species from 117 genera.

The classification of composites is still in the process of development and there are different opinions regarding the status of subfamilies and tribes. The subfamily Cichorioideae (=Lactucoideae) is placed at the end of the family by the majority

---

[1]This volume was made ready for the press before the now well-known political changes associated with the dissolution of the USSR and the changed status of the autonomous republics constituting Russia. Hence in the text we have mostly retained the names of politico-administrative units as they stood in 1989– 90.

of authors. However, A.L. Takhtajan in the book *Sistema magnoliofitov* [The System of Magnoliophyta] (1987: 266–272) prefers to place it at the beginning of the family and includes in it not only the tribe Lactuceae which is widely distributed in the region of our *Flora* but also the tribe Cardueae along with the tribes Echinopeae and Carlineae which were separated from it. Takhtajan relates the tribes Eupatorieae, Senecioneae, Calenduleae, Heliantheae, Tageteae, Inuleae, Anthemideae, and Astereae to another subfamily Asteroideae, represented in the European part of Russia and bordering regions. However, in *Flora of Russia: The European Part and Bordering Regions,* the accepted system of the composites is closer to the traditional, considering the fact that the subfamily Cichorioideae in its narrow interpretation has already been published in the eighth volume of this *Flora.* At the head of the system of the family and subfamily Asteroideae we place the tribe Heliantheae (including the tribes Tageteae and Ambrosieae closely related to it). It is characterized by such relatively primitive features as the frequent presence of bracts on the common receptacle, predominantly yellow color of the flowers, relatively less incised pappus of the achenes (except in the anomalous genus *Arnica*) and entire, opposite leaves. The tribes Eupatorieae and Cardueae, which are closest to the subfamily Cichorioideae, have naturally been placed at the end of the subfamily.

6   The family Compositae occupies the first place in the flora of the USSR as well as its European part, in terms of the number of species, although in economic importance it lags way behind such families as the Poaceae and Fabaceae. However, the composites include many economically important species, of which it is sufficient to recall the most important oleiferous plant— sunflower (*Helianthus annuus*), some salad plants (*Lactuca sativa, Artemisia dracunculus, Cichorium endivia*, and others), numerous ornamental plants from the genus of dahalias (*Dahlia*), aster (*Aster* and *Callistephus*), Chrysanthemums (*Chrysanthemum* and *Dendranthema*), daisies (*Bellis*), cone-flowers (*Rudbeckia*), calendulas (*Calandula*), zinnias (*Zinnia*), marigold (*Tagetes*), coreopses (*Coreopsis*), cosmos (*Cosmos*), strawflower (*Xerochrysum*), ageratum or floss-flower (*Ageratum*), and so on, whole range of important medicinal plants of the genera *Artemisia, Inula, Matricaria, Achillea, Gnapalium, Calendula, Rhaponticum,* and others. *Artemisia* species play an important role in the composition of steppe and semidesert phytocenoses. At the same time, this family includes a whole range of obnoxious weeds of fields and

plantations of various crops, in the first instance Canada thistle (*Cirsium arvense* s.l.) and field sow-thistle (*Sonchus arvensis* s.l.). Ecdemic but spreading fast in the European part of Russia and bordering regions are the weeds from the genera *Ambrosia* and *Cyclachaena* which have poisonous pollen. Exozoochorous weeds of the genera *Xanthium* and *Bidens* have spread out very extensively.

As in the earlier volumes, we give below the list of families included in the *Flora*. Also there is the map of the regions of the *Flora* with explanatory text. Illustrations (Plates I–XXXIII) have been prepared by artists O.V. Zaitseva (Plates I–VII, XIII–XVI), S.S. Loseva (Plates VIII–X, XXVII), and A.B. Nikolaeva (Plates XI, XII, XVII–XXVI, XXVIII–XXXIII).

Synonymy for all the taxa included in this volume has been verified by S.K. Czerepanov.

The alphabetic index of the Latin names of plants was prepared by L.S. Krasovskaya and a larger part of the work of preparing the manuscript for the press was handled by I.V. Sokolova.

Contents

List of Families Included in *Flora of Russia: The European Part and Bordering Regions*

1. **Lycopodiaceae**—I, 1974
2. **Huperziaceae**—I, 1974
3. **Selaginellaceae**—I, 1974
4. **Isoëtaceae**—I, 1974
5. **Equisetaceae**—I, 1974
6. **Ophioglossaceae**—I, 1974
7. **Osmundaceae**—I, 1974
8. **Onocleaceae**—I, 1974
9. **Athyriaceae**—I, 1974
10. **Aspidiaceae**—I, 1974
11. **Thelypteriadaceae**—I, 1974
12. **Aspleniaceae**—I, 1974
13. **Blechnaceae**—I, 1974
14. **Hemionitidaceae**—I, 1974
15. **Sinopteridaceae**—I, 1974
16. **Adiantaceae**—I, 1974
17. **Cryptogrammaceae**—I, 1974
18. **Hypolepidaceae**—I, 1974
19. **Polypodiaceae**—I, 1974
20. **Pteridaceae**—I, 1974
21. **Marsileaceae**—I, 1974
22. **Salviniaceae**—I, 1974
23. **Pinaceae**—I, 1974
24. **Cupressaceae**—I, 1974
25. **Taxaceae**—I, 1974
26. **Ephedraceae**—I, 1974
27. Magnoliaceae
28. Lauraceae
29. Aristolochiaceae
30. Nymphaeaceae
31. Ceratophyllaceae
32. Nelumbonaceae
33. Ranunculaceae
34. Berberidaceae
35. Papaveraceae
36. Hypecoaceae
37. Fumariaceae
38. Platanaceae
39. Ulmaceae
40. Moraceae
41. Cannabaceae
42. Urticaceae
43. Fagaceae
44. Betulaceae
45. Myricaceae
46. Juglandaceae
47. Phytolaccaceae
48. Nyctaginaceae
49. Molluginaceae
50. Portulacaceae
51. Basellaceae
52. Illecebraceae
53. Caryophyllaceae
54. Chenopodiaceae
55. Amaranthaceae
56. Polygonaceae
57. Plumbaginiaceae
58. Theligoniaceae
59. Paeoniaceae
60. Hypericaceae
61. Elatinaceae

Regions of *Flora of Russia: The European Part and Bordering Regions*

1. ARCTIC (A.)

1a. Fr.-Joz.	—	Franz-Jozef Land
1b. N.-Zem.	—	Novaya Zemlya Islands
1c. Arct.-Eur.	—	Murmansk Region north of the Nikel—Murmansk-Voron'y—Kanevka—Palka line, Nenets national district

2. NORTH (N.)

2a. Kar.-Murm.	—	Murmansk Region south of the above line, Karelia
2b. Dv.-Pech.	—	Arkhangelsk Region south of the polar circle, Vologda Region, Komi ASSR

3. BALTIC (B.)

3. Balt.	—	Lithuania, Latvia, and Estonia, Kaliningrad Region of Russia

4. CENTER (C.)

4a. Lad.-Ilm.	—	Leningrad, Pskov, Novgorod regions
4b. Up. Dniep.	—	Belorussia; Smolensk, Bryansk regions of Russia
4c. Up. Volg.	—	Tver, Yaroslavl', Kaluga, Moscow, Vladimir regions, Nizhny Novgorod Region west of the Oka and Volga rivers (above their confluence), Ivanovo Region west of the Volga River

| 4d. Volg.-Kam | — | Kostroma, Kirov, Perm regions, Marii and Udmurt ASSR; Ivanovo and Nizhny Novgorod regions north and east of the Volga River; Tataria north of the Volga and Kama rivers; Sverdlovsk and Chelyabinsk regions west of the Kushva—Nizhny Tagil—Sverdlovsk—Zlatoust—Magnitogorsk —Orsk line. |
| 4e. Volg.-Don | — | Tula, Ryazan, Orel, Tambov, Penza, Kursk, Voronezh regions; Chuvash and Morodov ASSR; Nizhny Novgorod Region south of the Oka and Volga rivers (below their confluence); Ulyanovsk, Kuibyshev Saratov regions and Tataria west of the Volga River |

9

5. WEST (W.)

5a. Carpa.	—	L'vov, Transcarpathian, Ivano-Frankovsk and Chernovitsy regions of the Ukraine
5b. Dniep.	—	Volyn', Rovno, Ternopol, Kamenets-Podolsk, Zhitomir, Vinnitsa, Kiev regions; Chernigov, Sumy, Poltava, Kirovograd, Dnepropetrovsk, Kharkov, Lugansk regions of the Ukraine
5c. Mold.	—	Moldavia; Izmail District of Odessa Region of the Ukraine
5d. Bl. Sea	—	Odessa (excluding Izmail District), Nikolaev, Kherson, Zaporozhe, Donetzk regions of the Ukraine

6. EAST (E.)

| 6a. Lower Don | — | Volgograd Region west of the Volga River; Rostov Region; Krasnodar Territory north of the Temryuk—Kropotkin line, Kalmykia |
| 6b. Trans-Volga | — | Saratov, Kuibyshev, Ul'yanovsk Region east of the Volga River; Tataria south of the Kama River; Bashkiria; Orenburg Region (west) |

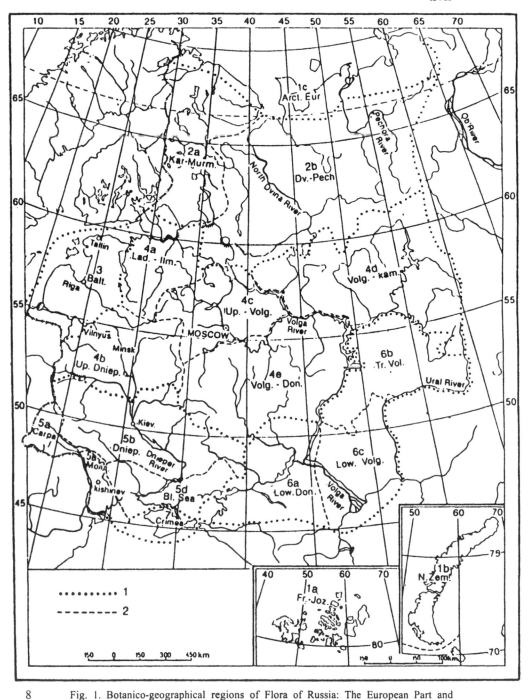

Fig. 1. Botanico-geographical regions of Flora of Russia: The European Part and Bordering Regions. 1—Regional boundaries; 2—Subregional boundaries.

8

Division 5. Magnoliophyta (=Angiospermae)

Class 7. Magnoliopsida (=Dicotyledones)

Order 65. ASTERALES

Includes only one family. A brief description of the family is given below.

FAMILY **155. ASTERACEAE** Dumort.
(**Compositae** Giseke)

The flowers are clustered in an inflorescence—heads surrounded by an involucre of bracts and subtending leaves. All the flowers in a single head are either bisexual (and then heads homogamous) or partly unisexual or sterile (and then heads heterogamous). Rarely the heads are homogamous, comprising only pistillate or only staminate flowers (in such cases plants may be monoecious or dioecious). Usually the flowers in large numbers are disposed on a flat, convex or conical receptacle of the head (for the sake of brevity called simply receptacle) in the axils of scalelike bracts or without them. The calyx tube is fused with the ovary and its lobes are usually modified into a tuft (pappus) comprising one or several rows of membranous scales or bristles, less often having the form of an entire or more or less toothed corona, or is altogether absent. The corolla is gamopetalous, of variable structure, usually tubular and then actinomorphic with (four) five teeth or lobes or ligulate and then zygomorphic with more or less long ligule with three to five teeth, less often tubular-infundibuliform, almost filiform or bilabiate, sometimes (in pistillate flowers) completely absent. Ligulate flowers—ray florets—are usually disposed on the periphery of the head, are of various colors and serve to attract pollinator insects, while the tubular, usually female flowers—disk florets—occupy the central part of the head; in the genera of the subfamily Cichorioideae, all flowers in a head are ligulate. Stamens (four)five, inserted in the corolla tube, alternating with its teeth; their filaments are usually free; the anthers are almost always connate in a tube, introrse, and open longitudinally in the upper part and often also at the base with more or less developed appendages. The gynoecium is syncarpous, comprising two fused carpels; ovary is inferior, unilocular, with one(two) anatropous ovules. The single-seeded indehiscent fruits, called achenes, usually apically bear a tuft (pappus) that facilitates their dispersal, rarely the pappus is absent. Predominantly perennial or annual herbs, less often semishrubs and shrubs; outside the limits of our *Flora*, as also trees with alternate or opposite leaves of various forms.

11   Nearly 1,000 genera and 30,000 species, distributed throughout the world, except for the snow- and ice-covered areas.

Literature: Cronquist, A. 1955. Phylogeny and taxonomy of the Compositae. *Amer. Midl. Natur.* **53**: 478–511.—Carlquist, S. 1976. Tribal inter-relationships and phylogeny of the Asteraceae. *Aliso*, **8**: 465–492.—Wagenitz, G. 1976. Systematics and phylogeny of the Compositae (Asteraceae). *Plant Syst. a. Evol.*, **125**: 29–46.—Wagenitz, G. 1977. New aspects of the systematics of Asteridae. *Plant Syst. a. Evol.*, Suppl. **1**: 375–395.—Cronquist, A. 1977. The Compositae revisited. *Brittonia*, **29**, 2: 137–154.—Jeffrey, C. 1978. Asterales. In: Heywood, V.H. Flowering Plants of the World. Oxford: 263–268.—Bolick, M.A. 1978. Taxonomic, evolutionary and functional considerations of Compositae pollen ultrastructure and sculpture. *Plant Syst. a. Evol.*, **130**, 3–4: 209–218. Grau., J. and H. Hopf. 1985. Das Endosperm der Compositae. *Bot. Jahrb.*, **107**, 1–4: 251–268.—Takhtajan A. 1987. Sistema magnoliofitov [The system of the Magnoliophyta]. Leningrad: 1–439.

1. All flowers in a head ligulate, gradually reduced from periphery to center; ligules at apex with five teeth not always distinct; achenes usually with pappus of scabrous

or plumose setae, rarely of scales or pappus absent ...
... Subfam. 2. **Cichorioideae** (cf. *Flora*, Vol. VIII. 1989).
+ Only peripheral flowers ligulate (but in cultivated species often all flowers), others tubular or all flowers in head tubular (and then peripheral often enlarged, tubular infundibuliform, deeply incised); ligules of peripheral flowers (if ligulate) with two or three not always distinct teeth; achenes with pappus of setae or scales with more or less distinct corona as membranous border at apex of achenes or corona and pappus absent (subfam. 1. Asteroideae) .... 2.
2. All leaves in basal rosette, entire, without spines; stalk of head—scape—without fully developed as also scaly subtending leaves. Heads solitary, with white or pink ligulate flowers .... 3.
+ Stems with one to many developed leaves, less often without fully developed leaves but bearing distant scaly leaves, sometimes heads subsessile in center of leaf rosette, and then leaves more or less incised, spiny................ 4.
3. Ovary and achenes without pappus, rarely with very short (much shorter than achene) pappus .... 77. **Bellis.**
+ Ovaries and achenes with pappus of scarbrous setae more than two times as long as achene ....
.... 79. **Bellidastrum.**
4. Stems with one or several heads appearing in spring before emergence of green leaves and bearing only strongly reduced scaly leaves; vegetative shoots with green leaves appearing only at the end of flowering or after flowering .... 5.
+ Flowering stems with one to many green but often reduced leaves; sometimes plants stemless .... 6.
5. Heads solitary, 20–25 mm in dia, with golden yellow flowers; leaf blade orbicular-cordate not longer than wide .... 33. **Tussilago.**
+ Heads usually more than two on stem, with whitish, pink-purple, or light yellow flowers, rarely in (in arctic *P. gmelinii*) solitary, and then flowers not golden yellow, and leaf blades longer than wide .... 34. **Petasites.**
6. Heads with only one bisexual flower, surrounded by involucre, clustered in very dense and large (more than 20 mm in dia) globose common inflorescence without subtending leaves at base; leaves spiny........ 92. **Echinops.**

4

+ Heads with more than two flowers, rarely one or two flowers and then not clustered in large globose common inflorescences .... 7.

12   7. Heads unisexual, but monoecious: pistillate usually in lower part of flowering branches, with only one or two flowers, without perianth or with weakly developed filiform-tubular perianth, staminate many-flowered with tubular flowers .... 8.

+ Heads usually bisexual, rarely unisexual, and then dioecious (pistillate and staminate on different plants) and all many-flowered .... 9.

8. All or only lower and middle leaves opposite, usually with pinnately divided, less often palmate; three-, to five-lobed blade; pistillate heads monochromatic, small, without spines .... 23. **Ambrosia.**

+ All leaves alternate, usually undivided, more or less toothed, rarely shallow-lobed; pistillate heads one- or two-flowered, at fruiting rather large (more than 10 mm long), more or less covered with spines and terminating in (one) two spines .... 24. **Xanthium.**

9. All or only lower and middle leaves opposite, rarely partly whorled; stem always developed .... 10.

+ All leaves alternate; sometimes plants stemless .... 28.

10. Ovaries and achenes with pappus of more than 10 long and thin setae. Perennials .... 11.

+ Ovaries and achenes without pappus, or with pappus of scales or two to five stiff bristles .... 12.

11. Leaves undivided, partly in basal rosette, cauline in one to three(four) pairs; heads one to seven, 25–80 mm in dia, with yellow or orange-yellow ligulate flowers ....
.... 18. **Arnica.**

+ Leaves cut into three to five lobes or segments, not forming basal rosette, cauline in many pairs; heads numerous, 1.5–2.5 mm in dia, usually with five pinkish or lilac tubular flowers, ligulate flowers absent ....
.... 90. **Eupatorium.**

12. Annuals, 30–100 cm high, throughout densely covered with glandular hairs and hence sticky; leaves entire, lower opposite, upper alternate; heads 15–20 mm in dia, with yellow ligulate flowers .... 16. **Madia.**

+ Plants eglandular or with glandular hairs only in inflorescence .... 13.

13. Ovaries and achenes with two to four stiff persistent bristles at apex, clingy because of downward directed spines; heads 4–30 mm in dia, with or without yellow ligulate flowers. Annuals, with undivided or more or less incised leaves .... 13. **Bidens.**

  + Ovaries and achenes without stiff bristles at apex or non-clinging bristles, rarely with two to four somewhat clingy but readily detaching bristles (and then heads considerably larger and usually with pink ligulate flowers) .... 14.

14. Heads 2–7 mm in dia, without ligulate flowers or with very short ligulate flowers. Annuals with undivided, rarely shallow three-lobed, rather broad leaves .... 15.

  + Heads more than 10 mm in dia, always with well-developed ligulate flowers (in garden forms often all flowers ligulate— heads "double") .... 18.

15. Peripheral flowers ligulate, white, with short three-lobed ligules .... 15. **Galinsoga.**

  + All flowers in head tubular .... 16.

16. Heads 2–5 mm in dia, in paniculate inflorescence; flowers pale green or yellowish-green. Plant (15) 50–150 (200) cm high; upper leaves usually alternate... 22. **Cyclachaena.**

  + Heads 3–7 mm in dia, in corymbose-paniculate or corymbose inflorescence; flowers yellow-azure, pink, or whitish; all leaves opposite .... 17.

17. Involucral bracts glandular-hairy, outer stellate, inner surrounding achene; flowers yellow ...... 8. **Sigesbeckia.**

  + Involucral bracts without glandular hairs or with very short glandular hairs, all adherent to flowers but not surrounding achenes; flowers light azure, pink, or whitish .... 91. **Ageratum.**

18(4). Upper cauline leaves at base connate in pairs by their margins, forming cupuliform involucre around stem; heads 40–80 mm in dia, yellow. Plants 50–250 cm high .... 1. **Silphium.**

  + Upper cauline leaves sessile or petiolate, but their bases not connate in pairs .... 19.

19. All or almost all (except uppermost) leaves pinnately lobed or pinnately divided .... 20.

  + All leaves undivided, entire or more or less toothed ..... 23.

20. Involucre comprising many bracts connate to more than half their length; flowers usually yellow, orange or reddish-brown. Ornamental annuals to 100 cm high ..... 20. **Tegetes.**

+ Involucre of two or more rows of free or only basally connate bracts ....21.

21. Terminal lobe of leaves broadly lanceolate to ovate, toothed; heads 50–200 mm in dia, of various colors. Perennials with tuberously thickened roots .... 12. **Dahlia.**

  + Terminal lobe of leaves narrow-linear to linear-lanceolate, not toothed or with few large teeth; heads to 100 mm in dia. Annuals ....22.

22. Outer involucral bracts deflected sideways; ligulate flowers yellow, orange, dark red or purple-reddish-brown; terminal lobe of leaves 1–5 mm wide .... 11. **Coreopsis.**

  + Outer involucral bracts adpressed to inner; ligulate flowers usually pink, less often white or yellow; terminal lobe of usually bipinnatipartite leaves 0.7–1.3 mm wide .... 14. **Cosmos.**

23(19). All or almost all (except the lowermost petiolate) leaves sessile, semiamplexicaul, broadly lanceolate to ovate, undivided or indistinctly toothed. Annuals ....24.

  + Leaves, except uppermost sessile, petiolate, often with short or auriculately expanded bases ....25.

24. All leaves opposite, sessile; heads solitary, 50–120 mm in dia; ligulate flowers of various colors, usually not yellow .... 2. **Zinnia.**

  + Uppermost leaves often alternate, lowermost petiolate; heads usually rather numerous on stem, 20–50 mm in dia; ligulate flowers yellow .... 10. **Guizotia.**

25. Heads 20–25 mm in dia; ligulate flowers orange or yellow, tubular dark violet, less often yellow. Annuals, 15–25 cm high, with strongly branched, usually procumbent stem; all leaves opposite .... 3. **Sanvitalia.**

  + Heads 25–100 mm in dia; all flowers yellow. Plant 30–250 cm high, with erect stem usually branched only in upper part ....26.

26. Ovaries and achenes glabrous, with very weakly developed corona at apex; ligulate flowers persisting with fruits; all leaves opposite .... 4. **Heliopsis.**

  + Ovaries and achenes more or less hairy, apically with one to four (five) often easily detaching bristles or scales; ligulate flowers not persisting with fruits; upper leaves usually alternate ....27.

27. Achenes strongly flattened, winged on two ribs, with two persistent, stiff bristles in sinus between wings;

heads 25–45 mm in dia. Annuals, 30–80 cm high ....
.... 9. **Ximenesia.**
+ Achenes somewhat flattened, not winged, with one to
four (five) readily detaching bristles or scales at apex;
heads 30–100 mm in dia. Perennials, with tuberously
thickened roots or annuals ....5. **Helianthus** (in part).
28(9). Involucral bracts less numerous, large (considerably longer
then yellow, narrow-ligulate, peripheral flowers), herbaceous, stellately spreading and spinescent. Annuals,
with undivided and entire, soft-hairy leaves; heads 25–
60 mm in dia... 38. **Pallenis.**
+ Involucral bracts not spinescent, if spinescent then peri-pheral
flowers not ligulate and leaves also often spiny.............. 29.
29. Ovaries and achenes hairy, with persistent pappus of 6–12
bristles deflected sideways, in lower half wide, scale-like
and then abruptly terminating in subulate tip; receptacle
with stiff scalelike bracts; heads 40–100 mm in dia, with
ligulate flowers of various colors; disk flowers usually
purple. Ornamental plants........................... 19. **Gaillardia.**
+ Ovaries and achenes glabrous or hairy, without pappus
or with pappus of filiform or scalelike bristles gradually
narrowed toward apex... 30.
30. Involucre of head 11–23 mm in dia, sticky outside but
without glandular hairs, imbricate, usually more or less
deflexed, herbaceous, nonspiny, linear-lanceolate or linear;
peripheral flowers ligulate, yellow; leaves with undivided,
usually more or less toothed blades......... 75. **Grindelia.**
+ Involucre not sticky outside or sticky from glandular hairs,
less often somewhat sticky without glandular hairs, and
then peripheral flowers not ligulate and not yellow .... 31.
31. Ovaries and achenes of all flowers without pappus, but
often with corona as undivided or more or less toothed
border at achene apex, or with pappus of one to four
bristles or one to five membranous scales.................. 32.
+ Ovaries and achenes with pappus of five or more scabrous
or plumose bristles, often more or less flattened or gradually
expanded toward base, pappus sometimes absent in all
or peripheral flowers; also with corona or additional pappus
of membranous scales besides typical pappus of bristles
.... 66.
32. Flowers pink, peripheral tubular but strongly enlarged,
with infundibuliform, deeply five-parted corolla, sterile,

inner tubular; heads rather large (30–50 mm in dia); involucral bracts apically with membranous appendages. Perennials, with undivided but often coarsely toothed or partly pinnatilobate leaves .... 33.

+ Flowers of various colors, peripheral ligulate, usually three-toothed, often with very short ligules, or all flowers in heads tubular, fertile .... 34.

33. All leaves entire; stems usually with one head; appendages of involucral bracts regularly ciliate; achenes obtuse at apex .... 109. **Phalacrachena.**

+ Lower leaves often coarsely toothed or pinnatilobate, less often all leaves undivided and entire; stems often with several heads; appendages of involucral bracts without cilia, undivided but often on margin more or less erose or toothed; achenes at apex acute ....

.... 110. **Centaurea** (sect. *Jacea*).

34. Receptacle covered with deciduous or persistent, often very narrow, scalelike, bracts, but sometimes (in *Anthemis cotula*) bracts present only in middle part of receptacle .... 35.

+ Receptacle glabrous, without scalelike bracts .... 47.

35. Peripheral flowers ligulate, with very short (1.5–2.5 mm) bilobate ligules, persisting with fruits, in axil of broadly obovate, inner involucral bracts and with two sterile tubular flowers on sides, enclosed by bracts; achenes falling with involucral bracts and sterile flowers; heads 4–8 mm in dia, in corymbose or corymbose-paniculate inflorescence. Grayish-tomentose semishrubs, 30–100 cm high ....

.... 21. **Parthenium.**

+ Peripheral ligulate flowers, if present, do not bear on sides two sterile tubular flowers enclosed by bracts; achenes not falling with inner involucral bracts and sterile flowers .... 36.

36. Small heads without ligulate flowers clustered in subglobose glomerule 5–10 mm in dia, in stem dichotomies and terminal on branches. Annuals, 4–15 cm high, grayish from dense pubescence, with undivided and entire leaves .... **Filago** (*F. filaginoides*).

+ Heads not clustered in glomerules .... 37.

37. Heads 3–10 (12) mm in dia (including ligules of peripheral flowers if present), usually numerous and in corymbose or corymbose-paniculate inflorescences; leaves pinnati-lobate to bipinnatipartite .... 38.

15

+ Heads more than 12 mm, less often 7–12 mm in dia, and then all leaves undivided, more or less toothed .... 39.

38. Peripheral flowers ligulate, but with short (0.5–3.6 mm) ligules, pistillate; disk flowers tubular, bisexual. Perennials, with rosulate vegetative branches .... 57. **Achillea.**

   + All flowers in head tubular, bisexual. Semishrubs or perennials, with very narrow pinnatilobate or pinnatipartite leaves .... 58. **Santolina.**

39. Outer involucral bracts herbaceous, but with more or less developed membranous border, distinctly shorter than innermost; leaves glabrous, soft-pubescent or more or less tomentose but never scabrous from short stiff hairs ..... 40.

   + Outer involucral bracts herbaceous, not shorter, often even longer than inner bracts; heads 25–100 mm and more in dia; leaves usually undivided and more or less scabrous from short stiff hairs, less often (in *Rudbeckia lacinniata*) incised and subglabrous .... 43.

40. Peripheral achenes strongly flattened, with two broad-winged ribs projecting above achene apex, others also strongly flattened but not winged. Predominantly cultivated annuals, with strong fragrance ....
   .... 59. **Anacyclus.**

   + Peripheral achenes not flattened or more or less flattened and then not winged .... 41.

41. Achenes strongly flattened, with two prominent ribs, without corona. Perennials, with undivided, more or less toothed leaves, rarely (in alpine *P. tenuifolia*) with strongly incised leaves .... 56. **Ptarmica.**

16 + Achenes not flattened or more or less flattened, and then apically with short corona, corona absent only in annuals and biennials; leaves always more or less incised.... 42.

42. Receptacular bracts acute or cuspidate; base of corolla tube without downward directed processes. Wild plants, with weak, less often strong and then unpleasant odor
   .... 54. **Anthemis.**

   + Receptacular bracts obtuse; base of corolla tube with downward directed processes covering the apex of ovary. Cultivated and occasionally escaped plants, with strong aromatic odor .... 55. **Chamaemelum.**

43(39). Ligulate flowers pink or purple, tubular reddish-reddish-brown; achenes glabrous, with corona 0.5–1.0 mm long more or less toothed .... 7. **Echinacea.**

+ Ligulate flowers yellow, tubular yellow, less often purple
.... 44.

44. Ovaries and achenes hairy at least in upper part, apically
with (one) two (three) readily detaching scales or with
two short bristles; lower leaves often opposite .... 45.

+ Ovaries and achenes glabrous, with or without very short,
unevenly toothed corona at apex; all leaves alternate ..... 46.

45. Ovaries and achenes apically with (one) two (three) readily
detaching lanceolate scales; achenes ribbed, not winged,
usually more or less hairy only in upper part. Annuals or
perennials .... 5. **Helianthus** (in part).

+ Ovaries and achenes apically with two persistent stiff
bristles; achenes strongly flattened, winged on two ribs,
hairy throughout. Annuals ....
.... 9. **Ximenesia** (cf. also couplet 27).

46. Receptacle strongly convex, obtusely conical; leaves more
or less incised or undivided, and then entire or only
upper with few teeth .... 6. **Rudbeckia.**

+ Receptacle weakly convex; leaves undivided, all toothed
.... 37. **Telekia.**

47(34). Heads 13–25 mm in dia, nodding, with subtending leaves
closely approximate to their bases; leaves undivided,
rather broad .... 41. **Carpesium.**

+ Heads not drooping, if nodding then considerably smaller
.... 48.

48. Ligulate flowers yellow of various shades or absent ... 49.

+ Ligulate flowers white, less often pink, always present ... 58.

49. Grayish-tomentose annuals, 5–20(30) cm high, with entire
leaves; heads small (3–4 mm in dia) clustered in twos or
threes in glomerule-like inflorescences 6–15 mm in dia;
all flowers tubular .... 53. **Bombycilaena.**

+ Plant not grayish-tomentose, less often grayish-tomentose,
and then perennials or their leaves more or less incised
.... 50.

50. Ovaries and achenes covered with rather large glandular
hairs and hence sticky; heads without ligulate flowers,
4–6 mm long; mature achenes 7–8 mm long, stellately
arranged. Perennials, 30–80 cm high, glandular-hairy in
inflorescence .... 42. **Adenocaulon.**

+ Ovaries and achenes glabrous, rarely short-hairy, but
without glandular hairs, not sticky .... 51.

51. Heads without ligulate flowers, to 6 mm in dia, often nodding, clustered in paniculate, less often racemose (to subspicate) inflorescences; achenes indistinctly ribbed, without corona; pollen grains smooth .... 74. **Artemisia.**

17   + Heads usually exceeding 6 mm in dia, with ligulate flowers, less often 5–8 mm in dia, without ligulate flowers, and then clustered in corymbose or paniculate-corymbose inflorescences, always erect; pollen grains spinose .... 52.

52. Perennials with more or less long, often branched rhizomes .... 53.

    + Annuals with slender root .... 54.

53. Leaves undivided, more or less toothed ....
.... 69. **Balsamita.**

    + Leaves more or less incised, pinnatilobate to bipinnati-partite .... 70. **Tanacetum.**

54. Heads without ligulate flowers, 4–12 mm in dia, usually rather numerous; corolla of tubular flowers four-toothed .... 61. **Lepidotheca.**

    + Heads with ligulate flowers, 2–70(80) mm in dia, often solitary; corolla of tubular flowers five-toothed .... 55.

55. All or only inner achenes with tubercles or wrinkles in addition to ribs; outer achenes larger, 7–15 mm long ... 56.

    + Achenes ribbed but without tubercles and wrinkles, all 1.8–4.5 mm long .... 57.

56. Achenes 6–15 mm long, rather strongly differing in form and size in the same head, but not of two distinctly different types, peripheral tuberculate, not winged; leaves undivided, entire or finely toothed .... 35. **Calendula.**

    + Achenes in the same head of two distinctly different types: outer 7–8 mm long, strongly flattened, obovate, not rugose, not tuberculate but with two broad wings; inner 3–4 mm long, conical-prismatic, winged, rugose and tuberculate; some leaves undivided and entire, some (usually lower) coarsely toothed ..... 36. **Dimorphotheca.**

57. Ovaries and achenes without corona, with three or two strongly raised, winged ribs in peripheral ligulate flowers; leaves pinnatilobate or pinnatipartite, less often some remotely coarsely toothed .... 64. **Chrysanthemum.**

    + Achenes of all flowers similar, with more or less truncate corona 1.2–1.8 mm long, and with eight or nine weak ribs; leaves undivided, serrate-toothed ....
.... 65. **Coleostephus.**

12

58(48). Achenes of all or only peripheral ligulate flowers trigonous, with three strongly raised, thick or somewhat winged ribs .... 59.

+ Achenes of all flowers not trigonous but often with 5–10 more or less distinct ribs .... 60.

59. Achenes of all flowers trigonous, with three strongly thickened subobtuse ribs and two dark colored glands in upper part of spine. Small perennating or annual wild herbs with leaves incised to very narrow lobes .... 62. **Tripleurospermum.**

+ Only achenes of peripheral flowers trigonous, with three winged ribs, without dark colored glands. Cultivated semishrubs, with leaves more or less incised to broader lobes .... 63. **Argyranthemum.**

60. Cauline leaves undivided, more or less toothed, sessile or subsessile, basal usually also undivided, rarely pinnatilobate, with two to five entire lobes on each side (in the Carpathians, plant 5–20 cm high, with very sparsely leafy stem and solitary heads); heads rather large and often solitary .... 61.

18 + Usually all, less often only middle and lower cauline leaves more or less incised, usually with toothed or incised segments or lobes of first order .... 63.

61. Leaves of basal rosette uniformly pinnatilobate, cauline reduced, undivided, two to five; all achenes with somewhat truncate corona 0.5–1.0 mm long .... 71. **Leucanthemopsis.**

+ All leaves undivided, toothed, less often blades of lower leaves somewhat lobed near base; achenes of disk flowers without corona or with corona of five subobtuse teeth . 62.

62. Peripheral flowers pistillate, fertile. Predominantly meadow plants, usually forming rosettes of basal leaves .... 66. **Leucanthemum.**

+ Peripheral flowers sterile. Plants of marshes, always without rosettes of basal leaves .... 67. **Leucanthemella.**

63. Annuals, with strong aroma; leaves strongly incised, with linear terminal lobes; heads 8–25 mm in dia, with ligules deflexed always from base; achenes without corona .... 60. **Matricaria.**

+ Perennials, not strongly aromatic, rarely (in cultivation) annuals with rather strong aroma, and then terminal lobes of leaves considerably broader .... 64.

64. Ovary and achenes apically with short but distinct corona 0.1–1.5 mm long;s achenes with 5–16 more or less raised ribs; ligulate flowers white .... 68. **Pyrethrum.**

+ Ovary and achenes without corona .... 65.

65. Achenes with five to eight distinct ribs; apical appendages of anthers oblong, broadly rounded. Arctic littoral plants, with somewhat fleshy leaves and white ligulate flowers .... 72. **Arctanthemum.**

+ Achenes with indistinct ribs; apical appendages of anthers lanceolate-ovate, subobtuse. Plants of pine forests, chalk and limestone outcrops or ornamentals (and then often with "double" heads); leaves not fleshy and ligulate flowers of various colors (often pink).............................
.... 73. **Dendranthema.**

66(31). Receptacles covered with scalelike or bristle-like bracts, less often with long hairs .... 67.

+ Receptacles without bracts and long hairs but edges of receptacle alveolae may be slightly raised forming more or less toothed membranous border............................. 89.

67. Receptacle covered with rather long hairs; heads 8–20 mm in dia, clustered in corymbose inflorescence; flowers usually reddish, peripheral pistillate with filiform, nonuniformly four-toothed corona. Perennials, 40–150 cm high, with undivided sessile leaves .... 43. **Karelinia.**

+ Receptacle covered with scalelike or bristle-like bracts; peripheral flowers tubular or tubular-infundibuliform .. 68.

68. Pappus bristles (at least in the majority of flowers of a head) plumose, their hairs more than 0.5 mm long.... 69.

+ Pappus bristles more or less scabrous from spinules or very short (to 0.2 mm) hairs .... 76.

69. Ovaries and achenes short-hairy; pappus bristles uniseriate, at base connate in groups, not detaching. Perennials or biennials, with spiny-toothed leaves .... 93. **Carlina.**

+ Ovaries and achenes glabrous .... 70.

70. Leaves glabrous on margins, hairy or scabrous from very short (to 0.3 mm), stiff hairs or papillae but without stiff bristles more than 0.6 mm long or spinules .... 71.

+ Leaves on margins with spinules or stiff bristles exceeding 0.6 mm .... 73.

71. Flowers yellow, peripheral strongly enlarged; leaves undivided, middle cauline decurrent on stem forming broad wings .... 114. **Chartolepis.**

14

+ Flowers pink, lilac, bluish, whitish; peripheral not enlarged
.... 72.

72. Heads 7–20 mm in dia, almost always many on stem, rarely solitary; leaves undivided, less often more or less incised, often decurrent on stem .... 97. **Saussurea.**

+ Heads 35–60 mm in dia, almost always solitary on stem; leaves undivided, not decurrent .... 107. **Rhaponticum.**

73. Heads 50–120 mm in dia, with bluish or bluish-violet flowers; involucral bracts strongly fleshy near base; receptacle strongly thickened. Cultivated plant with large, weakly spinose, more or less incised leaves ....
.... 104. **Cynara.**

+ Flowers pink, purple, yellowish-white, dirty yellow; involucral bracts not fleshy near base .... 74.

74. Heads 7–15 mm in dia, usually several clustered at apices of stem and its branches and covered with subtending leaves bearing yellowish spines 7–16 mm long; flowers pink. Annual plant of Crimea, with spiny-winged internodes
.... 101. **Picnomon.**

+ Heads not covered with subtending leaves, less often covered, and then subtending leaves not spiny, but with stiff bristles to 55 mm long .... 75.

75. Biennial or perennial plants to 200 cm high, sometimes stemless, with weakly prickly leaves lacking spines or with strongly prickly leaves, and then scabrous above from short stiff bristles .... 100. **Cirsium.**

+ Strongly prickly perennial plant of Crimea, 15–35 cm high; cauline leaves approximate, deeply pinnatipartite, tomentose beneath, glabrous and smooth above, with large yellowish spines at apices of narrow-lanceolate lateral lobes ....
.... 102. **Lamyra.**

76(68). Leaves more or less prickly on margins from prickles or stiff bristles more than 0.7 mm long, very rarely (in *Carthamus tinctorius*) lacking spines; and then leaflike involucral bracts spinescent .... 77.

+ Leaves glabrous and smooth on margins, more or less hairy or scabrous from very short (to 0.5 mm long), stiff papillae or spines; if involucral bracts spinescent, then leaf and often with membranous appendage besides prickles .... 81.

77. Internodes (at least some) prickly-winged; leaves more or less decurrent .... 99. **Carduus.**

+ Stems without wings and spines, not prickly; leaves often amplexicaul but not or scarcely decurrent .... 78.

78. Heads 40–60 mm in dia; cultivated and occasionally escaped plant, strongly prickly, with almost coriaceous and usually white-spotted leaves .... 102. **Silybum.**

+ Heads 10–35 mm in dia (excluding subtending leaves approximate to their base) .... 79.

79. Heads at base not covered with subtending leaves, more or less arachnoid-hairy; outer involucral bracts not leaflike and not longer than next row of bracts; flowers whitish or yellowish. Perennials, 10–30 cm high ... 96. **Cousinia.**

+ Heads with subtending leaves approximate to their bases, rather gradually merging with and longer than involucral bracts. Annuals or biennials, 30–80 cm high .... 80.

80. Achenes obtusely tetragonous, broadly conical, apically acute, with numerous (exceeding 30) strongly flattened, scalelike bristles of same length .... 115. **Carthamus.**

+ Achenes with 10–15 ribs, more or less terete, apically with very short toothed corona and pappus of 10 outer stiff bristles usually exceeding achenes, and 10 inner bristles, less than one-third as long as outer ....
.... 116. **Cnicus.**

81(76). All or almost all involucral bracts terminating in uncinate cusp; heads clingy, 12–35 mm in dia; perennials to 2 m high, with large undivided leaves .... 95. **Arctium.**

+ Involucral bracts apically not uncinate but often with deflexed apices; heads not clingy .... 82.

82. Heads 5–10 mm in dia, with 3–8(10) fertile flowers; ovaries and achenes with biseriate pappus; outer pappus exceeding achenes and comprising numerous scabrous bristles of same length, inner considerably shorter than outer, comprising 5–10 scalelike bristles of same length. Small biennials .... 83.

+ Heads with large number of flowers .... 84.

83. All leaves undivided and entire; ovaries and achenes subglabrous; scalelike bristles of inner pappus apically hairy .... 111. **Hyalea.**

+ Cauline leaves pinnatipartite; ovaries and achenes densely hairy; scalelike bristles of inner pappus glabrous ....
.... 112. **Crupina.**

84. Ovaries and achenes short-hairy, with pappus of 5–15 one-rowed scalelike bristles strongly expanded in lower

20

part and often more or less pink. Grayish- or white-tomentose annuals .... 94. **Xeranthemum.**

+ Ovaries and achenes glabrous or short-hairy with pappus of more numerous bristles not expanded in lower part ... 85.

85. Peripheral flowers strongly enlarged, tubular-infundibuli-form, sterile; disk florets tubular, bisexual, fertile ..... 86.

+ Peripheral flowers not enlarged or slightly enlarged, tubular, fertile, usually bisexual, less often pistillate, rather gradually merging with disk florets .... 87.

86. Ovaries and achenes with many-rowed pappus of bristles of various form; outer and middle bracts of involucre usually apically with membranous, undivided or more or less incised, sometimes prickly appendage, rarely without it. Perennials or biennials, with more or less hairy or glabrous leaves.... 110. **Centaurea** (cf. also couplet 33).

+ Ovaries and achenes with one-rowed pappus of bristles of uniform structure; outer and middle involucral bracts narrowly scabrous on margin, inner with lanceolate, membranous appendage. Annuals, with glabrous or subglabrous leaves .... 113. **Amberboa.**

87. Leaves pinnatipartite to undivided and entire, whitish- or grayish-tomentose beneath............................. 98. **Jurinea.**

+ Leaves of different form, glabrous or more or less hairy beneath, sometimes grayish from rather dense pubescence of crisped hairs but not tomentose .... 88.

88. Outer and middle involucral bracts without apical membranous appendage, but often with stiff cusp deflected sideways. Perennials with short rhizome and more or less incised, less often undivided, leaves ..... 106. **Serratula.**

+ All involucral bracts apically with membranous appendage, terminating in small soft cusp. Perennials with long and strongly branched rhizome; leaves undivided, lanceolate or linear-lanceolate .... 108. **Acroptilon.**

89(66). Tall (40–200 cm) biennial plants with highly prickly leaves and involucres; heads 40–70 mm in dia; flowers all tubular, pink, peripheral large and deeply five-parted .... 103. **Onopordum.**

+ Leaves and involucral bracts not prickly...................... 90.

90. Dioecious glabrous or subglabrous shrubs, 0.5–2.0 m high, with more or less toothed, less often entire leaves; heads 4–7 mm in dia, only with tubular, light yellow or whitish flowers .... 89. **Baccharis.**

+ Herbaceous plants, monoecious, less often dioecious .. 91.
91. Stems and leaves on both sides, less often only beneath, whitish- or grayish-tomentose; heads without ligulate flowers, to 8 mm in dia, clustered in dense corymbose, corymbose-capitate, glomerulate, spicate, or glomerular inflorescence, rarely (in dwarf, to 10 cm high plants) solitary; involucre imbricate in upper part or bracts almost entirely membranous. Plants to 50 (60) cm high, with undivided and entire leaves .... 92.
+ Stem and leaves glabrous or more or less hairy, but not tomentose, rarely tomentose and then either heads with ligulate flowers or involucres with herbaceous bracts almost lacking membranous border .... 100.
92. Annuals, with slender root .... 93.
+ Perennials, with creeping or reduced rhizome............. 96.
93. Involucre cup-shaped or subcylindrical, narrow, few-flowered, lanate-tomentose to half or more; heads clustered in more or less distant, glomerulate inflorescences, more or less spicate in large plants... 94.
+ Involucre bowl-shaped, broader, and many-flowered, more or less hairy only near base; heads clustered in corymbose-capitate inflorescences, in turn forming corymb or corymbose-panicles .... 95.
94. All involucral bracts acute or obtuse, but always without cusp .... 51. **Logfia**.
+ Middle involucral bracts with yellowish cusp 1.0–1.5 mm long .... 52. **Filago** (in part).
95. Subtending leaves present at base of glomerular inflorescences; involucre whitish or brownish ....
.... 50. **Filaginella**.
+ Subtending leaves remote from base of glomerular inflorescences; involucre often with yellowish tinge ....
....48. **Gnaphalium**.
96. Groups of densely clustered heads surrounded at base with rather large stellately spreading subtending leaves ....
.... 44. **Leontopodium**.
+ Common inflorescences lacking stellately spreading subtending leaves .... 97.
97. Heads forming spicate inflorescences, less often (in small arcto-alpine plants up to 10 cm high) solitary, silvery or more or less brownish; leaves often not tomentose above
.... 49. **Omalotheca**.

22

+ Heads forming rather dense corymb or corymbose-paniculate inflorescences .... 98.

98. Dioecious plants, with creeping rhizome and more or less pinkish or whitish heads .... 45. **Antennaria.**

+ Monoecious plants; flowers yellow .... 99.

99. Involucral bracts more or less deflected, almost entirely membranous. Plants with reduced rhizome ....
.... 46. **Helichrysum.**

+ Involucral bracts adpressed, herbaceous, with narrow membranous border only on margin. Plants with long, strongly branched rhizomes ....
.... 82. **Galatella** (section *Chrysocomella*).

100(91). Involucral bracts almost entirely coriaceous-membranous, inner strongly enlarged and usually colored, resembling dry ligules of peripheral flowers; heads rather large, one each at apices of stem and its branches. Cultivated ornamental plant .... 47. **Xerochrysum.**

+ Involucral bracts in large part herbaceous, inner not resembling ligules of peripheral flowers .... 101.

101. Pappus of achenes double, outer comprising row of scales more or less connate at base into corona, inner of numerous longer, scabrous bristles; cauline leaves undivided and usually subentire, sessile and basally amplexicaul .... 40. **Pulicaria.**

+ Pappus of one or many rows of bristles of similar structure
.... 102.

102. Pappus of strongly flattened and basally expanded bristles 2.5–4.0 mm long; mature achenes 2.5–4.0 mm long, black, rather densely hairy, with two or three ribs. Annual ornamental plant, 20–50 cm high, with sessile leaves; heads 40–50 mm in dia, with yellow or whitish ligulate flowers .... 17. **Layia.**

+ Pappus of thin, basally unexpanded (but often connate) bristles; body of achene not black .... 103.

103. Peripheral ligulate flowers yellow, less often orange, often absent and then disk flowers yellow, rarely creamish-yellow .... 104.

+ Peripheral ligulate flowers white, pink, lilac, or azure but disk flowers often yellow; if ligulate flowers absent, then disk flowers not yellow (usually whitish-pinkish, somewhat lilac, rarely yellowish-white)... 110.

104. All flowers in head tubular, bisexual; heads usually rather numerous, in corymbose or corymbose-paniculate

inflorescences; cauline leaves numerous, linear, entire. Perennials with creeping rhizome.....82. **Galatella** (in part).

+ Peripheral ligulate flowers present but often very small and inconspicuous, pistillate or sterile, rarely completely absent, and then plants annual with toothed leaves ..... 105.

105. Achenes of peripheral ligulate flowers without pappus or with pappus of one-rowed bristles; disk flowers with pappus of many-rowed bristles; heads large, solitary or two to eight on one stem; leaves more or less toothed, basal and lower cauline with long petioles and broad blades .... 30. **Doronicum.**

+ All achenes with pappus of similar structure........... 106.

106. Lower and often also middle leaves with long petioles and cordate or ovate blades deeply notched at base; heads 25–40 mm in dia, in racemose or compressed racemose-paniculate inflorescence; involucre of one row of inner bracts and one or two narrower outer bracts. Perennials, 30–130 cm high .... 28. **Ligularia.**

+ Leaf blades not cordate or reniform with deep notch at base .... 107.

107. Involucre of one row of bracts of same form and size, always appressed; heads 15–40 mm in dia, always with ligulate flowers; leaves undivided, entire or more or less toothed .... 27. **Tephroseris.**

+ Involucre many-rowed, comprising bracts of various lengths and forms, less often of same length, and then involucre at base with considerably smaller outer bracts, or some bracts deflected or bent sideways .... 108.

108. Involucre of one row of similar bracts, but at base with several (to 10) much smaller outer bracts considerably differing from them .... 25. **Senecio.**

+ Involucre of many rows of bracts, imbricate or bracts almost equal but in both cases outer gradually merging into inner... 109.

109. Heads solitary or less numerous, more than 20 mm in dia, with large ligulate flowers, less often more numerous and small, with scarcely conspicuous ligulate flowers, clustered in corymbose or corymbose-paniculate inflorescences, and then cauline leaves subentire, elliptical or broadly lanceolate, more or less tomentose .... 39. **Inula.**

+ Heads always numerous and small, to 15 (20) mm in dia, with peripheral ligulate flowers 0.6–8.0 mm long, clustered in racemose, paniculate, less often corymbose inflorescences; all or only lower and middle leaves distinctly toothed, rarely all entire, and then linear or linear-lanceolate .... 76. **Solidago.**

110(103). Ovaries and achenes glabrous; heads without ligulate flowers .... 111.

+ Ovaries and achenes short-hairy, rarely subglabrous, and then with well-developed ligulate flowers .... 114.

111. Stems with one to three shorter leaves and one head; basal leaves evergreen, with long petiole and reniform, almost coriaceous blade .... 32. **Homogyne.**

+ Stems with large number of leaves and more numerous heads in general inflorescences; basal leaves not evergreen .... 112.

112. Annual weeds, 20–200 cm high, with pinnatipartite or undivided and then oblong or lanceolate cauline leaves; flowers dirty yellowish-white, peripheral pistillate, filiform-tubular, inner tubular, bisexual; heads 5–8 mm in dia, clustered in corymbose or corymbose-paniculate inflorescences .... 26. **Erechites.**

+ Forest perennials, with undivided, rather broad leaves; all flowers in head tubular, bisexual .... 113.

113. Lower and middle cauline leaves with deltoid-hastate-blades basally narrowed; heads in paniculate, less often racemose inflorescences; flowers whitish .... 29. **Cacalia.**

+ Lower and middle cauline leaves with cordate blades, deeply notched at base; heads in corymbose-paniculate inflorescences; flowers pink .... 31. **Adenostyles.**

114. Peripheral flowers narrow and small, filiform or ligulate with ligules to 3 mm long; heads 2.5–8.0 mm in dia ..... 115.

+ Peripheral flowers ligulate with ligules more than 5 mm long; heads more than 8 mm in dia .... 118.

115. Outer involucral bracts more than half as long as inner .... 116.

+ Outer involucral bracts one-fourth to two-fifths as long as inner .... 117.

116. Peripheral flowers pistillate, in several rows, with short, obliquely truncate corolla tube; style far exserted from corolla tube. Late-flowering (usually in September) annuals .... 84. **Brachyactis.**

+ Peripheral flowers ligulate but very small and narrow, often with convolute ligules; flowers next to them with obliquely truncate corolla tube. Biennials or perennials flowering in end of spring and summer (May–August) .. .. 86. **Erigeron** (in part).

117. Plants glabrous or subglabrous, often spreadingly branched; flowers pinkish, light azure or pale violet; achenes 1.6–2.0 mm long .... 83. **Conyzanthus.**
   + Plants more or less hairy, usually branched in upper part; flowers whitish, peripheral often with pinkish or lilac tinge; achenes 0.8–1.5 mm long .... 88. **Conyza.**

118. Outer involucral bracts large, next row herbaceous, leaflike; heads 30–130 mm in dia, often "double" (all flowers ligulate). Ornamental annuals .... 78. **Callistephus.**
   + Outer involucral bracts not larger than next row, less often (in some species of *Aster*) larger than inner, and then heads considerably smaller .... 119.

119. Achenes red, of ligulate flowers with pappus of bristles 0.2–0.3 mm long, connate in lower part, in flowers with double pappus consisting of outer row of bristles 0.2–0.3 mm long, connate in lower part and inner row of bristles 1.6–3.0 mm long. Biennials or annuals .... 87. **Phalacroloma.**
   + Achenes of all flowers in head similar, but in peripheral ligulate flowers often not developed .... 120.

120. Peripheral ligulate flowers sterile, usually with undeveloped or scarcely branched rudimentary style .... 82. **Galatella** (in part).
   + Peripheral ligulate flowers pistillate, fertile, with well-developed biparted style .... 121.

121. Perennials, often with stoloniferous shoots .... 122.
   + Biennials or annuals without stolons .... 123.

122. Involucral bracts subacute or obtuse, usually imbricate and strongly unequal, rarely (in *A. alpinus*) almost all equal; peripheral ligulate flowers up to 10, usually in one row .... 80. **Aster.**
   + Involucral bracts gradually pointed, usually almost equal; peripheral ligulate flowers more than 10, often in one and one-half rows .... 86. **Erigeron** (in part).

123. Glabrous or subglabrous halophytes with distinctly fleshy leaves .... 81. **Tripolium.**
   + Short-hairy ecdemic plant with nonfleshy leaves .... 85. **Heteropappus.**

25   SUBFAMILY **1. A S T E R O I D E A E**

All flowers in heads tubular, (four) five-toothed, or peripheral flowers more or less enlarged, ligulate (and then with three teeth on ligule apex) or tubular-infundibuliform (and then usually with deeply parted, five-lobed corolla); sometimes the corolla of peripheral flowers is indistinctly ligulate or with obliquely truncate tube, very rarely (in few- or one-flowered pistillate heads) the corolla is reduced. The achenes have a pappus of scales or bristles, less often with membranous or coriaceous border as a corona or lacking both corona and pappus. These are herbs, less often shrubs (in the tropics also trees), with alternate or opposite leaves, almost always without laticiferous canals in stem, leaves, and roots.

Type: type genus of the family.

TRIBE 1. **HELIANTHEAE** Cass.[1]
(incl. Ambrosieae, Tageteae)

Heads heterogamous, rarely homogamous, sometimes unisexual, and then pistillate with one to a few flowers; peripheral flowers ligulate, yellow, rarely white or pink, sometimes absent; inner (disk florets) tubular, usually yellow. Receptacle with scaly bracts, less often bracts absent. Anthers at base obtuse or sagittate. Style branches seemingly truncate, often apically without stigmatic papillae. Pollen helianthoid type. Achenes with pappus of scales or scalelike bristles, rarely of thin scabrous bristles, often with undivided corona or both corona and pappus absent. Leaves opposite, less often alternate, usually with undivided, less often more or less incised blade.

Type: *Helianthus* L.

GENUS 1. *SILPHIUM* L.
1753, Sp. Pl.: 919; id. 1754, Gen. Pl., ed. 5: 391

Heads heterogamous, 4–8 cm in dia, clustered at apices of stem and its branches, often forming terminal paniculate inflorescences. Involucre broadly bowl-shaped, 16–24 mm long; involucral bracts less numerous (usually five), of same length

---

[1]Treatment by V.V. Protopopova (excluding the genus *Arnica* L.).

and form, ovate or oblong-ovate, herbaceous. Receptacle flat with scalelike bracts. Peripheral flowers ligulate, pistillate, yellow; disk florets bisexual, tubular, yellow. Achenes 10–12 mm long and 7–8 mm wide, strongly flattened and winged on two ribs, dark gray, glabrous, apically notched, without pappus and corona. Perennial plants 50–250 cm high, with erect stem and undivided opposite leaves.

Lectotype: *S. asteriscus* L.

About 15 species, distributed in North America, of which one rather extensively introduced in cultivation.

Literature: Gritsak, Z.I. 1965. Novoe mnogoletnee kormovoe rastenie—silfila pronzonnolistnaya. V. Kn.: *Novye Kormovo-silosnye Rasteniya* [A new perennial fodder plant—cup rosinweed. In: New Fodder and Silage Plants]. Minsk: 86–93.—Medvedev, P.F. 1971. Introduktsiya silfii pronzonnolistnoi v Leningradskoi Oblasti [Introduction of cup rosinweed in Leningrad Region]. *Rastit. Resursy*, **7**, 2: 290–295.

1. **S. perfoliatum** L. 1763, Sp. Pl., ed. 2: 1301; Dobrocz. 1962, Fl. URSR, **11**: 160; Tutin, 1976, Fl. Europ. **4**: 142.

Lower and middle leaves with long, somewhat winged petiole, upper sessile, connate in pairs at base and seemingly enclosing stem; blades ovate or oblong-ovate, remotely sinuate-toothed, more or less scabrous from short hairs.

Type: USA ("in Missisipi").

Center (Ladoga-Ilmen; Upper Dnieper; Upper Volga); *West* (Dnieper; Black Sea).—Occasionally cultivated as an ornamental or fodder plant and as an escape in gardens and parks.—*General distribution*: North America; cultivated in many other extratropical countries.—2n = 14–16.

GENUS 2. *ZINNIA* L.
1759, Syst. Nat., ed. 10, 2: 1189, 1221, 1377, nom. conserv.

Heads heterogamous, 5–12 cm in dia, usually solitary, on stalks thickened below involucre. Involucre bowl-shaped, 15–25 mm long; involucral bracts ovate or oblong, obtuse, imbricate, in many rows, herbaceous, with black border. Receptacle obtusely conical, with scalelike carinately folded bracts. Peripheral flowers ligulate, pistillate, of various colors (pink, lilac, purple or vari-colored); disk florets tubular, bisexual, yellow or orange. Achenes 5–14 mm long, 2 mm wide, more or less flattened, triquetrous to almost flat, with more or less raised (to narrow-winged) lateral ribs, dark gray or brown, glabrous or more or less scabrous, with pappus of two or three bristles or pappus absent. Annual or perennial plants with erect stem and undivided opposite leaves.

Type: *Z. peruviana* (L.) L.

About 20 species, predominantly in the tropics and subtropics and partly also in warm temperate regions of America; several species introduced into cultivation as ornamental plants.

Literature: Andrew, M.T. 1963. Taxonomy of *Zinnia*. *Brittonia*, **15**, 1: 1–25.

1. **Z. elegans** Jacq. 1789, Collect. Bot. **3**: 152; id. 1793, Icon. Pl. Rar. **3**: tab. 589; Vass. 1959, Fl. SSSR, **25**: 532.

Annual, 20–80 cm high, with simple or furcately branched stem; leaves sessile, oblong-ovate, entire, scabrous from very short stiff pubescence.

Type: Mexico ("in Mexico").

North; *Baltic*; *Center*; *West*; *East*; *Crimea*.—Cultivated as an ornamental plant.—*General distribution*: Mexico; cultivated in many other countries of both hemispheres.—2n = 24.

GENUS 3. *SANVITALIA* Lam.
1792, Journ. Hist. Nat. (Paris), **2**: 176, tab. 33

Heads heterogamous, 2.0–2.5 cm in dia, solitary terminal on stem and its branches. Involucre bowl-shaped; involucral bracts herbaceous, almost imbricate, in two or three rows, outer considerably exceeding inner. Receptacle more or less connate, with scalelike bracts. Peripheral flowers ligulate, pistillate, yellow or orange; disk florets tubular, bisexual, dark violet or yellow. Achenes 2.5–3.5 mm long, 1.5–2.0 mm wide, outer flattened, triquetrous-prismatic, yellowish-gray, glabrous, apically with three awnlike bristles, inner more flattened, with narrowly winged, usually short-ciliate lateral ribs, dark gray, with or without one or two awnlike bristles at apex. Annual plants with more or less branched, procumbent or ascending stems and undivided opposite leaves.

Type: *S. procumbens* Lam.

27 Seven species in the southwest of USA, Central and South America; some species occur in other countries as cultivated or ecdemic plants.

1. **S. procumbens** Lam. 1792, Journ. Hist. Nat. (Paris), **2**: 176, tab. 33; Vass. 1959, Fl. SSSR, **25**: 534.

Short-hairy plant 15–30(50) cm high; leaves oblong- or rhombic-ovate, entire, less often more or less toothed, on short petiole; ligulate flowers orange-yellow, tubular dark violet.

Type: Described from plants cultivated in the Paris Botanical Garden and originating from "South America" ("dans l'Amerique meridionale"), but possibly from Mexico.

Baltic; *Center*; *West*; *East*; *Crimea*.—Cultivated as an ornamental plant.—*General distribution*: North America (Mexico, Central America), South America; as cultivated or ecdemic plant in other warm temperate, and subtropical countries.—2n = 16.

GENUS 4. *HELIOPSIS* Pers.
1807, Syn. Pl. 2: 473

Heads heterogamous, 2–8 cm in dia, solitary terminal on stem and its branches. Involucre bowl-shaped or cup-shaped; involucral bracts oblong, herbaceous, of almost similar length, arranged in two or three rows. Receptacle obtusely conical, with scalelike bracts. Peripheral flowers ligulate, pistillate, yellow; disk florets tubular, bisexual, yellow. Achenes 4.0–4.5 mm long, about 1.5 mm wide, all more or less similar, cuneate-prismatic, with scarcely developed sharp edge—corona. Perennial plants with erect stems and opposite leaves with undivided more or less toothed blade on more or less long petiole.

Type: *H. laevis* (L.) Pers.

The genus includes 12 species, distributed in the temperate and tropical regions of America. Of these one is cultivated and as an escape in other extratropical countries.

Literature: Fischer, T.R. 1957. Taxonomy of the genus *Heliopsis* (Compositae). *Ohio Journ. Sci.* 57: 171–191.

1. **H. scabra** Dun. 1819, Mém. Mus. Hist. Nat. (Paris), 5: 56; Kotov, 1949, Bot. Zhurn. Akad. Nauk URSR, 6, 1: 75; Vass. 1959, Fl. SSSR, 25: 535.—*H. helianthoides* (L.) Sweet subsp. *scabra* (Dun.) Fischer, 1957, Ohio Journ. Sci. 57: 190; Hansen, 1976, Fl. Europ. 4: 143.

Stem 50–100 cm high, scabrous, more or less branched; leaf blades ovate or oblong-ovate, serrate-dentate, scabrous from short hairs; heads 20–30 mm in dia.

Type: Cultivated plant originating from the Missouri River ("Garden plants from seeds collected by Lambert on lower Missouri R.").

Baltic; *Center*; *West*; *East*.—Cultivated as an ornamental plant and occasionally an escape (in Kiev and Kherson regions of the Ukraine), found in gardens and parks, near habitations.— *General distribution*: North America, cultivated and as escape in other extratropical countries.—2n = 28.

GENUS 5. *HELIANTHUS* L.
1753, Sp. Pl.: 904; id. 1754, Gen. Pl., ed. 5: 386

28 Heads heterogamous, solitary terminal on stem and its branches, 4–70 cm in dia. Involucre more than 15 mm long; involucral bracts imbricate, in two to five rows but scarcely differing in length, lanceolate to broadly ovate, herbaceous, acuminate, adpressed or deflected sideways, often more or less covering base of peripheral flowers. Common receptacle flat or somewhat convex, with membranous scalelike bracts, often more or less covering base of flowers. Peripheral flowers ligulate, yellow; disk florets tubular, yellow or purple. Achenes oblong, usually tetraquetrous, more or less flattened from sides, with readily detaching pappus of one or two (three) oblong or lanceolate scales. Perennial or annual herbaceous plants with stem simple or branched in upper part; leaves alternate or opposite, entire or more or less toothed, usually short- and stiff-hairy, on long petiole.

Lectotype: *H. annuus* L.

The genus includes 70–110 species, distributed predominantly in North and partly also in South America. Some species are cultivated in many other extratropical countries or occur there as ecdemic weeds.

Literature: Heiser, C.B. 1948. Taxonomic and cytological notes on the annual species of *Helianthus*. *Bull. Torrey Bot. Club*, **75**, 5: 512–515.—Heiser, C.B. 1960. Notes on origin of two ornamental sunflowers *Helianthus multiflorus* L. and *H. laetiflorus* Pers. *Baileya*, **8**: 146–149.—Clevenger, S. and C.B. Heiser, 1963. *Helianthus laetiflorus* and *H. rigidus*—hybrids or species? *Rhodora*, **65**, 762: 121–133.—Heiser, C.B. 1969. The North American sunflowers (*Helianthus*). *Mem. Torrey Bot. Club*, **22**, 3: 1–298.— Barbaricz, A.I. and O.N. Dubovik, 1972. Novi vidi sonyashnika (*Helianthus* L.) na Ukraine (New species of sunflower [*Helianthus* L.] in the Ukraine). *Ukr. Bot. Zhurn.*, **29**, 5: 647–650.—Skvortsov, A.K. 1973. Novye dannye ob adventivnoi flore Moskovskoi oblasti, II [New data on the adventive flora of Moscow Region, II]. *Byull. Glavn. Bot. Sada*, **88**: 30–34.—Ignatov, M.S. 1986. Dopolnenie k

27

advenstivnoi flore Dal'nego Vostoka [Addendum to the adventive flora of the Far East]. *Bot. Zhurn.*, **71**, 8: 1133–1134.

1. Annuals; leaves usually alternate .... 2.
+ Perennials, with tuberous thickenings on stolons ...... 4.
2. Tubular flowers yellow; heads (10)15–40(70) cm in dia, usually solitary, less often many, and then the main head larger than others... 2. **H. annuus.**
+ Tubular flowers purple; heads 4–15 cm in dia, usually many, of which the main head not larger than others........... 3.
3. Involucral bracts lanceolate, without cilia at base; leaf blades broadly lanceolate or lanceolate-ovate, rarely lower leaves somewhat cordate .... 1. **H. petiolaris.**
+ Involucral bracts ovate or oblong-ovate, at base more or less ciliate along margin; leaf blades more or less ovate, lower leaves cordate .... 3. **H. lenticularis.**
4. Stems glabrous, less often weakly scabrous from very short hairs .... 5.
+ Stems short-hairy or scabrous from rather dense stiff hairs .... 6.
5. Leaves, except the uppermost, opposite, lanceolate, at base narrowed into indistinctly demarcated petiole 0.5–1.0 cm long, entire or weakly toothed; involucral bracts as long as tubular flowers or slightly exceeding them, adpressed, not leafy, broadly lanceolate, dorsally glabrous .... 4. **H. strumosus.**
+ Leaves alternate in upper part of stem, opposite in middle and below, lanceolate to broadly ovate, rounded at base or decurrent on petiole 2–5 cm long, deeply, but often weakly, toothed; involucral bracts distinctly exceeding tubular flowers, often recurved, leafy, linear-lanceolate, scatteredly hairy...................................5. **H. decapetalus.**
6. All leaves, or except uppermost opposite, lanceolate to lanceolate-ovate, entire to coarsely toothed, short-petiolate or subsessile; involucral bracts shorter than disk florets or as long, adpressed.. 7.
+ Leaves in upper and often middle part of stem alternate, oblong-ovate to ovate, more or less cordate at base, more or less toothed, with distinct petioles; involucral bracts exceeding disk florets, more or less deflected sideways.. 8.
7. Lower and middle leaves opposite, short-petiolate, oblong-lanceolate to ovate-lanceolate, with broadly cuneate base,

29

more or less serrate-dentate; upper leaves alternate, smaller, entire, subsessile; heads usually many; involucral bracts two- or three-rowed, ovate-lanceolate or oblong, dorsally more or less pubescent, rarely subglabrous, ciliate on margin; disk florets yellow .... 6. **H. laetiflorus.**

+ Almost all leaves (except two very small below head) opposite, subsessile, lanceolate to lanceolate-ovate, with narrow cuneate base, entire to more or less toothed; heads solitary; involucral bracts imbricate, ovate, dorsally glabrous, less often scatteredly pubescent, puberulent on margin; disk florets with yellow tube and dark red lobes, sometimes pubescent outside .... 7. **H. rigidus.**

8. Stems densely leafy; leaves to 15 cm wide, scatteredly hairy, with short stiff hairs on both sides; involucral bracts moderately hairy; tubers large, globose or ovoid, on very short stolons, clustered at base of stem ....

....8. **H. tuberosus.**

+ Stems moderately leafy; leaves to 8 cm wide, densely hairy beneath with thin soft hairs; involucral bracts densely hairy and ciliate; tubers small, pyriform or almost fusiform, on short stolons, not clustered at base of stem ....

....9. **H. subcanescens.**

1. **H. petiolaris** Nutt. 1821, Journ. Acad. Sci. Philadelphia, **2:** 115.

Type: North America ("... ad ripas fl. Arkansas").

Baltic (vicinity of Vilnius); *Center* (Volga-Kama: Udmurtia); *West* (Black Sea: vicinity of Odessa).—Endemic by roadsides.— *General distribution*: North America; ecdemic in other countries.— 2n = 54.

Note. A weed of quarantine significance. Spotted in 1963 in Odessa, later disappeared.

2. **H. annuus** L. 1753, Sp. Pl.: 904; Vass. 1959, Fl. SSSR, **25:** 542; Hansen, 1976, Fl. Europ. **4:** 141.

Type: Peru, Mexico ("in Peru, Mexico").

North; *Baltic*; *Center*; *West*; *East*; *Crimea*.—Cultivated as an oil crop or fodder plant and often found as incidental ecdemic plant by roadsides and in habitations.—*General distribution*: North America; widely cultivated and sometimes as escape in many other extratropical countries.—2n = 34.

Note. Introduced in Europe (Madrid) in the year 1510, in Russia in the 17th century.

3. **H. lenticularia** Dougl. ex Lindl. 1829, Bot. Reg. **15**: tab. 1265; V.V. Nikit. 1983, Sorn. Rast. SSSR: 364.—*H. annuus* L. subsp. *lenticularis* (Dougl. ex Lindl.) Cockerell, 1914, Science, **11**, 4: 284; Safonov, 1986, Novosti Sist. Vyssh. Rast. **23**: 252, subvar.

Type: West of North America ("n. W. coast of Am., and in the interior about the Columbia, and to the south of that river").

Baltic; *Center* (Volga-Kama); *East*.—As an ecdemic by roadsides, in habitations, fields, forest nurseries.—*General distribution*: North America; ecdemic in many other extratropical countries (in the former Soviet Union: south of Western and Eastern Siberia, Far East).—2n = 34.

Note. A weed of quarantine significance.

30   4. **H. strumosus** L. 1753, Sp. Pl.: 905; A. Skvortsov 1973, Byull. Gl. Bot. Sada Akad. Nauk SSSR, **88**: 34; Bochkin, 1989, Probl. Izhch. Advent. Fl.: 37.

Type: Canada ("in Canada").

Center (Upper Dnieper: southeast; Upper Volga: Moscow Region); *East* (Lower Don: Ilovya District); *Crimea*.—Cultivated as an ornamental plant, sometimes as escape, found in habitations, by roadsides.—*General distribution*: North America; cultivated and as escape in countries of Europe.—2n = 68, 102.

5. **H. decapetalus** L. 1753, Sp. Pl.: 905; Dobrocz. 1962, Fl. URSR, **11**: 176; Fodor and Yantso, 1980, Fl. Rastit. Ukr.: 192.

Type: Canada ("in Canada ex semine").

West (Carpathians: Uzhgorod; Dnieper: Kiev).—Cultivated as an ornamental plant, sometimes as escape, found near roads, in habitations, on banks of water bodies.—*General distribution*: North America; cultivated and as escape in other extratropical countries.—2n = 34, 60.

6. **H. laetiflorus** Pers. 1807, Syn. Pl. **2**: 476; Barbarich and Dubovik, 1972, Ukr. Bot. Zhurn. **29**, 5: 648; Hansen, 1976, Fl. Europ. **4**: 141.

Type: Described from cultivated plants of North American origin ("Garden specimens, originating from N. Am.").

Baltic (vicinity of Kaunas); *Center* (Upper Dnieper: Brest Region; Upper Volga: Moscow Region, possibly also Volga-Don); *West* (Carpathians: vicinity of Uzhgorod; Dnieper: vicinity of Kiev and Oster; Black Sea: vicinity of Odessa).—Cultivated as an ornamental plant and as escape, found in habitations, by roadsides.—*General distribution*: North America; cultivated and as escape in other countries (in the former Soviet Union: Far East).—2n = 102.

Note. Evolved through hybridization of *H. rigidus* × *H. tuberosus*, but at the present has established as a hybridogenic species, relatively more extensively distributed than *H. rigidus*.

7. **H. rigidus** (Cass.) Desf. 1829, Cat. Pl. Horti. Paris, ed. 3: 184; Dobrocz. 1962, Fl. URSR, **11**: 177; Ignatov, 1986, Bot. Zhurn. **71**, 8: 1133.—*Harpalium rigidum* Cass. 1818, Bull. Soc. Philom. Paris, 1818: 141.

Type: Described from plants grown in the Paris Botanical Garden and originating from North America ("...de l'Amerique septentrionale").

Baltic (vicinity of Kaunas); *West* (Carpathians: vicinity of Uzhgorod; Dnieper: vicinity of Kiev).—Cultivated as an ornamental plant, sometimes as escape, found in habitations, by roadsides.— *General distribution*: North America; cultivated and as escape in other extratropical countries (in the former Soviet Union: Far East).—2n = 102.

Note. Hybridizes with other species. In the vicinity of Uzhgorod the hybrid *H. subcanescens* × *H. ridigus* has been found.

8. **H. tuberosus** L. 1753, Sp. Pl.: 905; Vass. 1959, Fl. SSSR, **25**: 544; Hansen, 1976, Fl. Europ. **4**: 141; A. Skvorts. 1973, Byull. Gl. Bot. Sada Akad. Nauk SSSR, **88**: 34.—(Plate I, 1).

Type: Brazil ("in Brasilia").

Baltic; *Center*; *West*; *East*; *Crimea*.—Cultivated as a fodder, technical, food, and ornamental plant; sometimes as escape, found in habitations, by roadsides.—*General distribution*: North America; as a cultivated or more or less naturalised plant—Caucasus, Western and Eastern Siberia (south), Far East (south), Russian Central Asia; Scandinavia (south), Central and Atlantic Europe, Mediterranean, Asia Minor, Japan-China, India—Himalayas; South America, Australia, Africa.—2n = 102.

Note. Introduced in Europe in the 17th century; in Russia in the 19th century.

31   Plate I.

1—*Helianthus tuberosus* L.; 2—*Bidans radiata* Thuill.; 3—*B. tripartita* L., head, 3a—achene; 3b—receptacular scale; 4—*B. cernua* L., head, 4a—leaf, 4b—achene; 5—*B. frondosa* L., achene; 6—*B. tripartita* L. var. *orientalis* (Velen.) Stojan. et Steph.; 7—*Galinsoga parviflora* Cav., 7a— achene, 7b—receptacular scale; 8—*G. quadriradiata* Ruiz et Pavl., achene, 8a—receptacular scale.

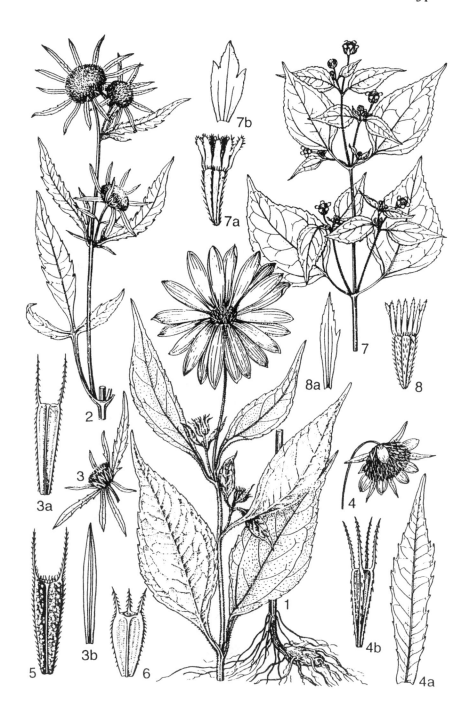

9. **H. subcanescens** (A. Gray) E.E. Wats. 1929, Pap. Mich. Acad. Sci. **9**: 430; Barbarich and Dubovik, 1972, Ukr. Bot. Zhurn. **29**, 5: 647; Alekseev and Makarov, 1977, Byull. Mosk. Obl. Isp. Prir. Otd. Biol. **82**, 6: 91.—*H. tuberosus* L. var. *subcanescens* A. Gray, 1884, Syn. Fl. North Am. **1**, 2: 280.

Lectsotype: USA, Minnesota ("Minnesota").

Baltic (vicinity of Kaunas); *Center* (Upper Volga: vicinity of Moscow; Upper Dnieper; Volga-Don); *West* (Carpathians; Dnieper).—Cultivated as an ornamental and fodder plant, sometimes as escape, found in habitations, by roadsides.—*General distribution*: North America; cultivated and as escape in other extratropical countries.—2n = 102.

Note. More extensive distribution of this species is not ruled out because often collectors confuse it with the previous species. For the Ukraine, *H. salicifolius* A. Dietr. is also reported, which is grown in the Kiev and L'vov botanical gardens.

GENUS 6. *RUDBECKIA* L.
1753, Sp. Pl.: 906; id. 1754, Gen. Pl., ed. 5: 387

Heads heterogamous, 5–10 cm in dia, solitary terminal at apices of stem and its branches. Involucre cup-shaped, biseriate, consisting of herbaceous, more or less pubescence bracts of almost same length. Receptacle obtusely conical, with scalelike bracts. Peripheral flowers sterile, yellow; disk florets tubular, bisexual, fertile, yellow or purple. Achenes tetragonous-prismatic, glabrous, almost black, lacking pappus but often with very short irregularly toothed corona. Perennial or biennial herbaceous plants, with erect, usually more or less branched stems, alternate (in the former Soviet Union also opposite) undivided, lower pinnatisect and upper ternate or trilobate leaves.

Lectotype: *R. hirta* L.

About 25 species, distributed in North America; some species cultivated as ornamental plants, sometimes as escape.

Literature: Perdue, R.E. 1957. Synopsis of *Rudbeckia* subgenus *Rudbeckia*. Rhodora, **59**, 708: 293–298.—Perdue, R.E. 1959. The somatic chromosomes of *Rudbeckia* and related genera of the Compositae. *Contr. Gray Herb.*, **185**: 129–162.

    1. Perennial, glabrous plants, 40–150 cm high; lower and middle leaves pinnatipartite, upper ternate or three-lobed, on shorter petiole, uppermost undivided, subsessile; outer

involucral bracts ovate-lanceolate; glabrous, as long as tubular flowers or shorter; achenes 4–5 mm long, with small irregularly toothed corona .... 1. **R. laciniata.**
+ Biennial, rather densely hairy plants, 20–50 cm high, with undivided, entire or more or less toothed leaves, lower petiolate, upper sessile; outer involucral bracts lanceolate, densely pubescent, exceeding tubular flowers; achenes 2.0–2.7 mm long, almost tetragonous, without corona.. .... 2. **R. hirta.**

1. **R. laciniata** L. 1753, Sp. Pl.: 906; Vass. 1959, Fl. SSSR, **25:** 540; Hansen, 1976, Fl. Europ. **4:** 141.

Type: North America ("in Virginia, Canada").
North (south); *Baltic*; *Center*; *West*; *East*; *Crimea.*—Cultivated as an ornamental plant and sometimes as escape found in habitations, by roadsides, on riverbanks.—*General distribution*: North America, cultivated and as escape in many other countries of both hemispheres.—2n = 36, 38, 54, 72, 76, 64–79, 102.

2. **R. hirta** L. 1753, Sp. Pl.: 907; Vass. 1959, Fl. SSSR, **25:** 539; Hansen, 1976, Fl. Europ. **4:** 141.

Type: North America ("in Virginia, Canada").
Baltic; *Center*; *West*; *East*; *Crimea.*—Cultivated as an ornamental plant and often as escape; found in habitations, by roadsides, on riverbanks, in fields and meadows.—*General distribution*: North America; cultivated and as escape in many other extratropical countries of both hemispheres (in the former Soviet Union: Far East).—2n = 38.

33   *Note.* In southern and southwestern regions of our *Flora*, occasionally cultivated as ornamental plants are three more species: *R. speciosa* Wender—perennial plant with multiradiate heads to 4 cm in dia and undivided, coarsely sinuate-toothed leaves; *R. triloba* L.—with small heads, clustered in corymbose inflorescence; and *R. umbrosa* Boynton ex Beadle—pubescent perennial with undivided, ovate, finely toothed leaves and large solitary heads.

GENUS 7. *ECHINACEA* Moench
1794, Meth. Pl.: 591

Heads heterogamous, to 10 cm in dia, on long stalks. Involucre hemispherical, two- to four (five)-seriate. Receptacle obtusely conical, with stiff lanceolate-subulate bracts exceeding tubular flowers. Peripheral flowers sterile, ligulate; disk florets bisexual, tubular. Achenes broadly cuneate, flattened-tetragonous, glabrous, dark brown, apically with irregularly toothed corona 0.5–1.0 mm long. Perennial herbaceous plants, with erect stems and alternate undivided but usually more or less toothed (to entire) leaves, of which the lower and middle petiolate and upper usually sessile.

Type: *E. purpurea* (L.) Moench.

The genus includes three to nine species, distributed in the eastern and central parts of North America; one of them is cultivated in Europe as an ornamental plant.

Literature: Denisova, E.K. 1961. K farmakognozii ekhinatsei (rudbekii) purpurnoi [On the pharmacognosy of purple coneflower (*Rudbeckia*)]. *Uch. Zap. Pyatigorsk. Farm Inst.*, **5**: 95–98.—McGregor, R.L. 1968. The taxonomy of the genus *Echinacea* (Compositae). *Univ. Kansas Sci. Bull.*, **48**, 4: 113–142.—Menshova, V.A. et al., 1957. Medonosnoe znachenie *Echinacea purpurea* (L.) Moench, introdutsirovonnoi na Ukrain [Importance of *Echinacea purpurea* (L.) Moench, introduced in the Ukraine for honey]. *Rast. Resursy*, **2**, 4: 612–616.

1. **E. purpurea** (L.) Moench, 1794, Meth. Pl.: 591; Vass. 1959, Fl. SSSR, **25**: 541.—*Rudbeckia purpurea* L. 1753, Sp. Pl. **2**: 907.

Plant 60–150 cm high, with more or less branched stem. Leaves scabrous, basal with ovate blade on long winged petiole, cauline lanceolate or oblong-ovate, sessile or subsessile. Heads nodding, with pink ligulate and reddish-reddish-brown tubular flowers.

Type: USA, states of Virginia and Carolina ("in Virginia, Carolina").

Baltic (south); *Center* (Upper Dnieper; Upper Volga; Volga-Don); *West*; *East*; *Crimea*.—Cultivated as an ornamental and medicinal plant.—*General distribution*: North America; cultivated in many other extratropical countries.—2n = 22.

GENUS 8. *SIGESBECKIA* L.
1753, Sp. Pl.: 900; id. 1754, Gen. Pl., ed. 5: 383

Heads heterogamous, about 5 mm in dia, stalked, clustered in corymbose inflorescence. Involucre campanulate, biseriate; outer bracts five, linear-spatulate, distinctly exceeding inner, patent, densely glandular-hairy outside and on margin, sticky; inner obovate, less densely hairy. Receptacle flat, with membranous bracts enclosing achenes. Flowers yellow; peripheral ligulate, pistillate; disk florets tubular, bisexual. Achenes elongate-cuneate, tetragonous, without pappus. Annual plants, with erect stem and opposite undivided leaves.

Lectotype: *S. orientalis* L.

Six species, distributed predominantly in the tropics and subtropics of both hemispheres, one of them ecdemic in many countries of the temperate zone and partly naturalized there.

34   *Literature*: Brummitt, R.K. 1967. Three species of *Sigesbeckia* in Britain. *Proc. Bot. Soc. Brit. Isles*, 7: 19–21.

1. **S. orientalis** L. 1753, Sp. Pl.: 900; Vass. 1959, Fl. SSSR, **25**: 536; Dobrocz. 1962, Fl. URSR, **11**: 166–168; Brummitt, 1976, Fl. Europ. **4**: 140.

Plant 30–100 cm high, appressed-hairy, with more or less branched stem; leaves ovate-rhombic, coarsely toothed, grayish from dense pubescence; inflorescence lax of small heads.

Type: China ("in China, Media ad pagos").

West (Dnieper: ecdemic near station of Semenovka, Chernigov Region); *Crimea* (southern coast).—As ecdemic plant of roadsides, in habitations, more or less weedy meadows, on gravel-beds and clayey slopes.—*General distribution*: Caucasus, Far East, Russian Central Asia; Central Europe (south of Romania), Mediterranean, Japan-China; North America, Australia.—2n = 30 (60).

GENUS **9.** *XIMENSIA* Cav.
1793, Icon. Descr. Pl. 2: 60, tab. 178

Heads heterogamous, 2.5–4.5 cm in dia, solitary at apices of stem and its branches. Involucre of head hemispherical; its bracts linear-lanceolate, acuminate, short-hairy, outer longest, herbaceous, adpressed. Common receptacle convex with deciduous scalelike bracts. All flowers fertile, yellow; peripheral ligulate, pistillate, disk, florets tubular, bisexual. Achenes oblong, strongly flattened, about 6 mm long, 3 mm wide, short-hairy, with four ribs, winged on two lateral ribs, apically with two short bristles in sinus between wings. Annual plants, with erect stem and alternate (except the lowermost opposite), undivided, irregularly toothed, leaves with non-winged or short-winged petioles.

Type: *X. encelioides* Cav.

Four species in the tropical and temperate regions of America. One ecdemic in Europe.

36

1. **X. encelioides** Cav. 1793, Icon. Descr. Pl. **2**: tab. 178; Dobrocz. 1958, Bot. Zhurn. Akad. Nauk URSR, **15**, 3: 87.—*Verbesina encelioides* (Cav.) Benth. et Hook. fil. ex A. Gray, 1876, in Brewer, Watson, and Gray, Bot. California, **1**: 350; Hansen, 1976, Fl. Europ. **4**: 142; Kondratyuk and others, 1987, Ukr. Bot. Zhurn. **44**, 3: 26.

Plant 30–80 cm high, with more or less branched stem, usually grayish from dense pubescence; leaf blades rhombic- or lanceolate-ovate.

Type: Cultivated plant originating from Mexico ("in Mexico, undo introducta in Regnum hortum Matritensem").

West (Dnieper: Kiev and Dnepropetrovsk regions; Black Sea).—As ecdemic weed of roadsides, in habitations, on riverbanks.—*General distribution*: North America; ecdemic in other countries and often naturalized (Caucasus; Atlantic and Central Europe, Africa).—2n = 34.

GENUS 10. *GUIZOTIA* Cass.
1829, Dict. Sci. Nat. **59**: 237, 247, 248, nom. conserv.

Heads heterogamous, 2–5 cm in dia, clustered in lax paniculate inflorescences. Involucre campanulate, biseriate. Receptacle convex or conical, covered with scalelike bracts. Flowers yellow, peripheral ligulate, pistillate, disk florets tubular, bisexual. Achenes small, cuneate, dark, without pappus. Annual herbs, with erect, branched stem and undivided opposite (or upper alternate), leaves.

Type: *G. abyssinica* (L. fil.) Cass.
About 12 species in tropical Africa.
Literature: Kanevskaya, G.S. 1971. Biologicheskie predposylki selektsii novoi maslychnoi kultury gizotsii—*Guizotia abyssinica* (L. fil.) Cass. [Biological basis of breeding a new oil crop of ramtil *Guizotia abyssinica* (L. fil.) Cass.]. Author's Abstr. Kand. [Ph.D.] Diss. Voronezh: 1–25.—Baage, J. 1974. The genus *Guizotia* (Compositae). A taxonomic revision. *Bot. Tidsskr.* **69**: 1–39.

1. **G. abyssinica** (L. fil.) Cass. 1829, Dict. Sci. Nat. **59**: 248; Vass. 1959, Fl. SSSR, **25**: 545; Hansen, 1976, Fl. Europ. **4**: 139.—*Polymnia abyssinica* L. fil. 1781, Suppl.: 383.—*Guizotia oleifera* DC. 1836, Prodr. **5**: 551.

Plant 30–150 cm high, with sessile semiamplexicaul (lower petiolate), lanceolate-ovate or lanceolate leaves and more or less large heads on more or less long pubescent stalks. Outer involucral bracts ovate or broadly oval, herbaceous, inner narrower, scaly.

Type: Ethiopia ("in Abbyssinia").

Baltic (Kaliningrad Region, ecdemic); *West* (Black Sea: in cultivation).—Cultivated as an oil crop and sometimes as escape—*General distribution*: Africa (mountains of Ethiopia); cultivated and as escape in Central Europe, Mediterranean, Asia (India—Himalayas); North America, Africa.—2n = 30.

Note. In Ethiopia introduced in cultivation in 2500 B.C.

GENUS 11. *COREOPSIS* L.
1753, Sp. Pl.: 907; id. 1754, Gen. Pl., ed. 5: 388

Heads heterogamous, 2.5–6.5 cm in dia, solitary terminal on stem and its branches, forming corymbose or paniculate inflorescence. Involucre of head hemispherical, biseriate, with basally connate bracts, outer herbaceous, inner more or less membranous, erect, longer. Peripheral flowers ligulate, pistillate but usually sterile; disk flowers with narrow tube and expanded five-toothed limb, bisexual, fertile. Annual or perennial plants with erect, more or less branched stems and opposite (sometimes upper alternate), pinnatisect, less often undivided leaves.

Lectotype: *C. lanceolata* L.

About 50 species in warm temperate, subtropical, and tropical regions of America; several species introduced into cultivation as ornamental plants.

Literature: Sherff, E.E. 1936. Revision of the genus *Coreopsis*. *Publ. Field Mus. (Bot.)*, **11**: 279–475.—Dress, W.J. 1965. Notes on the cultivated Compositae 8. *Thelesperma, Bidens, Coreopsis. Baileya*, **13**, 1: 20–42.

1. Heads 2.5–3.0 cm in dia; ligulate flowers yellow or orange with dark reddish-brown base, less often purple-reddish-brown or dark red; tubular predominantly dark red with reddish-brown border; achenes narrow, almost fusiform .... 1. **C. tinctoria.**
+ Heads to 6 cm in dia; ligulate flowers golden yellow; tubular dark yellow; achenes ellipsoidal or obovate, somewhat curved .... 2. **C. grandiflora.**

1. **C. tinctoria** Nutt. 1821, Journ. Acad. Sci. Philadelphia, **2**: 114; Vass. 1959, Fl. SSSR, **25**: 547.—*Calliopsis bicolor* Reichenb. 1823, Icon. Pl. Cult.: Pl. 114.—*C. tinctoria* (Nutt.) DC. 1836, Prodr. **5**: 565.

Type: USA, State of Arkansas ("in territorio Arkansano ad ripas flum. Rubri in pratis inundatis").

Baltic; *Center*; *West*; *East*; *Crimea.*—Cultivated as an ornamental plant, may be used as a dye plant.—*General distribution*: North America, cultivated in many other countries, sometimes escaped.—2n = 24.

2. C. grandiflora Hogg. ex Sweet, 1826, Brit. Fl. Gard.: tab. 175; Vass. 1959, Fl. SSSR, **25**: 547.—*C. heterophylla* Nutt. 1841, Trans. Amer. Philos. Soc. **2**, 7: 358.

Type: Described from garden plant, originating from the USA (New York).

Baltic; *Center*; *West*; *East*; *Crimea.*—Cultivated as an ornamental plant.—*General distribution*: North America; cultivated in many other countries.—2n = 26, 28.

Note. In the Ukraine, sometimes another North American species, *C. lanceolata* L; is cultivated, which differs by having lanceolate or broadly linear undivided leaves. (In the species described above, the leaves are pinnatipartite or bipinnatipartite.)

GENUS 12. *DAHLIA* Cav.
1791, Icon. Descr. Pl. **1**: 56.—*Georgina* Willd. 1804, Sp. Pl. **3**: 2124

Heads heterogamous, 4–20 cm in dia, solitary terminal on stem and its branches. Involucre bowl-shaped, two- to five-rowed; its bracts connate at base, green. Receptacle flat, fleshy, with large membranous bracts. Peripheral flowers pistillate, with ligulate corolla of various forms and color; disk flowers bisexual, with tubular corolla having five short teeth, yellow or reddish-reddish-brown. Perennial plants, with erect stems and opposite, petiolate, pinnati- or bipinnati- or tripinnatisect leaves.

Type: *D. pinnata* Cav.

The genus includes 27 species, growing in Central and South America, one of them is extensively cultivated in many other countries.

Literature: Sorensen, P.D. 1969. Revision of the genus *Dahlia* (Compositae, Heliantheae—Coreopsidinae). *Rhodora*, **71**, 786: 309–365.—Alshoeva, B.Ya. and others. 1985. Georginy [Dahalias]: 1–75.

1. D. pinnata Cav. 1791, Icon. Descr. Pl. **1**: 57, tab. 80; Vass. 1959, Fl. SSSR, **25**: 548.—*D. rosea* Cav. 1796, Icon. Descr. Pl. **3**: 33.—

D. variabilis Desf. 1829, Cat. Hort. Paris, **3**: 182, nom. nud.—*Georgina purpurea* Willd. 1804, Sp. Pl. **3**: 2124.—*G. rosea* (Cav.) Willd. 1804, ibid.—*G. variabilis* Willd. 1809, Enum. Pl. Hort. Berol. **1**: Pl. 93.

Plant 40–200 cm high, with tuberously thickened roots. Leaves pinnatisect, usually with five ovate or oblong-ovate, serrate-dentate leaflets. Heads 50–200 mm in dia, often only with ligulate flowers ("double").

Type: Described from plants grown in the Madrid Botanical Garden and originating from Mexico ("in Mexico").

North; *Baltic*; *Center*; *West*; *East*; *Crimea*.—Extensively culti-vated as an ornamental plant.—*General distribution*: Mexico and Central America; cultivated in many other countries.—2n = 64.

Note. In the circumscription of this species, often four smaller taxa are recognized. The key for their identification is given below:

1. Heads not "double", with ligulate and tubular flowers; ligulate flowers red or purple-lilac .... 2.
+ Heads usually "double" with only ligulate flowers of various colors, 30–200 mm in dia....................................
.... **D. cultorum** Thorsr. and Reis.
2. Ligulate flowers purple-lilac .... **D. merckii** Lehm.
+ Ligulate flowers red .... 3.
3. Inflorescences 80–100 cm in dia; stem strongly branched
.... **D. coccinea** Cav.
+ Inflorescences 100–120 cm in dia; stem weakly branched
.... **D. pinnata** Cav.

37

GENUS 13. *BIDENS* L.
1753, Sp. Pl.: 831; id. 1754, Gen. Pl., ed. **5**: 362

Heads homogamous or heterogamous, 4–30 mm in dia, solitary at apices of stem and its branches or clustered in racemose or corymbose inflorescence. Involucre bowl-shaped or cup-shaped, biseriate; outer herbaceous, inner with wide membranous border. Receptacle flat, with linear membranous bracts resembling inner involucral bracts and falling together with achenes. All flowers in a head yellow, bisexual, tubular or peripheral ligulate, rudimentary, sterile. Achenes obconical, trigonous, flattened or tetraquetrous, apically with two to four sticky bristles. Annual plants, with opposite (rarely upper alternate), undivided three- to five-parted or incised leaves.

Lectotype: *B. tripartita* L.

About 230 species in the tropical, subtropical, and warm temperate regions of both hemispheres, but predominantly in America.

Literature: Sherff, E.E. 1937. The genus *Bidens*. *Publ. Field Mus.* (*Bot.*), **16**, 1: 1–346.—Schumacher, A. 1942. Die fremden *Bidens*-Arten in Mitteleuropa. *Feddes Repert. Beih.* **131**: 42–93.—Voroschilov, V.N. 1949. Opisanie osoboi formy cheredy—*Bidens tripartita* var. *minor* Wimm. et Grab. [Description of special form of bur-marigold *Bidans tripartita* var. *minor* Wimm. et Grab.]. *Byull. Mosk. Obshch. Isp. prir., Otd. Biol.*, **2**: 56–60.—Kornass, Ya. 1960. O nakhodke *Bidens melanocarpus* Wiegand v Breste [On the discovery of *Bidens melanocarpus* Wiegand in Brest]. *Bot. Mat.* (*Leningrad*), **20**: 337–339.—Lhotsk, M. 1968. Karyologia und Karpobiologie der tschechoslowakischen vertreter der Gattung *Bidans*. *Rozpr. Ceskosl. Akad. Věd. Rada Mat. Prir. Věd.* **78**, 10: 1–85.—Mosyakin, S.L. 1988. Rid *Bidens* L. (Asteraceae) u flori URSR (Genus *Bidens* L. (Asteraceae) in the flora of the Ukraine). *Ukr. Bot. Zhurn.*, **45**, 6: 63–64.

1. Head nodding; achenes tetragonous, apically with four bristles; ligulate flowers present and rather large or absent; leaves sessile, undivided, serrate-dentate .... 1. **B. cernua.**
  + Heads erect; achenes strongly flattened, rarely tetragonous, apically with two to four(five) bristles; ligulate flowers absent, rarely present but very small and inconspicuous; leaves petiolate .... 2.
2. Outer involucral bracts more than 10; bracts linear or narrow-lanceolate, distinctly exceeding mature achenes (including awns); achenes smooth or somewhat rugose, hairs on their ribs retrorse; leaves three- to five-parted, rarely partly undivided .... 2. **B. radiata.**
  + Outer involucral bracts less than 10 .... 3.
3. Leaves undivided or shallow-pinnatilobate, petiolate (excluding the uppermost); achenes tetragonous, more or less sulcate, apically with four(five) stiff bristles, hairs on ribs of achenes spreading .... 5. **B. connata.**
  + Leaves three- to five-parted or incised; achenes strongly flattened .... 4.
4. Achenes densely warty; hairs on edges of achenes upward directed; bracts abruptly acuminate .... 4. **B. frondosa.**
  + Achenes not warty, more or less hairy; hairs on edges of achenes downward directed; bracts gradually acuminate .... 3. **B. tripartita.**

1. **B. cernua** L. 1753, Sp. Pl.: 832; Vass. 1959, Fl. SSSR, **25:** 555; Tutin, 1976, Fl. Europ. **4:** 140.—(Plate I, 4).

Type: Europe ("in Europa ad fontes et fossas").

North; *Baltic*; *Center*; *West*; *East*; *Crimea*.—On banks of water bodies and swampy meadows.—*General distribution*: Caucasus, Western and Eastern Siberia, Far East, Russian Central Asia; Scandinavia, Central and Atlantic Europe, Mediterranean, Dzhungaria-Kashgaria, Japan-China; North America.—2n = 24.

Note. Besides the type variety var. *cernua* with ligulate flowers, there is often found another variety, var. *discoidea* Ledeb., without ligulate flowers.

2. **B. radiata** Thuill. 1800, Fl. Paris, ed. **2:** 422; Vass. 1959, Fl. SSSR, **25:** 558; Tutin, 1976, Fl. Europ. **4:** 140; Baranova, 1987, Vestn. Leningr. Univ. Biol. **3:** 92; Mosyakin, 1988, Ukr. Bot. Zhurn., **45**, 6: 63.—(Plate I, 2).

Type: France ("St. Hubert").

North (Dvina-Pechora); *Baltic*; *Center*; *West* (Dnieper: Kiev and Chernigov regions); *East*.—On banks of water bodies, in swamps and swampy meadows.—*General distribution*: Western and Eastern Siberia, Far East, Russian Central Asia; Scandinavia, Atlantic and Central Europe, Mongolia, Japan-China.—2n = 28, 48.

3. **B. tripartita** L. 1753, Sp. Pl.: 832; Vass. 1959, Fl. SSSR, **25:** 557; Tutin, 1976, Fl. Europ. **4:** 140.—*B. orientalis* Velen, 1891, Fl. Bulg.: 250; Zefirov, 1954, Bot. Mat. (Leningrad), **16:** 373.—(Plate I, 3).

Type: Europe ("in Europae inundatis").

North; *Baltic*; *Center*; *West*; *East*; *Crimea*.—On banks of water bodies, in swampy meadows and forest glades, swampy forests, by roadsides and walkways, in habitations, sometimes in crops.—*General distribution*: Caucasus, Western and Eastern Siberia, Far Fast, Russian Central Asia; Scandinavia, Central and Atlantic Europe, Mediterranean, Asia Minor, Iran, Dzhungaria-Kashgaria, Mongolia, Japan-China, Himalayas; North America, Australia.—2n = 48, 72.

Note. A highly polymorphic species, in the circumscription of which several variants are identified. For Crimea, var. *orientalis* (Velen.) Stojan. et Steph. (Plate I, 6) was reported as a separate species, *B. orientalis* Velen., which is distinguished by undivided lower leaves, broader heads, large number of involucral bracts, and ruffus-brown color or the outer involucral bracts. Var. *minor*

Wimm. et Grab. (cf. Voroschilov, loc. cit.) has been described from very small plants.

4. **B. frondosa** L. 1753. Sp. Pl.: 832; Vass. 1959, Fl. SSSR, **25:** 550; Tutin, 1976, Fl. Europ. **4:** 140; A. Skvorts. 1982, Byull. Gl. Bot. Sada Akad. Nauk SSSR, **124:** 45; A. Skvorts. and others, 982, Byull. Mosk. Obshch. Isp. Prir. Otd. Biol., **87,** 3: 109; Tikhomir. 1987, Nauch. Dokl. Vyssh. Shkoy, Biol. Nauki, **6:** 78; Protopopova, 1987, Opred. Vyssh. Rast. Ukrainy: 331; Makarov and Ignatov, 1989, Byull. Gl. Bot. Sada Akad. Nauk SSSR, **127:** 42.—*B. melanocarpa* Wieg. 1899, Byull. Torry. Bot. Club, **26:** 405; Kornass, 1960, Bot. Mat. (Leningrad), **20:** 337; Lisitsyna and Artemenko, 1990, Byull. Mosk. Obshch. Isp. Prir., Otd. Biol. 95, 4: 110.—(Plate I, 5).

Type: North America ("in America septentrionali").

Baltic; *Center* (Upper Dnieper; Upper Volga; Volga-Kama: Izhevsk; Volga-Don); *West*; *East* (on Volga).—On banks of water bodies, by roadsides, in habitations.—*General distribution*: Caucasus, Far East; Scandinavia, Central and Atlantic Europe, Mediter-ranean, Japan-China; North America (Native place).—2n = 48.

Note. In Europe, for the first time reported in 1777 in the Oder Basin, in the former Soviet Union in 1959 in the Far East, in the territory of our Flora in 1960 in the vicinity of Brest. At present it is spreading fast. In Kiev Region, hybrids *B. frondosa* × *B. tripartita* (= *B.* × *garumnae* Jean. et Debr.) which differ from typical *B. frondosa* L. by winged petioles and indistinctly warty achenes, as also *B. frondosa* × *B. connata*, and *B. frondosa* × *B. cernua* were found.

5. **B. connata** Muehl. ex Willd. 1803, Sp. Pl.: **3,** 3: 1718; Tutin, 1976, Fl. Europ. **4:** 140; Mosyakin, 1988, Ukr. Bot. Zhurn. **45,** 2: 72.

Type: North America ("in Sylvis et agris Amer. bor.").

West (Dnieper: vicinity of Kiev).—On banks of water bodies, in swampy meadows, by roadsides.—*General distribution*: Central and Atlantic Europe; North America.

Note. European plants related to var. *fallax* (Warnst.) Sherff, differ from var. *connata* by longer leaf-like outer involucral bracts.

39   GENUS **14.** *COSMOS* Cav.
1791, Icon. Descr. Pl.: **1:** 9, tab. 14.—*Cosmea* Willd. 1803, Sp. Pl.:
3, 3: 2250

Heads heterogamous, 4–10 cm in dia, on long stalks, solitary or clustered in lax paniculate inflorescences. Involucre biseriate, with adpressed bracts. Peripheral flowers ligulate, sterile, large, usually pink, less often white or golden yellow to orange; disk florets bisexual, tubular, with yellow five-toothed corolla. Receptacle flat, with membranous bracts, attenuate in filiform tip, less often short and obtuse. Achenes apically attenuate in a beak bearing 2–4 readily detached anus, provided with small recurved spines, rarely awnless. Annual or perennial herbs, with erect stem and opposite, short-petiolate, bipinnatisect leaves.

Type: *C. bipinnatus* Cav.

The genus includes 26 species in the tropics and subtropics of America; two species are introduced in cultivation as ornamental plants.

Literature: Sherff, E.E. 1932. Revision of the genus *Cosmos* (family Compositae). *Publ. Field. Mus. (Bot.)*, **8**: 401–447.

1. **C. bipinnatus** Cav. 1791, Icon. Descr. Pl. **1**: 10; Vass. 1959, Fl. SSSR, **25**: 562.—*Cosmea bipinnata* (Cav.) Willd. 1804, Sp. Pl. **3**: 2250.

Annual glabrous plant, 80–160 cm high, with more or less branched stem. Leaf segments linear-filiform. Heads solitary or numerous, 5–10 cm in dia, with pink, less often white ligulate flowers. Involucre hemispherical; its outer bracts herbaceous, oblong-ovate, inner membranous, broader.

Type: Mexico ("in Mexico").

North (south); *Baltic*; *Center*; *West*; *East*; *Crimea*.—Cultivated as ornamental plants and sometimes escaped, found in habitations and by roadsides.—*General distribution*: North America, extensively cultivated in many other countries where it is often found as an escape.—2n = 24.

Note. C. sulphureus Cav. is much less often cultivated. It has smaller (4–6 cm in dia) heads, yellow ligulate flowers and broader leaf segments. Native place: Central and South America.

GENUS 15. *GALINSOGA* Ruiz. et Pav.
1794, Fl. Peruv. Chil. Prodr.: 110, tab. 24

Heads heterogamous, 3–5 mm in dia, usually rather numerous at apices of stems and branches. Involucre hemispherical or campanulate, biseriate; its bracts oval, membranous on margin, five, outer broader and usually longer than inner. Peripheral flowers four or five, pistillate, fertile, ligulate, white; disk flowers bisexual, fertile, tubular, yellow. Receptacle conical, with membranous bracts.

Achenes cuneate, five-angled or more or less compressed, finely pubescent, with pappus of scales, often absent in peripheral achenes. Annual plants, with more or less branched erect stems and opposite, petiolate, undivided leaves.

Lectotype: *G. parviflora* Cav.

Four species in the tropical and subtropical regions of America (from Mexico to Argentina), but as an ecdemic and often completely naturalized plant also in other regions of America, Europe, Asia, Africa and Australia.

40   *Literature*: Thellung, A. 1915. Über die in Mitteleuropa vorkommenden *Galinsoga*-Formen. *Allg. Bot. Zeitschr.* **21**: 1–16.—Lousley, J.E. 1950. The nomenclature of the British species of *Galinsoga*. *Watsonia*, **1**, 4: 238–241.—Kruberg, Yu.K. 1955. O poyavlenii v flore Leningrada galinsogi [On the presence of *Galinsoga* in the flora of Leningrad]. *Uchen. Zap. Leningr. Ped. Inst.*, **109**: 239–245.—Gusev, Yu.D. 1966. Rasselenie vidov *Galinsoga* v Leningradskoi oblasti [Dispersal of species of *Galinsoga* in Leningrad Region]. *Bot. Zhurn.*, **51**, 4: 577–578.—Shontz, N.N. and J.P. Shontz, 1970. *Galinsoga ciliata* (Raf.) Blake: its arrival and spread in the northeastern United States. *Rhodora*, **72**: 386–392.—Canne, J.M. 1977. A revision of the genus *Galinsoga* (Compositae: Heliantheae). *Rhodora*, **79**: 319–389.—Tikhomirov, V.N. 1982. Tri novykh vida dlya flory Ryazanskoi oblasti [Three new species in the flora of Ryazan Region]. *Nauchn. Dokl. Vyssh. Shkoly*, **12**: 52–53.

1. Plants subglabrous, only in upper part scatteredly pubescent; stem below heads entirely or almost entirely lacking glandular pubescence; bracts of middle disk flowers three-parted; pappus of peripheral achenes absent or weakly developed, of middle achenes with numerous ciliate scales .... 1. **G. parviflora.**
+ Plants moderately to densely pubescent from spreading hairs; stem below heads more or less densely glandular-hairy; bracts of middle disk flowers undivided; pappus of short-ciliate, apically incised scales fully developed in peripheral achenes but half as long as in middle achenes .... 2. **G. quadriradiata.**

1. **G. parviflora** Cav. 1794, Icon. Descr. Pl. **3**: 41; tab. 281; Vass. 1959, Fl. SSSR, **25**: 563; Tutin, 1976, Fl. Europ. **4**: 144; Kozhevnikova and Makhaeva, 1978, Bot. Zhurn. **61**, 4: 566; Efimova and others, 1981, Bot. Zhurn. **66**, 7: 1049; Safonov, 1982, Byull. Glavn. Bot. Sada Akad. Nauk SSSR, **124**: 49; Maitullin, 1984,

Byull. Glavn. Bot. Sada Akad. Nauk SSSR, **132**: 46; Saksonov, 1987, Bot. Zhurn., **72**, 10: 1403; Papchenkov and Dimitrieva, 1987, Bot. Zhurn., **72**, 4: 527.—(Plate I, 7).

Type: Described from plants cultivated in the Paris Botanical Garden from seeds received from Peru.

North (Dvina-Pechora: west and south); *Baltic*; *Center*; *West*; *East*; *Crimea.*—In habitations, by roadsides, in fields and plantations of various crops, in more or less weedy places.— *General distribution*: South America; as an ecdemic and often completely naturalized plant: Caucasus, Far East (south), Russian Central Asia; Scandinavia (south), Central and Atlantic Europe, Mediterranean, Asia Minor, Himalayas, Japan-China; North America, Australia, Africa.—2n = 16, 32, 36.

2. **G. quadriradiata** Ruiz et Pav. 1798, Syst. Veg. **1**: 198; Vass. 1959, Fl. SSSR, **25**: 564; Dobrocz. 1962, Fl. URSR, **11**: 193; Bogachev and others, 1962, Bot. Zhurn. **47**, 11: 1668; Bochkin and others, 1968, Byull. Glavn. Bot. Sada Akad. Nauk SSSR, **151**: 53.—*Adventita ciliata* Rafin. 1836, New Fl. North Amer. **1**: 67.—*Galinsoga ciliata* (Rafin.) Blake, 1922, Rhodora, **24**: 35; Vorosch. 1966, Fl. Sov. Daln. Vost.: 406; Tutin, 1976, Fl. Europ. **4**: 144; Gusev. 1976, Bot. Zhurn. **61**, 4: 568; Tikhomir, 1982, Nauchn. Dokl. Vyssh. Shkoly Biol. Nauki, **12**: 53; Bosek, 1983, Bot. Zhurn. **68**, 5: 673; Tikhomir and Khaitontsev, 1984, Nauchn. Dokl. Vyssh. Shkoly Biol. Nauki, **8**: 74; Maitullin, 1984, Byull. Glavn. Bot. Sada Akad. Nauk SSSR, **132**: 46; Voloskova, 1986, Nauchn. Dokl. Vyssh. Shkoly Biol. Nauki, **8**: 74.—(Plate I, 8).

Type: Peru ("Peru: Lima, Lima and Chancay").

Baltic; *Center*; *West*; *Crimea.*—In habitations, gardens and parks, by roadsides, in weedy places, sometimes in fields.— *General distribution*: North and South America; as an ecdemic and often naturalized plant: Caucasus, Far East (south); Scandinavia, Atlantic and Central Europe, Mediterranean, Himalayas, Japan-China, South Asia, Africa (southwest).—2n = 32.

GENUS 16. *MADIA* Molina
1781 (1782), Sag. Stor. Nat. Chili, **1**: 136, 354

Heads heterogamous, 15–20 mm in dia, solitary or in clusters in axils of upper leaves. Involucres subspherical, uniseriate; involucral bracts boat-shaped, oblong-ovate, carinate, enveloping peripheral achenes, pubescent from simple and glandular hairs.

46

Receptacle flat or somewhat convex, with one row of ovate or
41 broadly oval bracts enveloping achenes. Flowers yellow; peripheral
pistillate, ligulate; disk florets bisexual, with tubular, five-toothed,
corolla. Achenes glabrous, dark gray, maculate, with apical beak,
without pappus, peripheral strongly flattened, middle four- or
five-gonal, cuneate or cylindrical, more or less curved. Annual,
glandular-pubescent plant with erect stems and simple, undivided,
sessile leaves, lower opposite, upper alternate.

Lectotype: *M. sativa* Molina.

The genus includes nearly 20 species, distributed in North
and South America; one species is cultivated for obtaining essential
oil in other countries.

Literature: Hesse, V.F. 1968. Notes on *Madia sativa* and
related species. *Madroño*, **19**, 6: 210–215.

1. **M. sativa** Molina, 1781, Sag. Stor. Nat. Chili, **1**: 136; Vass.
1959, Fl. SSSR, **25**: 567.

Plant 30–100 cm high, covered with simple and glandular
hairs, with aromatic fragrance; stem simple or more or less
branched, densely leafy; leaves broadly lanceolate to linear-
lanceolate; heads one or many on short stalks in axils of upper
leaves considerably exceeding heads.

Type: Chile ("... il Regno del Chili").

West; *Crimea*.—Sometimes cultivated for essential oil.—
General distribution: North and South America; as a cultivated
plant in other warm temperate and subtropical countries.—2n = 32.

GENUS 17. *LAYIA* Hook. et Arnott ex DC.
1838, Prodr. **7**, 1: 294, nom. conserv.

Heads heterogamous, 4–5 cm in dia, on long stalk, solitary
terminal on stem and its branches. Involucre biseriate; involucral
bracts oblong, adpressed, outer exceeding inner or almost as
long, herbaceous, more or less covered with hairs and spinules.
Receptacle lacking bracts, glabrous. Peripheral flowers ligulate,
pistillate, yellow or white, 8–20; disk florets bisexual, with five-
toothed tubular corolla, yellow. Achenes 2.5–4.0 mm long, narrow-
cuneate, black, rather densely hairy, with two or three ribs; pappus
of numerous, basally expanded, scabrous, 2.5–4.0 mm long bristles.
Annual plants, with more or less branched or simple stem and
alternate, sessile, undivided or pinnatipartite leaves.

Type: *L. gaillardioides* (Hook. et Arnott.) DC.

The genus includes 15 species from the western regions of the United States of America. One of them is grown as an ornamental plant in other countries.

1. **L. elegans** Torr. et Gray, 1842, Fl. North Amer., **2**: 314; Vass. 1959, Fl. SSSR, **25**: 568.

Plant 20–50 cm high, with scatteredly hairy, erect or more or less ascending stems and more or less hairy, sessile, semiamplexicaul leaves; lower leaves pinnately divided to pinnatifid, upper linear-lanceolate or lanceolate, undivided or more or less toothed.

Type: California ("in California").

West; *East*; *Crimea*.—Grown as an ornamental plant.—*General distribution*: The USA (western regions); grown in other countries.—2n = 14.

42 GENUS **18.** *ARNICA* L.[1]
1753, Sp. Pl.: 884; id. 1754, Gen. Pl., 5: 376

Heads heterogamous, with numerous yellow or orange-yellow flowers, solitary or in corymbose inflorescence. Involucre bowl-shaped, biseriate, consisting of herbaceous bracts of similar length. Receptacle flat, hairy. Peripheral flowers ligulate, pistillate, central tubular, bisexual. Corolla more or less pubescent outside. Stamens with basally obtuse anthers and narrow-deltoid connective. Style branches thick, apically somewhat expanded with pointed tip, uniformly papillate. Achenes somewhat narrowed toward base, black or reddish-brown, with pappus of scabrous bristles. Perennial herbs, with erect stem. Leaves opposite, less often upper alternate, lanceolate or elliptical, entire or finely toothed.

Lectotype: *A. montana* L.

Includes 32 species in extratropical regions of the Northern Hemisphere.

Literature: Maguire, B. 1943. A monograph of the genus *Arnica. Brittonia*, **4**, 3: 386–510.

1. Leaves elliptical, subobtuse; heads one to seven, 4–8 cm in dia .... 1. **A. montana.**
+ Leaves lanceolate or sublinear, acuminate; heads solitary, less often two or three, 2.5–4.0 cm in dia .... 2. **A. angustifolia.**

[1]Treatment by G.Yu. Konechnaya.

48

1. **A. montana** L. 1753, Sp. Pl.: 884; Iljin, 1961, Fl. SSSR, **26:** 663; I1.K. Ferguson, 1976, Fl. Europ. **4:** 189.

Type: The Alps ("in Alpibus").

Baltic (Lithuania); *Center* (Upper Dnieper); *West* (Carpathians).—In pine forests, on forest edges and in glades, mountain meadows.—*General distribution*: Scandinavia (south), Central and Atlantic Europe, Mediterranean (north).—2n = 38.

2. **A. angustifolia** Vahl. 1816, in Hornem. Fl. Dan. **9:** 26: 5; I.K. Ferguson, 1976, Fl. Europ. **4:** 189.

Type: Greenland ("E. colon. Omenak et Godhavn Gronlandiac").

a. Subsp. **alpina** (L.) I.K. Ferguson, 1973, Journ. Linn. Soc. London, Bot. 67: 282; id. 1976, Fl. Europ. **4:** 189.—*A. montana* L. var. *alpina* L. 1753, Sp. Pl.: 884.—*A. alpina* (L.) Olin et Ladau, 1799, Diss. Arnica: 11, non Salisb. 1796; Iljin, 1961, Fl. SSSR, **26:** 657.—*A. fenno-scandica* Jursz. et Korobkov, 1987, Arkt. Fl. SSSR, **10:** 191.—Involucres and stem below head covered with short glandular and long simple hairs, glandular hairs inconspicuous.

Type: Northern Europe ("in pratis Europae frigidioris").

Arctic (Arctic Europe: Murmansk Region); *North* (Karelia-Murmansk).—In mountain meadows, meadow tundras.—*General distribution*: Scandinavia (north).—2n = 76.

b. Subsp. **iljinii** (Maguire) I.K. Ferguson, 1973, Journ. Linn. Soc. London, Bot. 67: 282; id. 1976, Fl. Europ. **4:** 189.—*A. alpina* (L.) Olin et Ladau subsp. *iljinii* Maguire, 1943, Brittonia, **4:** 411.—*A. iljinii* (Maguire) Iljin, 1961, Fl. SSSR, **26:** 658.—Involucres and stem below head covered with long glandular and long simple hairs; glandular hairs conspicuous.

Type: Mouth of the Yenisei River ("unterlauf des Jenissei, Ust Jenisseisky Port").

Arctic (Novaya Zemlya; Arctic Europe: Arkhangelsk Region); *North* (Dvina-Pechora).—In mossy and lichen tundras, on limestone outcrops, in meadows, forest glades and edges.—*General distribution*: Western and Eastern Siberia.—2n = 56.

GENUS 19. *GAILLARDIA* Foug.
1786, Observ. Phys. **29:** 55 ("*Gaillarda*"); id. 1788, Mém. Hist. Acad. Sci. Paris. 1786: 5

Heads heterogamous, 4–10 cm in dia, solitary, at apices of stem and its branches. Involucre saucer-shaped, many-seriate; involucral bracts almost equal, outer oblong-lanceolate, herbaceous, more or less adpressed. Receptacle convex, with subulate bracts. Peripheral flowers ligulate, one- to many-rowed, pistillate or sertile, of various colors; disk flowers fertile, bisexual, with tubular corolla and five-toothed limb, usually purple. Achenes conical, with five to nine ribs, hairy, with pappus of scaly bristles, expanded in lower half and then rather abruptly terminating in stiff, subulate tip. More or less hairy annuals, less often perennial plants, with erect usually more or less branched stem and alternate leaves, of which the lowest petiolate, undivided or pinnatilobate; middle and upper leaves sessile, oblong or lanceolate, entire or more or less toothed.

Type: *G. pulchella* Foug.

The genus includes 26 species in North and South America, of which 5 are cultivated as ornamental plants in other countries.

Literature: Biddulph, S.F. 1944. A revision of the genus *Gaillardia. Res. Stud. State Coll. Wash.*, **12**: 195–256.—Stoutamire, W.P. 1960. The history of cultivated Gaillardias. *Baileya*, **8**: 13–17.

1. Achenes 2.0–2.5 mm long; pappus 3–4 mm long; involucre 2.5–3.0 cm in dia; heads 4–6 cm in dia. Annuals..........
.... 1. **G. pulchella.**
+ Achenes 3–4 mm long; pappus 6–8 mm long; involucre 3–5 cm in dia; heads usually 6–10 cm in dia. Perennials
.... 2. **G. aristata.**

1. **G. pulchella** Foug. 1788. Mém Hist. Acad. Sci. Paris, 1786; 5; Dobrocz., 1962, Fl. URSR, **11**: 197; Hansen, 1976, Fl. Europ. **4**: 144.

Type: The USA, State of Lousiana ("... in Lousiana").

Baltic; *Center*; *West*; *East*; *Crimea*.—Cultivated as an ornamental plant.—*General distribution*: North America; cultivated in other countries.—2n = 34, 36, 68.

2. **G. aristata** Pursh, 1814, Fl. Amer. Sept. **2**: 573; Yashenko, 1985, Dekorat. Rast.: 398; Hansen, 1976, Fl. Europ. **4**: 144; Tuyanaev and Puzyrev, 1988, Gemerefity Vyatsko-Kamskogo mezhdurechya [Hemenophytes of the Vyatka-Kama interfluve]: 96.

Type: The USA, the Rocky Mountains ("in collibus siccis and Rocky Mountains").

Baltic; *Center*; *West*; *East*; *Crimea*.—Cultivated as an ornamental plant; sometimes as an escape.—*General distribution*: North America; cultivated and as an escape in other countries.— 2n = 34, 36, 68, 72.

Note. A hybridogenic species *G. × hybrida* Lort. is also found in cultivation.

50

GENUS **20.** *TAGETES* L.
1757, Sp. Pl.: 887; id. 1754, Gen. Pl. 5: 378

Heads heterogamous, 1.5–10.0 cm in dia. Involucre cup- or bowl-shaped, one-rowed with bracts connate over a large part. Common receptacle flat, without bracts. Peripheral flowers ligulate, pistillate, but usually sterile, yellow, brownish-red or orange; disk flowers bisexual, fertile, yellow, their corolla tubular with five-toothed limb. Achenes linear-oblong, narrowed toward base, with pappus of unequal scales. Annual plants, with erect, usually more or less branched stem and opposite or partly alternate pinnatipartite leaves.

Lectotype: *T. patula* L.

44 The genus includes 35–50 species in the tropical and subtropical regions of America, of which some are cultivated in other countries or found as ecdemics but often fully naturalized plants.

1. Stem usually branched from middle or above, with obliquely upward directed branches; stalks of heads strongly inflated in upper part; heads 3–10 cm in dia, with bowl-shaped involucre; ligulate flowers yellow .... 1. **T. erecta.**
   + Stem branched almost from base, with branches strongly deflected sideways; stalks of heads somewhat inflated in upper part; heads 1.5–4.0 cm in dia, with cup-shaped involucre; ligulate flowers dark orange-yellow or brownish-red .... 2.
2. Involucre 2–4 cm in dia; stem below involucre somewhat inflated; segments of pinnatisect leaves 5–7 mm wide, acutely toothed .... 2. **T. patula.**
   + Involucre 1.5–3.0 cm in dia; stem below involucre not inflated; segments of pinnatisect leaves 2.0–2.5 mm wide, finely acutely toothed .... 3. **T. tenuifolia.**

1. **T. erecta** L. 1753, Sp. Pl.: 887; Gorschk. 1959, Fl. SSSR, **25:** 571.

Type: Mexico ("... in Mexico").

North (south); *Baltic; Center; West; East; Crimea.*—Cultivated as an ornamental plant, also used for essential oil and as condiment.—*General distribution*: North America; cultivated in many other countries.—2n = 24.

2. **T. patula** L. 1753, Sp. Pl.: 887; Gorschk. 1959, Fl. SSSR, **25:** 571.

Type: Mexico ("in Mexico").

Baltic; *Center*; *West*; *East*; *Crimea*.—Cultivated as an ornamental, medicinal, essential oil, and condiment plant.—*General distribution*: North America (south); cultivated in many other countries.—2n = 48.

3. **T. tenuifolia** Cav. 1797, Icon. Descr. Pl. **4**: 31; tab. 352; DC. 1836, Prodr. **5**: 644.—*T. signata* Bartl. 1837, Index Sem. Horti Gotting.: 7; Dobrocz. 1962, Fl. URSR, **11**: 200.

Type: Peru ("... in Peru").

Baltic; *Center*; *West*; *East*; *Crimea*.—Cultivated as an ornamental plant.—*General distribution*: North and South America; cultivated in other countries.—2n = 36.

GENUS **21.** *PARTHENIUM* L.
1753, Sp. Pl.: 988; id. 1754, Gen. Pl., ed. 5: 378

Heads heterogamous, to 8 mm in dia, usually in corymbose or corymbose-paniculate inflorescences. Involucre bowl-shaped, 3–6 mm in dia, two- or three-seriate; outer involucral bracts herbaceous, narrow-lanceolate to ovate, inner broadly ovate, almost entirely membranous. Receptacle almost flat, with scalelike bracts. Peripheral flowers pistillate, five, ligulate, with very short two-lobed ligule, yellow, less often whitish; disk flowers tubular, with five-toothed corolla, bisexual, but usually sterile. Achenes 2.3–2.6 mm long, oblanceolate-ovate, strongly flattened, usually grayish, with two prominent and two weak ribs, apically with two persistent lanceolate scales; achenes falling with ligulate corolla enclosing them, with broad inner involucral bracts and two covering pales of two sterile tubular flowers. Semishrubs perennial, and annual herbaceous plants, containing latex; leaves alternate, undivided or more or less incised.

Lectotype: *P. hysterophorus* L.

45 About 15 species in North and South America, of which one is introduced into cultivation in other countries.

Literature: Rollins, R.C. 1950. The guayule rubber plant and its relatives. *Contr. Gray Herb.* **172**: 3–73.

1. **P. argentatum** A. Gray, 1884, Syn. Fl. North Amer. **1, 2**: 245; Smoljan, 1959, Fl. SSSR, **25**: 530.

Grayish-tomentose semishrub, 30–100 cm high, with strongly branched stems. Leaves undivided, and then usually coarsely toothed, pinnatilobate, or pinnatipartite.

Type: The USA, State of Texas ("Texas").

Crimea.—Cultivated as rubber-producing and ornamental plant.—*General distribution*: North America (south); cultivated in some other countries (in the former Soviet Union also in the Caucasus and Russian Central Asia).—2n = 36, 72.

GENUS 22. *CYCLACHAENA* Fresen.
1836, Index Sem. Hort. Frankof.: 4; id. 1838, in Schlecht. Linnaea (Literat.), **12**: 78.—*Iva* L. sect. *Cyclachaena* (Fresen.) A. Gray, 1884, Syn. Fl. North Am. **1, 2**: 245

Heads heterogamous, 2–5 mm in dia, clustered in large terminal paniculate inflorescences. Involucre consisting of five outer elliptical herbaceous and five inner obovate, almost undivided membranous bracts. Receptacle almost flat, with or without few scalelike bracts. Peripheral flowers pistillate, with strongly reduced (annular) perianth, fertile; inner bisexual with five-lobed corolla, undeveloped pistil and almost free anthers, sterile, their corolla pale greenish or yellowish-green. Achenes oblong-obovate or cuneate, flattened-trigonous, without pappus. Annual plants, with erect stem and predominantly opposite (except the uppermost) distant leaves with undivided, less often pinnatilobate blades on rather long petiole.

Type: *C. xanthiifolia* (Nutt.) Fresen.

Four species, distributed in North America, one of them ecdemic in many other extratropical countries, where it is fully naturalized.

1. **C. xanthiifolia** (Nutt.) Fresen. 1836, Index Sem. Hort. Frankof.: 4; Smoljan. 1959, Fl. SSSR, **25**: 515; Gusev, 1968, Bot. Zhurn. **53, 2**: 267; A. Skvorts. 1973, Byull. Glavn. Bot. Sada Akad. Nauk SSSR, **88**: 33; Kozhevnikova and Makhaeva, 1976, Bot. Zhurn. **61, 4**: 567; Schultz, 1976, Bot. Zhurn. **61, 10**: 1451; Alekseev and Makarov, 1977, Byull. Mosk. Obshch. Isp. Prir., Otd. Biol. **82, 6**: 90; Tuganaev and others, 1978, Bot. Zhurn., **63, 10**: 1512; Bosek, 1979, Bot. Zhurn., **64, 2**: 244; Rakov and Pchelkin, 1980, Bot. Zhurn., **65, 5**: 713; Ilminskikh and others, 1981, Bot. Zhurn., **66, 8**: 1222; A. Skvorts. 1982, Byull. Glavn. Bot. Sada Akad. Nauk SSSR, **124**: 49; Golulaev and Kosykh, 1982, Bot. Zhurn., **67, 9**: 1299; Majevski and Ivanov, 1984, Sostav i Persp. Issl. Fl. Sredn. Pol. Evrop. Chasti SSSR [Content and Prospects of Studies on the Flora of Central Belt of the European Part of the USSR]: 61; Matekeitite, 1985, Bot. Zhurn., **70, 10**: 1419; Voloskova, 1986, Nauchn. Dokl. Vyssh. Shkoly, Biol. Nauki, **8**: 74; Dimitriev, 1987, Novosti Sist. Vyssh. Rast., **24**: 225; Oktybreva and Tikhomir,

1987, Opred. Rast. Meshchery, **2**: 104.—*Iva xanthiifolia* Nutt. 1818, Gen. North Amer. Pl. **2**, 185; Hansen, 1976, Fl. Europ. **4**: 141; Liakavicius, 1982, Bot. Zhurn., **67**, 2: 233.—(Plate II, 1).

Usually weakly pubescent plants, 15–200 cm high; leaf blades ovate, more or less toothed, less often pinnatilobate; heads yellowish-green, 2–4 mm in dia.

Type: The USA, State of North Dakota ("in arid soils, near Fort Manden, etc., on the banks of the Missouri").

North (Karelia-Murman, Olonets Station); *Baltic*; *Center*; *West*; *East*; *Crimea.*—By roadsides, in habitations, on banks of water bodies, in weedy meadows, pastures, forest glades, sometimes in fields, in more northern regions found as a rare ecdemic plant, often not fruiting, southward becomes common.—*General distribution*: North America; ecdemic and introduced: Caucasus, Far East (south); Central and Atlantic Europe, Mediterranean, Asia Minor, Japan-China; South America, Australia.—2n = 28, 36.

GENUS **23.** *AMBROSIA* L.
1753, Sp. Pl.: 987; id. 1754, Gen. Pl., ed. 5: 425

Heads homogamous, but unisexual. Staminate flowers with infundibuliform, five-toothed corolla, forming small, hemispherical, more or less flattened heads with fused involucre, clustered in terminal spicate or racemose inflorescences; pistillate flowers without perianth, borne in lower part of the same inflorescence in axils of terminal or upper leaves. Receptacle of male heads with or without few scalelike bracts. Achenes without pappus, enclosed by involucre hardening at fruiting. Annuals, less often perennial plants with erect, less often procumbent stems and opposite petiolate leaves with more or less incised blades.

Lectotype: *A. maritima* L.

The genus includes 35–40 species, distributed (often as ecdemic, not naturalized plants) in almost all tropical and temperate regions of both hemispheres but predominantly in America.

Literature: Payne, W.W. 1964. A re-evaluation of the genus *Ambrosia* (Compositae). *Journ. Arnold Arbor.*, **45**: 401–439.—Wagner, W.H. 1959. An annotated bibliography of ragweed (*Ambrosia*). *Rev. Allergy Appl. Immuno.*, **13**: 353–403.—Basset, I.J. and C.W. Crompton, 1975. The biology of Canadian weeds. 2. *Ambrosia artemisiifolia* L. and *A. psilostachya* DC. *Canad. Journ. Pl. Sci.*, 55: 463–476.

1. Leaves pinnatisect or bipinnatisect; receptacle with bracts; involucre of male heads without ribs .... 2.

+ Leaves three- to five-lobed; receptacle without bracts; involucre of male heads with three to eight ribs .... 3.

2. Plant appressed-hairy, annual; leaves bipinnatisect, dark green and subglabrous above, grayish-green and short hairy beneath; involucre with five to eight spines surrounding them, glabrous in upper part ....

.... 1. **A. artemisiifolia.**

+ Rather densely hirsute, soboiferous perennials, with hairy tubercles; leaves pinnatisect, scabrous from rather stiff hairs and usually more stiff; involucre not surrounded by spines but often with four obtuse tubercles, pubescent in upper part .... 2. **A. psilostachya.**

3. Petioles cylindrical, not winged, densely setose; involucre of male heads with four to eight ribs .... 4. **A. aptera.**

+ Petioles expanded near base, narrow-winged, ciliate; involucre of male heads with three ribs .... 3. **A. trifida.**

1. **A. artemisiifolia** L. 1753, Sp. Pl.: 988; Smoljan. 1959, Fl. SSSR, **25:** 519; A. Skvorts. 1971, Tr. Bot. Sada Mosk. Univ. **7:** 63; Hansen, 1976, Fl. Europ. **4:** 142; Schultz. 1976, Bot. Zhurn. **61,** 10: 1452; Oktyabreva and others, 1978, Nauchn. Dokl. Vyssh. Shakoly, Biol. Nauki, **12:** 93; Rakov and Pchelkin, 1980, Bot. Zhurn. **65,** 5: 713; Voloskova, 1981, Nauchn. Dokl. Vyssh. Shkoly, Biol. Nauki, 6: 64; Liakavicius, 1982, Bot. Zhurn. **67,** 2: 233; Safonov, 1982, Byull. Glavn. Bot. Sada Akad. Nauk SSSR, 124: 49; Golubev and Kosykh, 1982, Bot. Zhurn. **67,** 9: 1299; Majevski and Ivanov, 1984, Sostav i Persp. Issl. Fl. Sredn. Pol. Evrop. Chasti SSSR: 61; Bosek, 1986, Bot. Zhurn. **71,** 1: 98; Oktyabreva and Tikhomir. 1987, Opred. Rast. Meshchery, **2:** 104; Lavrenko and Kustysheva, 1987, Tr. Akad. Nauk SSSR, **82:** 83.—(Plate II, 2).

Type: North America ("Virginia, Pensylvania").

48 *North* (Dvina-Pechora: south); *Baltic*; *Center*; *West*; *East*; *Crimea*.—As ecdemic but often fully naturalized plant, by roadsides, in habitations, weedy meadows and pastures, on banks of water bodies, in forest nurseries; predominantly in the forest-

47 Plate II.

1—*Cyclachaena xanthiifolia* (Nutt.) Fresen., la—head; 2—*Ambrosia artemisiifolia* L., 2a—head; 3—*A. bifida* L., 3a—head; 4—*A. psilostachya* DC. leaf, 4a—head; 5—*Xanthium spinosum* L., 5a—involucre of achenes; 6—*X. albinum* (Willd.) H. Scholz, leaf, 6a—involucre of achenes; 7—*X. strumarium* L., involucre of achenes.

steppe and steppe regions.—*General distribution*: North America, ecdemic and naturalized: Caucasus, Far East (south), Russian Central Asia; Central and Atlantic Europe, Mediterranean, Asia Minor, Iran, Japan-China; South America, Australia, Africa.—2n = 36.

2. **A. psilostachya** DC. 1836, Prodr. **5:** 526; Smoljan. 1959, Fl. SSSR, **25:** 519; Khvalina, 1965, Bot. Zhurn. **50,** 4: 532; Schulz, 1976, Bot. Zhurn., **61,** 10: 1452; Oktyabreva and Tikhomir, 1987, Opred. Rast. Meshchery: 104.—(Plate II, 4).

Type: Mexico ("in Mexico inter San-Fernando et Matamoros").
Baltic; *Center* (Upper Volga: vicinity of Moscow; Volga-Don); *East* (Lower Don; Trans-Volga; Lower Volga).—As ecdemic weed of roadsides, in habitations.—*General distribution*: North and Central America; ecdemic: Caucasus (Ciscaucasia).—2n = 72, 108, 144.

3. **A. trifida** L. 1753, Sp. Pl.: 987; Smoljan. 1959, Fl. SSSR, 25: 520; A. Skvorts. 1971, Tr. Bot. Sada Mosk. Univ. 7: 63; Schult, 1976, Bot. Zhurn., **61,** 10: 1452; Tarasov, 1978, Ukr. Bot. Zhurn., **35,** 2: 188; V'yunkova, 1983, Byull. Mosk. Obshch. Isp. Prir. Otd. Biol., **88,** 1: 133; Majevski and Ivanov, 1984, Sostav i Persp. Issl. Fl. Sredn. Pol. Evrop. Chasti SSSR: 61; Puzyrev, 1985, Bot. Zhurn., **70,** 2: 268; Ignatov and Makarov, 1985, Bot. Zhurn., **70,** 6: 852.—(Plate II, 3).

Type: North America ("ad ripas fl. Amer. bor. a Canada ad Georgiam").
Baltic; *Center*; *West* (Dnieper: Dnepropetrovsk; Moldavia; Black Sea); *East*.—By roadsides, in habitations, fields, on banks of water bodies.—*General distribution*: North America; ecdemic: Caucasus, Eastern Siberia (south), Far East (south); Central and Atlantic Europe, Mediterranean.—2n = 24.

4. **A. aptera** DC. 1836, Prodr. **5:** 527; Smoljan. 1959, Fl. SSSR, **25:** 520; Kotov, 1970, Ukr. Bot. Zhurn., **27,** 1: 1078.

Type: Mexico ("in Mexico circa Bejar locis inundatis").
Crimea (Alushta District, village of Pritsvetnoe).—As ecdemic weed in habitations, by roadsides, in plantations of various crops.—*General distribution*: North America (Mexico): ecdemic: Caucasus (western Transcaucasia); Mediterranean.

GENUS 24. *XANTHIUM* L.
1753, Sp. Pl.: 987; id. 1754, Gen. Pl., ed. 5: 424

Heads homogamous, but unisexual: staminate in upper part of common inflorescence, many-flowered with uniseriate involucre of free bracts and cylindrical receptacle with small scalelike bracts; pistillate solitary of in glomerules, in lower part of inflorescence, one- to two-flowered, with biseriate involucre; inner involucral bracts of pistillate heads connate, with one or two hard beaklike spines at apex and numerous spines allover involucre. Annual plants, with erect stem and alternate leaves with undivided, but toothed or more or less lobate blades on petioles.

Lectotype: *X. strumarium* L.

About 30 species, distributed mostly in North America; some species occur in warm temperate and subtropical regions of other countries, usually only as ecdemics but then fully naturalized.

Literature: Millspaugh, Ch.F. and E.E. Sherff. 1919. Revision of the North American species of *Xanthium. Publ. Field Mus. (Bot.)*, **204,** 12: 9–51.—Widder, F.J. 1923. Die Arten der Gattung *Xanthium.* Feddes Repert. Beih. **20**: 1–221.—Widder, F.J. 1925. Übersicht über die bischer in Europa beobachteten *Xanthium*-Arten und Bastarde. *Feddes Repert.* **44**: 273–303.—Wein, K. 1925. Beitrage zur Geschichte der Einfuhrung und Einburgerung einiger Arten von *Xanthium* in Europa. *Beih. Bot. Centralbl.* **42**, 2: 151–176.—Love, D. and P. Danserean, 1959. Biosystematic studies on *Xanthium*: taxonomic appraisal and ecological status. *Canad. Journ. Bot.*, **37**: 173–208.—Duguyan, D.K. 1960. O vidakh roda *Xanthium* L. flory SSSR [On species of the genus *Xanthium* in the flora of the USSR]. *Tr. Rostov. Otd. Vsesoyuz. Bot. Obshch.*, **1**: 1–57.—Protopopova, V.V. 1964. Novya danny of sistematicheskom sostave roda *Xanthium* L. na Ukraine [New data of the systematics of the genus *Xanthium* in the Ukraine]. *Ukr. Bot. Zhurn.*, **21,** 4: 78–84.—Hardtl, H. 1973. Über die Inhaltsstoffe einiger *Xanthium*-Arten. *Phyton.* **15**: 1–25.—Hardtl, H. 1974. Nachweis von Sippen bei *Xanthium albinum. Bot. Jahrb.*, **94**: 541–548.—Brande, A. 1976. Zur Ausbreitunggeschichte von *Xanthium* in südostlichen Europa. *Bot. Jahrb.*, **95**: 406–410.

   1. Leaves basally with erect, glabrous, yellow spines, dark green above, densely grayish-pubescent beneath; involucre **of pistillate heads at fruiting (involucre of achenes) light**-yellow, subglabrous, densely covered with glabrous uncinate spines, with subulate beak at apex .... 10. **X. spinosum.**

  + Leaves basally lacking spines, green on both sides, scatteredly pubescent; involucre of achenes greenish, brownish, reddish-brown, reddish-brown-yellow, with

58

pubescent spines and two uncinate beaks (very rarely one undeveloped) .... 2.

2. Plants relatively soft-hairy; involucre of achenes 10–17 mm long; beaks parallel, erect or somewhat bent inward, often unequal (very rarely one undeveloped) .... 3.

+ Plants covered with very short glandular hairs; involucre of achenes 16–28 mm long; beaks divergent at angle, almost equal, well-developed, curved, apically uncinate .... 5.

3. Involucre of achenes broadly ellipsoidal to subglobose (half and more as long as wide), 10–15 mm long and 6–8 mm wide, very densely covered with spines up to apex .... 2. **X. brasilicum.**

+ Involucre of achenes more than two times as long as wide .... 4.

4. Involucre of achenes narrow-ellipsoidal, gradually narrowed at both ends, 14–17 mm long and 4–7 mm wide, with occasional spines, absent at apex; spines 2–3 mm long, thickened at base; beaks more or less equal and approximate .... 1. **X. strumarium.**

+ Involucre of achenes obovate or elliptical, narrowed toward base, 12–14 mm long and 5–8 mm wide; beaks scarcely thickened at base, dissimilar, widest, sometimes one undeveloped or entirely absent .... 3. **X. sibiricum.**

5. Involucre of achenes narrow-ellipsoidal, three and more times as long as wide .... 6.

+ Involucre of achenes ovate, obovate or elliptical, always less than three times as long as wide .... 8.

6. Involucral spines yellow, densely covered with squarrose hairs .... 7.

+ Involucral spines reddish-brown, scatteredly hairy; involucre of achenes distinctly inequilateral (on one side thickened at base or below middle), densely glandular, 17–26 mm long, 5–8 mm wide, densely covered with more or less pubescent subulate or somewhat curved 4–7 mm long spines; beaks, stiff, falcately curved toward each other, 4–7 mm long .... 7. **X. pensylvanicum.**

7. Involucre of achenes bronze-yellow, elongate-ovate, 23–27 mm long, 6–8 mm wide, densely covered with uncinate 6–10 mm long spines equaling diameter of involucre or exceeding it .... 8. **X. italicum.**

+ Involucre of achenes yellowish-brown or reddish, inequilateral (with one side thickened at apex), 16–22 mm long and 5–9 mm wide; spines somewhat curved, apically often subulate, 2–3 mm long, distinctly shorter than diameter of involucre .... 4. **X. ripicola.**

8. Involucre of achenes elongate-fusiform, 20–28 mm long, 6–8 mm wide; beaks falcate, 5–9 mm long, not split; spines 5–7 mm long, more or less bent, uncinate ....
.... 6. **X. californicum.**

+ Involucre of achenes elongate-ovate, inequilateral, 15–26 mm long, 7–9 mm wide; beaks thickened, divergent at right angle, weakly uncinate at apex, each often split into two unequal parts; spines straight, more or less uncinate ..... 9.

9. Beaks 4–8 mm long; spines 3–5 mm long ....
.... 5. **X. albinum.**

+ Beaks 2.5–3.5 mm long; spines 1–2 mm long ....
.... 9. **X. palustre.**

50

SUBGENUS **1. *XANTHIUM***

Stem not spiny; leaves undivided or shallowly three-lobed; involucre of seeds with two distinct beaks.

Type: type species.

Section 1. Xanthium.

Plant grayish-green, soft-hairy; involucre of achenes mostly with straight parallel beaks, covered with short soft hairs.

Type: type species.

1. **X. strumarium** L. 1753, Sp. Pl.: 987; Smoljan. 1959, Fl. SSSR, **25**: 524; D. Love, 1976, Fl. Europ. **4**: 143.—(Plate II, 7).

Type: Europe, Canada, Virginia, Jamaica, Ceylon, Japan ("in Europa, Canada, Virginia, Jamaica, Zeylona, Japonia").

North (Dvina-Pechora: south); *Baltic*; *Center*; *West*; *East*; *Crimea.*—By roadsides, in habitations, weedy meadows and pastures, on banks of water bodies, sometimes in fields.—*General distribution*: Caucasus, Western Siberia (south), Far East (rarely), Russian Central Asia; Central and Atlantic Europe, Mediterranean, Asia Minor, Iran, Dzhungaria-Kashgaria, Japan-China; North and South America, Africa.—2n = 36.

Note. Many reports of this species relate to *X. albinum.*

2. **X. brasilicum** Vellozo, 1827, Fl. Flumin.: 10, tab. 23; Duguyan, 1960, Tr. Rostov. Otd. Vsesouz. Bot. Obshch. **1**: 54.

Type: Brazil (the above cited figure is the type).

Center (Volga-Kama: vicinity of the village of Kizner); *West* (Dnieper; Black Sea); *East* (Lower Don: Rostov Region).—On banks of rivers, sand and gravel-beds of seacoast, by roadsides, in habitations; as an ecdemic plant.—*General distribution*: Caucasus; Central and Atlantic Europe, Mediterranean; North and South America.—2n = 36.

3. **X. sibiricum** Patrin ex Widd. 1923, Feddes Repert. Beih. **20**: 32; Smoljan. 1959, Fl. SSSR, **25**: 524; Duguyan, 1960, Tr. Rostov. Otd. Vsesoyuz. Bot. Obshch. **1**: 55; Golubov and Kossykh, 1982, Bot. Zhurn. **67**, 9: 1299.—*X. japonicum* auct. non Widd.: Vorosch. 1985, Flor. Issl. v Razn. Raionakh SSSR: 194.

Type: Altai ("in Kamenogorsk").

Center (Volga-Kama: vicinity of Sarapul); *West* (Black Sea: Zaporozh'e Region); *East* (Lower Don: Rostov Region); *Crimea* (Yalta).—On banks of water bodies, in habitations, by roadsides; as an ecdemic plant.—*General distribution*: Caucasus (Transcaucasia), Western Siberia, Eastern Siberia, Far East, Russian Central Asia; Iran, India, Japan-China.

Section 2. Campylorrhyncha (Wallr. ex Widd.) Widd. 1967, Phyton, **12**: 185.—*Xanthium* subsect. *Campylorrhyncha* Wallr. ex Widd. 1923, Feddes Repert. Beih. **20**: 18.

Plant yellowish-green, covered with very short, somewhat stiff, squarrose hairs; involucre of achenes with curved or falcate beaks divergent at an angle, setose with mixture of glandular hairs.

Type: *X. albinum* (Widder) H. Scholz.

4. **X. ripicola** Holub, 1976, Folia Geobot. Phytotax. (Praha), **11**: 83.—*X. riparium* Lasch, 1856, Bot. Zeit. **14**: 412, non Itz and Hertsch, 1854.—*X. riparium* auct. non Itz et Hertsch: Smoljan. 1959, Fl. SSSR, **25**: 529.

Type: Germany ("Flu gebiet der Oder: pr. Driesen "Neomarchiae"").
Baltic; *Center* (Volga-Kama; Volga-Don); *West* (Dnieper; Black Sea).—By roadsides, in habitations, on banks of water bodies, in weedy meadows and pastures.—*General distribution*: Scandinavia (south), Central and Atlantic Europe; North America.—2n = 36.

Note. According to Widder's monograph of the genus, this species evolved from the North American *X. saccharatum* Wallr.

5. **X. albinum** (Widd.) H. Scholz, 1960, Verh. Bot. Ver. Brandenb. **47**: 98–100; Schultz, 1976, Bot. Zhurn. **61**, 10: 1452; Alekseev and Makarov, 1977, Byull. Mosk. Obshch. Isp. Prir., Otd. Biol. **82**, 6: 91; Ilminskikh and others, 1981, Bot. Zhurn., **66**, 8: 1222;

Oktyabreva and Tikhomir, 1987, Opred. Rast. Meshcher. **2**: 105; Tikhomir, 1987, Nauchn. Dokl. Vyssh. Shkoly, Biol. Nauki, **6**: 77.—*X. riparium* Itz. et Hertsch. var. *albinum* Widd. 1923, Feddes Repert. Beih. **20**: 105.—*X. californicum* auct. non Greene: Smoljan. 1959, Fl. SSSR, **25**: 529, p. p.—*X. cavanillesii* auct. non Schouw. ex Didr.: Protopopova, 1964, Ukr. Bot. Zhurn. **21**, 4: 80, fig. 4.—*X. speciosum* auct. non Kearn.: Protopopova, 1964, ibid.: 82, fig. 7.—*X. occidentale* auct. non Bertol.: Kotov, 1965, Vizn. Rosl. Ukr., ed. 2: 670.—(Plate II, 6).

Type: Germany ("Dresden, a. Uferd Elbe").

North (Dvina-Pechora: south); *Baltic*; *Center*; *West*; *East*; *Crimea*.—By roadsides, on banks of water bodies, in habitations, weedy meadows and sometimes in fields.—*General distribution*: Caucasus, Far East (south), Russian Central Asia; Central Europe, Mediterranean, Asia Minor; North America.—2n = 36.

Note. The species most common in the European part of Russia and bordering regions, occurs quite abundantly in the river floodplains.

6. **X. californicum** Greene, 1899, Pittonia, **4**: 62; Smoljan. 1959, Fl. SSSR, **25**: 526, p. p.

Type: North America ("Middle California, especially about San Francisco Bay").

Baltic (rarely); *Center* (Upper Dnieper; Upper Don); *West* (Moldavia; Black Sea); *East* (Lower Don); *Crimea* (vicinity of Simferopol and Yalta).—By roadsides, in habitations, weedy meadows, and pastures, on banks of water bodies.—*General distribution*: Caucasus, Russian Central Asia; North America.—2n = 36.

Note. Large volume of literature on the distribution of this species relates to *X. albinum.*

7. **X. pensylvanicum** Wallr. 1842, Beitr. Bot. **1**: 236; Duguyan, 1960, Tr. Rostov. Otd. Vsesoyuz. Bot. Obshch. **1**: 50.

Type: Pensylvania ("in Pensylvania: without locality invading the marshes of paludosis"), North Carolina ("....from Asheville").

West (Moldavia; Black Sea: Odessa, Nikolaev and Kherson regions); *East* (Lower Don; Lower Volga); *Crimea*.—By roadsides, on sandy riverbanks, in habitations.—*General distribution*: North America.—2n = 36.

Note. Hybrids of *X. pensylvanicum* × *X. ripicola* and *X. pensylvanicum* × *X. italicum* are reported in Volgograd Region and Kalmyk ASSR.

8. **X. italicum** Moretti, 1822, Giorn. Fis. Pavia. **5**: 326; Protopopova, 1964, Ukr. Bot. Zhurn. **21**, 4: 79.

Type: Italy ("sulla riva del Po").

West (Moldavia; Black Sea: Odessa Region, on the Danube).—On banks of water bodies, by roadsides.—*General distribution*: Central and Atlantic Europe, Mediterranean.—2n = 36.

9. **X. palustre** Greene, 899, Pittonia, **4**: 63; Duguyan, 1960, Tr. Rostov. Otd. Vsesoyuz. Bot. Obshch. **1**: 51.

Type: California ("California: suisum Marsh").

52   *East* (Lower Don: vicinity of Rostov; Lower Volga: Astrakhan Region).—On sandy and gravelly riverbanks.—*General distribution*: North America.

Note. Predisposition to hybridization and polymorphism of the involucre in the subgenus *Xanthium* has led to great confusion in botanical literature. Some species were cited for the European part of Russia and Bordering Regions under incorrect nomenclature. Thus *X. albinum* was sometimes related to species in our Flora which were never found, sometimes to *X. italicum*, *X. californicum*, or to *X. strumarium*. Under *X. californicum* often four species were included: *X. californicum*, *X. albicum*, *X. pensylvanicum*, and *X. ripicola*. For a long time species not only from the section *Xanthium* but also section *Campylorrhyncha* were related to *X. strumarium*.

SUBGENUS **2. *ACANTHOXANTHIUM*** (DC.) Widd. 1965, Phyton, **11**: 71.—*Xanthium* sect. *Acanthoxanthium* DC. 1836, Prodr, **5**: 523.—*Acanthoxanthium* (DC.) Fourr. 1869, Ann. Soc. Linn. Lyon, N.S. 17: 110.

Leaf bases with yellow triparted spines; leaves sinuate-pinnatipartite; involucre of achenes straw-yellow, predominantly with one straight subulate beak.

Type: *X. spinosum* L.

10. **X. spinosum** L. 1753, Sp. Pl. 987; Smoljan. 1959, Fl. SSSR, **25**: 523; D. Löve, 1976, Fl. Europ. **4**: 143; Rakov and Pchelkin, 1980, Bot. Zhurn. **65**, 5: 713; Ilminskikh and others, 1981, Bot. Zhurn. **66**, 8: 1222; Puzyrev, 1986, Bot. Zhurn. **71**, 2: 260; Dimitriev, 1987, Novosti Sist. Vyssh. Rast. **24**: 225; Oktyabreva and Tikhomir, 1987, Opred. Rast. Meschchery, **2**: 106.—*Acanthoxanthium spinosum* (L.) Fourr. 1869, Ann. Soc. Linn. Lyon, NS, **17**: 110.— (Plate II, 5).

Type: Portugal ("in Lusitania").

Baltic; *Center* (Upper Dnieper: south; Upper Volga; Volga-Kama: south; Volga-Don); *West*; *East*; *Crimea*.—In habitations, by roadsides, in weedy meadows and pastures, on banks of water bodies.—*General distribution*: Caucasus, Western Siberia (southwest), Far East (south), Russian Central Asia; Atlantic and Central Europe, Mediterranean, Asia Minor, Iran; North and South America, Australia, Africa.

TRIBE 2. **SENECIONEAE** Cass.[1]

Heads heterogamous, less often homogamous; peripheral flowers ligulate, less often filiform-tubular, yellow, white, reddish-pinkish and so on, less often absent; inner flowers (disk flowers) tubular, usually yellow. Receptacle lacking bracts, glabrous, rarely with hairs. Anthers at base obtuse, sagittate or auriculate. Style branches usually seemingly truncate, less often acuminate, apically often papillate. Pollen helianthoid type. Achenes with pappus of thin, scabrous bristles. Leaves alternate, with undivided, less often more or less incised blades.

Type: *Senecio* L.

GENUS **25.** *SENECIO* L.
1753, Sp. Pl.: 866; id. 1754, Gen. Pl., ed. 5: 373

Heads heterogamous or homogamous, with numerous yellow flowers, solitary or clustered in corymbose, racemose, or paniculate inflorescence. Involucre campanulate or cylindrical, biseriate; involucral bracts of inner row wide with scarious border, outer narrower, lacking membranous border, usually shorter than inner, occasionally as long as or exceeding. Receptacle glabrous, flat. All flowers tubular, bisexual or peripheral ligulate, pistillate; central tubular, bisexual. Stamens with trigonal or ovate connectives, obtuse bases of anthers, and downward expanded antheropodia (neck of anthers). Style branches truncate, with crown of hairs at apex, stigmatic surface as two strips on inner side of each branch. Achenes cylindrical, with 10 longitudinal, more or less distinct ribs, pubescent or glabrous, grayish, olive or light brown; pappus white or brownish, consisting of scabrous bristles. Annual, biennial or perennial herbs, rarely semishrubs or shrubs; leaves

53

[1]Treatment by G.Yu. Konechnaya.

64

alternate, with undivided, toothed, less often entire blade or
pinnatisect to various degree.

Lectotype: *S. vulgaris* L.

About 3,000 species, distributed in all the continents except
Antarctica, but predominantly in the extratropical countries and
montane regions of the tropics.

Literature: Alexander, J.C.M. 1979. The Mediterranean species
of *Senecio* sections *Senecio* and *Delphinifolius Notes Roy. Bot.
Gard. Edinb.*, **37**, 3: 387–428.—Konechnaya, G.Yu. 1981. Kariologo-
anatomicheskie priznaki vidov roda *Senecio* s. l. (Asteraceae) v
svazi s ikh sistematikoi [The karyological-anatomical characters
of species of the genus *Senecio* s. l. (Asteraceae) in connection
with their systematics]. *Bot. Zhurn.*, **66**, 6: 834–842.

1. Annuals; outer involucral bracts less than one-fourth as
   long as inner; achenes very narrow, almost acicular .... 2.
 + Biennials or perennials; outer involucral bracts not shorter
   or not less than one-third as long as inner; achenes
   cylindrical or broadly cylindrical .... 7.
2. Ligulate flowers absent in head...................................... 3.
 + Ligulate flowers present, but often very short and
   inconspicuous .... 4.
3. Leaves coarsely toothed or pinnatilobate, with broadly
   toothed lobes .... 1. **S. vulgaris.**
 + Leaves pinnatisect, with linear, undivided or indistinctly
   toothed lobes, narrowed toward apex .... 2. **S. dubius.**
4. Plants sticky from glandular pubescence; ligulate flowers
   very small.. 3. **S. viscosus.**
 + Plants glabrous or arachnoid-hairy .... 5.
5. Involucre 2–4 mm in dia; ligulate flowers very small, their
   limb about 1 mm long..............................4. **S. sylvaticus.**
 + Involucre more than 4 mm in dia; ligulate flowers one
   and one-half to two times as long as involucre .... 6.
6. Plants usually arachnoid-hairy; leaf segments broad,
   toothed ....5. **S. vernalis.**
 + Plants glabrous; leaf segments narrow, undivided or with
   sharp teeth..6. **S. noeanus.**
7(1). All or only upper cauline leaves pinnatilobate to pinnatisect
   or lyrate... 8.
 + All leaves undivided, more or less toothed, rarely entire
   ....20.
8. Plants with single head; leaf segments linear................
   .... 17. **S. carpaticus.**

+ Plants with many heads, aggregated in corymbose or corymbose-paniculate inflorescence .... 9.

9. Perennial, nonrosulate plants .... 10.

+ Biennial or perennial plants, with well-developed rosette of basal leaves often withering by flowering .... 12.

10. Cultivated, but sometimes as escape, white-tomentose plants; achenes glabrous or subglabrous........................
.... 15. **S. cineraria.**

+ Wild plants, with green or only on lower surface grayish-tomentose leaves; achenes densely hairy .... 11.

11. Leaves bipinnatisect to narrow segments but lower sometimes pinnatilobate, glabrous or sparsely floccose-tomentose... 18. **S. erucifolius.**

+ Leaves pinnatilobate or pinnatisect to wide segments, usually tomentose beneath........... 19. **S. grandidentatus.**

54   12(9). Entire plant silver-gray from dense pubescence ....
.... 16. **S. carniolicus.**

+ Plant glabrous or weakly pubescent .... 13.

13. Basal and lower cauline leaves undivided, cordate.......
.... 14. **S. subalpinus.**

+ Basal leaves lyrate, pinnatisect, or undivided, cuneate ... 14.

14. All achenes glabrous or peripheral glabrous, middle covered with short papillate hairs................................ 15.

+ Peripheral achenes (of ligulate flowers) glabrous, inner (of tubular flowers) covered with rather long 1–8 mm hairs .... 16.

15. Lateral segments of leaves at acute angle to rachis limb of ligulate flowers 8–10 mm long .... 12. **S. aquaticus.**

+ Lateral segments of leaves at right angle to rachis; limb of ligulate flowers to 5 mm long .... 13. **S. erraticus.**

16. Cauline leaves twice or thrice divided to linear segments about 1 mm wide. Heads in corymbose-paniculate inflorescence, not approximate after flowering. Plants of sands, 20–100 cm high.................... 7. **S. borysthenicus.**

+ Cauline leaves divided in broader (3–6 mm) lobes .... 17.

17. Heads (20) 30–50 (60), but smaller; involucre (4)5–7(8) mm wide. Plants 20–100 cm high .... 18.

+ Heads (4)8–20(25), rather large, in corymbose inflorescence, not approximate after flowering, (7)8–13(15) mm wide. Plants (10)20–50(70) cm high .... 19.

18. Common inflorescence during flowering rapidly turning corymbose-paniculate, with spreading branches; heads

at different levels and not approximate after flowering. Plants of sands, of terraces above floodplain in steppe and forest-steppe zones.................. 8. **S. andrzejowskyi.**

+ Common inflorescence with heads more or less at same level, usually somewhat approximate after flowering. Widely distributed plant... 9. **S. jacobaea.**

19. Plants of steppe slopes of southeastern European part of Russia and bordering regions, 20–70 cm high, usually with 10–25 heads; involucre 7–10 mm in dia................. .... 10. **S. ferganensis.**

+ Plants of Crimean yailas, 10–30 cm high, usually with 4–15 heads; involucre 10–15 mm in dia ... 11. **S. tauricus.**

20(7). Plants with rosette of basal leaves; cauline leaves fewer (to 12) ....21.

+ Plants without leaf rosette, with numerous lanceolate or oblong-ovate cauline leaves ....23.

21. Common inflorescence racemose or paniculate-racemose; heads with 13 ligulate flowers............ 26. **S. paucifolius.**

+ Common inflorescence corymbose; heads with five to eight ligulate flowers..22.

22. Heads with five ligulate flowers; involucre 4–6 mm in dia .... 24. **S. schvetzovii.**

+ Heads with eight ligulate flowers; involucre 7–10 mm in dia.. 25. **S. umbrosus.**

23(20). Stem slender, not hollow. Plant glabrous or pubescent, with erect hairs; leaves sometimes arachnoid-hairy beneath; ligulate flowers 5–8(13) in head...................24.

+ Stem thick, hollow. Plant usually tomentose or arachnoid-hairy; ligulate flowers 13 and more in each head......27.

24. Plants with long creeping rhizome; leaves serrate-dentate .... 23. **S. fluviatilis.**

+ Plants with short (to 3 cm) rhizomes; leaves with erect or inclined teeth ....25.

25. Ligulate flowers 8(13) in head; involucre 6–8(12) mm in dia.. 20. **S. nemorensis.**

+ Ligulate flowers five in each head; involucre 2–5 mm in dia ....26.

26. Outer involucral bracts as long as or exceeding inner; upper and middle cauline leaves sessile or with broadly winged petiole...................................... 21. **S. jacquinianus.**

+ Outer involucral bracts half as long as inner; all leaves petiolate or sessile and narrow-cuneate....22. **S. ovatus.**

27(23). Leaves beneath glabrous or arachnoid-hairy, but not white-
tomentose... 27. **S. paludosus.**
  + Leaves white-tomentose beneath, also finely tomentose
    above ....28. **S. tataricus.**

SUBGENUS 1. *SENECIO*

Annuals, with leaves variously incised. Outer involucral bracts less than one-fourth as long as inner. Achenes very narrow, almost acicular, with 10 round, less often oval strands of sclerenchyma in transverse section.

Type: lectotype of genus.

1. **S. vulgaris** L. 1753, Sp. Pl.: 867; Schischk. 1961, Fl. SSSR, **26:** 780; Chater and Walters, 1976, Fl. Europ. **4:** 204.—(Plate III, 7).

Type: Europe ("in Europae cultis ruderatis, succulentis").
North; *Baltic*; *Center*; *West*; *East*; *Crimea.*—In fields and plantations of different crops, by roadsides and pathways, in habitations.—*General distribution*: Caucasus, Western and Eastern Siberia, Far East, Russian Central Asia; Scandinavia, Central and Atlantic Europe, Mediterranean, Asia Minor, Iran, Dzhungaria-Kashgaria, Mongolia, Himalayas, Japan-China; ecdemic in North and South America, Australia and Africa.—2n = 20, 40.

2. **S. dubitalis** C. Jeffrey et G.L. Chen, 1984, Kew. Bull. **39,** 2: 427.—*S. dubius* Ledeb. VII–XII, 1833, Fl. Alt. **4:** 112, non Beck, V–VI, 1833; Schischk. 1961, Fl. SSSR, **26:** 781.—*S. vulgaris* auct.: Chater and Walters, 1976, Fl. Europ. **4:** 204, p. p.

Type: Eastern Kazakhstan ("in deserto songorokirghisico ad rivarum Tschaganka").
Center (Volga-Kama: east); *East* (Lower Don: southeast; Lower Volga).—On banks of water bodies, sands and gravelbeds, in solonetzic meadows.—*General distribution*: Western and Eastern Siberia (south), Far East (west), Russian Central Asia; Mongolia, Tibet, Himalayas, India.

3. **S. viscosus** L. 1753, Sp. Pl.: 868; Schischk. 1961, Fl. SSSR, **26:** 779; Chater and Walters, 1976, Fl. Europ. **4:** 204.—(Plate III, 8).

Type: Europe ("in Europae pagis, urbibus").
North; *Baltic*; *Center*; *West.*—On coastal sands, sandy banks of rivers and lakes, by roadsides, in habitations, predominantly ecdemic plant.—*General distribution*: Far East (ecdemic);

Scandinavia, Central and Atlantic Europe, Mediterranean; North America (ecdemic).—2n = 40.

4. **S. sylvaticus** L. 1753, Sp. Pl.: 868; Schischk. 1961, Fl. SSSR, **26**: 782; Chater and Walters, 1976, Fl. Europ. **4**: 204.— (Plate III, 5).

Type: Europe ("in Europae borealis sylvis ceduis").

North (Karelia-Murman); *Baltic*; *Center*; *West.*—By roadsides, in habitations, sometimes in pine forests, in sandy forest glades, predominantly as an ecdemic plant.—*General distribution*: Caucasus; Scandinavia, Central and Atlantic Europe, Mediterranean; North and South America (ecdemic).—2n = 40.

5. **S. vernalis** Waldst. et Kit. 1802, Descr. Icon. Pl. Rar. Hung. **1**: 23; Schischk. 1961, Fl. SSSR, **26**: 783; Chater and Walters, 1976, Fl. Europ. **4**: 204.

Type: Yugoslavia ("ad sepes vincarum et in aggeribus in Comitatu Syrmeinsi").

56 *North* (ecdemic); *Baltic* (ecdemic); *Center* (in northern regions, only as ecdemic); *West*; *Crimea.*—In sandy meadows and forest glades, on coastal sands, gravel-beds, stony slopes, by roadsides, sometimes in fields.—*General distribution*: Caucasus, Russian Central Asia; Scandinavia, Central and Atlantic Europe, Mediterranean, Asia Minor.—2n = 20, 40.

6. **S. noeanus** Rupr. 1856, Bull. Phys.-Math. Acad. Sci. Petersb. **14**: 231; Schischk. 1961, Fl. SSSR, **26**: 787; Safonov, 1986, Novosti Sist. Vyssh. Rast. **23**: 253.—*S. gallicus* auct.: Chater and Walters, 1976, Fl. Europ. **4**: 203, p. p.—(Plate III, 6).

Type: Iraq ("Kutt").

East (Lower Volga); *Crimea.*—On sandy and gravelly banks of rivers and lakes, in steppes and semisteppes, on solonchaks.—*General distribution*: Caucasus, Russian Central Asia; Asia Minor, Iran.

Note. The species is described from plants raised in 1853 from seeds collected by Noe from Iran.

---

57   Plate III.

1—*Senecio numorensis* L., 1a—head; 2—*S. ovatus* (Gaertn., Mey et Scherb.) Willd., head, 2a—middle cauline leaf; 3—*S. fluviatilis* Wallr., head, 3a— middle cauline leaf; 4—*S. jacquinianues* Reichenb., head, 4a—upper leaf; 5—*S. sylvaticus* L., head, 5a—leaf; 6—*S. neoanus* Rupr., head, 6a—leaf; 7—*S. vulgaris* L., head, 7a—leaf; 8—*S. viscosus* L., head, 8a—leaf.

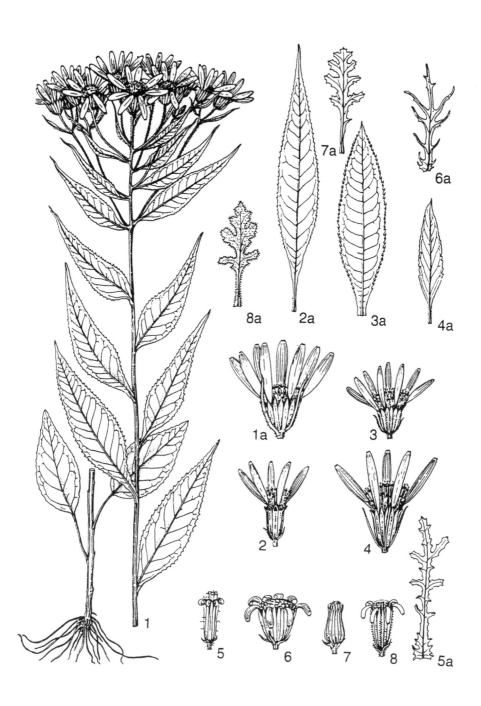

SUBGENUS **2. *JACOBAEA* Cass.**
1822, Dict. Sci. Nat. **24:** 110; Reichenb. 1841, Repert. Herb.: 87

Perennial or biennial plants. Leaves undivided or pinnatisect to various degree. Outer involucral bracts not shorter or not less than one-third as long as inner. Achenes terete, with 10 oval or oblong strands of sclerenchyma in transverse section.

Lectotype: *S. jacobaea* L.

Section 1. Jacobaea (Cass.) Dumort. 1827, Fl. Belg.: 65, p. p.; Schischk. 1961, Fl. SSSR, **26:** 710; Chater and Walters, 1976, Fl. Europ. **4:** 201.—*Senecio* subgen. *Jacobaea* Cass. 1822, Dict. Sci. Nat. **24:** 110.

Biennial or perennial plants; stem solid; leaves coarsely toothed to twice- or thrice-pinnatisect; ligulate flowers usually about 13, occasionally absent; achenes glabrous or pubescent.

Lectotype: *S. jacobaea* L.

Subsection 1. Jacobaea Konechn. 1981, Bot. Zhurn. **66,** 6: 841.

Leaves pinnatisect, lyrate, less often undivided, toothed, leaf segments and teeth subobtuse; secretory canals absent in pericarp.

Type: type of section.

7. **S. borysthenicus** (DC.) Andrz. ex Czern. 1859, Consp. Pl. Charcov.: 32; Gruner, 1868, Bull. Soc. Nat. Moscou, **41,** 2: 423; Schischk. 1961, Fl. SSSR, **26:** 718.—*S. praealtus* Bertol. β. *borysthenicus* DC. 1838, Prodr. **6:** 351.—(Plate IV, 4).

Type: Ukraine ("in pratis Podoliae, Cherson, Russiae circ. Borysth.").

West (Dnieper; Moldavia; Black Sea); *Crimea.*—On coastal and river sands, sandy glades in pine forests.—*General distribution*: Mediterranean (Romania).

O 8. **S. andrzejowskyi** Tzvel. 1986, Novosti Sist. Vyssh. Rast. **23:** 254.

Type: Voronezh Region ("Novokhopersk District, Khoper Preserve, terraces above floodplain of the Khoper River, sandy meadow near the village of Varvarino").

Center (Volga-Don: south and east); *West* (Dnieper: east; Black Sea); *East.*—On sandy terraces above the floodplain of the river.—Endemic.

9. **S. jacobaea** L. 1753, Sp. Pl.: 870; Schischk. 1961, Fl. SSSR, **26:** 715; Chater and Walters, 1976, Fl. Europ. **4:** 201.—(Plate IV, 3).

Type: Europe ("in Europae pascuis").

North (Dvina-Pechora: ecdemic); *Baltic*; *Center*; *West*; *East*; *Crimea.*—In dry bottom meadows, forest glades and edges, pine forests, by roadsides.—*General distribution*: Caucasus, Western and Eastern Siberia (south), Russian Central Asia; Scandinavia (south), Central and Atlantic Europe (ecdemic).—2n = 40, 80.

10. **S. ferganensis** Schischk. 1961, Fl. SSSR, **26**: 881, 720; Tzvel. 1986, Novosti Sist. Vyssh. Rast. **23**: 255.

58   Type: Russian Central Asia ("Asia Media, jugum Alaicum, in faucibus fluminis Czigirczik, in clivo montis supra ripam").

Center (Volga-Don: east); *West*: (Dnieper: east); *East.*—In steppes, in steppefied forest glades.—*General distribution*: Russian Central Asia.

Note. It is mentioned in the original description of this species that all achenes are glabrous, although only its peripheral achenes are glabrous.

○ 11. **S. tauricus** Konechn. 1985, Novosti Sist. Vyssh. Rast. **22**: 230.—(Plate IV, 5).

Type: Crimea ("Crimea, Nikita yaila").

Crimea.—On grassy patches and stony slopes of yaila.— Endemic.

12. **S. aquaticus** Hill, 1761, Veg. Syst., **2**: 120; Huds, 1762, Fl. Angl.: 317; Schischk. 1961, Fl. SSSR, **26**: 788.—*S. aquaticus* Hill subsp. *aquaticus*; Chater and Walters, 1976, Fl. Europ. **4**: 202.— (Plate IV, 1).

Type: Great Britain ("native of our ditch fides").

Baltic; *Center* (Ladoga-Ilmen; Upper Dnieper).—In more or less wet meadows, forest glades and edges.—*General distribution*: Scandinavia (south), Central and Atlantic Europe.—2n = 40.

13. **S. erraticus** Bertol. 1810, Rar. Ital. Pl. **3**: 62; Schischk. 1961, Fl. SSSR, **26**: 707.—*S. aquaticus* Hill. var. *barbareifolius* Wimm. et Grab. 1829, Fl. Siles, **2**, 2: 151.—*S. barbareifolius* (Wimm. et Grab.) Reichenb. 1831, Fl. Germ. Excurs. **2**: 244; Konechn. 1981, Bot. Zhurn. **66**, 11: 1614.—*S. aquaticus* subsp. *barbareifolius* (Wimm. et Grab.) Walters, 1976, Journ. Linn. Soc. London (Bot.), **71**, 4: 266; Chater and Walters, 1976, Fl. Europ. **4**: 202.—(Plate IV, 2).

Type: Italy ("Sarzanae huc illuc erraticus, praecipue ad canaliculos campestres").

Baltic (south); *West* (Carpathians; Dnieper: west).—In damp meadows, ditches, on banks of water bodies.—*General distribution*:

Caucasus; Central and Atlantic Europe, Mediterranean, northern Africa; South America (ecdemic).—2n = 20, 40.

14. **S. subalpinus** Koch, 1834, Flora (Regensb.), **17:** 614; Schischk. 1961, Fl. SSSR, **26:** 726; Chater and Walters, 1976, Fl. Europ. **4:** 201.—*Cineraria cordifolia auriculata* Jacq. 1774, Fl. Austr. 2: 47, tab. 177.—(Plate V, 5).

Type: Austria ("in Austria").

West (Carpathians).—In meadows and forest glades in middle and upper mountain zones.—*General distribution*: Central Europe.—2n = 40.

15. **S. cineraria** DC. 1838, Prodr. **6:** 355; Schischk. 1961, Fl. SSSR, **26:** 720.—*Othonna maritima* L. 753, Sp. Pl.: 925.—*Cineraria maritima* (L.) L. 1763, Sp. Pl., ed. 2: 1244.—*Senecio bicolor* (Willd.) Tod. subsp. *cineraria* (DC.) Chater, 1974, Journ. Linn. Soc. London (Bot.), **68:** 273; Chater and Walters, 1976, Fl. Europ. **4:** 144.—*S. bicolor* auct. non (Willd.) Tod.: Czer. 1981, Sosud. Rast. SSSR: 95.—(Plate IV, 9).

Type: Southern Europe ("ad Maris inferi littora").

North; *Baltic*; *Center*; *West*; *East*; *Crimea*.—Extensively cultivated as an ornamental plant, as an escape in southern Crimea, on coastal slopes.—*General distribution*: Mediterranean; cultivated in many other countries.—2n = 40.

Note. In cultivation this species is known as "*Cineraria maritima*".

16. **S. carniolicus** Willd. 1800, Sp. Pl.: **3,** 3: 1803; Schischk. 1961, Fl. SSSR, **26:** 748.—*S. incanus* L. subsp. *carniolicus* (Willd.) Br. Bl. 1913, Neue Denkschr. Schweiz. Naturf. Ges. **48:** 300; Chater and Walters, 1976, Fl. Europ. **4:** 194.—(Plate IV, 8).

Type: Yugoslavia ("in alpibus Styriae, Carnioliae").

West (Carpathians).—On rocks and taluses in subalpine zone.— *General distribution*: Central Europe.—2n = 120.

Note. I have not seen the authentic herbarium specimens of this species from the territory of the former Soviet Union.

---

59   Plate IV.

1—*Senecio aquaticus* Hill, la—achene; 2—*S. erraticus* Bertol., middle cauline leaf; 3—*S. jacobaea* L., middle cauline leaf, 3a—head, 3b— peripheral achene, 3c—middle achene; 4—*S. borysthenicus* (DC.) Andrz. ex Czern., middle cauline leaf; 5—*S. tauricus* Konechn., head; 6—*S. erucifolius* L., middle cauline leaf, 6a—achene; 7—*S. grandidentatus* Ledeb., middle cauline leaf; 8—*S. carniolicus* Willd.; 9—*S. cineraria* DC., cauline leaf; 9a—achene.

3

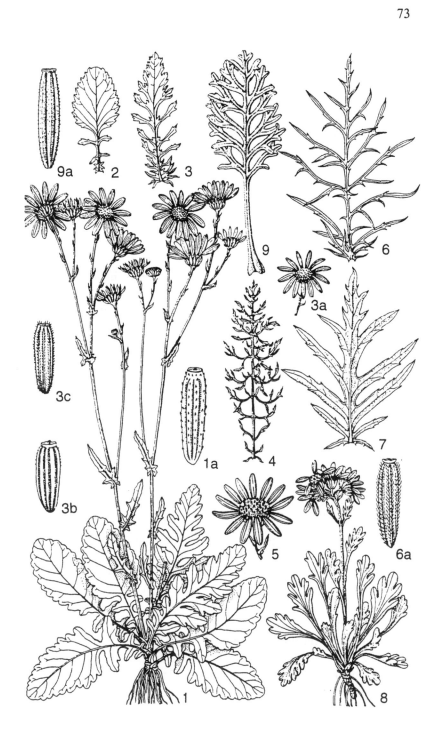

60   17. **S. carpaticus** Herbich, 1831, Add. Fl. Galic.: 43; Schischk.
1961, Fl. SSSR, **26:** 750.—*S. abrotanifolius* L. subsp. *carpaticus*
(Herbich) Nym. 1879, Consp. Fl. Eur. **2:** 356; Chater and Walters,
1976, Fl. Europ. **4:** 203.—(Plate V, 4).

Type: Poland ("in alpibus supra lacum Morskie Oko ad nives").
West (Carpathians).—In meadows, on stony slopes and rocks
in upper mountain zone.—*General distribution*: Central Europe
(Carpathians).—2n = 40.
Subsection 2. Erucifolii Rouy, 1913, Fl. Fr. **8:** 333, s. str.;
Konech. 1981, Bot. Zhurn. **66,** 11: 1615.

Leaves pinnatisect to varying degrees, their segments
acuminate, lanceolate; secretory canals present in pericarp.
Lectotype: *S. erucifolius* L.

18. **S. erucifolius** L. 1753, Sp. Pl.: 869; Schischk. 1961, Fl.
SSSR, **26:** 70; Chater and Walters, 1976, Fl. Europ. **4:** 202, p. p.—
S. tenuifolius Jacq. 1775, Fl. Austr. **3:** 42; Katina, 1987, Opred.
Vyssh. Rast. Ukr.: 344.—(Plate IV, 6).

Type: Europe ("in aggeribus Europae temperatae").
North (Dvina-Pechora: Pechora Basin); *Center* (Upper Dnieper;
Upper Volga; Volga-Kama; Volga-Don); *West*; *East*.—In meadows,
forest glades and edges, in steppes and thinned-out forests,
sometimes as an ecdemic of roadsides.—*General distribution*:
Western and Eastern Siberia; Scandinavia (south of Sweden),
Central and Atlantic Europe, Mediterranean, Asia Minor; North
America (ecdemic).—2n = 40.

19. **S. grandidentatus** Ledeb. 1845, Fl. Ross. **2:** 636; Schischk.
1961, Fl. SSSR, **26:** 711.—*S. erucifolius* auct. non L.: Chater and
Walters, 1976, Fl. Europ. **4:** 202, p. p.—(Plate IV, 7).

Type: Caucasus ["in provinciis caucasicis (in insula Sara m.
Caspii)"].
Center (Volga-Don: south and east); *West* (Dnieper; Black
Sea); *East*; *Crimea*.—On banks of water bodies, in more or less
wet meadows and forest glades.—*General distribution*: Caucasus,
Western Siberia (Kurgan Region); Iran.
Section 2. Pseudooliganthi Sof. 1957, Izv. Akad. Nauk
AzerbSSR, **1:** 88; Schischk. 1961, Fl. SSSR **26:** 742.—*Senecio* C.
Doria Reichenb. 1831, Fl. Germ. Excurs. **2:** 244, p. p.—*Senecio*
sect. *Doria* (Reichenb.) Godr. 1851, in Gren. and Godr. Fl. Fr. **2:**
117, p. p.; Chater and Walters, 1976, Fl. Europ. **4:** 195, p. p.

Perennial, nonrosette plants; stem solid; leaves undivided; ligulate flowers, 5, 8, less often 13 and more in one head; achenes glabrous.

Lectotype: *S. nemorensis* L.

20. **S. nemorensis** L. 1753, Sp. Pl.: 870, p. p. (excl. pl. Germ.); Schischk. 1961, Fl. SSSR, **26**: 742, p. p.—*S. otoglossus* DC. 1838, Prodr. **6**: 353, p. p.—*S. nemorensis* L. var. *polyglossus* Rupr. 1845, Beitr. Pflanzenk. Russ. Reich., **2**: 43.—*S. nemorensis* L. subsp. *nemorensis*; Chater and Walters, 1976, Fl. Europ. **4**: 196, p. min. p.—(Plate III, 1).

Type: Siberia ("Siberiae nemoribus").

North (Karelia-Murman: Murmansk Region; Dvina-Pechora); *Center* (Volga-Kama); *East* (Trans-Volga).—In forests, forest glades and edges.—*General distribution*: Western and Eastern Siberia, Russian Central Asia.—2n = 40.

21. **S. jacquinianus** Reichenb. 1825, Iconogr. Bot. Pl. Crit. **3**: 801; id. 1831, Fl. Germ. Excurs. **2**: 245.—*S. nemorensis* L. 1753, Sp. Pl.: 870, quoad pl. Germ.; Jacq. 1768, Obs. Bot. **3**: 15, tab. 65, 66; Schischk. 1961, Fl. SSSR, **26**: 742, p. p.—*S. nemorensis* subsp. *nemorensis*; Chater and Walters, 1976, Fl. Europ. **4**: 196, p. max. p.

Type: Hungary ("planta sylvestri Pannonica Austriacaque").

Center (Volga-Kama); *West* (Carpathians; Dnieper: West); *East* (Trans-Volga).—In forests on open lands, near streams.—*General distribution*: Central Europe.—2n = 40.

Note. The figure from the work of Jacquin, 1768, op. cit.: tab. 65, 66, drawn from Hungarian plants, is considered the type of the species.

22. **S. ovatus** (Gaertn., Mey. et Scherb.) Willd. 1803, Sp. Pl.: **3**, 3: 2004; L'Herborg, 1985, Willdenowia, **15**: 183.—*Jacobaea* 62 *ovata* Gaertn., Mey. et Scherb. 1801, Fl. Wett. **3**, 1: 212.—*S. fuchsii* C.C. Gmel. 1808, Fl. Bad. **3**: 444; Schischk. 1961, Fl. SSSR, **26**: 745.—*S. nemorensis* L. subsp. *fuchsii* (C.C. Gmel.) Čelak. 1871, Prodr. Fl. Böhm. 241; Chater and Walters, 1976, Fl. Europ. **4**: 196.—*S. sarracenicus* auct. non L.: Minder. 1962, Fl. URSR, **11**: 380.—(Plate III, 2).

Type: Germany ("in lichten Waldern und an deren Randern. In der Hanauer Fasanerie; im Bachköbler Wald, am Kohlbrunnen und im alten Wald; an dem Reichenbach hinter Falkenstein; an einem Bächlein am Rüdlinghayner Wäldchen am Oberwald; bei dem Taufstein im Oberwald; auf dem Münsterberg zwishcen Orb und Salmünster").

Baltic (Hiiumaa Island); *West* (Carpathians).—In forests, forest glades and edges, in scrublands.—*General distribution*: Scandinavia (south of Sweden), Central and Atlantic Europe.—2n = 40.

Note. This species was collected on Hiiumaa Island in 1874 by Winkler. The report needs verification.

23. S. fluviatalis Wallr. 1840, Linnaea, **14**: 646; Schischk. 1961, Fl. SSSR, **26**: 745; Chater and Walters, 1976, Fl. Europ. **4**: 195.—*S. sarracenicus* L. 1753, Sp. Pl.: 871, nom. ambig.—(Plate III, 3).

Type: Germany ("der Flusse der südlichen Harzgränze").

North (Dvina-Pechora: southwest); *Baltic*; *Center* (Ladoga-Ilmen: southeast; Upper Dnieper; Upper Volga; Volga-Kama; Volga-Don); *West* (Carpathians; Dnieper); *East* (Trans-Volga).—On banks of water bodies.—*General distribution*: Caucasus (southern Transcaucasia), Western and Eastern Siberia, Russian Central Asia; Central and Atlantic Europe.—2n = 40.

Section 3. Doria (Reichenb.) Godr. 1851, in Gren. and Godr. Fl. Fr. **2**: 117, p. min. p.; Chater and Walters, 1976, Fl. Europ. **4**: 195, p. p.—*Senecio* C. *Doria* Reichenb. 1831, Fl. Germ. Excurs. **2**: 244, p. min. p.—*Senecio* sect. *Crociserides* auct. non Rouy: Schischk. 1961, Fl. SSSR, **26**: 726, p. p.

Perennial, semirosette plants; stem solid; leaves undivided; heads with 5, 8, less than 13 ligulate flowers; achenes glabrous or pubescent.

Type: *S. doria* L.

24. S. schvetzovii Korsh. 1898, Mém. Acad. Sci. Petersb. **7**, 1: 519; Schischk. 1961, Fl. SSSR, **26**: 733.—*S. doria* L. subsp. *doria*; Chater and Walters, 1976, Fl. Europ. **4**: 196, p. p.

Type: Bashkiria ("in gub. Ufa: SW, prope Dawlekanowo").

Center (Volga-Don: south and east); *West* (Dnieper; Moldavia; Black Sea); *East.*—In damp, often more or less alkaline meadows, on outcrops of chalk, abandoned lands, in steppes, thinned-out forests, scrublands.—*General distribution*: Western Siberia, Russian Central Asia; Mediterranean (Romania).—2n = 40.

---

61   Plate V.

1—*Senecio paucifolius* S.G. Gmel., 1a—achene; 2—*S. paludosus* L., 2a—achene; 3—*S. tataricus* Less., achene, 3a—leaf; 4—*S. carpaticus* Herbich; 5—*S. subalpinus* Koch, head, 5a—cauline leaf, 5b—rosulate leaf.

78

25. **S. umbrosus** Waldst. et Kit. 1806, Descr. Icon. Pl. Rar. Hung. **3**: 232; Schischk. 1961, Fl. SSSR, **26**: 734.—*S. doria* L. subsp. *umbrosus* (Waldst. et Kit.) Soó, 1946, Erdész Kisérl. **46**: 282; Chater and Walters, 1976, Fl. Europ. **4**: 196.

Type: Czechoslovakia ("in sylvis acerosis ad thermas ucskienses Comitatus Liptoviensis, et non procul vetusta arce Arva in Provincia hujus nominis").

West (Carpathians; Dnieper: west).—In meadows, forest glades and edges.—*General distribution*: Central Europe, Mediterranean.—2n = 40.

26. **S. paucifolius** S.G. Gmel. 1770, Reise Russland, **1**: 171, tab. 38, fig. 1; Schischk. 1961. Fl. SSSR, **26**: 733.—*S. kirghisicus* DC. 1838, Prodr. **6**: 362; Paschkov, 1985, Fl. Nizhn. Dona: 101.— *S. doria* L. subsp. *kirghisicus* (DC.) Chater, 1974, Journ. Linn. Soc. London (Bot.), **68**, 4: 276; Chater and Walters, 1976, Fl. Europ. **4**: 196.—(Plate V, 1).

Type: Rostov Region ("in Cimlia").

Center (Volga-Kama: east; Volga-Don: south and east); *West* (Dnieper: east; Black Sea); *East*.—In saline meadows, steppes.— *General distribution*: Western Siberia.

63   **Section 4. Crociserides** (DC.) Rouy, 1903, Fl. Fr. **8**: 325; Schischk. 1961, Fl. SSSR, **26**: 726, p. min. p.—*Senecio* sect. *Crociseris* (Reichenb.) Minder. 1962, Fl. URSR, **11**: 375, p. p.; Chater and Walters, 1976, Fl. Europ. **4**: 196.

Perennial plants; stem hollow; leaves undivided; heads with 13–21 ligulate flowers; achenes glabrous or pubescent.

Lectotype: *S. scopolii* Hoppe et Hornsch. ex Bluff et Fingerh.

27. **S. paludosus** L. 1753, Sp. Pl.: 870, s. str.; Schischk. 1961, Fl. SSSR, **26**: 735; Chater and Walters, 1976, Fl. Europ. **4**: 196.— (Plate V, 2).

Type: Europe ("in Europae paludibus maritimis").

Baltic; *Center* (Ladoga-Ilmen; Upper Dnieper); *West* (Dnieper: west; Moldavia).—On banks of water bodies, in inundated meadows.—*General distribution*: Scandinavia, Central Europe, Mediterranean.—2n = 40.

28. **S. tataricus** Less. 1834, Linnaea, **9**: 192; Schischk. 1961, Fl. SSSR, **26**: 736.—*S. paludosus* L. 1753, Sp. Pl.: 870, p. p.; Chater and Walters, 1976, Fl. Europ. **4**: 196, p. p.—*S. auratus* DC. 1838, Prodr. **6**: 348; Mikhailovskaja, 1967, Opred. Rast. Beloruss.: 607.—(Plate V, 3).

Type: Orenberg Region ("ad fl. Ilek pr. Ilezkaja Saschtschita ad fl. Ural infra Ilezki Gorodok").

North (Dvina-Pechora: basin of the North Dvina River); *Center*; *West* (Carpathians; Dnieper; Black Sea); *East.*—On banks of water bodies and in inundated meadows.—*General distribution*: Western Siberia.

GENUS 26. *ERECHTITES* Raf.
1817, Fl. Ludov.: 65

Heads heterogamous, with numerous white, reddish or pinkish-violet flowers, clustered in corymbose or racemose-paniculate inflorescence. Involucre cup-shaped, biseriate; involucral bracts narrow-ovate, outer one-fourth to one-third as long as inner. Receptacle flat. Peripheral flowers filiform-tubular, pistillate; disk flowers tubular, bisexual. Stamens with sagittate anthers and oblong-deltoid connective. Style branches with conical apical appendage. Achenes fusiform, with 10 ribs; pappus white or violet-pink, of several rows of scabrous bristles, many times as long as achene. Annual or perennial herbaceous plants with erect stem; leaves alternate with undivided or more or less incised blades.

Type: *E. praealta* Raf. [=*E. hieracifolia* (L.) Raf. ex DC.].

The genus includes five species, distributed in America; two of them are ecdemic in many other countries.

Literature: Belcher, R. 1956. A revision of the genus *Erechtites* (Compositae) with inquiries into *Senecio* and *Arrhenechtites*. *Ann. Missouri Bot. Gard.*, **43**: 1–85.

1. **E. hieracifolia** (L.) Raf. ex DC. 1838, Prodr. **6**: 294; Pojark. 1961., Fl. SSSR, **26**: 683; Tutin, 1976, Fl. Europ. **4**: 191.—*Senecio hieracifolius* L. 1753, Sp. Pl.: 866.

Annual plant, 30–150 cm high; leaves oblong-elliptical or lanceolate, undivided, more or less toothed or pinnatilobate with acute lobes, lower leaves on winged petioles, middle and upper sessile; flowers whitish.

Type: North America ("in America septentrionali").

West (Carpathians).—As ecdemic plant of habitations, by roadsides.—*General distribution*: North and South America; ecdemic in other countries.—2n = 40.

64   GENUS **27. *TEPHROSERIS*** (Reichenb.) Reichenb.
1841, Repert. Herb.: 87; id. 1842, Fl. Saxon.: 146.—*Cineraria* L. β.
Tephroseris Reichenb. 1831, Fl. Germ. Excurs. **2**: 241.

Heads 1–4 cm in dia, heterogamous, less often homogamous, many-flowered, solitary or clustered in corymbose inflorescence. Involucre uniseriate, comprising linear or lanceolate bracts of equal length. Receptacle glabrous, flat. Peripheral flowers ligulate, pistillate or absent; disk flowers tubular, bisexual. Corolla yellow, orange or reddish-violet. Stamens with narrow-deltoid connective, basally obtuse anthers and cylindrical indehiscent antheropodia. Style branches truncate with apical crown of hairs, without hairs in peripheral flowers; stigmatic surface as a broad strip on inner side of style branches. Achenes 2–6 mm long, terete, with five coarser and between them five finer ribs or with 10–15 similar ribs, yellowish-brown, with pappus of scabrous bristles. Perennial or biennial herbaceous plants with erect stem; leaves alternate with undivided, more or less toothed or entire, rarely pinnatilobate or pinnatisect leaf blades.

Lectotype: *T. campestris* (Retz.) Reichenb. (=*T. integrifolia* (L.) Holub).

About 100 species in the Arctic and temperate zone of Eurasia and North America.

Literature: Cufodontis, G. 1933. Kritische Revision von *Senecio* sectio *Tephroseris*. *Feddes Repert.* (*Beih.*), **70**, 1: 1–266.—Holub, J. 1973. New names in Phanerogamae. 2. *Folia Geobot. Phytotax.* (*Praha*), **8**: 155–179.—Konechanaya, H.L. 1981. Kariologo-anatomicheskie priznaki vidov roda *Senecio* s. 1. (Asteraceae) v syazi s ikh sistematikoi [Karyological anatomical characters of species of the genus *Senecio* s. 1. (Asteraceae) in connection with their systematics]. *Bot. Zhurn.*, **66**, 6: 836–842.

1. Plants glandular-pubescent; pappus with fruits two to three times as long as involucre .... 1. **T. palustris.**
+ Plants glabrous or tomentose; pappus as long as involucre or somewhat longer .... 2.
2. All leaves undivided .... 3.
+ Lower and middle cauline leaves pinnatilobate or pinnatisect .... 2. **T. heterophylla.**
3. Heads solitary, rarely two or three; involucre violet; rosulate leaves oblong-elliptical, usually glabrous ....
.... 3. **T. atropurpurea.**

+ Heads 3–20, clustered in corymbose or umbellate inflorescence... 4.

4. Heads 2–3 cm in dia; involucre 1.5–2.0 cm in dia; ligulate flowers often orange. Plants somewhat dwarf, with rather thick, 0–20 cm high stem........................ 4. **T. tundricola.**

+ Heads 1.0–1.5 cm in dia; involucre 0.6–1.0 cm in dia .... 5.

5. All flowers yellow; involucre usually green................. 7.

+ All or only peripheral flowers orange, red or violet; involucre violet.. 6.

6. Plants weakly pubescent; common inflorescence lax; heads on long (1–5 cm) stalks.........................5. **T. aurantiaca.**

+ Plant white-tomentose; common inflorescence dense, capitate; heads on short (0.5–1.0 cm) stalks ....6. **T. capitata.**

7. Rosulate leaves linear-lanceolate, toothed, four to five times as long as wide .... 7. **T. papposa.**

+ Rosulate leaves oval, ovate, or spatulate, one and one-half to three times as long as wide .... 8.

8. Rosulate leaves cordate or spatulate, with truncate or rather abruptly narrowed base, petiolate .... 8. **T. rivularis.**

+ Rosulate leaves oval, ovate or oblong-ovate, gradually narrowed into petiole .... 9.

9. Plant of Crimea, white-tomentose throughout vegetative period; rosulate leaves oval, almost entire...................... .... 10. **T. jailicola.**

+ Plant glabrous or floccose-tomentose, sometimes white-tomentose only in spring .... 10.

10. Plant biennial, usually pubescent; rosulate leaves ovate or oblong-ovate; involucre 0.7–1.0 cm in dia .... 9. **T. integrifolia.**

+ Plant perennial, usually glabrous; rosulate leaves narrow-elliptical; involucre 0.5–0.7 cm in dia .... 10. **T. igoschinae.**

Section 1. Eriopappus (Dumort.) Holub, 1973, Folia Geobot. Phytotax. (Praha), **8:** 173.—*Cineraria* L. sect. *Eriopappus* Dumort. 1827, Fl. Belg.: 65.—*Senecio* L. sect. *Eiopappus* (Dumort.) Schischk. 1961, Fl. SSSR, **26:** 752; Chater and Walters, 1976, Fl. Europ. **4:** 201.

Glandular-hairy biennial plant; stem hollow; leaves undivided, irregularly toothed, less often pinnatilobate; achenes glabrous; pappus many times as long as achene.

Type: *Cineraria palustris* [=*T. palustris* (L.) Reichenb.].

1. **T. palustris** (L.) Richenb. 1842, Fl. Saxon.: 146.—*Othonna palustris* L. 1753, Sp. Pl.: 924.—*Cineraria palustris* (L.) L. 1763, Sp. Pl.: 1243.—*C. congesta* R. Br. 1824, Journ. Voy. N.W. Pass. (Suppl. App.): 279.—*Senecio congestus* (R. Br.) DC. 1838, Prodr. **6**: 363; Chater and Walters, 1976, Fl. Europ. **4**: 201.—*S. arcticus* Rupr. 1845, Beitr. Pflanzenk. Russ. Reich.: 44; Schischk. 1961, Fl. SSSR, **26**: 752.

Type: Europe ("in Europae paludibus maritimis").

Arctic; *North* (Dvina-Pechora); *Baltic*; *Center*; *West* (Dnieper; Black Sea); *East* (Lower Don; Trans-Volga).—On clayey banks of water bodies, in swamps and swampy tundra.—*General distribution*: Western and Eastern Siberia, Far East, Russian Central Asia (northwest); Scandinavia, Central and Atlantic Europe.—2n = 48.

Section 2. Aurei (Rydb.) Konechn. 1981, Bot. Zhurn. **66**, 6: 840.—*Senecio* L. Aurei Rydb. 1900, Bull. Torrey Bot. Club. **27**: 173.

Glabrous or weakly pubescent perennial plants; stem solid; leaves lyrate, pinnatilobate or pinnatisect; achenes glabrous; pappus two times as long as achene.

Lectotype: *Senecio aureus* L. [=*T. aurea* (L.) Konechn.].

2. **T. heterophylla** (Fisch.) Konechn. 1981, Bot. Zhurn. **66**, 6: 840.—*Cineraria heterophylla* Fisch. 1812, Mém. Soc. Nat. Moscou, **3**: 79.—*Senecio resedifolius* Less. 1831, Linnaea, **6**: 243; Schischk. 1961, Fl. SSSR, **26**: 750; Chater and Walters, 1976, Fl. Europ. **4**: 203.—*Packera resedifolia* (Less.) A. et D. Löve, 1976, Bot. Not. (Lund), **128**: 521.—(Plate VI, 4).

Type: Siberia ("Sibiria").

Arctic; *North* (Dvina-Pechora); *Center* (Volga-Kama: south).— In moss-lichen tundra, mountains in alpine zone.—*General distribution*: Western and Eastern Siberia, Far East; Mongolia; North America (northwest).—2n = 46.

Section 3. Tephroseris.

More or less tomentose, biennial or perennial herbaceous plant; stem hollow; leaves undivided, toothed or entire; achenes glabrous or pubescent; pappus two to three times as long as achene.

Type: lectotype of genus.

3. **T. atropurpurea** (Ledeb.) Holub, 1973, Folia Geobot. Phytotax. (Praha), **8**: 173.—*Cineraria atropurpurea* Ledeb. 1815, Mém. Acad. Sci. Petersb. **5**: 574.—*Senecio atropurpureus* (Ledeb.) B. Fedtsch.

1911, in B. Fedtsch. and Fler. Fl. Evrop. Rossii: 992; Schischk. 1961, Fl. SSSR, **26**: 771.

Type: Siberia ("in Sibiria").

66   *Arctic.*—In various tundra.—*General distribution*: Western and Eastern Siberia, Far East; North America (Arctic).—2n = 48.

4. **T. tundricola** (Tolm.) Holub, 1973, Folia Geobot. Phytotax. (Praha), **8**: 174.—*Senecio tundricola* Tolm. 1928, Compt. Rend. Acad. Sci. URSS, 1928: 266; Schischk. 1961, Fl. SSSR, **26**: 776.— *S. integrifolius* (L.) Clairv. subsp. *tundricola* (Tolm.) Chater, 1974, Journ. Linn. Soc. London (Bot.), **68**, 4: 276; Chater and Walters, 1976, Fl. Europ. **4**: 199.—*Tephroseris integrifolia* subsp. *tundricola* (Tolm.) B. Nord. 1978, Opera Bot. (Lund), **44**: 45.

Type: Siberia ("Siberia arctica a sinu Inisei usque ad ostia Kolymae fluminis et regio alpina montium Altaicorum et Sajanensium").

Arctic (Arctic Europe).—In mountains.—*General distribution*: Western and Eastern Siberia, Far East.

5. **T. aurantiaca** (Hoppe ex Willd.) Griseb. et Schenk, 1852, Arch. Naturg. (Berlin), **18**, 1: 342.—*Cineraria aurantiaca* Hoppe ex Willd. 1803, Sp. Pl.: **3**, 3: 2081.—*Senecio besserianus* Minder, 1956, Ukr. Bot. Zhurn. **13**, 3: 58; Schischk. 1961, Fl. SSSR, **26**: 768.—*S. integrifolius* (L.) Clairv. subsp. *aurantiacus* (Hoppe ex Willd.) Briq. et Cavill. 1916, in Burnat, Fl. Alp. Marit. **6**: 42; Chater and Walters, 1976, Fl. Europ. **4**: 199.

Type: Southern Europe ("in subalpinis Galliae, Italiae, Carinthiae").

West (Carpathians; Dnieper: west).—In meadows, forest glades and edges.—*General distribution*: Central Europe, Mediterranean.

6. **T. capitata** (Wahlenb.) Griseb. ex Schenk, 1852, Arch. Naturg. (Berlin), **18**, 1: 342.—*Cineraria capitata* Wahlenb. 1814, Fl. Carpat.: 271.—*Senecio capitatus* (Wahlenb.) Steud. 1841, Nomencl. Bot. **2**: 559; Schischk. 1961, Fl. SSSR, **26**: 768.—*S. integrifolius* (L.) Clairv. subsp. *capitatus* (Wahlnb.) Cuf. 1933, Feddes Repert. **70**: 14; Chater and Walters, 1976, Fl. Europ. **4**: 199.—*Tephroseris integrifolia* (L.) Holub. subsp. *capitata* (Wahlenb.) B. Nord. 1978, Opera Bot. (Lund): **44**: 45.

Type: Carpathians ("in alpium Scepusiensium").

West (Carpathians).—In alpine meadows.—*General distribution*: Central Europe (Carpathians).—2n = 96.

7. **T. papposa** (Reichenb.) Schur, 1866, Enum. Pl. Transs.: 344.—*Cineraria papposa* Reichenb. 1824, Icon. Bot. Pl. Crit. **2:** 13.—*Senecio papposus* (Reichenb.) Less. 1831, Linnaea, **6:** 244; Schischk. 1961, Fl. SSSR, **26:** 760; Chater and Walters, 1976, Fl. Europ. **4:** 201.—(Plate VI, 5).

Type: Ukraine ("inter Krzywczyce et Kamienopol").

West (Carpathians).—In cultivated meadows and on stony slopes of upper mountain zone.—*General distribution*: Central Europe, Mediterranean.

8. **T. rivularis** (Waldst. et Kit.) Schur, 1866, Enum. Pl. Transsilv.: 347.—*Cineraria rivularis* Waldst. et Kit. 1812, Descr. Icon. Pl. Rar. Hung. **3:** 265.—*Senecio rivularis* (Waldst. et Kit.) DC. 1838, Prodr. **6:** 359; Schischk. 1961, Fl. SSSR, **26:** 765; Chater and Walters, 1976, Fl. Europ. **4:** 200.—(Plate VI, 6).

Type: Hungary ("in vallibus Mantrae umbrosis: velut infra Kis-Kut...").

Center (Upper Dnieper); *West* (Carpathians).—In meadows.—*General distribution*: Central Europe.

9. **T. integrifolia** (L.) Holub, 1973, Folia Geobot. Phytotax. (Praha), **8:** 173.—*Othonna integrifolia* L. 1753, Sp. Pl.: 925.—*Senecio campestris* (Retz.) DC. 1838, Prodr. **6:** 361; Schischk. 1961, Fl. SSSR, **26:** 753.—*Tephroseris campestris* (Retz.) Reichenb. 1842, Fl. Saxon.: 147.—*Senecio integrifolius* (L.) Clairv. subsp. *integrifolius*; Chater and Walters, 1976, Fl. Europ. **4:** 199, p. p.— *S. czernjaevii* Minder, 1953, Ukr. Bot. Zhurn. **13,** 3: 55; Schischk. 1961, op. cit.: 754.

Type: Europe and Siberia ("in Alpibus Pyrenaicis, Helveticis, Austriacis, Sibiricis").

North; *Baltic* (Estonia); *Center*; *West*; *East*.—In meadows, forest glades and edges, on steppe slopes, sometimes in pine forests.—*General distribution*: Central and Atlantic Europe.—2n = 48.

O 10. **T. jailicola** (Juz.) Konechn. 1981, Bot. Zhurn. **66,** 6: 838.—*Senecio jailicola* Juz. 1953, Bot. Mat. (Leningrad), **15:** 399; Schischk. 1961, Fl. SSSR, **26:** 757.—*S. integrifolius* (L.) Clairv. subsp. *integrifolius*; Chater and Walters, 1976, Fl. Europ. **4:** 199, p. min. p.

---

67   Plate VI.

1—*Homogyne alpina* (L.) Cass.; 2—*Petasites hybridus* (L.) Gaertn., Mey. et Scherb., inflorescence, basal leaf; 3—*P. frigidus* (L.) Fries; 4—*Tephroseris heterophylla* (Fisch.) Konechn.; 5—*T. papposa* (Reichenb.) Schur, leaf, 5a—head; 6—*T. rivularis* (Waldst. et Kit.) Schur, leaf, 6a—achene.

Type: Crimea ("Babugan-Yaila, not far from Gavrell-Bogaz, open slopes and cultivated meadows").

Crimea.—In cultivated meadows and on stony slopes of yaila.—Endemic.

11. **T. igoschinae** (Schischk.) B. Nord. 1978, Opera Bot. (Lund), **44:** 44.—*Senecio igoschinae* Schischk. 1961, Fl. SSSR, **26:** 885, 763.—*S. papposus* auct. non (Reichenb.) Less.: Chater and Walters, 1976, Fl. Europ. **4:** 201, p. p.

Type: Urals ("Montes uralenses australes, prope pag. Machmutovo non procul a flumine Belaja").

Arctic (Arctic Europe); *North*; *Center* (Volga-Kama: east).— In cultivated meadows, on stony slopes and rocks, in stony tundras.—*General distribution*: Western Siberia.

GENUS 28. *LIGULARIA* Cass.
1816, Bull. Soc. Philom. Paris, 1816: 198, nom. conserv.

Heads 2–4 cm in dia, heterogamous, many-flowered, yellow, clustered in racemose or racemose-paniculate inflorescence. Involucre cylindrical or campanulate, biseriate; outer bracts one to three, small and narrow. Receptacle glabrous, flat. Peripheral flowers ligulate, pistillate; style branches with pointed tip, uniformly short-hairy outside; inner flowers tubular, bisexual, staminate, with deltoid or narrow-ovate connective; anthers obtuse at base with cylindrical antheropodia. Style branches apically rounded, short-hairy outside, with crown of longer hairs at apex. Achenes 3–6 mm long, terete, ribbed, glabrous, with whitish or brownish pappus of scabrous bristles. Perennial herbs with short rhizome and erect stem; leaves alternate with deltoid-cordate or oblong-obovate blades on long petiole.

Type: *L. sibirica* (L.) Cass.

The genus includes 150 species in the extratropical regions of Eurasia.

1. Rosulate leaves oblong-ovate; pappus considerably shorter than achenes .... 5. **L. carpatica.**
+ Rosulate leaves deltoid-cordate; pappus as long as or longer than achene .... 2.
2. Plants to 50 cm high; inflorescence short (of 5–15 heads); rosulate leaves 2–8 cm in dia...................................... 3.
+ Plants 50–150 cm high, with larger inflorescence and leaves.. 4.

3. Heads with five or six ligulate flowers ..... 3. **L. arctica.**
+ Heads with seven or eight ligulate flowers ....
.... 4. **L. bucovinensis.**
4. Leaves glabrous beneath or hairy only on principal veins
.... 1. **L. sibirica.**
+ Leaves hairy beneath .... 2. **L. lydiae.**
Section 1. Ligularia.
Pappus as long as achene or longer; leaves deltoid-cordate.

Type: type species.

1. **L. sibirica** (L.) Cass. 1823, Dict. Sci. Nat. **26:** 402; Pojark. 1961, Fl. SSSR, **26:** 807; Chater, 1976, Fl. Europ. **4:** 205, p. p.—*Othonna sibirica* L. 1753, Sp. Pl.: 924.—(Plate VII, 4).

Type: Siberia ("in Sibiria").

70   *North*; *Center* (Ladoga-Ilmen: east; Upper Volga: north; Volga-Kama); *East* (Trans-Volga).—In wet meadows, key and transitional bogs.—*General distribution*: Western and Eastern Siberia, Far East, Russian Central Asia (east).—2n = 60.

2. **L. lydiae** Minder. 1957. Ukr. Bot. Zhurn., **14,** 2: 48; Pojark. 1961, Fl. SSSR, **26:** 808.—*L. sibirica* auct. non (L.) Cass.: Chater, 1976, Fl. Europ. **4:** 205, p. p.

Type: Leningrad Region ("URSS pars Europaea dit. Leningradensis, meridiem versus ab opp. Gatczina, prope st. Voskressenskoje, fl. Sujdae ripa sinistra, in dumetis paludosis copiose").

North; *Baltic*; *Center* (Ladoga-Ilmen; Upper Dnieper: north; Upper Volga; Volga-Don).—In spring-fed bogs, swampy forests scrublands, on banks of water bodies.—*General distribution*: Central Europe.

3. **L. arctica** Pojark. 1961, Fl. SSSR, **26:** 891, 817.—*L. sibirica* auct. non (L.) Cass.: Chater, 1976, Fl. Europ. **4:** 205, p. p.

Type: Arkhangelsk Region ("Rossia septentrionalis, peninsula Kanin, in deeclivi boreali jugi Pae-hoi").

Arctic (Arctic Europe); *North*—In bogs, swampy meadows, stony and scrub tundras.—*General distribution*: Western Siberia (northwest).

4. **L. bucovinensis** Nakai, 1944, Journ. Jap. Bot. **20:** 135; Pojark. 1961, Fl. SSSR, **26:** 818.—*L. sibirica* auct. non (L.) Cass.: Chater, 1976, Fl. Europ. **4:** 205, p. p.

Type: Northeast of Romania.

88

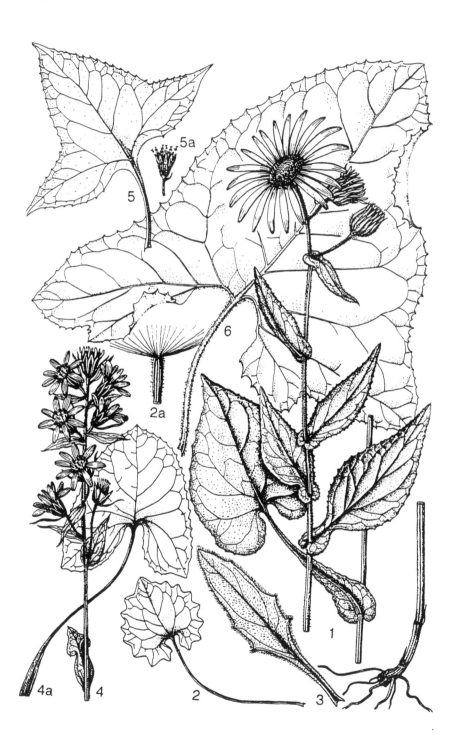

West (Carpathians; Dnieper: west).—In swampy meadows, bogs, wet forests.—*General distribution*: Central Europe (Carpathians).

Section 2. Senecillis (Gaertn.) Kitam. 1942, Compos. Japon. **3**: 187; Pojark. 1961, Fl. SSSR, **26**: 836.—*Senecillis* Gaertn. 1791, Fruct. Semen. **2**: 453.

Pappus one-fourth to half as long as achene; leaves oblong-ovate.

Type: *L. glauca* (L.) O. Hoffm.

5. **L. carpatica** (Schott, Nym. et Kotschy) Pojark. 1961, Fl. SSSR, **26**: 848.—*Senecillis carpatica* Schott, Nym. et Kotschy, 1854, Analect. Bot.: 5.—*Ligularia glauca* auct. non (L.) O. Hoffm.: Chater, 1976, Fl. Europ. **4**: 205.—*L. bucovinensis* auct. non Nakai: Czopik, 1977, Vizn. Rosl. Ukr. Karp.: 303.

Type: Transylvania ("Carpathos Transylvaniae inhabitat").

West (Carpathians; Dnieper: west).—On dry meadow slopes.—*General distribution*: Central Europe.

Note. Besides the above listed species is found in cultivation *L. hodgsonii* Hook. (=*L. clivorum* Maxim.)—a tall plant with green or purple leaves and corymbose inflorescence of large heads.

GENUS **29.** *CACALIA* L.
1753, Sp. Pl.: 834; id. 1754, Gen. Pl., ed. 5: 362

Head homogamous, with (1)3–20 bisexual, tubular whitish or yellowish flowers, clustered in racemose- or corymbose-paniculate inflorescence. Involucre cylindrical, one- or two-rowed (with small subtending leaves approximate to base of heads). Receptacle flat, glabrous. Stamens with sagittate anthers and oblong or narrow-deltoid connective. Style branches obtuse, with crown of hairs at apex. Achenes terete, slightly narrowed toward base, ribbed, glabrous, with pappus of scabrous bristles. Perennial herbaceous plants, with short rhizomes and erect stems; leaves alternate, with deltoid-hastate blades and rather long petiole.

Lectotype: *C. hastata* L.

About 50 species in extratropical Asia and north of Europe.

69 Plate VII.
1—*Doronicum austriacum* Jacq.; 2—*D. carpaticum* (Griseb. et Schenk) Nym., leaf, 2a—achene; 3—*D. clusii* (All.) Tausch, leaf; 4—*Ligularia sibirica* (L.) Cass., common inflorescence, 4a—leaf; 5—*Cacalia hastata* L., leaf, 5a—head; 6—*Petasites spurius* (Retz.) Reichenb., leaf.

71   *Literature*: Pojarkova, A.I. 1960. Kriticheskie zametki o rode *Cacalia* L. s. l. [Critical notes on the genus *Cacalia* L. s. l.]. *Bot. Mat.* (*Leningrad*), **20**, 370–391.—Koyama, H. 1963. Taxonomic studies on the tribe Senecioneae of Eastern Asia. II. Enumeration of the species of Eastern Asia. *Mem. Coll. Sci. Kyoto Univ. Biol.*, **2**, 2: 137–183.

1. **C. hastata** L. 1753, Sp. Pl.: 835; Pojark. 1961, Fl. SSSR, **26**: 687; Chater, 1976, Fl. Europ. **4**: 206.—*Hasteola hastata* (L.) Pojark. 1960, Bot. Mat. (Leningrad), **20**: 381.—*Koyamacalia hastata* (L.) H. Robinson et R.D. Brettell, 1973, Phytologia, **27**, 4: 272.—(Plate VII, 5).

Stems 40–150 cm high; leaf blades triangularly hastate, more or less toothed, sometimes entire; common inflorescence racemose-paniculate; heads drooping, with 4–20 whitish flowers; achenes 4–6 mm long.

Type: Siberia ("in Sibiria").

North (Dvina-Pechora); *Center* (Volga-Kama); *East* (Trans-Volga: southern Urals).—In forests, forest glades and edges, on banks of water bodies.—*General distribution*: Western and Eastern Siberia, Arctic, Far East; Mongolia, Japan-China.—2n = 60.

GENUS 30. *DORONICUM* L.
1753, Sp. Pl.: 885; id. 1754, Gen. Pl., ed. 5: 377

Heads heterogamous, with numerous flowers, solitary or clustered in corymbose inflorescence. Involucre cup-shaped, biseriate, comprising herbaceous bracts of similar length. Receptacle flat, glabrous. All flowers yellow, peripheral ligulate, pistillate (sometimes with sterile stamens); inner flowers tubular, bisexual. Stamens with obtuse or short-sagittate anthers and deltoid crown of papillate hairs. Achenes more or less terete, with 10 similar ribs, yellowish-brown or reddish-brown, glabrous or pubescent, all with pappus of scabrous bristles, or peripheral without pappus. Perennial herbaceous plants with erect stems, more or less long, often tuberously thickened rhizomes at base of stems, and alternate undivided leaves.

About 40 species, in mountains of warm temperate regions of Eurasia and northern Africa.

Lectotype: *D. pardalianches* L.

Literature: Cavillier, F. 1911. Nouvelles études sur le genre *Doronicum. Annu. Cons. Jard. Geneve*, 13–14: 195–368.

1. Plants without rosulate basal leaves; cauline leaves numerous, distant; heads 3–20 .... 1. **D. austriacum.**

   + Plants with rosulate basal leaves; cauline leaves not numerous; heads solitary, less often three to five ..... 2.

2. All leaves oblong with cuneate or round-cuneate base; rosulate and lower cauline leaves gradually narrowed into petiole, middle and upper sessile .... 3.

   + Rosulate and lower cauline leaves cordate, petiolate, middle and upper sessile, panduriform to oblong .... 4.

3. Rhizome tuberculate, with short (1.0–1.5 cm) stolons; stem 20–80 cm high; leaves glabrous or short-hairy, entire, less often small-toothed .... 5. **D. hungaricum.**

   + Rhizome not thickened, short; stem 10–30 cm high; leaves long- and soft-hairy, entire or with coarse upward directed teeth .... 6. **D. clusii.**

4. All achenes with pappus; rhizome horizontal, branched, uniformly thickened; rosulate leaves rounded-cordate with acute tip and acute teeth .... 2. **D. carpaticum.**

   + Peripheral achenes not pappose; rhizome tuberously thickened at stem base, stoloniferous; leaves cordate, subobtuse, and with obtuse teeth .... 5.

5. Rhizome lanate at stem base; cauline leaves one or two; rosulate, leaves 1.5–6.0 cm long .... 3. **D. orientale.**

   + Rhizome glabrous; cauline leaves two to six; rosulate leaves 6–12 cm long .... 4. **D. pardalianches.**

1. **D. austriacum** Jacq. 1774, Fl. Austr. **2:** 18; Gorschk. 1961, Fl. SSSR, **26:** 676; I.K. Ferguson, 1976, Fl. Europ. **4:** 190.—(Plate VII, 1).

Type: Austria ("ad margines sylvarum subalpinarum").

West (Carpathians).—In forest glades and edges, on banks of streams in forest zone of mountains.--*General distribution*: Central and Atlantic Europe, Mediterranean.—2n = 60.

2. **D. carpaticum** (Griseb. et Schenk) Nym. 1865, Syll. Fl. Europ. Suppl.: 1; Grosch. 1961, Fl. SSSR, **26:** 675.—*Aronicum scorpioides* DC. var. *carpaticum* Griseb. et Schenk, 1852, Arch. Naturg. (Berlin), **18,** 1: 342.—*D. orientale* auct. non. O. Hoffm.: I.K. Ferguson, 1976, Fl. Europ. **4:** 190, p. p.—*D. columnae* auct. non Ten.: Fodor. 1974, Fl. Zakarp.: 140; Ferguson, 1976, Fl. Europ. **4:** 190, p. p.—(Plate VII, 2).

Type: Carpathians ("....südlichen Karpaten, z.b. am. Szurul.").

West (Carpathians); in other regions (*Baltic*; *Center*) cultivated as an ornamental plant.—In damp meadows and forest glades, on stony slopes, banks of streams, in middle and upper mountain zones.—*General distribution*: Central Europe.

3. **D. orientale** O. Hoffm. 1803, Comment. Soc. Phys. Med. Univ. Mosq.: **1**: 8; Gorsch. 1961, Fl. SSSR, **26**: 677; I.K. Ferguson, 1976, Fl. Europ. **4**: 190; Kosykh, 1981, Bot. Zhurn. **66**, 9: 1327.

Type: Georgia ("circa Zehet in Iberia").

Crimea (northern slope of Crimean mountains, valley of the Pisara River).—In broad-leaved mountain forests.—*General distribution*: Caucasus; Central Europe, Mediterranean, Asia Minor.—2n = 60.

4. **D. pardalianches** L. 1753, Sp. Pl.: 885; Gorschk. 1961, Fl. SSSR, **26**: 679; Fodor, 1974, Fl. Zakarp.: 140; I.K. Ferguson, 1976, Fl. Europ. **4**: 191.

Type: Mountains of Middle Europe ("in Alpibus Helvetiae, Pannoniae, Vallesiae").

Baltic; *Center*; *West*; *East*; *Crimea*.—Cultivated as an ornamental plant; for Carpathians (Fodor, op. cit.) reported as wild in forests, on banks of rivers.—*General distribution*: Scandinavia (south), Central and Atlantic Europe, Mediterranean; cultivated in other extratropical countries.—2n = 60.

5. **D. hungaricum** Reichenb. fil. 1854, Icon. Fl. Germ. **16**: 34; Geideman, 1975, Opred. Vyssh. Rast. Mold. SSR: 497; I.K. Ferguson, 1976, Fl. Europ. **4**: 191.—*D. longifolium* Griseb. et Schenk, 1852, Arch. Naturg. (Berlin), **18**: 341, non Reichenb. 1831; Gorschk. 1961, Fl. SSSR, **26**: 680.

Type: Hungary ("in arenosis montium Hungariae").

West (Carpathians; Dnieper: southwest; Moldavia; Black Sea: west).—In thinned-out forests, forest glades and edges.—*General distribution*: Central Europe (south east), Mediterranean.

6. **D. clusii** (All.) Tausch, 1828, Flora (Regensb.), **11**: 178; Gorschk. 1961, Fl. SSSR, **26**: 674; I.K. Ferguson, 1976, Fl. Europ. **4**: 191.—*Arnica clusii* All. 1770–1773, Mél. Philos. Math. Soc. Roy. Turin. (Misc. Taur.), **5**: 70.—(Plate VII, 3).

Type: Italy.

West (Carpathians: Chernogora Range).—On stony slopes and rocks in upper mountain zone.—*General distribution*: Central and Atlantic Europe, Mediterranean.—2n = 60, 120.

Note. In the former Soviet Union, apparently, this species is represented only by subspecies *D. clusii* subsp. *stiriacum* (Tausch) Vierh. (Soják, 1983, *Sborn. Narodn. Muz. Praze*, **39B**, 1: 54).

73   GENUS **31.** *ADENOSTYLES* Cass.
1816, Dict. Sci. Nat. 1, Suppl.: 59.—*Cacalia* L. 1754, Gen. Pl., ed. 5: 362, p. p.

Heads homogamous, with 8–30, bisexual, tubular, tetramerous flowers, clustered in rather dense corymbose-paniculate inflorescence. Involucre campanulate or cup-shaped, biseriate, outer row comprising one or many small and narrow bracts. Receptacle flat, glabrous. Corolla violet or sordid pink. Stamens with anthers obtuse at base and oblong-ovate connective. Style branches elongate, pointed, uniformly covered with papillae. Achenes almost terete, glabrous, with eight ribs; pappus longer than achenes, of white scabrous bristles. Perennial herbaceous plants with erect stem; leaves alternate, cordate-reniform or deltoid, lower more or less long-petiolate, upper sessile.

Lectotype: *A. viridis* Cass. (=*A. alpina* (L.) Bluff et Fingerh.).

Three to six species in the mountains of Europe and Asia Minor.

Literature: Wagenitz, G. 1983. Die Gattung *Adenostyles* Cass. (Compositae—Senecioneae). *Phyton.* **23**, 1: 141–159.

1. **A. alliariae** (Gouan) A. Kerner, 1871, Oesterr. Bot. Zeitschr. **21**: 12; Tamamsch. 1959, Fl. SSSR, **25**: 22; Tutin, 1976, Fl. Europ. **4**: 189.—*Cacalia alliariae* Gouan, 1773, Ill. Observ. Bot.: 65.

Stem 40–150 cm high; leaves tomentose beneath, subglabrous above; heads two- to five-flowered; achenes about 3 mm long.

Type: Pyrenees mountains ("in Pyrenaeis").

West (Carpathians).—In mountain forests and subalpine meadows.—*General distribution*: Central Europe, Mediterranean.— 2n = 38.

GENUS **32.** *HOMOGYNE* Cass.
1816, Bull. Soc. Philom. Paris, 1816: 198

Heads heterogamous, with numerous dirty lilac flowers, solitary. Involucre cup-shaped, biseriate, outer row consisting of two to four small, narrow bracts. Receptacle flat, glabrous. Peripheral flowers pistillate, with short (one-third to half as long as pistil), narrow-tubular corolla, obliquely truncate and bidentate at apex, central bisexual, tubular, with five-toothed corolla. Stamens with short-sagittate anthers and oblong-deltoid connective. Style branches terete, rounded at tip, uniformly papillate. Achenes fusiform, glabrous, yellowish-brown, with five ribs; pappus of scabrous bristles. Perennial herbaceous plants, with creeping stems rooting at nodes and erect leafy scapes. Leaves alternate, basal reniform-orbicular, more or less toothed, long-petiolate, cauline leaves fewer, sessile, lower reniform with broad sheath, upper ovate.

Lectotype: *H. alpia* (L.) Cass.

Three species in the mountainous regions of Europe.

1. **H. alpina** (L.) Cass. 1816, Bull. Soc. Philom. Paris, 1816: 198; Kuprian, 1961, Fl. SSSR, **26:** 655; Tutin, 1976, Fl. Europ. **4:** 188.—*Tussilago alpina* L. 1753, Sp. Pl.: 865.—(Plate VI, 1).

Plant 10–50 cm high, more or less covered with ochreous entangled hairs; leaves almost coriaceous; achenes 4–6 mm long, with 6–10 mm long pappus.

Type: Europe ("in Alpibus Helvetiae, Austriae, Bohemiae").

West (Carpathians).—In alpine meadows, forest edges and glades in middle and upper mountain zones.—*General distribution*: Central Europe, Mediterranean.—2n = 140.

74   GENUS **33.** *TUSSILAGO* L.
          1753, Sp. Pl.: 865; id. 1754, Gen. Pl., ed. 5: 372

Heads heterogamous, with numerous yellow flowers, solitary. Involucre biseriate; outer bracts fewer, shorter than inner. Receptacle flat, glabrous. Peripheral flowers filiform, pistillate, many-rowed, fertile; pistil with short flat lobes [of stigma]. Inner flowers tubular, bisexual, but sterile; pistil with capitate stigma; stamens with short-sagittate anthers and oblong-deltoid connective. Achenes narrow-cylindrical, yellowish-brown, with 5–10 weakly raised ribs; pappus of scabrous bristles. Perennial herbaceous plants, with creeping branched rhizomes. Scapes arising in early spring, 8–40 cm high, erect with numerous scale-like, oblong-ovate, greenish or more or less purple-violet leaves. Green leaves developing post-flowering, all basal, orbicular-cordate, 5–25 cm in dia, sinuate-dentate, white-tomentose beneath, subglabrous above.

Type: *T. farfara* L.

A monotypic genus.

1. **T. farfara** L. 1753, Sp. Pl.: 865; Kuprian. 1961, Fl. SSSR, **26:** 641; Tutin, 1976, Fl. Europ. **4:** 186.

Type: Europe ("in Europae argillosis subutus humidis").

Arctic; *North*; *Baltic*; *Center*; *West*; *East*; *Crimea*.—On banks of water bodies at gulley bottoms, precipitous, predominantly clayey slopes, in more, or less wet, weedy areas in forests and by roadsides and pathways.—*General distribution*: Caucasus, Western and Eastern Siberia, Far East, Russian Central Asia; Central and Atlantic Europe, Mediterranean, Asia Minor, Iran, Himalayas, Japan-China; North America (ecdemic), Africa (north).—2n = 60.

GENUS 34. *PETASITES* Mill.
1754, Gard. Dict., ed. 4

Heads heterogamous, less often homogamous, many-flowered, usually clustered in corymbose or racemose-paniculate inflorescence, less often solitary. Involucre broadly cylindrical or campanulate, biseriate; inner involucral bracts lanceolate or linear, with scarious border equal; outer bracts smaller and narrower, fewer. Receptacle flat, glabrous. Usually dioecious plant; female plants with numerous peripheral ligulate or filiform flowers, obliquely truncate at apex, pistillate; inner flowers tubular, bisexual, sterile, less numerous (one to four) or absent. In male plants, peripheral flowers less numerous (sometimes absent), ligulate or filiform, pistillate, sterile; inner numerous, tubular, bisexual. Corolla yellowish, white or pink-purple. Stamens with basally obtuse anthers and ovate or deltoid connective. Pistil of bisexual flowers thickened in upper part, with short-deltoid, less often longer cylindrical branches. Achenes narrow-cylindrical, yellowish-brown or brick red-reddish-brown, with fine ribs; pappus of scabrous bristles. Perennial herbaceous plants, with creeping rhizome. Scapes developing in spring 10–60 cm high, with alternate, ovate or oblong-linear leaves greenish, purple or whitish-yellow. Basal green leaves developing postflowering, orbicular or deltoid-cordate, less often elliptical or ovate, lobate or toothed.

Lectotype: *P. major* Mill. (= *P. hybridus* (L.) Gaertn., Mey. et Scherb.).

75 About 20 species, distributed in extratropical regions of the Northern Hemisphere.

Literature: Toman, J. 1972. A taxonomic survey of the genera *Petasites* and *Endocellion. Folia Geobot. Phytotax. (Praha)*, **7:** 381–406.

1. Stem 10–20 cm high, with one, sometimes three, heads; basal leaves elliptical or ovate, cuneate, toothed .... 6. **P. sibiricus.**

+ Heads more numerous, clustered in corymbose or racemose-paniculate inflorescences; basal leaves cordate .... 2.

2. Basal leaves large, deltoid-cordate, with acute lobes at base, white-tomentose beneath or on both sides, sometimes subglabrous; cauline leaves oblong-linear, to 25 cm long, greenish-white; flowers yellowish- or pinkish-white .... 3. **P. spurius.**

+ Basal leaves uniformly lobed or toothed; cauline leaves usually ovate, less often oblong-lanceolate, to 15 cm long .... 3.

3. Basal leaves coriaceous, glabrous; flowering shoots greenish-yellow; peripheral flowers ligulate in male inflorescence, yellow .... 4. **P. radiatus.**

+ Basal leaves pubescent at least beneath, rarely glabrous, but sometimes thin and not coriaceous .... 4.

4. Basal leaves deltoid-cordate, uniformly lobed or coarsely toothed, 5–15 (25) cm in dia. Common inflorescence corymbose; involucre glandular-pubescent, purple-violet; peripheral flowers in male inflorescences ligulate, yellow; cauline leaves oblong-lanceolate, 3–15 cm long, purple .... 5. **P. frigidus.**

+ Basal leaves large, 20–70 cm in dia. Common inflorescence racemose-paniculate or racemose .... 5.

5. Basal leaves broadly deltoid or orbicular-cordate, unevenly bidentate, grayish-tomentose beneath; common inflorescence racemose-paniculate, two to four times as long as wide; flowers purple-pink, less often pale yellow; involucral bracts obtuse, somewhat arachnoid-hairy; pistils of bisexual flowers capitate or ovately thickened, with very short-deltoid lobes .... 1. **P. hybridus.**

+ Basal leaves orbicular-cordate, acutely bisinuat-dentate, white-tomentose beneath, less often almost glabrous (usually first leaves subglabrous, later white-tomentose), hairy only on veins; common inflorescence racemose, one and one-half to two times as long as wide; flowers yellowish-white; involucral bracts acute, dorsally glandular-hairy; pistils of bisexual flowers with rather long cylindrical branches with pointed tip .... 2. **P. albus.**

SUBGENUS 1. *PETASITES*

Peripheral pistillate flowers in head of androgynous (male) plants filiform or tubular, with obliquely truncate corolla, less numerous or absent. Common inflorescence consisting of many heads.

Type: lectotype of genus.

1. **P. hybridus** (L.) Gaertn., Mey. et Scherb. 1801. Fl. Wett. **3**: 184; Kuprian. 1961, Fl. SSSR, **26**: 643; Dingwall, 1976, Fl. Europ. **4**: 187.—*Tussilago hybrida* L. 1753, Sp. Pl.: 866.—(Plate VI, 2).

76   Type: Western Europe ("in Germania, Hollandia").

a. Subsp. **hybridus**.—Common inflorescence, as also stem and cauline leaves, purple-pink.

Baltic; *Center* (Ladoga-Ilmen; Upper Dnieper; Upper Volga; Volga-Don); *West*; *Crimea*.—On banks of water bodies, in more or less wet meadows, damp forests, cultivated as an ornamental plant and escaped.—*General distribution*: Caucasus; Central and Atlantic Europe, Mediterranean.—2n = 60.

b. Subsp. **ochroleucus** (Boiss. et Heut) Šourek, 1962, Rozpr. Československ. Akad. Věd. **72, 5**: 26.—*P. ochroleucus* Boiss. et Huet, 1856, in Boiss. Diagn. Pl. or. Ser. **2**, 3: 5.—Stem and flowers yellowish-white; cauline leaves and involucral bracts pinkish.

Type: Turkey ("in Armenia circa Zazalar hane").

Crimea (southern coast).—On banks of rivers and streams.—*General distribution*: Caucasus; Mediterranean (east), Asia Minor, Iran.

2. **A. albus** (L.) Gaertn. 1791, Fruct. Sem. Pl. **2**: 406; Kuprian. 1961, Fl. SSSR, **26**: 644; Dingwall, 1976, Fl. Europ. **4**: 187.—*P. kablikianus* auct. non Tausch ex Bercht.: Katina, 1987, Opred. Vyssh. Rast. Ukr.: 341.

West (Carpathians).—On forested banks of rivers and streams.—*General distribution*: Caucasus; Central and Atlantic Europe, Mediterranean.—2n =60.

Note. For the Carpathians, one more species, *P. kablikianus* Tausch ex Bercht. has been recorded, which is of hybrid origin (*P. hybridus* × *P. albus*). In the structure of flowers and inflorescence it is similar to *P. albus* and in the form of the leaves, to *P. hybridus*. But, its leaves are usually glabrous beneath or pubescent only on veins. However, it must be borne in mind that the first one or two leaves of vegetative shoots of all these species are usually glabrous or pubescent only on veins and only later the

98

larger leaves have pubescence typical of the species. In my opinion, the Carpathian plants of *P. kablikianus* examined by me relate to *P. albus.*

SUBGENUS **2.** *NARDOSMIA* (Cass.) Peterm. 1848, Deutschl. Fl.: 277

Peripheral pistillate flowers in heads of androgynous (male) plants ligulate, with 1–9 mm long limb, numerous in many rows. Common inflorescence with fewer heads, less often heads solitary.

Type: *Nardosmia denticulata* Cass. (= *Petastites fragrans* (Vill.) C. Presl).

3. P. spurius (Retz.) Reichenb. 1831, Fl. Germ. Excurs. **1:** 279; Kuprian. 1961, Fl. SSSR, **26:** 643; Dingwall, 1976, Fl. Europ. **4:** 188.— *Tussilago spuria* Retz. 1779, Observ. Bot. **1:** 29.—(Plate VII, 6).

Type: Sweden ("ad Bofarja juxta Ringsiön in scania").
North (Dvina-Pechora); *Baltic*; *Center*; *West* (Dnieper; Moldavia; Black Sea); *East.*—On sandy banks of rivers and lakes.— *General distribution*: Western Siberia, Russian Central Asia (northwest); Central and Atlantic Europe.—2n = 60.

4. P. radiatus (J.F. Gmel.) Toman, 1972, Folia Geobot. Phytotax. (Praha), **7:** 388; Dingwall, 1976, Fl. Europ. **4:** 187.—*Tussilago radiata* J.F. Gmel. 1792, Syst. Nat., ed. 13, **2**, 2: 1226.—*T. laevigata* Willd. 1800, Sp. Pl.: **3**, 3: 1969.—*Nardosmia laevigata* (Willd.) DC. 1836, Prodr. **5:** 205; Kuprian. 1961, Fl. SSSR, **26:** 650.

Type: Siberia (" ad Irtim fluvium et hinc ad Ieniseam usque... in alucis pigris Irtis, Obi, et Lenisieae fluviorum...").
Arctic (Novaya Zemlya; Arctic Europe); *North*; *Center* (Volga-Kama); *East* (Trans-Volga).—On stony and gravelly banks of rivers and streams.—*General distribution*: Western and Eastern Siberia.

5. P. frigidus (L.) Fries, 1846, Summa Veg. Scand.: 182; Dingwall, 1976, Fl. Europ. **4:** 188.—*Tussilago frigida* L. 1753, Sp. Pl.: 865.—*Nardosmia angulosa* Cass. 1825, Dict. Sci. Nat. **34:** 188; Kuprian. 1961, Fl. SSSR, **26:** 648.—*N. frigida* (L.) Hook. 1833, Fl. Bor. Amer. **1:** 307; Kuprian. op. cit.: 649.—(Plate VI, 2).

77　　Type: Europe and Siberia ("in Alpium Lapponiae, Helvetiae, Sibiriae").
Arctic; *North*; *Center* (Ladoga-Ilmen: north and east; Upper Volga; Volga-Kama).—In swampy tundras, on banks of swamps and water bodies in swampy forests.—*General distribution*: Eastern

and Western Siberia, Arctic, Far East; Scandinavia, Mongolia; North America.—2n = 60.

6. **P. sibiricus** (J.F. Gmel.) Dingwall, 1975, Journ. Linn. Soc. London (Bot.), **71**: 273; id. 1976, Fl. Europ. **4**: 188.—*Tussilago sibirica* J.F. Gmel. 1792, Syst. Nat., ed. 13, **2**, 2: 1224.—*Nardosmia gmelini* Turcz. ex DC. 1838, Prodr. **7**: 271; Kuprian. 1961, Fl. SSSR, **26**: 653.

Type: Yakutia ("ad Bielae fluvii tractum in itinere a Jacutia Ochotium versus").

North (Dvina-Pechora: northern Urals).—In stony tundra.— *General distribution*: Arctic, Eastern Siberia, Far East; Mongolia.

TRIBE 3. **CALENDULEAE** Cass.

Heads heterogamous; peripheral flowers ligulate, usually yellow or orange, less often of other color; inner flowers (disk flowers) tubular, usually yellow. Receptacle without bracts, usually glabrous. Anthers more or less sagittate. Style branches in bisexual flowers flattened, truncate, and papillate at tip. Pollen of helianthoid type. Achenes without pappus, rarely with weakly developed pappus of scabrous bristles, usually heteromorphic. Leaves alternate, undivided.

Type: *Calendula* L.

GENUS **35.** *CALENDULA* L.[1]
1753, Sp. Pl.: 921; id. 1754, Gen. Pl., ed. 5: 393

Heads heterogamous, 7–70 mm in dia, terminal on stem and its branches, many-flowered. Involucre 7–15 mm in dia; involucral bracts one or two rows, lanceolate, herbaceous, acute. Receptacle flat, glabrous. Flowers yellow or orange, peripheral ligulate, one and one-half to three times as long as involucre, pistillate, fertile; inner tubular, bisexual but often sterile, their corolla five-toothed. Anthers sagittate, with caudate appendages. Achenes without pappus or crown of hairs, heteromorphic, of three kinds: peripheral curved, dorsally with spinules of spines; middle with longitudinal ribs, wing-like lateral processes and a median process, saccate, carinate; inner smaller, almost annular tuberculate or rugose. Annual herbs with alternate, undivided, usually oblong leaves.

Lectotype: *C. officinalis* L.

[1]Treatment by S.S. Ikonnikov.

100

About 20 species, distributed predominantly in the countries of the Mediterranean and southwestern Asia. One species, *C. officinalis* L., is widely cultivated in other extratropical countries and sometimes found as an escape.

Literature: Lanza, D. 1923. Monografio del genero *Calendula* L. *Atti Reale Acad.* (*Palermo*), ser. 3, **12**: 1–166.—Bach, H. 1953, Heterocarpie bei *Calendula* (Entwicklung, Organstellung, Abhängigkiet von äuseren Einflussen). *Flora*, **140**, 2: 326–344.— Meusel, H. and H. Ohle. 1966, Zur Taxonomie und Cytologie der Gattung *Calendula*. *Österr. Bot. Zeitscher.*, **113**: 191–210.—Harza, R.R. 1970. Chromosome studies in *Calendula*. *Bull. Bot. Soc. Bengal*, **24**, 1–2: 95–100.—Ohle, H. 1974. Beitrage zur Taxonomie der Gattung *Calendula*. 2. Taxonomische Revision der Südeuropäischen perennierenden *Calendula*-Sippen. *Feddes Repert.* **85**, 4: 245–283.

1. Heads 1–2 cm in dia; ligulate flowers one and one-half times as long as involucre, yellow; peripheral achenes dorsally with large spines, largest of them 1.5–2.0 cm long; leaves 3–8 cm long and 0.5–1.4 cm wide. Plants to 20 cm high .... **2. C. arvensis.**

78 + Heads 3–5(7) cm in dia; ligulate flowers two to three times as long as involucre, orange, less often yellow; peripheral achenes dorsally with longitudinal rows of spinules, largest of them 2.0–2.5 cm long; leaves 7–15 cm long, and 1–5 cm wide. Plants 20–60 cm high ....
.... **1. C. officinalis.**

1. **C. officinalis** L. 1753, Sp. Pl.: 921; Vassil. 1961, Fl. SSSR, **26**: 860; Meikle, 1976, Fl. Europ. **4**: 207.

Type Europe ("in Europae arvis").

North; *Baltic*; *Center*; *West*; *East*; *Crimea*.—Widely cultivated as an ornamental and medicinal plant; often as ecdemic or escape.—. *General distribution*: Mediterranean; cultivated in many other extratropical countries where it is often found as an escape.—2n = 28, 32.

2. **C. arvensis** L. 1763, Sp. Pl., ed. 2: 1303; Vassil. 1961, Fl. SSSR, **26**: 859; Meikle, 1976, Fl. Europ. **4**: 207.

Type: Ethiopia ("in Aethiopia").

Baltic (ecdemic in the vicinity of Tartu); *West* (Dnieper: ecdemic in Kiev and Dnepropetrovsk); *Crimea* (Southern coast).—By roadsides, in habitations, on open stony and clayey slopes, in forest glades.—*General distribution*: Caucasus (ecdemic); Central

Europe (South), Mediterranean, Africa (north); ecdemic in other extratropical countries.—2n = 36, 44.

GENUS **36.** *DIMORPHOTHECA* Moench[1]
1794, Meth. Pl.: 588, nom. conserv.

Heads 30–50 mm in dia, heterogamous, many-flowered, terminal on stem and its branches. Involucre 20–25 mm in dia; involucral bracts in one or two rows, lanceolate, 8–12 mm long, gradually narrowed. Receptacle flat, glabrous. Peripheral flowers ligulate, pistillate, white, whitish, yellow or yellowish-orange, one and one-half times as long as involucre; inner staminate, yellow with purple tips. Achenes without pappus or crown, heteromorphic: some strongly flattened, smooth, broadly elliptical to almost round, 7–8 mm long, acute at base and with goffered margin; others almost cylindrical, tuberculate, 3–5 mm long. Annual herbs, with alternate simple leaves, some toothed.

Type: *D. pluvialis* (L.) Moench.
Seven species in Africa; cultivated in many other countries.
Literature: Nordlindh, T. 1943. Studies in Calenduleae. 1. Monographs of genera *Dimorphotheca, Castalis, Osteospermum, Gibbaria,* and *Chrysanthemoides.* Lund: 1–432.

1. Strongly flattened achenes almost round, 7–8 mm long and about 7 mm wide; subcylindrical achenes 5–6 mm long and 1.8–2.0 mm wide; ligulate flowers whitish or yellow .... 1. **D. pluvialis.**
+ Strongly flattened achenes round-deltoid, about 7 mm long and 5.0–5.5 mm wide; subcylindrical achenes light colored, about 4 mm long and 1.0–1.2 mm wide; ligulate flowers yellowish-orange .... 2. **D. hybrida.**

1. **D. pluviatlis** (L.) Moench, 1794, Meth. Pl.: 585.—*Calendula pluvialis* L. 1753, Sp. Pl.: 921.—*Dimophotheca annua* Less. 1832, Syn. Gen. Compos.: 257.

Type: Ethiopia ("in Aethiopia").
Baltic; *Center* (Ladoga-Ilmen; Upper Volga); *West.*—Cultivated as an ornamental plant.—*General distribution*: Africa; cultivated in many other countries.—2n = 18.

78   2. **D. hybrida** (L.) DC. 1837, Prodr. **6:** 70.—*Calendula hybrida* L. 1763, Sp. Pl., ed. 2: 1304.—*Dimorphotheca auranthiaca* auct.

[1]Treatment by S.S. Ikonnikov.

non DC.: Babarich, 1962, Fl. URSR, **11**: 413, in obs.; Wissjul. 1965, Vizn. Rosl. Ukr., ed. 2: 655; Katina, 1987, Opred. Vyssh. Rast. Ukr.: 346.

Type: South Africa, Cape of Good Hope ("ad Caputem Bonae"). *Baltic*; *Center*; *West*.—Cultivated as an ornamental plant.— *General distribution*: Africa (south); cultivated in many other countries.—2n = 18.

TRIBE 4. **INULEAE** Cass.[1]

Heads heterogamous, less often homogamous; peripheral flowers ligulate, filiform-tubular or absent; inner flowers (disk flowers) tubular. Receptacle with or without bracts. Anthers usually sagittate and with long caudate appendages. Style branches flattened, obtuse, truncate or subacute, apically with or without papillae. Pollen helianthoid type. Achenes usually with pappus of scabrous bristles, rarely pappus absent. Leaves alternate, undivided, often entire.

Type: *Inula* L.

GENUS **37.** *TELEKIA* Baumg.[1]
1816, Enum. Stirp. Transsilv. **3**: 149

Heads heterogamous, solitary at apices of stem and its branches, subtended by highly reduced leaves. Involucre of 5– 10 rows of imbricate bracts; bracts of each row strongly differing from those in other rows. Receptacle convex, with acute scale- like bracts. Flowers yellow, peripheral pistillate, ligulate, three times as long as involucre, in two whorls; disk flowers bisexual, tubular, five-toothed, many-whorled. Achenes somewhat flattened, longitudinally ribbed; peripheral terete-triquetrous, somewhat curved, without pappus; inner terete-tetraquetrous, with pappus as very short finely toothed corona. Perennial plants, with erect stem and alternate undivided leaves.

Type: *T. speciosa* (Schreb.) Baumg.
Two species, distributed in Europe and Asia Minor, one of them cultivated and as an escape in many other extratropical countries.

1. **T. speciosa** (Schreb.) Baumg. 1816, Enum. Strip. Transsilv. **3**: 150; Golubk. 1959, Fl. SSSR, **25**: 510; Tutin, 1976, Fl. Europ.

---

[1]Treatment by I.A. Gubanov.

4: 138.—*Buphthalmum speciosum* Schreb. 1766, Icon. Descr. Pl., Dec. **1:** 11.

Plant 50–200 cm high; leaves to 30 cm long, ovate or broadly lanceolate, middle and upper sessile, lower petiolate; rhizome 5–8 cm in dia.

Type: Turkey ("in Cappadocia").

West (Carpathians; Dnieper: west).—In forest glades and edges, on banks of rivers and streams, in thinned-out forests; cultivated as an ornamental plant in gardens and parks and often as an escape; stable populations of escaped plants are known in *Baltic* and *Center* (Ladoga-Ilmen; Upper Dnieper; Upper Volga).—*General distribution*: Caucasus; Central Europe, Mediterranean, Asia Minor.—2n = 20.

80
GENUS 38. *PALLENIS* (Cass.) Cass.[1]
1822, Dict. Sci. Nat. **23:** 566, nom. conserv.

Heads heterogamous, solitary at apices of stem and branches. Involucre two- or three-rowed; involucral bracts considerably exceeding flowers, spinescent, stellately spreading. Receptacle convex, with acute scalelike bracts. Flowers yellow; peripheral pistillate, with narrow three-toothed ligules, two whorled; disk flowers bisexual, tubular, five-toothed, in many whorls. Achenes heteromorphic: peripheral ovate, somewhat flattened, winged, with small depression at tip, and with stiff hairs on upper margin, without pappus; inner smaller, narrowly tetraquetrous, flattened, not winged, with pappus of one row of short bristles connate at base into a toothed corona. Annual plants, with alternate, oblong, undivided and entire leaves.

Type: *P. spinosa* (L.) Cass.
Four species, distributed predominantly in northern Africa; only one of them enters southern Europe and southwestern Asia.

1. **P. spinosa** (L.) Cass. 1825, Dict. Sci. Nat. **37:** 276; Golubk. 1959, Fl. SSSR, **25:** 508; Tutin, 1976, Fl. Europ. **4:** 139.—*Buphthalmum speciosum* L. 1753, Sp. Pl.: 903.—(Plate VIII, 1).

Plant 15–40 cm high; rhizome to 25 mm in dia.

Type: Europe ("in G. Narbonensi, Hispania, Italia, ad margines agrorum").

Crimea.—In dry meadows, by roadsides, in habitations, on edges of fields.—*General distribution*: Caucasus, Russian Central

[1]Treatment by I.A. Gubanov.

104

Asia; Central and Atlantic Europe, Mediterranean, Asia Minor, Iran.—2n = 10.

GENUS 39. *INULA* L.[1]
1753, Sp. Pl.: 881; id. 1754, Gen. Pl., ed. 5: 375

Heads heterogamous, many-flowered, clustered in corymbose inflorescence or solitary terminal on stem and its branches. Involucre cup-shaped, consisting of several rows of herbaceous bracts, of diverse form, imbricate. Receptacle flat or weakly convex, glabrous or more or less glandular on margin of alveoli. Flowers yellow, rarely (in *I. conyza*) reddish or brownish, peripheral in one whorl, usually distinctly exceeding involucre, pistillate, ligulate (ligules with three apical teeth), less often filiform-tubular, almost not exceeding involucre; disk flowers in many whorls, bisexual, infundibuliform-tubular, five-toothed. Anthers sagittate. Stigma lobes apically expanded, obtuse, ciliate. All achenes similar, prismatic or terete, tetraquetrous or ribbed; pappus of one row of numerous long scabrous bristles, often connate at base. Perennial, less often biennial herbaceous plants, with erect stems; leaves alternate with undivided blades.

Lectotype: *I. helenium* L.

About 100 species, distributed in Eurasia and Africa, predominantly in the temperate zone.

Literature: Beck, G. 1881. Inulae Europae. Die Europaischen *Inula*-Arten. *Denkschr. Akad. Wiss. (Math.-Naturw. Wien).* **44:** 283–339.——Avetisyan, V.E. 1955. Novoe v morfologii roda *Inula* L. [New data on the morphology of the genus *Inula* L.]. *Izv. Akad. Nauk. ArmSSR, Ser. Biol. i Sel'skokhoz. Nauk,* **8,** 6: 105–106.—Avetisyan, V.E. 1958. Kavkazskie predstaviteli roda *Inula* L. [The Caucasian members of the genus *Inula* L.]. *Tr. Bot. Inst. Akad. Nauk ArmSSR,* **11:** 3–72.—Buikq, R.A. 1959. Biologiya *Inula helenium* L. (devyasila vysokogo) i nakoplenie v nem efirnogo masla [Biology of the elecampane inula—*Inula helenium* L. and the accumulation of essential oil in it]. *Bot. Zhurn.,* **44,** 12: 1741–1747.—Nyàràdy, E.J. 1964. Genus *Inula* L. *Flora RPR,* **9:** 164–291.—Belyanina, N.B. and K.V. Kiseleva, 1980. *Inula caspia* (Asteraceae)—novyi vid dlya flory Kryma [*Inula caspia* (Asteraceae)—a new species in the flora of Crimea]. *Bot. Zhurn.,* **65,** 10: 1469–1470.

82

---

[1]Treatment by I.A. Gubanov.

Plate VIII.

1—*Pallenis spinosa* (L.) Cass.; 2—*Inula conyza* DC., 2a—head; 3—*Pulicaria dysenterica* (L.) Bernh.; 4—*P. vulgaris* Gaertn.

1. Plant 60–250 cm high, with fleshy root, thick stem (to 1.2 cm in dia) and large elliptical leaves (lower 10–50 cm long and 5–25 cm wide); heads 6–8 cm in dia; inner involucral bracts expanded above, obtuse...................... .... 1. **I. helenium.**

  + Plants much shorter, with slender stem and smaller leaves; root not fleshy; heads to 5 cm in dia; inner involucral bracts acute.. 2.

2. Peripheral flowers with conspicuous ligules, distinctly exceeding involucre, considerably larger than inner flowers .... 3.

  + Peripheral flowers with short ligules, almost not exceeding involucre and inner flowers or ligules altogether absent [and then] flowers filiform-tubular, little different from inner flowers.. 9.

3. Leaves with five to seven parallel veins, linear or linear-lanceolate, 0.3–0.6 mm wide, with absolutely flat margin, crowded on stem .... 3. **I. ensifolia.**

  + Leaves with pinnate venation, lanceolate or oblong, more than 0.5 cm wide, often not with flat margin (with five teeth, projections and cilia) .... 4.

4. Involucral bracts with dense long squarrose bristles; stem covered with stiff squarrose hairs, especially below inflorescence; leaves also with stiff hairs, particularly on margin and veins..2. **I. hirta.**

  + Involucral bracts glabrous, with small cilia on margin or more or less covered with soft hairs but always without squarrose bristles; stem and leaves glabrous or pubescent but not with stiff hairs .... 5.

5. Heads small (to 1 cm in dia), numerous, clustered in dense corymbs; peripheral flowers only slightly exceeding involucre .... 4. **I. germanica.**

  + Heads larger (1–5 cm in dia), solitary or less numerous in lax corymbs; peripheral flowers considerably exceeding involucre .... 6.

6. Ovary and achenes glabrous; leaves rather stiff, often almost coriaceous. Plant glabrous or weakly pubescent .... 5. **I. salicina.**

  + Ovary and achenes pubescent; leaves softer.............. 7.

7. Biennial plants, pubescent only in upper half of stem; peduncles and involucral bracts with long basally thickened hairs or hairs on tubercles; leaves narrow,

acuminate, sessile, with occasional stiff hairs on margin; heads 2–3 cm in dia; peripheral flowers exceeding involucre by one-third their length, inner flowers exceeding pappus .... 6. **I. caspica.**

+ Perennial plants, more or less pubescent with long sericeous hairs or even white-tomentose; leaves broader, lower elliptical or broadly lanceolate; heads 3.0–4.5 cm in dia; peripheral flowers two times as long as involucre, inner as long as pappus .... 8.

8. Outer involucral bracts distinctly shorter than inner, with dense lanate pubescence; leaves and stem densely pubescent; lower and partly middle leaves distinctly petiolate, only some upper leaves semiamplexicaul (that too not in all plants); achenes about 3 mm long, two-fifths as long as pappus .... 7. **I. oculus-christi.**

+ All involucral bracts almost similar in size and pubescence; leaves and stem less densely pubescent; majority of leaves sessile, semiamplexicaul, only lowermost leaves may be petiolate (that too not in all plants!); achenes about 1 mm long, one-fifth to one-fourth as long as pappus....
.... 8. **I. britannica.**

9(2). Cauline leaves decurrent on stem; stem winged; peripheral flowers yellow, their ligules to 3 mm long. Perennials...
.... 10. **I. thapsoides.**

+ Cauline leaves sessile but not decurrent on stem; stem not winged; peripheral flowers reddish, their ligules about 1 mm long. Biennials .... 9. **I. conyza.**

Section 1. Inula.—*Corvisartia* Mérat, 1812, Nouv. Fl. Paris: 328.—*Inula* sect. *Corvisartia* (Mérat) DC. 1836, Prodr. **5**: 463.

Peripheral flowers ligulate, without staminodes, two to two and one-half times as long as involucre; receptacle densely glandular on edges of alveoli; anther appendages long-fimbriate; pappus bristles connate at base.

Type: lectotype of genus.

1. **I. helenium** L. 1753, Sp. Pl.: 881; Gorschk. 1959, Fl. SSSR, **25**: 440; P.W. Ball and Tutin, 1976, Fl. Europ. **4**: 134.

Type: Europe ("in Anglia, Belgio").

North; *Baltic*; *Center*; *West*; *East*; *Crimea*.—In deciduous and mixed forests, scrublands, forest glades and edges, on banks of water bodies, sometimes in meadow steppes; in more northern regions predominantly as an escape or ecdemic plant; in the mountains ascending to subalpine zone; cultivated as an ornamental and medicinal plant.—*General distribution*: Caucasus, Western Siberia, Russian Central Asia; Scandinavia, Central and Atlantic Europe, Mediterranean, Asia Minor, Dzhungaria-Kashgaria; cultivated and escaped in some other extratropical countries.— $2n = 20$.

Section 2. Enula Duby, 1828, Bull. Gall. **1**: 267.—*Enula* Neck. 1790, Elem. Bot. **1**: 4, nom. illeg.—*Inula* subsect. *longeligulatae* G. Beck, 1881, Denkschr. Akad. Wiss. (Math.-Naturw. Wien), **44**: 288.

Peripheral flowers ligulate, without staminodes, one and one-half to two times as long as involucre; receptacle glabrous; anther appendages fimbriate; pappus bristles free or connate at base.

Type: *I. salicina* L.

2. **I. hirta** L. 1753, Sp. Pl.: 883; Gorschk. 1959, Fl. SSSR, **25**: 448; P.W. Ball and Tutin, 1976, Fl. Europ. **4**: 134.—(Plate IX, 1).

Type: Europe ("in Bavaria, Gallia, Genevae").
North (Karelia-Murmansk: near the Khibiny Station, ecdemic); *Center* (Upper Dnieper; Upper Volga; Volga-Kama; Volga-Don); *West*; *East*.—In steppes, steppefied meadows, scrublands, forest glades and edges, deciduous and pine forests, on chalk and limestone outcrops.—*General distribution*: Caucasus, Western Siberia (south), Russian Central Asia; Central and Atlantic Europe, Mediterranean, Asia Minor.—$2n = 16$.

3. **I. ensifolia** L. 1753, Sp. Pl.: 883; Gorschk. 1959, Fl. SSSR, **25**: 449; P.W. Ball and Tutin, 1976, Fl. Europ. **4**: 135.—(Plate X, 5).

Type: Austria ("in Austria inferiore").
Center (Upper Dnieper: southeast; Volga-Don); *West*; *East* (Lower Don); *Crimea*.—In steppes, steppefied meadows, forest glades and edges, scrublands, on chalk and limestone outcrops.— *General distribution*: Caucasus; Central and Atlantic Europe, Mediterranean, Asia Minor.—$2n = 16$.

Note. Sometimes a hybrid is found: *I. ensifolia* × *I. germanica* L. (= *I.* × *hybrida* Baumg.) with indistinct parallel venation of leaves.

4. **I. germanica** L. 1753, Sp. Pl.: 883; Gorschk. 1969, Fl. SSSR, **25**: 449; P.W. Ball and Tutin, 1976, Fl. Europ. **4**: 134.—(Plate X, 3).

Type: Europe ("in Misnia, Pannonia, Sibiria").
86 *Center* (Upper Dnieper: south; Volga-Kama; Volga-Don); *West* (Dnieper; Moldavia; Black Sea); *East*; *Crimea*.—In steppes, steppefied meadows, scrublands, forest glades and edges, on chalk and limestone outcrops.—*General distribution*: Caucasus,

84

Plate IX.

1—*Inula hirta* L.; 2—*I. salicina* L.; 3—*I. salicina* L. subsp. *aspera* (Poir.)
Hayek; 4—*I. salicina* L. subsp. *sabuletorum* (Czern. ex Lavr.) Soják; 5—
I. thapsoides (Bieb.) Spreng.

85

Plate X.
1—*Inula britannica* L.; 2—*I. caspica* Blum.; 3—*I. germanica* L.; 4—*I. oculus-christi* L., lower part of plant; 5—*I. ensifolia* L.

Western Siberia, Russian Central Asia; Central Europe, Mediterranean, Asia Minor, Iran.—2n = 16.

Note. Sometimes a hybrid is found: *I. germanica × I. salicina* (= *I. × media* Bieb.) which is distinguished by less numerous larger heads and subglabrous leaves.

5. **I. salicina** L. 1753, Sp. Pl.: 882; Gorschk. 1959, Fl. SSSR, **25**: 454; P.W. Ball and Tutin, 1976, Fl. Europ. **4**: 134.—(Plate IX, 2).

Type: Europe ("in Europae borealis pratis uliginosis, asperis").

a. Subsp. **salicina.**—Heads 2.5–4.0 cm in dia; leaves oblong-lanceolate, 4–10 cm long, 1–3 cm wide, often deflected from stem at right angle.

North; *Baltic*; *Center*; *West*; *East*; *Crimea.*—In pine, deciduous and mixed forests, forest glades and edges, forest fellings, scrublands, meadows, sometimes along railroads and highways.— *General distribution*: Caucasus, Western and Eastern Siberia, Far East, Russian Central Asia; Scandinavia, Central and Atlantic Europe, Mediterranean, Asia Minor, Iran, Dzhungaria-Kashgaria, Japan-China.—2n = 16.

b. Subsp. **aspera** (Poir.) Hayek, 1931, Prodr. Fl. Penins. Balcan. **2**: 602; P.W. Ball and Tutin, 1976, Fl. Europ. **4**: 134.—*I. aspera* Poir. 1813, in Lam. Encycl. Méth. Bot. Suppl. **3**: 154; Gorschk. 1959, Fl. SSSR, **25**: 452; Ignatev and others, 1990, Flor. Issled. v Mosk. Obl.: 88.—Heads 2–3 cm in dia; leaves oblong-lanceolate, 2–5 cm long, 0.6–2.5 cm wide, usually more or less upward directed (Plate IX, 3).

Type: Paris Botanical Garden ("On la cultive an Jardin des Plantes de Paris").

Center (Upper Volga: ecdemic in Moscow; Volga-Kama: south; Volga-Don); *West* (Dnieper; Black Sea); *East*; *Crimea.*—In steppes, dry and alkaline meadows, steppe scrublands, on chalk and limestone outcrops, less often in open dry forests and forest edges. Predominantly steppe plant.—*General distribution*: Caucasus, Western Siberia, Russian Central Asia; Mediterranean, Asia Minor, Iran.

c. Subsp. **sabuletorum** (Czern. ex Lavr.) Soják, 1972, Čas. Nár. Muz. odd. Přir. Praha, **140**, 3–4: 131.—*I. sabuletorum* Czern. ex Lavr. 1925–1926, Index Sem. Horti Bot. Charkov: 7: Gorschk. 1959, Fl. SSSR, **25**: 453.—Heads 1–2 cm in dia; leaves narrow-lanceolate, 5–9 cm long, 0.5–1.2 cm wide, divergent from stem at acute or almost right angle (Plate IX, 4).

Type: Ukraine ("Ucraina meridionalis (distr. Cherson... et orientalis (prov. Charjkov)").

112

North (Dvina-Pechora: ecdemic in the vicinity of Syktykar);
West (Dnieper; Black Sea); *East* (Lower Don; Lower Volga).—On
sandy river terraces above the floodplain, in sandy steppes.—
General distribution: Caucasus, Russian Central Asia.

Note. Subspecies of *I. salicina* s. l. are linked with all possible
transients and their delimitation often presents serious difficulties.
The situation also becomes complex because of frequent
hybridization with other species (cf. above).

6. **I. caspica**[1] Blum. 1822, in Ledeb. Index Sem. Hort. Acad.
Dorpat.: 10; Gorschk. 1959, Fl. SSSR, **25**: 460; P.W. Ball and
Tutin, 1976, Fl. Europ. **4**: 135; Belyanina and Kiseleva, 1980, Bot.
Zhurn. **65,** 10: 1469.—(Plate X, 2).

Type: Coast of the Caspian Sea ("ad mare Caspium").

Center (Volga-Kama: ecdemic in Sarapul); *East* (Lower Volga);
Crimea (Kerch Peninsula).—In wet saline meadows, coastal tall
grasses, rather rare.—*General distribution*: Caucasus, Western
Siberia, Russian Central Asia; Iran, Dzhungaria-Kashgaria.—2n = 16.

Note. Apparently hybridizes with *I. britannica.*

7. **I. oculus-christi** L. 1753, Sp. Pl.: 881; Gorsch. 1959, Fl.
SSSR, **25**: 461; P.W. Ball and Tutin, 1976, Fl. Europ. **4**: 135.—
(Plate X, 4).

87 Type Austria ("in Austria").

Center (Upper Dnieper: south; Volga-Don); *West* (Dnieper;
Moldavia; Black Sea); *East*; *Crimea.*—In steppes, steppefied
meadows, scrublands.—*General distribution*: Caucasus, Russian
Central Asia; Central Europe, Mediterranean, Asia Minor, Iran.—
2n = 30, 32.

8. **I. britannica** L. 1753, Sp. Pl.: 882; Gorschk. 1959, Fl. SSSR,
25: 465; P.W. Ball and Tutin, 1976, Fl. Europ. **4**: 135.—(Plate X, 1).

Type: Europe ("in Lutetia, Bavaria, Scania").

North; *Baltic*; *Center*; *West*; *East*; *Crimea.*—In forest glades
and edges, meadows, on banks of water bodies, in thinned-out
forests, scrublands, steppes, by roadsides, in habitations.—*General
distribution*: Caucasus, Western and Eastern Siberia, Far East,
Russian Central Asia; Scandinavia, Central and Atlantic Europe,
Mediterranean, Asia Minor, Iran, Dzhungaria-Kashgaria, Mongolia,
Japan-China.—2n = 16, 24, 32.

Note. Highly polymorphic species, possibly represented by
several ecogeographic races with different chromosome numbers.

[1]In the original description "caspica" and not "caspia."

Section 3. Breviligulatae (G. Beck) Gorschk. 1959, Fl. SSSR, **25**: 473.—*Inula* subsect. *Breviligulatae* G. Beck, 1881, Denkschr. Akad. Wiss. (Math.-Naturw. Wien), **44**: 291, 329.—*Inula* subgen. *Conyzoides* Kirschl. 1870, Fl. Voges Rhen.: 341.

Peripheral flowers with small ligules or ligules altogether absent ([flowers] filiform-tubular), with one to four staminodes, less often without them; receptacle glabrous; anther appendages acuminate; pappus bristles unevenly connate at base.

Type: *I. conyza* DC.

9. **I. conyza** DC. 1836, Prodr. **5**: 464; P.W. Ball and Tutin, 1976, Fl. Europ. **4**: 136.—*Conyza squarrosa* L. 1753, Sp. Pl.: 861, non *Inula squarrosa* L. 1753.—*Conyza vulgaris* Lam. 1778, Fl. Fr. **2**: 73, nom. illeg.—*Inula vulgaris* (Lam.) Trevis. 1842, Prosp. Fl. Euganea, 29, nom. illeg.; Gorschk. 1959, Fl. SSSR, **25**: 473. — (Plate VIII, 2).

Type: Europe ("in Germaniae, Belgii, Angliae, Galliae").

West (Dnieper: west; Moldavia); *Crimea* (south).—In forest glades and edges, deciduous forests.—*General distribution*: Caucasus; Central and Atlantic Europe, Asia Minor.—2n = 32.

10. **I. thapsoides** (Bieb.) Spreng. 1810, Index Sem. Horti Halens.: 16; Gorschk. 1959, Fl. SSSR, **25**: 474, cum auct. comb. DC.: P.W. Ball and Tutin, 1976, Fl. Europ. **4**: 136.—*Conyza thapsoides* Bieb. 1803, in Willd.[1] Sp. Pl.: **3**, 3: 1949.—(Plate IX, 5).

Type: Coast of the Caspian Sea ("ad mare Caspium").

Crimea (vicinity of Yalta).—In wet shady forests.—*General distribution*: Caucasus; Asia Minor, Iran.—2n = 16.

GENUS 40. *PULICARIA* Gaertn.[2]
1791, Fruct. Sem. Pl. 2: 461

Heads heterogamous, with two types of yellow flowers; peripheral pistillate, ligulate, three-toothed, in one whorl; inner bisexual, tubular, five-toothed, numerous. Involucre of five or six rows of imbricate bracts. Receptacles glabrous. All achenes similar, oblong, slightly flattened, longitudianlly ribbed, short-pubescent;

---

[1]While publishing the description of this species Willdenow cited the then unpublished *Fl. Taur. Cauc.*, about which he could learn only from its author Marschall Bieberstein. Hence it may be considered that the description of *Conyza thapsoides* was compiled by Bieberstein and published by Willdenow with his permission ("in litt.").

[2]Treatment by I.A. Gubanov.

pappus double: outer row or short scarious scales, connate at base into a crown, inner of scarious bristles. Annual or perennial herbs with unpleasant odor; leaves alternate, mostly sessile, undivided.

88 Type: *P. vulgaris* Gaertn.

About 50 species, distributed in Eurasia and Africa, predominantly in the countries of the Mediterranean.

1. Leaves cordate, mostly amplexicaul, all sessile; peripheral flowers one and one-half to two times as long as involucre .... 2. **P. dysenterica.**
+ Leaves rounded at base, not amplexicaul, lower short-petiolate; peripheral flowers not exceeding involucre (less often only slightly exceeding). Annuals .... 1. **P. vulgaris.**

1. **P. vulgaris** Gaertn. 1791, Fruct. Sem. Pl. 2: 461; Ratcliffe, 1976, Fl. Europ. **4:** 137.—*Inula prostrata* Gilib. 1781, Fl. Lithuan. **1:** 205, nom. illeg.—*Pulicaria prostrata* (Gilib.) Aschers, 1864, Fl. Prov. Brandenb. **1:** 304, nom. illeg.; Golubk. 1959, Fl. SSSR, **25:** 489.—(Plate VIII, 4).

Type: Figure in original description (Gaertn. op. cit.: 137, tab. 173, fig. 7); locality not mentioned.

Baltic; *Center* (Upper Dnieper; Upper Volga: south; Volga-Kama; Volga-Don); *West*; *East*; *Crimea.*—In damp meadows, on banks of water bodies, by roadsides, in habitations.—*General distribution*: Caucasus, Western Siberia, Far East, Russian Central Asia; Scandinavia, Central and Atlantic Europe, Mediterranean, Asia Minor, Iran, Dzhungaria-Kashgaria, Mongolia.—2n = 18.

2. **P. dysenterica** (L.) Bernh. 1800, Syst. Verz. Erfurt.: 153; Golubk. 1959, Fl. SSSR, **25:** 490; Ratcliffe, 1976, Fl. Europ. **4:** 137; Grigorjevskaya, 1990, Bot. Zhurn. **75**, 3: 434.—*Inula dysenterica* L. 1753, Sp. Pl.: 882.—*Pulicaria uliginosa* Stev. 1836, in DC. Prodr. **5:** 478; Golubk. 1959, Fl. SSSR, **25:** 491.—(Plate VIII, 3).

Type: Europe ("in Europae fossis").

Center (upper Dnieper: reported for the vicinity of Rogachev on the Dnieper, probably in error; Volga-Don: southeast); *West* (Moldavia; Black Sea); *Crimea.*—In damp meadows, on edges of swamps, banks of water bodies.—*General distribution*: Caucasus, Russian Central Asia; Central and Atlantic Europe, Mediterranean, Asia Minor, Iran.—2n = 18, 20.

Note. Populations with more numerous but smaller (to 18 mm in dia) heads are sometimes identified as a separate species.— *P. uliginosa* Stev.

GENUS **41.** *CARPESIUM* L.[1]
1753, Sp. Pl.: 859; id. 1754, Gen. Pl., ed. 5: 369

Heads heterogamous, drooping, solitary at apices of stem and its branches, surrounded by one or two subtending leaves, resembling cauline leaves in form and pubescence. Involucre of several rows of imbricate bracts; outer bracts deflected, densely hairy. Receptacle glabrous, flat. All flowers tubular, yellow; peripheral pistillate, three- to five-toothed, many-rowed; disk flowers bisexual, five-toothed, somewhat broader than peripheral. Achenes fusiform-triquetrous, weakly sulcate longitudinally, apically narrowed in beak, terminating in saucer-shaped areola thickened at margin, without pappus. Perennial, more or less pubescent plants, with erect stem and alternate oblong-lanceolate undivided leaves.

Lectotype: *C. cernuum* L.

The genus includes nine species, distributed in the temperate and subtropical regions of Eurasia, but predominantly in eastern Asia.

1. **C. cernuum** L. 1753, Sp. Pl.: 859; Golubk. 1959, Fl. SSSR, **25:** 501; Tutin, 1976, Fl. Europ. **4:** 138.

Plant 30–50 cm high; heads 13–25 mm in dia; achenes 4–5 mm long.

Type: Italy ("in Italia").

89   *West* (Carpathians; Dnieper: West).—In forests, scrublands, by roadsides, in ditches.—*General distribution*: Caucasus, Far East, Russian Central Asia; Central Europe, Mediterranean, Asia Minor, Dzhungaria-Kashgaria, Tibet, Himalayas, Japan-China.— $2n = 40$.

GENUS **42.** *ADENOCAULON* Hook.[2]
1830, Bot. Misc. **1:** 19

Heads heterogamous, clustered in lax terminal paniculate inflorescence. Involucre of five to seven bracts, pendent at fruiting,

---

[1]Treatment by I.A. Gubanov.
[2]Treatment by N.N. Tzvelev.

one-rowed, with almost equal bracts. Receptacle flat or some-what convex, without pales. Peripheral flowers pistillate, in one whorl, their corollas whitish, tube short and with four- or five-parted limb; disk flowers in many rows, staminate, their corollas with narrow long tube and five-lobed limb. Anthers with very short appendage at base. Styles of peripheral flowers with two broad branches; in disk flowers more or less reduced, almost not lobed. Achenes 7–8 mm long, clavate-cylindrical, weakly ribbed, stellately deflected from receptacle, without pappus, as also inflore-scence more or less covered with stalked glands and hence sticky. Perennial herbaceous plants, with erect stem and alternate leaves with undivided but usually toothed blade and winged petiole.

Type: *A. bicolor* Hook.

The genus includes four or five species in America and eastern Asia; occasionally introduced in other warm temperate and tropical countries.

1. **A. adhaerescens** Maxim. 1849, Mém. Prés. Acad. Sci. Petersb. Div. Sav. **9:** 152 (Prim. Fl. Amur); Golubk. 1959, Fl. SSSR, **25:** 506; Ignatov and others, 1988, Bot. Zhurn. **73,** 3: 440.

Plant 10–60 cm high; lower leaves orbicular-reniform, 3–8 cm long, middle orbicular-deltoid, 8–12 cm long; involucre 2.5–3.5 mm long; corolla of peripheral flowers 2.0–2.5 mm long, of disk flowers 2.5–3.0 mm long.

Type: Far East ("am untern Amur, von Borbi an... bis zur Ussuri-Mündung ... Chungar ... Dshare ... Sasurgu ...").

Center (Upper Volga: ecdemic and naturalized in Moscow Botanical Garden).—In coastal gray alder groves.—*General distribution*: Far East; Japan-China.

GENUS 43. *KARELINIA* Less.[1]
1834, Linnaea, **9:** 187

Heads heterogamous, 0.8–2.0 cm in dia, clustered in groups of two to nine in dense terminal corymbose inflorescences. Involucre 10–15 mm long, 6–15 mm in dia, obconical or campanulate; involucral bracts six- or seven-rowed, imbricate, outer ovate, coriaceous, inner oblong-lanceolate and linear, almost membranous, all light reddish-brown, with short appressed hairs on outside and cilia on margin predominantly in upper part. Flowers reddish or pinkish, numerous; peripheral flowers pistillate, fertile with filiform irregularly

---

[1]Treatment by L.I. Krupkina.

four-toothed corolla. Inner bisexual but functionally staminate, 10–20, with regular five-toothed corolla. Achenes 1–2 mm long and 0.3 mm wide, terete, faceted, somewhat curved, dark reddish-brown, glabrous, with whitish appendage at apex; pappus 12–15 mm of scabrous bristles, flexuous at base, with a tuft at apex in bisexual flowers. Perennial herbs, with glabrous (or subglabrous) erect, strongly branched stem 40–150 cm high; leaves alternate, sessile, undivided and entire, semiamplexicaul.

90

Type: *K. caspia* (Pall.) Less.
A monotypic genus.

1. **K. caspia** (Pall.) Less. 1834, Linnaea, **9:** 187; Smoljan. 1959, Fl. SSSR, **25:** 296; Tikhomir. 1976, Fl. Europ. **4:** 121.—*Serratula caspia* Pall. 1773, Reise, **2:** 743, tab. Z.—*Pulchea caspia* (Pall.) O. hoffm. ex Pauls. 1903, Kjoeb. Viddensk. Meddel. 1903: 147.

Type: Northern Ciscaspia ("in depressis, salsis versus mare Caspium, inde a statione Baksai").

East (Lower Volga: near the Caspian seacoast and on the Ural River).—On seacoast, banks of rivers and saline lakes, saltlicks and alkaline meadows.—*General distribution*: Russian Central Asia; Asia Minor, Iran (north), Dzhungaria-Kashgaria (northeast), Mongolia (northwest), Tibet (northwest).—2n = 20.

GENUS 44. *LEONTOPODIUM* (Pers.) R. Br.[1]

1817, Trans. Linn. Soc. (London), **12:** 124.—*Gnaphalium* L. + + *Leontopodium* Pers. 1806, Syn. Pl. **2:** 422

Heads usually heterogamous, 5–8 mm in dia, densely crowded in groups of 5–12 and all together enclosed by rather large tomentose subtending leaves, forming with them stellate inflorescences, 2.0–3.5 cm in dia. Involucre 3–5 mm long, cup-shaped; involucral bracts two- or three-rowed, imbricate, outer oblong-ovate, roughly half as long as oblong-lanceolate or oblong inner, more or less lanate, membranous at tip and on margin. Receptacle flat, smooth, without pales. Peripheral flowers usually pistillate, with narrow-tubular, three- or four-toothed yellowish corolla; inner bisexual but usually functionally staminate with more deeply incised, four- or five-toothed, tubular corolla, often also with sterile flowers having strongly developed nectaries or some heads with flowers of one sex. Achenes 1–5 mm long, 0.3–0.4 mm wide, very short-hairy; pappus 3.5–4.0 mm long, consisting

---

[1]Treatment by S.S. Ikonnikov.

of one row of scabrous bristles. More or less tomentose perennial plants with erect stem and usually condensed vegetative shoots; leaves alternate, undivided and entire, linear to oblong-spatulate.

Type: *Gnaphalium leontopodium* L. (= *Leontopodium alpinum* Cass.).

About 40 species, distributed in the mountainous regions of Europe and Asia.

Literature: Handel-Mazzetti, H. 1927. Systematische Monographie der Gattung *Leontopodium. Beih. Centralbl.* **44,** 2: 1–178.—Sokolowsky-Kulczycka, A. 1959. Apomiksia u *Leontopodium alpinum* Cass. [Apomixis in *Leontopodium alpinum* Cass.] *Acta Biol. Cravov (Bot.),* **2,** 1: 53–61.

1. **L. alpinum** Cass, 1822, Dict. Sci. Nat. **25:** 474; Grub. 1959, Fl. SSSR, **25:** 360; Dobrocz. 1962, Fl. URSR, **11:** 90; Tutin, 1976, Fl. Europ. **4;** 132, p. p. (quoad subsp. *alpinum*).—*Gnaphalium leontopodium* L. 1753, Sp. Pl.: 855.—(Plate XI, 1).

Plant 3–15 (20) cm high; subtending leaves white tomentose, linear-lanceolate, 6–20 (22) mm long and 2–5 mm wide.

Type: Mountains of Central Europe ("in Alpibus Helvetiae, Vallesiae, Carinthi, Austriae...").

West (Carpathians).—On limestone rocks, stony talus and slopes from 1,700 and 1,800 m above sea.—*General distribution*: Central Europe (mountains), Mediterranean (mountains).—2n = 52.

92 *Note.* This species has become quite rare in the Carpathians and is included in the Red Data Book of Wild Species of the Flora of the USSR needing protection (1975: 34; 1981, ed. 2: 53).

GENUS **45. *ANTENNARIA*** Gaertn.[1]
1791, Fruct. Sem. Pl., 2: 410

Heads unisexual, homogamous, 4–10 mm in dia, usually clustered in corymbose or corymbose-paniculate inflorescences.

---

91   Plate XI.

1—*Leontopodium alpinum* Cass., 1a—head, 1b—pistillate flower, 1c—staminate flower; 2—*Antennaria carpatica* (Wahlenb.) Bluff and Fingerh., 3—*A. dioica* (L.) Gaertn., 3a—head with pistillate flowers, 3b—head with staminate flowers, 3c—pistillate flower, 3d—staminate flowers; 4—*Gnaphalium luteo-album* L., 4a—head, 4b—inner bisexual flower, 4c—outer pistillate flower, 4d—achene; 5—*Helichrysum arenarium* (L.) Moench, 5a—head, 5b—flower, 5c—achene; 6—*H. graveolens* (Bieb.) Sweet, head.

---

[1]Treatment by S.S. Ikonnikov.

120

Involucre 5–13 mm long, cup-shaped; involucral bracts imbricate, in several rows, membranous in upper part, white, pink or brownish. Receptacle somewhat convex, without pales. Pistillate and bisexual flowers on different plants: pistillate with filiform, tubular, five-toothed, 5–7 mm long corolla, ovary glabrous and smooth or covered with very fine papillae, and pappus of numerous uniformly thick and weakly scabrous, 3–5 mm long bristles connate at base; bisexual (functionally staminate) flowers with broadly tubular, five-toothed corolla, 3–5 mm long, with more or less undeveloped ovary and pappus of less numerous, clavate bristles 3–7 mm long. Achenes oblong, 0.5–1.8 mm long, glabrous and smooth, with 3–5 mm long pappus. Perennial dioecious plants, with erect or ascending stems and condensed vegetative shoots; leaves alternate, undivided and entire, more or less tomentose like entire plant, linear to spatulate.

Type: *A. dioica* (L.) Gaertn.

Over 200 species, distributed in extratropical regions of the Northern Hemisphere (excluding northern Africa).

Literature: Porsild, A.E. 1950. The genus *Antennaria* in northern Canada. *Canad. Field Natur.*, **64**: 1–25.—Porsild, A.E. 1965. The genus *Antennaria* in eastern Arctic and subarctic America. *Bot. Tidsskr.* **61**: 22–55.—Petrovskii, V.V. 1983. Rod *Antennaria* Gaertn. (Asteraceae) na severo-vostoke Azii [The genus *Antennaria* Gaertn. in northeastern Asia]. *Novosti Sist. Vyssh. Rast.* **23**: 181–197.—Bayer, R.J. 1988. Typification of western North American *Antennaria* Gaertn. (Asteraceae: Inuleae): sexual species of sections *Alpinae, Dioicae* and *Plantaginifolia. Taxon*, **37**, 2: 292–298.

1. Plants lacking stoloniferous shoots; basal leaves oblanceolate to linear, 20–90 mm long and 3–7 mm wide ... 2.
+ Plant with aerial stoloniferous shoots; basal leaves obovate to oblanceolate-spatulate, 10–35 mm long and 1.5–10.0 mm wide .... 3.
2. Staminate flowers creamish with purple anthers; basal leaves glabrous above or subglabrous, white-tomentose beneath; pappus at mautiry almost as long as style. Plants of the Carpathians .... 4. **A. carpatica.**
+ Staminate flowers purple with yellowish anthers; basal leaves more or less lanate-tomentose, grayish on both sides; pappus at maturity considerably (by more than 1.75 mm) exceeding style. Plants of the Arctic ....
.... 5. **A. lanata.**

3. Basal leaves densely tomentose beneath, usually whitish or almost white; upper part of involucral bracts white, pink or greenish-brown, petalloid, of various forms from obovate to lanceolate .... 4.
+ Basal leaves glabrous on both sides or subglabrous beneath, green; upper part of involucral bracts dark, brownish or greenish-brown, not petalloid, lanceolate ..
.... 3. **A. porsildii.**
4. Upper part of involucral bracts white or pink, oblong or broadly obovate, obtuse .... 1. **A. dioica.**
+ Upper part of involucral bracts greenish-brown or brownish, subacute .... 2. **A. alpina.**

93

Section 1. Antennaria.—*Antennaria* sect. *Catipes* DC. 1838, Prodr. **6**: 269, p. p.; Boriss. 1959, Fl. SSSR, **25**: 328.

Rhizome horizontal, slender, with aerial stoloniferous shoots; basal leaves usually spatulate; involucral bracts white, pink or brownish in upper part.

Type: type species.

1. **A. dioica** (L.) Gaertn. 1791, Fruct. Sem. Pl. **2**: 410; Boriss. 1959, Fl. SSSR, **25**: 329; Dobrocz. 1962, Fl. URSR, **11**: 86; Halliday, 1976, Fl. Europ. **4**: 131; Petrovskii 1987, Arkt. Fl. SSSR, **10**: 87.— *Gnaphalium dioicum* L. 1753, Sp. Pl.: 850.—(Plate XI, 3).

Type: Europe ("in Europae apricis aridis").

Arctic (Arctic Europe); *North*; *Baltic*; *Center*; *West* (Carpathians; Dnieper; Black Sea); *East*; *Crimea* (mountains).—In pine forests, on forest glades and edges, open slopes, dry meadows, stony slopes and rocks, particularly on sandy and sandy-loam, soils.— *General distribution*: Caucasus (Ciscaucasia), Western and Eastern Siberia, Arctic, Far East, Russian Central Asia (north); Scandinavia, Atlantic Europe, Mediterranean, Asia Minor, Dzhungaria-Kashgaria, Mongolia.—2n = 28.

2. **A. alpina** (L.) Gaertn. 1791, Fruct. Sem. Pl. **2**: 410; Boriss, 1959, Fl. SSSR, **25**: 336; Orlova, 1966, Fl. Murm. Obl. **5**: 204; Halliday, 1976, Fl. Europ. **4**: 132; Petrovskii, 1987, Arkt. Fl. SSSR, **10**: 88.—*Gnaphalium alpinum* L. 1753, Sp. Pl.: 856.

Type: Lapland ("in Lapponiae").

Arctic (Arctic Europe; Rybachi Peninsula).—On slopes of coastal terraces.—*General distribution*: Scandinavia.—2n = 70, 84, 85.

3. **A. porsildii** Elis Ekman, 1927, Svensk. Bot. Tidskr. **21**: 51; Halliday, 1976, Fl. Europ. **4**: 132.

Type: Danmark Island east of Greenland ("Danmarks Insel, ca. 70°30′ ").

Arctic (Arctic Europe: Murmansk Region, Rybachi Peninsula).— On more or less wet slopes of coastal terraces.—*General distribution*: Scandinavia; North America (eastern Greenland).— $2n = 63, 70$.

Section 2. Urolepis Boriss. 1960, Bot. Mat. (Leningrad), **20**: 291; id. 1959, Fl. SSSR, **25**: 337, descr. ross.

Rhizome vertical or oblique, rather thick, more or less branched in upper part, but without stolons; leaves linear to oblanceolate; involucral bracts brownish.

Type: *A. carptica* (Wahlenb.) Bluff et Fingerh.

4. A. carpatica (Wahlenb.) Bluff et Fingerh. 1825, Comp. Fl. Ferm. **2**: 348; Boriss. 1959, Fl. SSSR, **25**: 338 (cum auct. comb. R. Br.); Dobrocz. 1962, Fl. URSR, **11**: 89; Czerep. 1973, Svod. Dopl. Izm."Fl. SSSR": 56; Halliday, 1976, Fl. Europ. **4**: 132.—*Gnaphalium carpaticum* Wahlenb. 1814, Fl. Carp.: 258.—(Plate XI, 2).

Type: Europe, the Carpathians ("Hab. locis declivibus madidis rupestribus ad summa cucumena alpium carpaticarum rarius ad cacumen Rohats ... ad cacumen Raczkowa ... ad cacumine Kriwani ... in jugo Kahlbacher-grat ... in cacumine Lomnitzense ... et Thorichtegern Kesmarkensium...").

West (Carpathians).—On stony slopes, taluses, cultivated meadows in alpine zone.—*General distribution*: Central Europe (mountains), Mediterranean (mountains).—$2n = 56$.

5. A. lanata (Hook.) Greene, 1898, Pittonia, **3**: 288; Soják, 1983, Sborn. Nar. Muz. Praze, Rada B. Přir. Vědy, **39**, 1: 54; Chrtek and Pouzar, 1985, Preslia, **57**, 3: 196.—*A. carpatica* (β) *lanata* Hook. 1834, Fl. Bor.-Amer. **1**: 329.—*Gnaphalium carpaticum* var. (β) Wahlenb. 1826, Fl. Suec. **2**: 515.—*Antennaria villifera* Boriss. 1960, Bot. Mat. (Leningrad), **20**: 292; Boriss. 1959, Fl. SSSR, **25**: 340, descr. ross.; Halliday, 1976, Fl. Europ. **4**: 132; Petrovskii, 1987, Arkt. Fl. SSSR, **10**: 87.

Type: Sweden ("in lateribus praeruptus aqua alpium Lulensium a Virihjaur meridiem versus positarum pone Kåbrinoivii, ad pållaure alibique satis copiosa; etiam in alpibus supra Alten Finmarkiae").

Arctic (Novaya Zemlya; Arctic Europe: Kolguev, Vaigach, Pai-Khoi Range, Bolshezemelsk and Karsk tundra).—In meadows, on coastal slopes, banks of streams and rivulets.—*General distribution*: Eastern Siberia, Arctic; Scandinavia (north).—$2n = 38, 42$.

123

GENUS **46.** *HELICHRYSUM* Mill.[1]
1754, Gard. Dict. Abridg. ed. 4, nom. conserv.

Heads homogamous, less often heterogamous, 3–40(50) mm in dia, clustered in more or less dense corymbose inflorescences, usually lacking subtending leaves. Involucre 3–20 mm long, cup-shaped; involucral bracts imbricate, in four to six rows, obtuse or subacute, almost entirely membranous, often with brownish or yellow tinge, outer broadly ovate, one-fifth to half as long as oblong or lanceolate innermost bracts. Receptacle flat or somewhat convex without pales, but sometimes edges of alveoli projected as teeth or lobes. Flowers usually all bisexual, tubular, with light yellow, yellow, or light brown, five-toothed corolla, less often also with a few peripheral pistillate flowers; pistillate flowers narrow-tubular, with three to five, not always distinct, corolla teeth. Achenes usually 0.8–1.3 mm long and 0.3–0.4 mm wide, narrow-ellipsoidal-cylindrical or subcylindrical, but somewhat faceted with two to four, not always distinct veins, covered with scattered tubercular papillae, less often glabrous; pappus 2.5–4.0 mm long, of scabrous, almost equal bristles, connate at base in a ring, falling together or in fascicle. Grayish- or white-tomentose perennial plant, less often semishrub, 10–40(50) cm high with erect, less often ascending or procumbent, leafy stems; leaves alternate with undivided and entire blades.

Type: *H. orientale* (L.) Gaertn.
Nearly 500 species in the warm temperate and subtropical regions of both hemispheres, and partly also in the tropical mountainous regions.
Literature: Kirpicznikov, M.E. 1960. Obzor vidov roda *Helichrysum* Mill. corr. Pers. proizyastayushchikh v SSSR [Review of speices of the genus *Helichrysum* Mill. corr. Pers. occurring in the USSR]. *Bot. Mat. (Leningrad)*, **20**: 314–336.

1. Cultivated plants with stems more or less partly woody; all or almost all leaf blades broadly ovate to broadly elliptical, rather abruptly joined to petiole winged in upper part; ovary and achenes glabrous; involucre yellowish-white or light brownish............................ 6. **H. petiolare.**
+ Wild plants with nonwoody stems; leaf blades linear or elliptical, only lower leaves gradually narrowed in petiole,

---
[1]Treatment by N.N. Tzvelev.

others sessile; ovary and achenes covered with scattered tubercular papillae .... 2.

2. Outermost involucral bracts oblong or oblong-lanceolate, nearly half as long as longest inner bracts; peripheral pistillate flowers usually few. Alpine plants of Crimea, with rather long rhizomes .... 5. **H. graveolens.**

+ Outermost involucral bracts ovate or broadly ovate, one-fourth to one-third as long as longest inner bracts; all flowers in a head bisexual. Plants of plains and low mountains, with very short rhizomes .... 3.

3. Sideways deflected upper membranous part of middle involucral bracts 2.0–2.5 mm long and 1.5–2.0 mm wide .... 4. **H. buschii.**

+ Sideways deflected upper membranous part of middle involucral bracts 0.8–1.7 mm long and 0.8–1.6 mm wide... 4.

4. Entire plant silvery-gray from adpressed and rather dense tomentum; membranous upper part of middle involucral bracts 1.3–1.5 mm long and 0.8–1.0 mm wide, but considerably longer than wide .... 3. **H. tanaiticum.**

+ Plants grayish-tomentose from less dense (seemingly floccose) tomentum; membranous upper part of middle involucral bracts 0.8–1.6 mm long, and 1.2–1.7 mm wide, but often wider than long .... 5.

5. Upper and middle cauline leaves usually not linear, with margins firmly or loosely revolute; heads 4–6 mm long and as much wide, in rather dense corymbs .... 1. **H. arenarium.**

+ Upper and, often also middle, cauline leaves linear, 1–2 mm wide, with revolute margins; heads 3.5–4.0 mm long and as much wide, in rather lax corymbs .... 2. **H. corymbiforme.**

Section 1. Helichrysum. Receptacle with scarcely raised edges of alveoli; achenes with scattered tubercular papillae. Perennial herbaceous plants.

Type: type species.

1. **H. arenarium** (L.) Moench, 1794, Meth. Pl.: 575; Kirpicz. 1959, Fl. SSSR, **25:** 410; Clapham, 1976, Fl. Europ. **4:** 130.— *Gnaphalium arenarium* L. 1753, Sp. Pl.: 854.—(Plate XI, 5).

Type: Europe (" in Europae campis arenosis").

Baltic; *Center* (Ladoga-Ilmen: south; Upper Dnieper; Upper Volga: south; Volga-Kama: south; Volga-Don); *West*; *East*; *Crimea.*—

On dry, predominantly sandy and sandy-loam meadows and forest glades, stony slopes and rocks, in pine forests and steppes.—*General distribution*: Caucasus, Western and Eastern Siberia (south), Russain Central Asia; Scandinavia (south), Central and Atlantic Europe, Mediterranean, Mongolia.—2n = 14(?), 28.

○ 2. **H. corymbiforme** Opperm. ex Katina, 1952, Ukr. Bot. Zhurn. **9**, 3: 87; Katina, 1950, Vizn. Rosl. URSR: 528, descr. ucrain.; Dobrocz. 1962, Fl. URSR, **11**: 105.

Type: Kherson Region ("Golopristan district, Solonoozernaya summer house, birch grove on yagorlitsa Creek").

West (Black Sea: west).—On sands less often gravel-beds of seacoast and in the lower reaches of the Dnieper River.—Endemic.

Note. Possibly this species is also found in the Black Sea regions of Bulgaria and Romania, from where it is reported under the name *H. arenarium* subsp. *ponticum* (Velen.) Clapham (= *H. arenarium* var. *ponticum* Velen.).

3. **H. tanaiticum** P. Smirn. 1940, Byull. Mosk. Obshch. Isp. Prir., otd. Biol. **49**, 2: 87.—*H. arenarium* auct. non (L.) Moench: Kirpicz. 1959, Fl. SSSR, **25**: 410, p. min. p.

Type: Basin of the Don River, Golubaya River ("upper reaches of the Golubaya River, large bouldery slope of the chalk hill").

East (Lower Don).—On chalk and limestone outcrops.—*General distribution*: Caucasus (Ciscaucasia).

○ 4. **H. buschii** Juz. 1955, Spisok Rast. Gerb. Fl. SSSR, **13**: 97; Kirpicz. 1959, Fl. SSSR, **25**: 442, in adnot.; Clapham, 1976, Fl. Europ. **4**: 31, in adnot.

Type: Crimea, vicinity of Bakhchesarai ("in Tauria, iter Bachczisaraj et Czerkes-Kermen").

96 *Crimea.*—On limestone outcrops in the foothill and lower mountain zone.—Endemic.

Note. It is quite likely that this species evolved through hybridization of *H. arenarium* × *H. graveolens* as suggested by the author of the species. However, presently it is enough stabilized. In this state, it is common in the vicinity of Bakhchisarai where *H. graveolens* is absent.

5. **H. graveolens** (Bieb.) Sweet, 1827, Hort. Brit., ed. **1**: 223; Kirpicz. 1959, Fl. SSSR, **25**: 491; Clapham, 1976, Fl. Europ. **4**: 131.—*Gnaphalium graveolens* Bieb. 1808, Fl. Taur.-Cauc. **2**: 299; id 1819, ibid. **3**: 567.—(Plate XI, 6).

Type: Crimea, Mt. Chatyrdag ("in summo montis Tchatyrdagh Tauriae").

Crimea (Yaila).—In cultivated meadows, on stony slopes and rocks.—*General distribution*: Caucasus; Asia Minor.

Section 2. Lepiscline (Cass.) DC. 1837, Prodr. **6**: 197.— *Lepiscline* Cass. 1818, Bull. Soc. Philom. Paris, 1818: 31.

Receptacle with alveolar edges raised as teeth and lobes; achenes glabrous. Semishrubs.

Type: *H. cymosum* (L.) Less.

6. **H. petiolare** Hilliard et B. Burtt, 1973, Notes. Roy. Bot. Gard. Edinb. **32**: 357; Clapham, 1976, Fl. Europ. **4**: 131.—*H. petiolatum* auct. non (L.) DC.: Dobrocz. 1962, Fl. URSR, **11**: 109.

Type: South Africa.

West; *Crimea*.—Cultivated as an ornamental (predominantly border) plant; occasionally as an escape.—*General distribution*: Africa (south); cultivated and as escape in many warm-temperate and subtropical countries.

GENUS 47. *XEROCHRYSUM* Tzvel.[1]

1990, Novosti Sist. Vyssh. Rast. **27**: 151.—*Helichrysum* L. sect. *Xerochlaena* (DC.) Benth. 1867, Fl. Austr. **3**: 613, p. p.

Heads heterogamous or homogamous, 20–50 mm in dia, solitary at apices of stem and its branches. Involucre 12–20 mm long, cup-shaped; involucral bracts imbricate, in four to six rows, almost entirely coriaceous-membranous, light brown, yellow, silvery or reddish; outer broadly ovate, usually obtuse, one-sixth to one-fourth as long as oblong innermost, usually with apical cusp. Receptacle somewhat convex, without bracts, with somewhat raised alveoli. Flowers tubular, usually light colored, few outer often pistillate, others bisexual, with five-toothed corolla. Achenes 2.2–2.5 mm long and 0.6–0.7 mm wide, tetraquetrous-prismatic, glabrous, with four fine veins, almost not narrowed toward base, with narrow costricted areola at apex bearing pappus; pappus 6–8 mm long, of scabrous bristles connate at base into ring, falling wholly or partly. Weakly pubescent annual or perennial plants, 30–120 cm high, with erect leafy stem and alternate leaves; leaves undivided and entire.

Type: *X. bracteatum* (Vent.) Tzvel.

---

[1]Treatment by N.N. Tzvelev.

The genus includes 8–10 species in Australia (predominantly southern part); the highly polymorphic species *X. bracteatum* is extensively cultivated in other countries also.

1. **X. bracteatum** (Vent.) Tzvel. 1990, Novosti Sist. Vyssh. Rast. **27**: 151.—*Xeranthemum bracteatum* Vent. 1803, Jard. Malm **2**: tab. 2.—*Helichrysum bracteatum* (Vent.) Andrews, 1805, Bot. Reposit. **6**: subtab. 428; Kirpicz. 1959, Fl. SSSR, **25**: 406; Dobrocz. 1962, Fl. URSR, **11**: 110; Clapham, 1976, Fl. Europ. **4**: 131.

Type: Australia.

97   *Baltic*; *Center*; *West*; *East*; *Crimea*.—Cultivated as an ornamental (particularly for dry bouquettes) plant, sometime ecdemic.—*General distribution*: Australia; in cultivation and as ecdemic in many other warm-temperate and tropical contries of both hemispheres.—2n = 24, 28.

Note. There are numerous garden varieties that differ by the color of the involucral bracts (most often light brown, silvery or yellow) and size of the heads.

GENUS **48. *GNAPHALIUM* L.**[1]
1753, Sp. Pl.: 850, s. str.; id. 1754, Gen. Pl., ed. 5: 368, s. str.

Heads heterogamous, 3–5 mm in dia, clustered in rather dense corymbose inflorescences, without leaves, less often with one or two leaves at base. Involucre 3.5–4.0 mm long, cup-shaped; involucral bracts imbricate, in many rows, obtuse, almost entirely membranous, with yellowish, less often brownish tinge, outer ovate, half to two-thirds as long as inner oblong bracts. Receptacle almost flat, with somewhat raised edges of alveoli. Peripheral flowers numerous, pistillate, narrow-tubular, reddish at tip, with three or four corolla teeth, inner flowers bisexual, tubular, usually yellowish, less numerous, with (four) five corolla teeth. Achenes 0.5–0.7 mm long and 0.2 mm wide, somewhat flattened, with two or three thin veins, covered with scattered tubercular papillae; pappus 2.0–2.5 mm long, of 10–20 scabrous, almost equal bristles, somewhat connate at base, falling together or in fascicle. Annual plants with whitish-gray tomentum; stem erect, leafy, usually branched only in inflorescence; leaves alternate, undivided and entire.

Lectotype: *G. luteo-album* L.

---

[1]Treatment by N.N. Tzvelev.

128

About 50 species in extratropical countries and mountainous regions of the tropics. Often this genus is clubbed with the next. However, in the structure of inflorescence, heads, and achenes it is considerably closer to the genus *Helichrysum* Mill. than to *Filaginella* Opiz.

Literature: cf. the literature to the genus *Omalotheca* Cass. and "Addendum" at the end of the book.

1. **G. luteo-album** L. 1753, Sp. Pl.: 891; Kirpicz. 1959, Fl. SSSR, **25**: 395; Holub, 1976, Fl. Europ. **4**: 128.—(Plate XI, 4).

Plant 5–30(40) cm high; middle and upper leaves oblong to linear.

Type: Europe ("in Helvetia, G. Narbonensi, Hispania, Lusitania").

Baltic (south); *Center* (Upper Dnieper); *West*; *East* (Lower Don).—In wet depressions on sands, on banks of water bodies, in fields, and by roadsides.—*General distribution*: Caucasus, Russian Central Asia; Atlantic and Central Europe, Mediterranean, Asia Minor, Iran.—2n = 14.

GENUS 49. *OMALOTHECA* Cass.[1]

1828, Dict. Sci. Nat. **61**: 218.—*Gnaphalium* L. subgen. *Omalotheca* (Cass.) Endl. 1838, Gen. Pl.: 447

Heads heterogamous, 4–7 mm in dia, clustered in racemose, spicate, or narrow paniculate inflorescences, less often (in very small plants) solitary. Involucre 5–8 mm long, cup- or bowl-shaped; involucral bracts imbricate, in two to four rows, outer ovate or lanceolate-ovate, one-fourth to half as long as the longest, usually oblong-linear inner bracts, all membranous over a large part. Receptacle almost flat, without pales and raised edges of alveoli. Peripheral flowers numerous, pistillate, narrow-tubular (often almost filiform), with two to four not always distinct corolla teeth, with strongly exserted style, light brownish or pale yellowish, often almost colorless; inner flowers bisexual, 3–8, tubular, with (four) five corolla teeth and light brownish or yellowish corolla. Achenes 1.0–1.5 mm long and 0.2–0.3 mm wide, more or less flattened, with two, not always conspicuous, veins, scatteredly very short-hairy; pappus 2.7–4.0 mm long, of almost equal scabrous bristles, basally more or less connate or free. Perennial plants, more or less covered with grayish tomentum, with erect or ascending flowering shoots and condensed vegetative shoots; leaves alternate, undivided, and entire.

97

---

[1]Treatment by N.N. Tzvelev.

Type: *O. supina* (L.) DC.

About 100 species predominantly in extratropical countries of both the hemispheres and in the mountainous regions of the tropics.

Literature: Kirpicznikov, M.E. and L.A. Kuprianova. 1950. Morfologo-geograficheskie i palinologicheskie dannye k poznaniyu rodov podtriby gnafalievykh [Morphological-geographical and palynological data for interpretation of genera of the subtribe of cudweeds]. *Tr. Bot. Inst. Akad. Nauk SSSR*, ser. 1, **9**: 7–37.—Kirpicznikov, M.E. 1960. Konspekt vidov rodov *Gnaphalium* L. (emend), *Synchaeta* Kirp. i *Omalotheca* Cass. obitayushchikh v SSSR [Conspectus of species of the genera *Gnaphalium* L. (emend.), *Synchaeta* Kirp., and *Omalotheca* Cass. occurring in the USSR]. *Bot. Mat. (Leningrad)*, **20**: 296–313.

1. Plants 10–50 cm high, with simple or many-headed tap root and erect, rather thick, stems; outer involucral bracts one-fourth to one-third as long as the longest inner bracts; pappus bristles connate at base, all falling together; heads usually more than 10 on stem............................ 2.
+ Plants 2–12(15) cm high, with long rhizome and slender, often ascending, stems, densely caespitose; outer involucral bracts one-third to half as long as the longest inner bracts; pappus bristles not connate at base, falling singly; heads 1–10 on stem.. 3.
2. Plants appressed-tomentose; cauline leaves distant, 8–18, usually strongly reduced above, lower 3–8 mm, upper 2–3 mm wide; involucral bracts with light brown or whitish membranous part .... 1. **O. sylvatica.**
+ Plants floccose-tomentose; cauline leaves distant, 3–9, slightly reduced above, lower 8–15 mm, upper 4–8 mm wide; involucral bracts with dark brown membranous part ....2. **O. norvegica.**
3. Outer involucral bracts one-third to two-fifths as long as the longest inner bracts; hairs on achenes rather sparse and squarrose, not forming crown at base of pappus ..
.... 3. **O. hoppeana.**
+ Outer involucral bracts two-fifths to half as long as the longest inner bracts; hairs on achenes rather dense, appressed, seemingly forming crown at base of pappus .... 4. **O. supina.**

Section 1. Gamochaetopsis F.W. Schultz, 1861, Arch. Flore (Journ. Bot.): 311; Holub, 1976, Fl. Europ. **4**: 126.—*Synchaeta*

Kirp. 1950, Tr. Bot. Inst. Akad. Nauk SSSR, ser. **1**, 9: 33.—
Gnaphalium L. subgen. *Synchaeta* (Kirp.) Kirp. 1959, Fl. SSSR,
25: 397.

Pappus bristles connate at base, falling entirely; outer involucral
bracts one-fourth to one-third as long as the longest inner bracts;
heads on rather thick stalk, usually more than 10.

Type: *O. sylvatica* (L.) Sch. Bip. et F.W. Schultz.

1. **O. sylvatica** (L.) Sch. Bip. et F.W. Schultz, 1861, in F.W.
Schultz, Arch. Flore (Journ. Bot.): 311; Holub, 1976, Fl. Europ. **4**:
126.—*Gnaphalium sylvaticum* L. 1753, Sp. Pl.: 856; Kirpicz. 1959,
Fl. SSSR, **25**: 397.—*Synchaeta sylvatica* (L.) Kirp. 1950, Tr. Bot.
Inst. Akad. Nauk SSSR, ser. **1**, 9: 33.—(Plate XII, 1).

Type: Europe ("in Europae sylvis arenosis").

100   *Arctic* (Arctic Europe: vicinity of Pechenga, south of
Bolshezemsk tundra); *North*; *Baltic*; *Center*; *West*; *East*; *Crimea*
(mountains).—In forest glades and forest edges, dry meadows,
on edges of fields, in pine and thinned-out deciduous forests, by
roadsides.—*General distribution*: Western and Eastern Siberia,
Far East (ecdemic), Russian Central Asia (north); Scandinavia,
Central and Atlantic Europe, Mediterranean, Asia Minor; North
America.—2n = 36.

Note. Apparently, forms hybrids with the subsequent species—
O. × *traunsteineri* (Murr.) Dostal, to which may also be related the
plants of the northern populations of *O. sylvatica* with 8–12 distant
cauline leaves and brownish ·membranous part of involucral bracts.

2. **O. norvegica** (Gunn.) Sch. Bip. et F.W. Schultz, 1861, in
F.W. Schultz, Arch. Flore (Journ. Bot.): 311; Holub, 1976, Fl.
Europ. **4**: 126.—*Gnaphalium norvegicum* Gunn., 1766, Fl. Norveg.
2: 105; Kirpicz. 1959, Fl. SSSR, **25**: 399; Rebristaya, 1987, Arkt.
Fl. SSSR, **10**: 101, map. 34.—*G. sylvaticum* L. subsp. *norvegica*
(Gunn.) Mela et Cajand, 1906, Suomen Kasvio: 572.—*Synchaeta
norvegica* (Gunn.) Krip. 1960, Bot. Mat. (Leningrad), **20**: 312.

Type: Norway ("in praedio Engan ad Roraas; in sylva inter
Tronaes et Qvaefjorden norlandiae et passim alibi in norlandia...").

---

99   Plate XII.
1—*Omalotheca sylvatica* (L.) Sch. Bip. et F.W. Schultz. la—head, lb—
achene; 2—*O. supina* (L.) DC.; 3—*Filaginella uliginosa* (L.) Opiz, 3a—
head, 3b—achene, 3c—flower; 4—*Filago pyramidata* L., 4a—head; 5—
F. vulgaris Lam., 5a—head; 6—*Logfia minima* (Smith) Dummort.; 7—*L.
arvensis* (L.) Holub, 7a—head.

Arctic (Arctic Europe); *North* (Karelia-Murman: north; Dvina-Pechora: north and east); *Center* (Volga-Kama: peaks of the North Urals); *West* (Carpathians: high mountains).—In cultivated meadows, river sands and gravel-beds, on stony slopes and rocks, sometimes in inundated meadows.—*General distribution*: Western and Eastern Siberia, Russian Central Asia (mountains); Scandinavia, Central and Atlantic Europe (mountains), Mediterranean (mountains), Dzhungaria-Kashgaria; North America.—2n = 56.

Section 2. Omalotheca.

Pappus bristles not connate at base and falling singly; outer involucral bracts one-third to half as long as the longest inner bracts; heads 1–10 on very slender stalk.

Type: type species.

3. **O. hoppeana** (Koch) Sch. Bip. et F.W. Schultz, 1861, in F.W. Schultz, Arch. Flore (Journ. Bot.): 311; Holub, 1976, Fl. Europ. **4**: 127.—*Gnaphalium hoppeanum* Koch, 1843, Syn. Fl. Germ., ed. 2, **1**: 399; Fodor, 1974, Fl. Zakarp.: 134.

Type: Mountains of central Europe ["in alpibus, subalpinis et montibus altioribus (Kärnthen, Steyermark, Oestreich u höhere Mährische Gebirge ...")].

West (Carpathians: reported for high mountains).—On cultivated meadows, stony slopes and rocks, predominantly on limestones.— *General distribution*: Central Europe (mountains), Mediterranean (mountains).

4. **O. supina** (L.) DC. 1838, Prodr. **6**: 245; Kirpicz. 1960, Bot. Mat. (Leningrad), **20**: 313; Holub, 1976, Fl. Europ. **4**: 127.— *Gnaphalium supinum* L. 1768, Syst. Nat. ed. 12, **3**: 234; Kirpicz. 1959, Fl. SSSR, **25**: 402; Rebristaya, 1987, Arkt. Fl. SSSR, **10**: 99, map. 32.—(Plate XII, 2).

Type: The Alps in Switzerland and Italy ("in Alpibus Helveticis, Italicis").

Arctic (Arctic Europe); *North* (Karelia-Murman: Kola Peninsula; Dvina-Pechora: between the mouth of North Dvina and Kanin, middle reaches of the Pechora River, Cispolar and North Urals); *West* (Carpathians: high mountains)—On cultivated meadows, stony slopes and rocks, usually relatively more acidic.—*General distribution*: Caucasus, Western Siberia (altai), Eastern Siberia (mountains), Arctic, Russian Central Asia (mountains); Scandinavia, Central and Atlantic Europe (mountains), Mediterranean (mountains),

Asia Minor (mountains), Dzhungaria-Kashgaria (mountains); North America.—2n = 28.

GENUS **50.** *FILAGINELLA* Opiz[1]
1854, Abh. Konigl. Bohm. Ges. Wiss., ser. 5, **8** (Gesch.): 52

101 Heads heterogamous, 3–6 mm in dia, densely clustered in basally leafy glomerular inflorescences at apices of stem and its branches. Involucre 2.0–2.5 mm long; involucral bracts imbricate, in two or three rows, outer ovate, nearly half as long as the longest inner and usually broadly lanceolate bracts, all bracts membranous over longer part. Receptacle almost flat, without pales, with somewhat raised margins of alveoli. Peripheral flowers numerous, pistillate, narrow-tubular, with two to four, not always distinct, corolla teeth and strongly exserted style, light brownish or almost colorless; inner flowers 3–10, bisexual, tubular, with five light brownish or pale yellowish corolla teeth. Achenes 0.5–0.6 mm long and 0.2 mm wide, narrow-ellipsoidal with two or three fine veins, glabrous or very short-hairy; pappus 1.0–1.3 mm long, of fewer (6–10), basally free and rather fragile, almost equal, scabrous bristles. Annual plants, with grayish or whitish tomentum, 5–30(40) cm high, with erect, often strongly branched leafy stem; leaves alternate, undivided and entire.

Lectotype: *F. uliginosa* (L.) Opiz.

About 50 species in the extratropical countries of both hemispheres.

Literature: cf. literature for the preceding genus.

1. Ovary and achenes glabrous; plant grayish-tomentose with rather thin (below inflorescence, 0.3–0.8 mm thick) stem weakly tomentose in lower part .... 4. **F. uliginosa.**
 + Ovary and achenes covered with very short scattered hairs .... 2.
2. Plant grayish-tomentose, with rather thin (below inflorescence, 0.3–0.8 mm thick) stem weakly tomentose (to subglabrous) in lower and often middle part; inner involucral bracts subacute, usually with light brownish membranous part ....3. **F. pilularis.**
 + Plant covered with more appressed and more dense, almost whitish, tomentum, with thicker (below inflorescence, 1–2 mm thick) stem densely tomentose to base; inner involuscral bracts acuminate... 3.

---

[1]Treatment by N.N. Tzvelev.

3. Leaves linear-elliptical, rather abruptly acuminate; membranous part of inner involucral bracts with brownish tinge, rather abruptly acuminate; basal leaves withering early .... 1. **F. rossica.**
+ Leaves linear, gradually acuminate; membranous part of inner involucral bracts without brownish tinge, more gradually acuminate; basal leaves often persisting at flowering .... 2. **F. kasachstanica.**

1. **F. rossica** (Kirp.) Tzvel. 1990, Novosti Sist. Vyssh. Rast. **27:** 150.—*Gnaphalium rossicum* Kirp. 1958, Bot. Mat. (Leningrad), **19:** 349; id. 1959, Fl. SSSR, **25:** 389; Privalova and Ryndina, 1972, Opred. Vyssh. Rast. Kryma: 471.—*Filaginella uliginosa* (L.) Opiz subsp. *rossica* (Kirp.) Holub, 1976, Bot. Journ. Linn. Soc. (London), **71:** 271; id. 1976, Fl. Europ. **4:** 128.

Type: Penza Region ("Kuznetsk district, near Elyuzan station, upland bank of the Kadada River").

Center (Upper Volga: on the Oka River; Volga-Kama: south and east; Volga-Don); *West*; *East*; *Crimea* (vicinity of village of Magarach, probably ecdemic).—On coastal sands and gravel-beds, edges of fields, in steppe depressions, by roadsides and pathways.—*General distribution*: Caucasus, Western Siberia (south), Eastern Siberia (southwest), Russian Central Asia (north).

2. **F. kasachstanica** (Kirp.) Tzvel. 1990, Novosti Sist. Vyssh. Rast. **27:** 151.—*Gnaphalium kasachstanicum* Kirp. 1960, Bot. Mat. (Leningrad), **20:** 305; id. 1959, Fl. SSSR, **25:** 391, descr. 102 ross.—*Filaginella uliginosa* (L.) Opiz subsp. *kasachstanica* (Kirp.) Holub, 1976, Bot. Journ. Linn. Soc. (London), **71:** 271; id. 1976, Fl. Europ. **4:** 128.

Type: Akmolinsk Region ("Atbassar district").

East (Lower Volga: northeast).—In solenetz and solonchak meadows, steppe depressions.—*General distribution*: Western Siberia (south), Eastern Siberia (southwest), Russian Central Asia.

3. **F. pilularis** (Wahlenb.) Tzvel. 1990, Novosti Sist. Vyssh. Rast. **27:** 150.—*Gnaphalium pilulare* Wahlenb. 1812, Fl. Lapp.: 205, tab. 13; Tzvel. 1979, Novosti Sist. Vyssh. Rast. **16:** 204.—*G. uliginosum* L. subsp. *pilulare* (Wahlenb.) Nym. 1879, Consp. Fl. Eur.: 382.—*G. sibiricum* Kirp. 1960, Bot. Mat. (Leningrad), **20:** 302; id. 1959, Fl. SSSR, **25:** 390, descr. ross.—*Filaginella uliginosa* (L.) Opiz subsp. *sibiricum* (Kirp.) Holub, 1976, Bot. Journ. Linn. Soc. (London), **71:** 271; id. 1976, Fl. Europ. **4:** 128.

Type: Lapland ("ad Hietasuvando Lapponiae Kemensis copiose").

Arctic (Arctic Europe: vicinity of village of Gremikha near the mouth of the Pechora River, on the Usa River); *North* (Karelia-Murmansk: near Lake Imandra and Sandal, on the Varzuga and Kem rivers; Dvina-Pechora: valleys of larger rivers); *Center* (Ladoga-Ilmen: on the Volkhov River and Lake Ilmen; Upper Volga: on the Volga, Mologa, Kostroma rivers; Volga-Kama).—On clayey and sandy shoals, coastal buffs, by roadsides and pathways.—*General distribution*: Western and Eastern Siberia, Far East; Scandinavia (Lapland).

4. **F. uliginosa** (L.) Opiz, 1854, Abh. Konigel. Böhm. Ges. Wiss., ser, 5, **8** (Gesch.): 52; Holub, 1976, Fl. Europ. **4**: 127; *Gnaphalium uliginosum* L. 1753, Sp. Pl.: 856; Kirpicz. 1959, Fl. SSSR, **25**: 391.—*G. tomentosum* Luce, 1823, Topogr. Nachr. Ins. Oesel.: 275.—(Plate XII, 3).

Type: Europe ("in Euroape paludibus, ubi aquae stagnant").

North; *Baltic*; *Center*; *West* (Carpathians; Dnieper).—On shoals and gravel-beds, coastal buffs, in meadows, fields, forest glades and edges, by roadsides and pathways.—*General distribution*: Far East (south); Scandinavia, Central and Atlantic Europe, Mediterranean; North America.—2n = 14.

GENUS 51. *LOGFIA* Cass.[1]
1819, Bull. Soc. Philom. Paris, 1819: 143.—*Oglifa* Cass. 1819, ibid.: 143

Heads heterogamous, 2–5 mm in dia, clustered in groups of 2–12 (rarely solitary) in lax glomerules in axils of cauline leaves and in dichotomies of stem or in terminal racemose or paniculate inflorescences. Involucre cup-shaped; involucral bracts 15–20, in many rows; outer smaller, linear, middle and inner linear-lanceolate or oblong-lanceolate, longitudinally folded, all, except innermost, lanate-tomentose, obtuse or acute but always awnless, 2.5–3.5 mm long and 0.8–1.0 mm wide; at fruiting all involucral bracts more or less stellately divergent. Receptacle more or less convex, glabrous. All flowers tubular, outer pistillate, filiform, yellowish, 15–20; inner bisexual, three to five, with faintly purple tinge and four-toothed corolla. Achenes light reddish-brown, somewhat dimorphic: outer slightly curved, more or less papillate, oblong, 0.8–1.0 mm long

---

[1]Treatment by L.I. Krupkina.

and about 0.4 mm wide, without pappus; inner straight, oblong-obovate, 0.5–0.6 mm long and 0.2–0.3 mm wide; pappus 2.0–3.5 mm long, of white, scabrous, readily falling bristles. Annual plants, tomentose or lanate, with erect or somewhat ascending, 5–40(70) cm high stem; leaves alternate, sessile, (4)10–20 mm long and (0.5)1.0–4.0 mm wide, oblong-lanceolate or oblong-linear, entire.

Type: *Filago gallica* L. = (*Logfia gallica* (L.) Coss. et Germ.).

103 About 15 species, distributed almost throughout Europe (excluding the northernmost regions), northern Africa, and temperate regions of Asia to northwestern Mongolia, as also in the southwest of North America.

Literature: cf. [the literature for] *Filago* L.

1. Plants white-lanate-tomentose, with simple or more or less branched stem and more or less deflected, to 20 mm long leaves; heads 2.5–6.0 mm long, in glomerules of 3–12, sometimes solitary .... 1. **L. arvensis.**

+ Plants grayish-lanate-tomentose, with simple or dichotomously branched stem and more or less appressed, to 10 mm long leaves; heads 2.5–3.5 mm long, in glomerules of three to seven, sometimes solitary .... 2. **L. minima.**

1. **L. arvensis** (L.) Holub, 1975, Notes Roy. Bot. Gard. Edinb. **33**: 432; id. 1976, Fl. Europ. **4**: 123.—*Filago arvensis* L. 1753, Sp. Pl.: Add. post Indicem; Smoljan. 1959, Fl. SSSR, **25**: 322; Wagenitz, 1980, in Rech. fil. Fl. Iran, **145**: 24.—*F. montana* auct. non L.: smoljian. 1959 op. cit.: 323, p. p.—(Plate XII, 7).

Type: Europe ("in Europae campis sabulosis").

North (south of Karelia-Murmansk and Dvina-Pechora); *Baltic*; *Center*; *West*; *East*; *Crimea*.—In dry meadows and glades, usually with sandy or sandy loam soil, in fields and plantations of various crops, by roadsides.—*General distribution*: Caucasus, Western Siberia (southwest), Russian Central Asia; Scandinavia (south), Central and Atlantic Europe, Mediterranean, Asia Minor, Iran, Mongolia (northwest); ecdemic in other countries.—2n = 28.

2. **L. minima** (Smith) Dumort. 1827, Fl. Belg.: 68; Holub, 1976, Fl. Europ. **4**: 124.—*Gnaphalium minimum* Smith, 1800, Fl. Brit. **2**: 873.—*Filago montana* L. 1753, Sp. Pl.: Add. post Indicem. p. p.—*F. minima* (Smith) Pers. 1807, Syn. Pl. **2**: 422; Smoljian. 1959, Fl. SSSR, **25**: 324.—(Plate XII, 6).

Type: Great Britain ("in arenosis et glareosis frequens").

Baltic; *Center* (Ladoga-Ilmen: west and southwest; Volga-Don: west and north); *West* (Carpathians: Transcarpathia; Dnieper: west and north).—In dry sandy places, meadows and forest glades with sandy soil and in pine forests.—*General distribution*: Scandinavia (south), Central and Atlantic Europe, Mediterranean.—2n = 28.

GENUS 52. *FILAGO* L.[1]

1753, Sp. Pl. **2**: 927, 1199, 1230; id. 1754, Gen. Pl., ed. 5: 397, nom. conserv.—*Gifola* Cass. 1819, Bull. Soc. Philom. Paris, 1819: 247

Heads heterogamous, small, clustered in groups of (5)10–30(40) in dense, subspherical glomerules 5–123 mm in dia, at apices of stem and its branches. Involucre 3.5–6.0 mm long and 1.6–2.5 mm in dia, subcylindrical or cup-shaped; involucral bracts in many rows, outer linear-lanceolate or lanceolate, short-acuminate, lanate-tomentose outside; middle lanceolate, greenish or reddish, sparsely rather long-hairy, with erect or arcuate, yellowish, awn 1.0–1.5 mm long; inner obtuse or short-acuminate, membranous, glabrous, shorter than others; all bracts at fruiting somewhat approximate, falling singly from achene. Receptacle elongate, sometimes subglobose, without or with membranous bracts. All flowers tubular, yellowish, red, sometimes some inner pistillate with filiform, two- to four-toothed corolla, 20–30 or five to seven; inner bisexual (one) two to four (seven), with four-toothed corolla, sterile or functionally staminate. Achenes 0.5–1.0 mm long and 0.2–0.5 mm wide, terete or oblong-obovate, covered with whitish papillae or short hairs, reddish-brown; peripheral achenes usually without pappus, others with about 3 mm long pappus, of readily
104 falling scabrous bristles, less often all achenes without pappus. Grayish-tomentose annuals, with erect stem, dichotomously branched in upper part or from base, (2)4–30(40) cm high; leaves alternate, sessile, linear-lanceolate, lanceolate or oblong-spatulate, entire.

Type: *F. pyramidata* L.

About 20 species, distributed in Central and Atlantic Europe as also in the southwestern and Russian Central Asia, extending in the east to north India.

Literature: Holub, J. and J. Chrtek, 1962. Zur nomenklatur des Gattungsnamen *Filago* L. *Taxon*, **11**, 6: 195–201.—Chrtek, J. and J. Holub, 1963. Bemerkungen zur taxonomie und Nomenklatur der Gattungen *Evax* und *Filago*. *Presila* **35**: 1–17.—Wagenitz, G.

[1]Treatment by L.I. Krupkina.

1965. Zur Systematik und Nomenklatur einiger Arten von *Filago* L. emand. Gaertn. subgen. *Filago* (*Filago germanica*—Gruppe). *Willdenowia*, **4**, 1: 37–59.—Myrzakurov, P. 1968. K sistematike podtriby Filagininae O. Hoffm. semeistva slozhnotsvetnykh [On the systematics of the subtribe Filagininae O. Hoffm. of the family Compositae]. *Bot. Mat.* (*Alma-Ata*), **5**: 34–41.—Wagenitz, G. 1969. Abgrenzung und Gliederung der Gattung *Filago* L. s. l. (Compositae—Inulae). *Willdenowia*, **5**, 3: 395–444.—Wagenitz, G. 1970. Uber die Verbreitung einiger *Filago*-Arten. *Feddes Repert.* **81**, 1–5: 107–117.

1. All flowers without pappus; stem 4–15 cm high, with one to eight glomerules of heads; ovary and achenes short-hairy .... 3. **F. filaginoides.**
+ Only outermost flowers in axils of inner involucral bracts without pappus, others with pappus; stem to 40 cm high, with up to 20 glomerules of heads; ovary and achenes covered with whitish papillae .... 2.
2. Leaves oblong-lanceolate or lanceolate, broadest in lower part, upper not exceeding length of glomerule of heads; involucral bracts with straight awns, inner more or less reddish .... 1. **F. vulgaris.**
+ Leaves oblong-spatulate, broadest in upper part, upper distinctly exceeding length of glomerule of heads; involucral bracts with bent awns, not reddish ....
.... 2. **F. pyramidata.**

Section 1. Filago.

Achenes more or less covered with whitish papillae, only outermost achenes in head without pappus, others with pappus; receptacle without pales.

Type: type species.

1. **F. vulgaris** Lam. 1778, Fl. Fr. **2**: 61; Holub, 1976, Fl. Europ. **4**: 121.—*Gifola vulgaris* Cass. 1820, Dict. Sci. Nat. **18**: 531.—*Filago germanica* L. 1763, Sp. Pl. **2**: 1311, non Huds. 1762; Smoljian. 1959, Fl. SSSR, **25**: 316.—(Plate XII, 5).

Type: Europe ("in Europa").

Center (Upper Dnieper: vicinity of Mogilev); *West* (Carpathians: Transcarpathia and Chernovitsy regions; Moldavia); *Crimea* (southern coast).—On dry sandy, stony, or clayey slopes, in meadows and forest glades.—*General distribution*: Caucasus, Russian Central Asia; Scandinavia (south), Central and Atlantic Europe, Mediterranean, Asia Minor, Iran; ecdemic in other countries.—2n = 28.

2. **F. pyramidata** L. 1763, Sp. Pl. **2:** 1199; Holub, 1976, Fl. Europ. **4:** 122.—*F. germanica* Huds. 1762, Fl. Angl.: 328, non L. 1763.—*Gifola pyramidata* (L.) Dumort. 1829, Fl. Belg.: 69.—*Filago spathulata* C. Presl, 1822, Delic. Prag.: 99; Smoljian. 1959, Fl. SSSR, **25:** 321.—(Plate XII, 4).

Type: Spain ("Hispania").

Crimea (rarely).—On dry stony, weakly turfaceous slopes.— *General distribution*: Caucasus, Russian Central Asia; Central and Atlantic Europe, Mediterranean, Asia Minor, southern Asia (northwest).—2n = 28.

Section 2. Filaginoides (Smoljian.) Wagenitz, 1969, Willdenowia, **5,** 3: 417.—*Evax* Gaertn. sect. *Filajinoides* Smoljian. 1959, Fl. SSSR, **25:** 313.

Achenes short-hairy, all without pappus; receptacle with membranous pales.

105   Type: *F. filaginoides* (Kar. et Kir.) Wagenitz.

3. **F. filaginoides** (Kar. et Kir.) Wagenitz, 1969, Willdenowia, **5,** 3: 417; Holub, 1976, Fl. Europ. **4:** 123.—*Evax filaginoides* Kar. et Kir. 1842, Bull. Soc. Nat. Moscou, **15:** 379; Smoljian. 1959, Fl. SSSR, **25:** 313.

Type: Eastern Kazakhstan ("in arenosis Songariae prope Kusu-Kerpetsch ad fl. Ajagis").

East (Lower Volga: Astrakhan Region).—In semi-deserts, on sandy and clayey places on banks of water bodies.—*General distribution*: Russian Central Asia.

GENUS 53. *BOMBYCILAENA* (DC.) Smoljian.[1]
1955, Bot. Mat. (Leningrad), **17:** 450.—*Micropus* L. sect. *Bombycilaena*
DC. 1836, Prodr. **5:** 460

Heads heterogamous, 3–4 mm in dia, few-flowered, clustered in groups of two or three in globose glomerules, 6–15 mm in dia, at apices of branches, in dichotomies of stem, and in axils of upper leaves. Involucral bracts in two (three) rows; outer bracts four or five, to 1.5 mm long, membranous, herbaceous at base, oblong-lanceolate, sparsely rather long-hairy outside; inner bracts four to eight, helmet-shaped, laterally compressed, with oblique beak, densely lanate outside, enclosing female flowers and falling with achenes. Receptacle subcylindrical, glabrous. All flowers

---

[1]Treatment by L.I. Krupkina.

140

tubular, white or light reddish; peripheral pistillate, fertile, five to nine, with filiform two-toothed corolla, laterally fused with ovary; inner flowers bisexual but functionally staminate, (one) three to five, with five-toothed, broadly tubular corolla and anthers with short, filiform, basal appendage. Achenes obovate, flattened, 1.3–1.5 mm long and 1 mm wide, glabrous and smooth, light reddish-brown, gray, or greenish, without pappus. Lanate-tomentose annuals with erect or procumbent stem, often branched from base, 5–20(30) cm high; leaves alternate, sessile, undivided and entire, oblong-lanceolate or oblong-linear.

Type: *B. erecta* (L.) Smoljian.

Two species, distributed in Central and Southern Europe, northern Africa, and southwestern Asia.

1. Glomerules of heads 8–10 mm in dia, more or less appressed grayish- or whitish-lanate, surrounded by subtending leaves, usually exceeding them; inner involucral bracts five to eight, with fruits 2–3 mm long; leaves 1.4–2.5 mm wide, more or less undulate............................ 1. **B. erecta.**
+ Glomerules of heads 10–16 mm in dia, sparsely reddish-brown-lanate, surrounded by subtending leaves, usually not exceeding them; inner involucral bracts 2–5 mm wide, flat .... 2. **B. discolor.**

1. **B. erecta** (L.) Smoljian. 1955, Bot. Mat. (Leningrad), **17:** 450; id. 1959, Fl. SSSR, **25:** 302; Holub, 1976, Fl. Europ. **4:** 125; Dobrocz. 1987, Opred. Vyssh. Rast. Ukr.: 324.—*Micorpus erectus* L. 1753, Sp. Pl. 2: Add. post Indicem.; Wagenitz, 1965, in Hegi, Ill. Fl. Mitteleur., 2, Aufl. 6, 3, 2: 109.

Type: Spain ("Hispania").
Crimea.—In dry open sandy and stony places, sometimes in fields and plantations.—*General distribution*: Caucasus; Central Europe, Mediterranean, Asia Minor, Iran.

2. **B. discolor** (Pers.) Lainz, 1973, Bol. Inst. Estud. Astur. (Supl. Cienc.), **16:** 194.—*Micropus discolor* Pers. 1807, Syn. Pl. **2:** 423.—*M. bombycinus* Lag. 1816, Gen. Sp. Pl. Nov.: 32.—*M. erectus* L. subsp. *bombycinus* (Lag.) Rouy, 1903, Fl. Fr. **8:** 170.—*Bombycilaena bombycina* (Lag.) Sojak, 1962, Novit. Bot. Prag. 1962: 50.

Type: Spain ("in collibus aridis circa Matritum alibique in Hispania").
Crimea (vicinity of Balaklava).—In dry open sandy and stony slopes.—*General distribution*: Mediterranean, Asia Minor, Iran.

TRIBE 5. **ANTHEMIDEAE** Cass.[1]

Heads homogamous or heterogamous; peripheral flowers ligulate, often absent; inner flowers (disk flowers) tubular. Receptacle with or without scaly pales. Anthers usually without basal appendage. Style branches seemingly truncate and apically papillate. Pollen predominantly anthemoid type. Achenes with undivided or more or less incised corona at apex, often altogether absent. Leaves alternate, with undivided or more or less incised leaves.

Type: *Anthemis* L.

GENUS **54.** *ANTHEMIS* L.
1753, Sp. Pl.: 893; id. 1754, Gen. Pl., ed. 5: 381

Heads usually heterogamous, 12–40(50) mm in dia, solitary at apices of stem and branches, but sometimes rather numerous. Involucre saucer-shaped, less often cup-shaped, 3–14 mm long; involucral bracts imbricate. Receptacle somewhat convex to obtusely conical but with scaly pales throughout, less often only in middle part. Peripheral flowers ligulate, pistillate, less often sterile, yellow or white, sometimes absent; disk flowers tubular, bisexual, regular, with yellow five-toothed corolla. Achenes 1.5–3.0 mm long and 0.6–1.5 mm wide, conical-cylindrical with 5–9 more or less distinct ribs to more or less flattened and then with two strongly developed lateral ribs and 3–15 more or less distinct veins, at apex with short (less than half as long as remaining part of achene) crown or without it. Perennial or annual plants, 8–100 cm high with usually erect, leafy stems and alternate, more or less incised leaves.

Type: *A. arvensis* L.

About 150 species, distributed in Europe, north Africa, and western Asia but also found as ecdemic in other extratropical countries. It is divided into three very distinct subgenera that fully deserve the rank of separate genera.

Literature: Kuzmanov, B. 1981. A cytotaxonomic study of Bulgarian *Anthemis* species. *Candollea*, **36**, 1: 1976.—Skvortsov, A.K. and V.S. Dolgacheva, 1984. Estestvennaya mezhvidovaya gibridizatsiya u pupavok (*Anthemis*, Asteraceae) v gorakh Kryma.

---

[1]Treatment by N.N. Tzvelev, except for the genus *Artemisia* L.

142

Mikroevolyutsiya. Materialy Vsesoyuznoi konferensii po problem evolyusii [Natural interspecific hybridization in chemomiles (*Anthemis*, Asteraceae) in mountains of Crimea. In "Microevolution". Proceedings of the All-Union Conference on the Problems of Evolution]. Moscow: 36–37.

1. Ligulate flowers yellow .... 2.
+ Ligulate flowers white, rarely yellowish-white (pale colored) .... 4.
2. Cauline leaves less numerous, petiolate, without lateral lobes approximate to base; primary lobes two- to five-parted, intermediate lobes absent; bracts with shorter, thin and nonstiff cusp. Perennials of chalk outcrops, with more or less basally woody shoots, some shoots terminating in a crown of leaves .... 1. **A. trotzkiana.**
+ Cauline leaves numerous, all with lateral lobes approximating base; primary lobes usually pinnatipartite or pinnatilobate, less often uniformly toothed; intermediate lobes small, usually present; bracts with thick stiff cusp. Herbaceous perennials without rosulate shoots .... 3.
3. Stems solitary or two to six, 20–80 cm high, almost always more or less branched, bearing many larger (25–40 mm in dia) heads; disk of head 12–15 mm in dia; ligules of peripheral flowers 7–10 mm long ....9. **A. tinctoria.**
+ Stems rather many (usually exceeding five), 12–30 cm high, usually simple, less often more or less branched (as a result of nibbling by cattle); heads usually 15–25 mm in dia; disk 8–12 mm in dia; ligules of peripheral flowers 4–7 mm long .... 10. **A. monantha.**
4. Annuals, less often biennials, with slender readily uprootable tap root. Widely distributed plants .... 5.
+ Perennials with thicker root. Plants of Crimea, Carpathians, and extreme south of Ukraine.. 10.
5. Receptacle strongly convex, without bracts in lower part, with very narrow, narrow-linear bracts above; peripheral flowers sterile; achenes 1.7–2.0 mm long, not flattened, with eight or nine fine ribs, rounded at apex. Weakly pubescent plants, 10–30(40) cm high, with pleasant smell .... 6.
+ Receptacle more or less convex, with lanceolate or oblong bracts throughout, often abruptly terminating in cusp; peripheral flowers pistillate, fertile. Plants with mild pleasant smell .... 7.

6. Achenes smooth or almost smooth; terminal lobes of leaves 0.2–0.5 mm wide; base of corolla weakly thickened .... 7. **A. lithuanica.**

+ Achenes tuberculate on ribs; terminal lobes of leaves 0.4–1.0 mm wide; base of corolla strongly thickened ... . 8. **A. cotula.**

7. Bracts oblong, abruptly narrowed in stiff cusp, almost as long as expanded part; heads 30–45 mm in dia; receptacle hemispherical; achenes 2.3–2.6 mm long, dorsally flattened, with strongly raised lateral ribs and a crown about 0.1 mm long, weakly pubescent plants of Crimea, 20–100 cm high .... 15. **A. altissima.**

+ Bracts gradually pointed or terminating in a non-stiff cusp considerably shorter than their broader part ..... 8.

8. Achenes 2.0–2.5 mm long, dorsally flattened, with two strongly raised, acute, lateral ribs and about 0.1 mm-long crown; receptacle hemispherical; bracts broadly lanceolate, rather gradually pointed in short cusp; primary leaf lobes pectinately divided in similar [secondary] lobes............ ....14. **A. austriaca.**

+ Achenes 1.8–2.2 mm long, dorsally not flattened, obtusely tetraquetrous-terete with seven to nine obtuse ribs; receptacle strongly convex, conical; leaf lobes irregularly pinnately divided or toothed .... 9.

9. Achenes with obtuse or subobtuse at apex; bracts lanceolate, entire, gradually pointed. Biennial plant, often branched from base or wintering annual more or less covered with semiappressed hairs .... 5. **A. arvensis.**

+ Achenes with acute margin on one side at apex or up to 0.1 mm-long, crown; bracts rather abruptly narrowed in short cusp, more or less toothed on both sides of it; stems usually branched above base. Annual spring plant, with rather profuse, somewhat lax pubescence .... 6. **A. ruthenica.**

10(4). Cauline leaves less numerous; their primary lobes two- to five-parted or lobed, intermediate lobes absent, terminal lobe abruptly narrowed in very short (often scarcely visible even on magnification) cusp about 0.1 mm long. Plants more or less woody at base, with rosulate basal leaves .... 11.

+ Cauline leaves rather numerous; their primary lobes crenate-pinnatipartite, pinnate, or toothed; intermediate lobes often present; terminal lobe more or less gradually narrowed in cusp 0.2–0.3 mm long. Plants of Crimea, not woody at base, without rosulate basal leaves .... 13.

11. Involucral bracts with dark brown membranous border; receptacle hemispherical. Alpine plants of the Carpathians, 10–30 cm high .... 2. **A. carpatica.**

  + Involucral bracts with whitish or light brown membranous border; receptacle conical. Plants of Crimea .... 12.

12. Heads 20–30 mm in dia; involucral bracts with light brown membranous border; ligules of peripheral flowers 7–9 mm long; achenes 2.5 mm long, with crown 0.5–0.7 mm long. Plants of limestone outcrops and allied rocks, 15–40 cm high .... 3. **A. tranzscheliana.**

  + Heads 12–20 mm in dia; involucral bracts with whitish membranous border; ligules of peripheral flowers 4–7 mm long; achenes 2 mm long, with crown 0.2–0.5 mm long. Plants of schist outcrops, 8–20 cm high ....
.... 4. **A. sterilis.**

13. Involucral bracts with dark brown membranous border heads 30–50 mm in dia; ligules of peripheral flowers white, 13–17 mm long. Plants of yailas and upper part of forest zone of Crimean mountains .... 12. **A. jailensis.**

  + Involucral bracts with whitish or light brown membranous border; heads 20–35 mm in dia; ligules of peripheral flowers 7–14 mm long. Plants of lower altitudes .... 14.

14. Middle mountain plants with fewer stems, 30–70 cm high, relatively weakly pubescent from base; ligules of peripheral flowers white or yellowish-white ...... 11. **A. dumetorum.**

  + Lower mountain plants usually with many stems, 20–40 cm high, more or less grayish from base from profuse pubescence; ligules of peripheral flowers white ....
.... 13. **A. dubia.**

SUBGENUS **1.** *ANTHEMIS*

Receptacle hemispherical or conical. Bracts all over receptacle, lanceolate or oblong, gradually pointed towards apex or more or less gradually narrowed in short nonstiff cusp. Peripheral flowers sterile. Achenes dorsally flattened or somewhat flattened, without or with short crown, obtusely tetraquetrous. Leaves without intermediate lobes.

Type: type species of genus.

Section 1. Hiorthia (DC.) Fernandes, 1975, Journ. Linn. Soc. London (Bot.), **70**, 1: 6; id. 1976, Fl. Europ. **4**: 148.—*Anacyclus* L. sect. *Hiorthia* DC. 1837, Prodr. **6**: 17.—*Anthemis* L. sect. *Rumata* Fed. 1961, Fl. SSSR, **26**: 865, 16.

109   Type: *A. orientalis* (L.) Degen.

1. **A. trotzkiana** Claus, 1847, in Bunge, Delect. Sem. Hort. Dorpat.: 3, in obs.; Claus, 1852, Fl. Mestn. Privolzh. Stran.: 287; Fed. 1961, Fl. SSSR, **26**: 29; Fernandes, 1976, Fl. Europ. **4**: 148.

Type: Saratov Region, vicinity of Khvalynsk ("Unico loco in colle cretaceo prope Chwalynsk observata").
Center (Volga-Don: on the Volga River, downstream environs of Syzran); *East* (Lower Don: on the Volga River; Trans-Volga: south and east).—On chalk outcrops, less often on limestone outcrops.—*General distribution*: Russian Central Asia (northwest).

2. **A. carpatica** Waldst. et Kir. ex Willd. 1803, Sp. Pl.: **3**, 3: 2179; Fed. 1961, Fl. SSSR, **26**: 20; Fernandes, 1976, Fl. Europ. **4**: 148.

Type: Carpathians, probably Mt. Pope Ivan of Marmarosh? ("in alpibus Carpaticis").
West (Carpathians: Mt. Pope Ivan of Marmarosh).—On stony slopes and rocks in upper mountain zone.—*General distribution*: Central Europe (eastern Alps, Carpathians); Mediterranean (Pyrenees, Balkans).—2n = 36.

3. **A. tranzscheliana** Fed. 1961, Fl. SSSR, **26**: 866, 27; Fernandes, 1976, Fl. Europ. **4**: 150.

Type: Crimea, Mt. Karadagh ("Crimea, vicinity of Karadagh Scientific Station, on rocks below the Karagach summit").
Crimea (vicinity of Simferopol, mountains east of Alushta).— · On rocks and stony slopes.—Endemic.

4. **A. sterilis** Stev. 1856, Bull. Soc. Nat. Moscou, **29**, 2: 379; Fed. 1961, Fl. SSSR, **26**: 25; Fernandes, 1976, Fl. Europ. **4**: 150.—*A. fruticulosa* auct. non Bieb. 1798: Bieb. 1808, Fl. Taur.-Cauc. **2**: 329, quoad pl. taur.; Fedtsch. and Fler. 1910, Fl. Evr. Ross.: 968.—*A. saxatilis* auct. non DC.: Bieb. 1819, Fl. Taur.-Cauc. 3: 581, 648.

Type: Crimea, vicinity of village of Morskoe ("Legi inter pagos litoris meridionalis Capsichor et Uskut in sylvis e Junipero excelsa solo arido; in lapidosis circa Kutlak (MB)").

146

Crimea (between Alushta and Sudak).—On coastal shale outcrops.—Endemic.

Section 2. Anthemis.

Annual or biennials, lacking vegetative shoots with crown of leaves; receptacle subobtusely conical; achenes subobtuse on upper margin or acute on one side, without crown.

Type: type species.

5. **A. arvensis** L. 1753, Sp. Pl.: 894; Fed. 1961, Fl. SSSR, **26:** 35; Fernandes, 1976, Fl. Europ. **4:** 153.—(Plate XIII, 2).

Type: Europe, Sweden ("in Europae praesertim Sueciae agris").

North (Karelia-Murmansk: south; Dvina-Pechora: southwest); *Baltic*; *Center* (Ladoga-Ilmen; Upper Dnieper; Upper Volga: rarely; Black Sea: west). *Crimea* (mountains and southern coast).—In dry meadows and forest glades, by roadsides, in fields and habitations.—*General distribution*: Caucasus; Scandinavia, Atlantic and Central Europe, Mediterranean, Asia Minor: ecdemic in other extratropical countries.—2n = 18.

6. **A. ruthenica** Bieb. 1808, Fl. Taur.-Cauc. **2:** 330; Fed. 1961, Fl. SSSR, **26:** 36; Fernandes, 1976, Fl. Europ. **4:** 153; Galinis, 1980, Lietuv. TSR Fl. **6:** 87.—*A. arvensis* L. var. *ruthenica* (Bieb.) Schmalh. 1887, Fl. Sr. Yuzhn. Ross. **2:** 61.

Type: Crimea and the Ukraine ("in Tauriae et Ucraniae collibus apricis sterilibus"); lectotype from the vicinity of Nikolaev ("prope Nikolajew").

North (Karelia-Murmansk: ecdemic); *Baltic* (south); *Center* (Ladoga-Ilmen: ecdemic, rarely; upper Dnieper: ecdemic; Upper Volga:) ecdemic, rarely; Volga-Kama; Volga-Don); *West* (Carpathians: ecdemic in Lvov; Dnieper; Moldavia; Black Sea); *East* (Lower Don); *Crimea*.—On riverine and coastal sands, in dry meadows, on steppe slopes, outcrops of chalk and limestone, in fields, thinned-out pine forests, by roadsides, in habitations.—*General distribution*: Caucasus; Central Europe (east), Mediterranean (east).

110　　　　　　　　　Plate XIII.

1—*Anthemis tinctoria* L.: la—ligulate flower; lb—tubular flower, lc—achene; 2—*A. arvensis* L.: 2a—ligulate flower, 2b—tubular flower, 2c—achene; 2d—involucral bract; 3—*Ptarmica tenuifolia* (Schur) Schur, 3a—ligulate flower, 3b—tubular flower, 3c—involucral bract; 4—*P. vulgaris* Blakw. ex DC., 4a—achene.

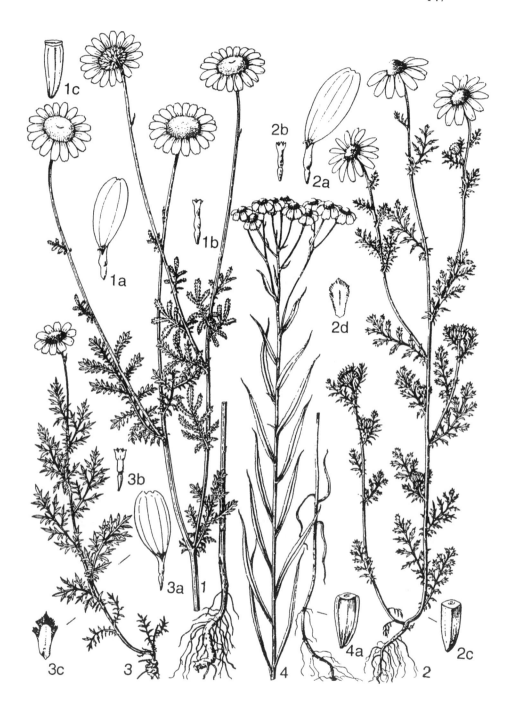

SUBGENUS **2.** *MARUTA* (Cass.) Tzvel. comb. nova
Maruta Cass. 1823, Dict. Sci. Nat. **29:** 174.—*Anthemis* sect. *Maruta*
(Cass.) Griseb. 1846, Spicil. Fl. Rumel. **2:** 205

Receptacle subobtusely conical. Bracts very narrow, narrow-linear, usually absent in lower part of receptacle. Peripheral flowers sterile. Achenes not flattened, with eight or nine weak, often tuberculate, ribs, apically with obtuse margin, without crown. Leaves without intermediate lobes.

Type: *A. cotula* L.

7. **A. lithuanica** (DC.) Trautv. 1883, Tr. Peterb. Bot. Sada, **8:** 392; Bess. ex DC. 1837, Prodr. **6:** 13, in syn.; Fed. 1961, Fl. SSSR, **26:** 64; Fernandes, 1976, Fl. Europ. **4:** 155.—*Maruta cotula* (L.) DC. var. *lithuanica* DC. 1837, op. cit.: 13.

Type: Lithuania ("in Lithuania meridionalis").

Baltic; *Center* (Ladoga-Ilmen: ecdemic in Leningrad and Staryi Russ; Upper Volga: south, rarely. Upper Dnieper; Volga-Don: rarely); *West* (Carpathians; Dnieper).—In more or less weedy meadows, fields, by roadsides, in habitations.—*General distribution*: Central Europe (east).

Note. Apparently, the more mesophytic race of *A. cotula* s. l. and is related through transitions to *A. cotula* s. str.

8. **A. cotula** L. 1753, Sp. Pl.: 894; Fed. 1961, Fl. SSSR, **26:** 63; Fernandes, 1976, Fl. Europ. **4:** 155.—*Maruta cotula* (L.) DC. 1837, Prodr. **6:** 13.

Type: Europe ("in Europae ruderatis, praecipue in Ucrania").

North (Karelia-Murmansk: ecdemic near the Khibiny station and Voknavolok; Dvina-Pechora: southwest); *Baltic*; *Center* (Ladoga-Ilmen: ecdemic; Upper Volga; Upper Dnieper; Volga-Kama; Volga-Don); *West*; *East* (Lower Don: rarely); *Crimea*.—In dry meadows, fields, by roadsides, in habitations.—*General distribution*: Caucasus; Scandinavia (south), Central Europe, Mediterranean, Asia Minor, ecdemic in other countries.—2n = 18.

Note. In the basin of the Oster River in Tula Region, a hybrid *A. cotula* × *A. tinctoria* = *A.* × *bollici* Sch. Bip. (Litvinov, 1917, in Majevski, *Fl. Sredn. Ross.,* ed. 5: 301) has been reported.

SUBGENUS **3.** *COTA* (J. Gay) Rouy

1903, Fl. Fr. **8:** 229.—*Cota* J. Gay, 1845, in Guss. Fl. Sic. Syn. **2:** 866.—
Anthemis sect. *Cota* (J. Gay ex Guss.) Rupr. 1860, Fl. Ingr.: 589

Receptacle hemispherical or weakly convex. Bracts lanceolate
or oblong, rather abruptly narrowed in stiff cusp, all over receptacle.
Peripheral flowers fertile. Achenes more or less flattened dorsally,
with two raised acute lateral ribs and 3–12 considerably finer
veins, without tubercles, with very short crown at apex. Leaves
usually with small intermediate lobes.

Lectotype: *A. tinctoria* L.

9. **A. tinctoria** L. 1753, Sp. Pl.: 896; Fed. 1961, Fl. SSSR, **26:**
39; Fernandes, 1976, Fl. Europ. **4:** 155.—*Cota tinctoria* (L.) J.
Gay, 1845, in Guss. Fl. Sic. Syn. **2:** 867.—*A. markhotensis* auct.
non Fed.: Dobrocz. 1966. Vizn. Rosl. Ukr.: 675; Fernandes, 1976,
Fl. Europ. **4:** 156.—(Plate XIII, 1).

Type: Europe ("in Sueciae, Germaniae apricis pratis siccis").

a. Subsp. **tinctoria.**—Leaves relatively weakly pubescent,
green, with rather broad (0.8–1.6 mm) and usually flat, terminal
lobes or segments; heads on the average larger, usually 30–40
mm in dia, bright yellow.

Arctic (ecdemic in vicinity of Murmansk and Vorkuta); *North*;
Baltic; *Center* (Ladoga-Ilmen; Upper Volga; Volga-Kama; Volga-
Don: west and north); *West* (Carpathians; Dnieper: Moldavia;
Black Sea: Kherson Region); *Crimea* (mountains and Kerch
Peninsula, rarely).—In dry meadows, forest glades and forest
edges, in scrublands, fields, by roadsides, in habitations.—*General*
112 *distribution*: Scandinavia, Atlantic and Central Europe, Mediter-
ranean; ecdemic in other extratropical countries.—2n = 18.

b. Subsp. **subtinctoria** (Dobrocz.) Soó, 1966, Acta Bot. Acad.
Sci. Hung. **12,** 3–4: 366; Smejkal, 1970, Folia Fac. Sci. Nat. Univ.
Purkyn. Brun. **11,** 3: 114; Fernandes, 1976, Fl. Europ. **4:** 156.—*A.
subtinctoria* Dobrocz. 1961, Ukr. Bot. Zhurn., **18,** 2: 67; Fed.
1961, Fl. SSSR, **26:** 40.—*A. zephyrovii* Dobrocz. 1961, op. cit.: 70,
fig. 1; Fed. 1961, op. cit.: 45; Fernandes, 1976, op. cit.: 156.—
Cota tinctoria (L.) J. Gay subsp. *subtinctoria* (Dobrocz.) Holub,
1974, Folia Geobot. Phytotax. (Praha), **9:** 270.—Leaves and stem
on the average more profusely hairy, grayish-green, with narrower
(0.3–1.0 mm), often convolute, terminal lobes or segments; heads
usually 25–30 mm in dia, on the average lighter colored.

150

Type: Donetzk Region of the Ukraine ("Khomutovskaya step preserve").

North (ecdemic); *Baltic* (predominantly in south); *Center*; *West*; *East* (Lower Volga; Trans-Volga); *Crimea.*—In dry meadows and forest glades, steppe slopes, sands, outcrops of chalk and limestone, in fields and by roadsides.—*General distribution*: Caucasus, Western Siberia (south), Eastern Siberia (southwest), Russian Central Asia; Central Europe (south and east), Mediterranean, Asia Minor, ecdemic in other countries.

Note. Possibly represents only the more xeromorphic populations of the preceding subspecies. Usually light yellow flowers are reported for subsp. *subtinctoria.* However, the natural populations of this species seen by me showed no differences in the color of flowers. I relate *A. zephyrovii* described from Kerch Peninsula to subsp. *subtinctoria,* for which smaller sizes of the whole plant and heads are reported. This is also found in other parts of the range of the subspecies.

○ 10. **A. monantha** Willd. 1803, Sp. Pl. **3,** 3: 2187; Fed. 1961, Fl. SSSR, **26:** 43; Fernandes, 1976, Fl. Europ. **4:** 156.—*A. tinctoria* L. var. *monantha* (Willd.) DC. 1837, Prodr. **6:** 11.—*A. cretacea* Zefir. 1954, Bot. Mat. (Leningrad), **16:** 371; Fed. 1961, op. cit.: 43; Fernandes, 1976, op. cit.: 156.—*A. parviceps* Dobrocz. et Fed. ex Klok. 1974. Novosti Sist. Vyssh. Nizsh. Rast. (Kiev), 1974: 115, fig. 8; Fernandes, 1976, op. cit.: 156.—*Cota cretacea* (Zefir.) Holub, 1974, Folia Geobot. Phytotax. (Praha), **9:** 270.

Type: Crimea ("ex Tauria merid.").

Crimea (predominantly southern coast and lower mountains).—On stony slopes and rocks, forest glades, in scrublands usually on outcrops of limestone and allied rocks.—Endemic.

Note. Possibly this species deserves only subspecies rank of *A. tinctoria* s. l., which does not differ very clearly from the smaller plants of subsp. *tinctoria.* Usually, the one-headed stems of *A. monantha,* because of cattle-grazing, become branched and bear several heads. Apparently, based on such a branched plant, *A. cretacea* was described from the vicinity of Bakhchisarai (chalk talus on the southern slope of Mangup-Kale). In addition, *A. cretacea* has bright yellow, and not pale yellow, flowers as in *A. monantha,* although this difference is difficult to detect in the herbarium material. The differences between *A. monantha* described from the vicinity of Yalta (slopes of the Mt. Darsan, 20.VIII.1962, Dobroczaeva) and *A. parviceps* are not clear to me.

11. **A. dumetorum** Sosn. 1927, Vestn. Tifl. Bot. Sada, Ser. 2, **3**: 160; Fed. 1961, Fl. SSSR, **26**: 47; Dobrocz. 1962, Fl. URSR, **11**: 221; id. 1965, Vizn. Rosl. Ukr.: 676; Fernandes, 1976, Fl. Europ. **4**: 157.

Type: Caucasus ("in collibus siccis praesertim in fruticetis atque ad silvarum margines totius Caucasi (300–700 m)").
Crimea (mountains and foothills).—On stony slopes and taluses, forest glades and forest edges, in scrublands.—*General distribution*: Caucasus; Asia Minor.
Note. The not very uncommon yellowish-white color of ligulate flowers in this species is apparently explained by hybridization of *A. dumetorum* × *A. tinctoria* s. 1.

O 12. **A. jailensis** Zefir. 1957, Bot. Mat. (Leningrad), **18**: 251; Fed. 1961, Fl. SSSR, **26**: 47; Fernandes, 1976, Fl. Europ. **4**: 157.— *Cota jailensis* (Zefir.) Holub, 1974, Folia Geobot. Phytotax. (Praha), **9**: 270.

113   Type: Crimea, Babugan-yaila ("State forest preserve, steppe slope adjoining yaila").
Crimea (yaila and upper part of forest zone).—On stony slopes and rocks, in cultivated meadows and forest glades, middle mountain zone (including the yaila).—Endemic.

O 13. **A. dubia** Stev. 1856, Bull. Soc. Nat. Moscou, **29**, 2: 380; Fed. 1961, Fl. SSSR, **26**: 51; Fernandes, 1976, Fl. Europ. **4**: 157.— *Cota dubia* (Stev.) Holub, 1974, Folia Geobot. Phytotax. (Praha), **9**: 270.

Type: Crimea, vicinity of Simferopol and the Alma River ("Circa Sympheropolin et ad Almam in collibus margaceis").
Crimea (predominantly lower mountains).—On stony slopes, taluses, rocks, sands and gravel-beds, forest glades and edges.—Endemic.
Note. In *Flora Europaea* (Fernandes, 1976: 156), one more closely related Balkans-Asia Minor species—*A. parnassica* (Boiss. et Heldr.) Fernandes (=*Cota parnassica* Boiss. et Heldr.)—has been reported for Crimea. It differs from *A. dubia* by having larger (12–14 mm long), but often absent, ligulate peripheral flowers, narrower (about 1 mm) inner involucral bracts, and achenes with longer (0.5–0.6 mm) crown.

14. **A. austriaca** Jacq. 1778, Fl. Austr. **5**: 22; Fed. 1961, Fl. SSSR, **26**: 59; Fernandes, 1976, Fl. Europ. **4**: 158; Dobrocz. 1987, Opred. Vyssh. Rast. Ukr.: 334.—*Cota austriaca* (Jacq.) J. Gay, 1845, in Guss. Fl. Sic. Syn. **2**: 866.

Type: Austria ("Austria").

Center (Ladoga-Ilmen: vicinity of Ivangorod); *West* (Carpathians; Dnieper: west; Moldavia); *Crimea* (mountains and Kerch Peninsula).—In dry meadows, on sands and gravel-beds, stony slopes, by roadsides, in habitations, fields.—*General distribution*: Caucasus; Central Europe, Mediterranean (east), Asia Minor.—2n = 18.

15. **A. altissima** L. 1753, Sp. Pl.: 893; Fed. 1961, Fl. SSSR, **26:** 58; Fernandes, 1976, Fl. Europ. **4:** 157; Dobrocz. 1987, Opred. Vyssh. Rast. Ukr.: 334.—*Cota altissima* (L.) J. Gay, 1845, in Guss. Fl. Sic. Syn. **2:** 867.

Type: Mediterranean Region ("in Italiae, Hispaniae, G. Narbonensis agris").

West (Black Sea: vicinity of Kherson, probably ecdemic); *Crimea* (southern coast).—On open stony and clayey slopes, gravel-beds, by roadsides and in habitations.—*General distribution*: Caucasus, Russian Central Asia (southwest); Atlantic and Central Europe (south), Mediterranean, Asia Minor.—2n = 18.

GENUS **55.** *CHAMAEMELUM* Mill.
1754, Gard. Dict. Abridg., ed. **4:** 1, sine pag.

Heads usually heterogamous, 18–25 mm in dia, solitary or many at apices of stem and its branches. Involucre saucer-shaped, 3–6 mm in dia; involucral bracts imbricate. Receptacle hemispherical, with obtuse scaly bracts. Peripheral flowers pistillate, ligulate, white; disk flowers bisexual, regular, with tubular, yellow, five-toothed corolla, in lower part thickened and more or less covered with downward directed processes from tip of ovary. Achenes 1.0–1.5 mm long and about 0.5 mm wide, prismatic-cylindrical, with three(four) finer ribs, without crown. Perennial, less often annual, plants (5)10–30(40) cm high with usually erect, leafy stems and alternate, more or less incised leaves.

Type: *C. nobile* (L.) All.

Five species in southern Europe and northern, Africa, predominantly in the western Mediterranean Region but, two species are sometimes cultivated as medicinal or aromatic plants; ecdemic in other extratropical countries.

1. **C. nobile** (L.) All. 1785, Fl. Pedem. **1:** 185; Tzvel. 1961, Fl. SSSR, **26:** 67; Tutin, 1976, Fl. Europ. **4:** 165.—*Anthemis nobilis* L. 1753, Sp. Pl.: 894; Galinis, 1980, Lietuv. TSR Fl. **6:** 91.

114 Leaves oblong, 1–6 cm long and 4–15 mm wide. Perennial.

Baltic (south); *Center* (Upper Dnieper); *West* (all regions, rarely); *Crimea*.—Occasionally cultivated as a medicinal and aromatic plant and found as escape or ecdemic in habitations, by roadsides.— *General distribution*: Atlantic and Central Europe, Mediterranean, Asia Minor; ecdemic in other countries.—2n = 18.

Note. As ecdemic, there is another annual Mediterranean species of this genus—*C. mixtum* (L.) All. (=*Anthemis mixtum* L.)—which is distinguished by less incised (usually pinnate) leaves and convolute, carinate bracts falling together with achenes. It is already reported for the Crimean mountains (vicinity of Sevastopol) under the name *Ormenis mixta* (L.) DC. (Schmalh. 1897, *Fl. Sr. Yuzhn. Ross.*: **2,** 63).

GENUS 56. *PTARMICA* Mill.

1754, Gard. Dict. Abridg., ed. **4**: 3; DC. 1837, Prodr. **6**: 19.—*Achillea* L. sect. *Ptarmica* (Mill.) Koch, 1837, Syn. Fl. Germ.: 372.—*Achillea* L. subgen. *Ptarmica* (Mill.) Rouy, 1903, Fl. Fr. **8**: 241

Heads heterogamous, 7–30 mm in dia, usually more or less numerous and clustered in corymbose inflorescences, rarely solitary at apices of stem and its branches. Involucre saucer-shaped or cup-shaped, 3–7 mm long; involucral bracts imbricate. Receptacle more or less convex, with scaly bracts. Peripheral flowers pistillate, ligulate, white, 6–15(20) with ligules 2.6–8.0(9.0) mm long, longer than wide. Disk flowers bisexual, regular, their corolla tubular, whitish or yellowish, five-toothed. Achenes 1.3–2.5 mm long and 0.6–1.0 mm wide, strongly flattened, cuneate-oblong, without crown and pappus, with two strongly raised lateral ribs, often also with one to three indistinct veins. Perennial plants (5)10–130(150) cm high, lacking vegetative shoots with crown of leaves, usually with erect leafy stems and alternate, undivided, less often more or less incised leaves.

Type: *P. vulgaris* Blakw. ex DC. (=*Achillea ptarmica* L.).

About 50 species, distributed in extratropical regions of Eurasia and in the northwest of North America.

Literature: Heimerl, A. 1854, Monographia sectionis *Ptarmica Achilleae* generis. *Denkschr. Akad. Wiss. Math.-Naturw. Kl. (Wien)*, **48**: 1–80.—Uotila, P. 1978. The distribution and history of *Achillea salicifolia* in Finland. *Memoranda Soc. Fauna Fl. Fenn.* **54**: 21–28.—Uotila, P. 1980. Hybridization of *Achillea salicifolia* and *A. ptarmica*, Compositae. *Acta Bot. Fenn.* **16**: 374–382.—Samutina,

154

M.L. 1983. O sistematike vidov roda *Achillea* L. sektsii *Ptarmicae* (Mill.) Koch, rasprostranennykh v evropeiskoi chasti SSSR [On the systematics of the species of the genus *Achillea* section *Ptarmica* (Mill.) Koch, growing in the European part of the USSR]. *Vestn. Leningr. Gos. Univ.,* **21:** 95–97.—Klokov, M.V., and I.L. Krytzka, 1984. Sistema roda *Ptarmica* Mill. i *Achillea* L. [Classification of the genera *Ptarmica* Mill. and *Achillea* L.]. *Ukr. Bot. Zhurn.,* **41,** 3: 1–11.—Sytnik, K.M. (ed.) 1984. Tysyachelistnik [The Yarrows]. Kiev, 1–272.

1. Heads solitary, 14–25(30) mm in dia; ligules of peripheral flowers 5–8(10) mm long; involucral bracts with broad, dark brown, membranous border; leaves bi- or tri-pinnately divided into linear or linear-lanceolate segments. Alpine plants of the Carpathians, 5–20 cm high.. 1. **P. tenuifolia.**

+ Heads in corymbose inflorescences, usually smaller; ligulate peripheral flowers 2.6–6.0 mm long; leaves undivided, more or less toothed.................................... 2.

2. Leaves apically rounded or somewhat emarginate. Alpine plants of the Carpathians, (10)15–40(50) cm high; involucral bracts with dark brown membranous border.. 2. **P. lingulata.**

+ Leaves apically acuminate. Plants of lower mountains and plains, 20–130(150) cm high; involucral bracts with light brown or whitish membranous border.................. 3.

3. Heads 12–17 mm wide, usually 5–25 in inflorescence; involucre usually 5.0–6.5 mm in dia; outermost involucral bracts more than two-thirds as long as the longest inner bracts; ligulate flowers 8–13, with ligule 4–6 mm long; leaves glabrous or subglabrous, without punctate glands.. 3. **P. vulgaris.**

+ Heads 7–12 mm wide, usually 20–40 in inflorescence; involucre 3.5–4.5 mm in dia; outermost involucral bracts half as long as the longest inner bracts; ligulate flowers six to eight, with ligule 2.5–4.5 mm long; leaves usually more or less hairy on both sides, less often glabrous, with or without punctate glands .... 4. **P. salicifolia.**

Section 1. Anthemoideae (DC.) Klok. et Krytzka, 1984, Tysyachelistniki: 166.—*Ptarmica* Anthemoideae* DC. 1837, Prodr. **6:** 19.—*Achillea* L. ser. *Anthemoideae* (DC.) Botsch. 1961, Fl. SSSR, **26:** 123.

115

Small alpine plants, often with solitary heads; leaves undivided to tri-pinnately divided; involucre more or less saucer-shaped, wider than long.

Lectotype: *P. barrelieri* (Ten.) DC.

1. **P. tenuifolia** (Schur) Schur, 1866, Enum. Pl. Transsilv.: 327, non *Achillea tenuifolia* Lam. 1783; Klok. and Krytzaka, 1984, Tysyachelistniki: 198.—*Anthemis tenuifolia* Schur, 1851, Verhendl. Siebenb. Ver. Natürw. **2**, 10: 171.—*Achillea schurii* Sch. Bip. 1856, Oesterr. Bot. Wochenbl. **6**: 300; Botsch. 1961, Fl. SSSR, **26**: 123.—*A. oxyloba* (DC.) Sch. Bip. subsp. *schurii* (Sch. Bip.) Heimerl, 1884, Monogr. Ptarm.: 25; Richards. 1976, Fl. Europ. **4**: 160.—(Plate XIII, 3).

Type: The Carpathians.

West (Carpathians: Chivchin mountains, Svidovets Range).—On limestone rocks and stony slopes in upper mountain zone.—*General distribution*: Central Europe (the Carpathians).—2n = 18.

Note. For Transcarpathia, S.S. Fodor (1974, *Fl. Zakarp.*: 138) reports one more species—*P. serbica* Nym. (=*P. ageratifolia* (Sibth. et Smith) Nym. var. *sérbica* Nym. Hayek) as a plant grown in gardens. This 15–40 cm high mountain plant, grayish-green from dense pubescence, is multicauline from base with fewer undivided, more or less crenate-toothed (to entire) cauline leaves and two to five heads at stem apex; the ligules of its peripheral flowers are 5–7 mm long.

Section 2. Lingulatae (Reichenb. fil.) Klok. et Krytzka, 1984, Tysyachelistniki: 165.—*Achillea* L. sect. *Ptarmica* (Mill.) Koch b. *Lingulatae* Reichenb. fil. 1854, Icon. Fl. Germ. Helv. **16**: 65.—*Achillea* L. ser. *Lingulatae* Botsch. 1961, Fl. SSSR, **26**: 105, descr. ross.

Alpine plants (10)15–40(50) cm high, with heads in corymbose inflorescences; leaves undivided, toothed, obtuse; involucre cup-shaped, slightly longer than wide.

Type: *P. lingulatae* (Waldst. et Kit.) DC.

2. **P. lingulata** (Waldst. et Kit.) DC. 1937, Prodr. **6**: 24; Kondr. 1962, Fl. URSR, **11**: 234.—*Achillea lingulata* Waldst. et Kit. 1799, Descr. Icon. Pl. Rar. Hung. **1**: 2, tab. 2; Botsch. 1961, Fl. SSSR, **26**: 105; Richards. 1976, Fl. Europ. **4**: 161.

Type: The Carpathians ("in alpibus orientalibus Hungariae: in Szathmariensi Rosaly, Transylvanica Guttin, Marmarosiesi Pop Ivan").

West (Carpathians).—In cultivated meadows and on stony slopes of upper mountain zone.—*General distribution*: Central Europe (the Carpathians), Mediterranean (Balkans).— 2n = 18.

Section 3. Ptarmica.

Plants of plains and low-mountains, 20–130(150) cm high, with heads in corymbose inflorescences; leaves undivided, toothed, acute; involucre cup-shaped, slightly longer than wide.

Type: type species of genus.

3. **P. vulgaris** Blakw. ex DC. 1837, Prodr. **6**: 23; Kondr. 1962, Fl. URSR, **11**: 231.—*Achillea ptarmica* L. 1753, Sp. Pl.: 898; Botsch. 1961, Fl. SSSR, **26**: 108; Richards. 1976, Fl. Europ. **4**: 162.—(Plate XIII, 4).

116   Type: Europe ("in Europa temperata").

Arctic (Arctic Europe: ecdemic in vicinity of Murmansk, on Rybachi Peninsula, in vicinity of Vorkuta); *North*; *Baltic*; *Center* (Ladoga-Ilmen; Upper Dnieper; Upper Volga; Volga-Kama: west and north; Volga-Don: west and north, vicinity of Tambov); *West* (Carpathians; Dnieper: north and west).—In meadows, forest glades and edges, on coastal sands and in gravel-beds, in willow groves, to lower mountain zone.—*General distribution*: Scandinavia, Central and Atlantic Europe, Mediterranean (rarely); ecdemic in Western Siberia and North America.—2n = 18.

4. **P. salicifolia** (Bess.) Serg. 1964, in Kryl. Fl. Zap. Sib. **12**, 2: 3484.—*Achillea salicifolia* Bess. 1812, Suppl. Cat. Pl. Volhyn. Cremen: 3; Botsch. 1961, Fl. SSSR, **26**: 112; Uotila, 1979, Ann. Bot. Fenn. **16**, 4: 381.—*A. ptarmica* L. subsp. *salicifolia* (Bess.) Aschers. et Graebn. 1899, Fl. Nordostdeut. Flachl.: 721.— *A. borysthenica* Klok. 1950, Vizn. Rosl. URSR: 543, descr. ucrain.— *Ptarmica borysthenica* Klok. et Sakalo, 1954, Bot. Mat. (Leningrad), **16**: 354.—*P. salicifolia* subsp. *borysthenica* (Klok. et Sakalo) Tzvel., 1987, Arkt. Fl. SSSR, **10**: 106.

Type: Valley of the Dniester River ("Dnestr").

a. Subsp. **salicifolia**.—Involucre 2.7–3.5 mm long and 3–4 mm wide; involucral bracts with light-colored membranous border; ligules of peripheral flowers 2.5–3.0 mm long; leaves rather densely hairy, with numerous punctate glands.

Center (Volga-Don: south and southeast); *West* (Dnieper; Moldavia; Black Sea); *East*; *Crimea* (mountains).—In meadows, swamps, on banks of water bodies, in willow groves.—*General distribution*: Caucasus (Ciscaucasia), Western Siberia (south), Russian Central Asia.

Note. P. borysthenica Klok. et Sakalo, described from the Dnieper (Trukhanov Island), can be clubbed with *P. salicifolia.*

b. Subsp. **cartilaginea** (Ledeb. ex Reichenb.) Tzvel. comb. nova.—*Achillea cartilaginea* Ledeb. ex Reichenb. 1832, Fl. Germ. Excurs. **3**: 849; Botsch. 1961, Fl. SSSR, **26**: 114; Richards. 1976, Fl. Europ. **4**: 162, p. p.—*Ptarmica cartilaginea* (Ledeb. ex Reichenb.) Ledeb. 1845, Fl. Ross. **2**, 2: 530.—*Achillea ptarmica* L. subsp. *cartilaginea* (Ledeb. ex. Reichenb.) Heimerl, 1884, Denkschr. Akad. Wiss. Math.-Naturw. Kl. (Wien), **48**: 174.— Involucre 3.3–4.0 mm long and 3.5–5.0 mm wide; involucral bracts with more or less brownish membranous border; ligules of peripheral flowers 3.0–4.5 mm long; leaves more or less hairy, with punctate glands, but often without them.

Type: Lithuania ("Litthauen", but, apparently, type from vicinity of Tartu ("Dorpat, in humidis").

Arctic (Arctic Europe: east and ecdemic in Murmansk); *North*; *Baltic*; *Center*; *West* (Dnieper).—In meadows, swamps, on banks of water bodies, in willow groves, thinned-out forests.—*General distribution*: Western and Eastern Siberia, Far East (ecdemic); Scandinavia, Central Europe.—2n = 18.

c. Subsp. **septentrionalis** (Serg.) Tzvel. comb. nova.—*P. cartilaginea* subsp. *septentrionalis* Serg. 1949, in Kryl. Fl. Zap. Sib. **11**: 2728.—*Achillea septentrionalis* (Serg.) Botsch. 1961, Fl. SSSR, **26**: 115.—*A. salicifolia* subsp. *septentrionalis* (Serg.) Uotila, 1979, Ann. Bot. Fenn. **16**, 4: 381.—*Ptarmica salicifolia* var. *septentrionalis* (Serg.) Tzvel. 1987, Arkt. Fl. SSSR, **10**: 106.— Like the preceding subspecies, but leaves glabrous or weakly hairy, on the average broader than in other subspecies, without punctate glands.

Type: Western Siberia (Lectotype: "Vicinity of Tyumen, inundated meadows near the confluence of the Balda River with the Tyshma River, 4.VII.1921, V. and L. Larinova").

North (Dvina-Pechora); *Center* (Upper Volga: east; Volga-Kama; Volga-Don: on the Volga River); *East* (Lower Don: northeast; Trans-Volga: north; Lower Volga: on the Volga River).—In meadows, river sands and gravel-beds, swamps, scrublands, on banks of water bodies.—*General distribution*: Western Siberia.

GENUS **57**. *ACHILLEA* L.
1753, Sp. Pl.: 896; id. 1754, Gen. Pl., ed. 5: 382

Heads heterogamous, 3–8(10) mm in dia, clustered in corymbose inflorescences. Involucre cup-shaped or bowl-shaped, 2.0–5.5 mm long; involucral bracts imbricate. Peripheral flowers pistillate, ligulate, (2)3–7(10), white, yellow, or pink, ligules 0.5–3.6 mm long, often wider than long; disk flowers bisexual, regular, with tubular corolla, yellow, less often whitish, with five teeth. Achenes 0.7–2.1 mm long and 0.3–0.9 mm wide, strongly flattened, cuneate-oblong, without crown and pappus, with two raised whitish lateral ribs. Perennial herbaceous plants, 10–120 cm high, with vegetative shoots bearing crown of leaves, usually erect, leafy flowering shoots, and alternate, more or less incised leaves.

Lectotype: *A. millefolium* L.

About 150 species in the extratropical countries of the Northern Hemisphere; most numerous in the countries of the eastern Mediterranean and southwestern Asia.

Literature: Afanasiev, S.S. 1959. Nomenklaturnye zametki o nekoto-rykh vidakh roda *Achillea* L. [Nomenclatural notes on species of the genus *Achillea* L.]. *Bot. Mat. (Leningrad)*, **19**: 360–366.—Dubovik, O.N. 1974. O vidovoi samostoyatel' nosti tysyachilistnika karpatskogo [On the species independence of the Carpathian yarrow]. *Novosti Sist. Vyssh. Rast. (Kiev)*, 1974: 92–98.—Dabrowska, J. 1982. Systematic and geographic studies of the genus *Achillea* L. in Poland with special reference to Silesia. *Acta Univ. Wratisl. Pr. Bot.* **24**: 1–222.—Zimann, S.M., M.V. Klokov, and L.I. Krytzka, 1983. Porivnyalono-morfologichnii analiz vidiv roda *Achillea* L. [Comparative-morphological analysis of species of genus *Achillea* L.]. *Ukr. Bot. Zhurn.,* **40**, 5: 90–96.—Khandzhan, N.S. 1983. Anatomiya semyanok nekotorykh vidov roda *Achillea* L. (Asteraceae) [Achene anatomy of some species of the genus *Achillea* L. (Asteraceae)]. *Bot. Zhurn.,* **68**, 3: 346–351.—Klokov, M.V. and L.I. Krytzka, 1984. Sistema rodiv *Ptarmica* Mill. i *Achillea* L. [System of the genera *Ptarmica* Mill. and *Achillea* L.]. *Ukr. Bot. Zhurn.,* **41**, 3: 1–11.—Sytnik, K.E. (ed.) 1984. Tysyachelistniki [The Yarrow]. Kiev, 1–272.

1. Leaves 10–40 mm long and 2.0–4.5 mm wide, pectinate-pinnatipartite; primary lobes lanceolate or linear, undivided and entire, rarely some with one or two lateral teeth; vegetative shoots elongate, with numerous distant leaves similar to leaves on flowering shoots; heads in dense corymbs;

involucre 2–3 mm wide; ligules of peripheral flowers yellowish-white, 1–2 mm long............26. **A. ochroleuca.**

+ Leaves with more less toothed, lobate or divided, primary lobes, rarely uppermost leaves with undivided primary lobes; leaves on condensed vegetative shoots less numerous and approximate, in crown, usually larger than leaves on flowering shoots,,, 2.

2. Ligulate flower yellow ....3.

+ Ligulate flowers white or pink, rarely yellowish-white ...... 11.

3. Receptacle narrow and long-conical, as long as involucre; heads 5.7–6.5 mm long and one and one-half to two times as long as involucre; involucral bracts gradually merging with usual bracts. At least some leaves 2–5 cm wide, their primary lobes usually more than 2 mm wide. Plants (30)50–100(120) cm high...................... 1. **A. filipendulina.**

+ Receptacle almost flat to short-conical, considerably shorter than involucre; involucral bracts distinct from usual bracts; all leaves narrower, their primary lobes to 1.5 mm wide at base. On the average shorter plants.............................. 4.

4. Outer involucral bracts narrow-lanceolate, more than two-thirds as long as longest inner; heads on very short and thick stalks, clustered in dense corymbs; leaves 5–10 mm wide, all with lateral lobes at base, densely hairy. Plants (15)25–50(70) cm high................................2. **A. coarctata.**

118
+ Outer involucral bracts deltoid to lanceolate, less than two-thirds as long as longest inner; general inflorescences usually less dense .... 5.

5. Whole plant glabrous............................ 8. **A. glaberrima.**

+ Whole plant grayish from dense pubescence.............. 6.

6. Larger cauline leaves with at least solitary small intermediate lobes in middle part or with teeth between primary lobes; ligulate flowers light yellow ....
....25. **A. micranthoides.**

+ All leaves lacking intermediate lobes or teeth, less often with intermediate lobes or teeth in upper part; ligulate flower usually yellow, less often light yellow (in hybrids with species of section *Achillea*)................................. 7.

7. Lower cauline leaves with long petiole, without auricles at base; middle cauline petiolate and with auricles at base; terminal lobes terminating in distinct (even at low magnification) spine 0.2–0.3 mm long. Plants of sands, 10(15)–40 (50) cm high.. 8.

+ Lower cauline leaves, as also middle, with auricles at base of relatively short petioles; terminal lobes terminating in spine 0.05–0.2 mm long. Predominantly steppe and petrophilous plants .... 9.

8. Involucre 2.0–3.3 mm long and 1.7–2.6 mm wide; general inflorescences relatively lax, with rather numerous heads on stalk 2–6 mm long and 0.2—0.4 mm thick. Predominantly steppe or petrophilous plants ....
.... 4. **A. micrantha.**

+ Involucre 3.2–3.5 mm long and 2.4–2.8 mm wide; general inflorescences more dense, but usually with fewer heads on stalk 1–4 mm long and 0.5–0.6 mm thick ....
.... 5. **A. birjuczensis.**

9. Plant 20–50 cm high; lower leaves often 8–15(20) mm wide, their terminal lobes usually linear, rather gradually narrowed in spinule 0.1–0.2 mm long; heads rather small and numerous; involucre 2.7–3.2 mm long and 1.6–2.5 mm wide .... 3. **A. biebersteinii.**

+ Plants 10–25(35) mm high; lower leaves usually to 8 mm wide, their terminal lobes lanceolate, oblong, or obovate, rounded at apex and abruptly narrowed in scarcely visible spinule even at magnification, 0.05–0.1 mm long; heads large but less numerous; involucre 3.2–3.6 mm long and 2.7–3.2 mm wide .... 10.

10. Stem and leaves appressed-hairy; terminal lobes of all leaves oblong or obovate .... 6. **A. leptophylla.**

+ Stem and also leaves squarrosely sparsely hairy; terminal lobes of lower leaves usually lanceolate, narrower; corymbs on the average more dense, with heads on shorter and thicker stalks .... 7. **A. taurica.**

11(2). Involucral bracts with rather broad, membranous, dark brown border; ligulate flowers white or pink. Arctic, subarctic, or alpine plants .... 12.

+ Involucral bracts with narrower, light brownish or whitish membranous border. Predominantly extra-arctic plants of plains and lower mountains .... 14.

12. Primary lobes of cauline leaves deeply pinnately divided; larger of terminal lobes linear-lanceolate or lanceolate, 0.2–0.4 mm wide, gradually narrowed in cusp. Plants of the northeastern European part of Russia ....
.... 20. **A. nigrescens.**

+ Primary lobes of cauline leaves pinnately divided or pinnatilobate; larger terminal lobes oblong or obovate, 0.4–1.2 mm wide, abruptly narrowed in cusp .... 13.

13. Cauline leaves usually 6–12, with solitary intermediate lobes between their bases and primary lobes. Plants of the Carpathians .... 17. **A. carpatica.**

+ Cauline leaves 4–8, without intermediate lobes between their bases and primary lobes. Plants of the north of the European part of Russia .... 18. **A. apiculata.**

14. Small intermediate lobes or teeth present between bases of primary lobes .... 15.

+ Intermediate lobes or teeth absent between bases of primary lobes (excluding the uppermost part of leaves) .... 20.

15. Heads rather large; involucre 4.5–5.5 mm long and 3–4 mm wide; ligules of peripheral flowers 2.2–3.6 mm long. Plants of forest or forest edges, 25–100 cm high ...... 16.

+ Heads smaller, usually 2.5–3.0 mm long and 2.0–2.6 mm wide; ligules of peripheral flowers 0.5–1.2 mm long. Plants of forest edges or meadow-steppes, 15–70 cm high .... 18.

16. Ligules of peripheral flowers 2.8–3.6 mm long, pink; cauline leaves oblong, with approximate primary lobes .... 10. **A. subtanacetifolia.**

+ Ligules of peripheral flowers 2.2–2.8 mm long, white, rarely light pink; cauline leaves lanceolate or linear-lanceolate with distant primary lobes .... 17.

17. Basal leaves lanceolate-elliptical, rapidly narrowed toward base, four to six times as long as wide, their primary lobes relatively longer and narrower (oblong-lanceolate) with distant and rather narrow secondary lobes; tertiary lobes oblong, coarsely narrow-toothed; cauline leaves with narrow and distant basal primary lobes, their rachis narrowly winged (usually 1–2 mm wide)... 9. **A. stricta.**

+ Basal leaves linear-lanceolate or linear, gradually narrowed toward base, six to eight times as long as wide; their primary lobes smaller and broader (usually oblong) with approximate and rather broader secondary lobes; tertiary lobes short, as teeth of various length; cauline leaves with approximate and broad primary lobes, their rachis 2–3 mm wide .... 11. **A. distans.**

18. Ligulate flowers white. Widely distributed plant, with appressed-hairy leaves; intermediate lobes of leaves usually numerous .... 23. **A. nobilis.**

+ Ligulate flowers yellowish-white (pale yellow). Plants of the southwestern European part of Russia .... 19.

19. Cauline leaves oblong, more than 1 cm wide, with rather numerous intermediate lobes. Plants sparsely hairy .... 24. **A. neilreichii.**

+ Cauline leaves broadly linear, usually 6–12 mm wide, with solitary intermediate lobes. Plants appressed-hairy .... 25. **A. micranthoides** (cf. also couplet 6).

20(14). Terminal lobes of leaves 0.1–0.3(0.4) mm wide, linear-lanceolate or linear, gradually narrowed into spinules .... 21.

+ Terminal lobes of leaves (0.3)0.4–1.5(2.0) mm wide, oblong, lanceolate, or lanceolate-ovate, rather abruptly narrowed in spinule .... 23.

21. Common inflorescence rather lax; involucre 3.5–4.2 mm long and 2.6–3.0 mm wide, their bracts with light brownish membranous margin; ligules of peripheral flowers white or pink, 1.6–2.3 mm long, usually longer than wide. Plants of the eastern European part of Russia .... 19. **A. asiatica.**

+ Common inflorescence dense; involucre 2.7–3.6 mm long and 1.5–2.5 mm wide, membranous margin of their bracts not brownish; ligules of peripheral flowers 0.8–1.7 mm long, usually wider than long. Plants of south of the European part of Russia .... 22.

22. Involucre 3.2–3.6 mm long and 2.0–2.5 mm wide; leaves on the average wide, some of them 8–14 mm wide; lowermost secondary lobes usually almost half as long as primary lobes; terminal lobes linear-lanceolate. 2n = 36 .... 21. **A. stepposa.**

+ Involucre 2.8–3.3 mm long and 1.5–2.2 mm wide; leaves· usually narrow, to 8(10) mm wide; lowermost secondary lobes usually nearly two-thirds as long as primary lobes; terminal lobes usually narrow-linear. 2n = 18 .... 22. **A. setacea.**

23. Upper and also some middle cauline leaves with broadly winged rachis, their primary lobes one and one-half to three times as long as width of winged rachis and usually weakly incised, with few wide (deltoid-lanceolate or deltoid) lateral lobes, segments or teeth; stems in lower part weakly appressed-hairy to subglabrous; ligulate flowers white, less often pink .... 24.

+ Upper and middle cauline leaves with narrowly winged (to almost wingless) rachis, their primary lobes three to six times as long as width of winged or wingless rachis and usually more strongly incised .... 25.

24. Basal and lower cauline leaves 5–15 mm wide; heads in dense corymbs, rather small; involucres 3.2–4.3 mm long; ligules of peripheral flowers 1.2–2.0 mm long. Plants 25–80 cm high .... 12. **A. euxina.**

+ Basal and lower cauline leaves 15–70 mm wide; heads in rather lax corymbs, large; involucre 4–5 mm long; ligules of peripheral flowers 2–3 mm long. Plants 35–130 cm high .... 13. **A. inundata.**

25. Plants 25–90 cm high, grayish from dense pubescence; stems to base, also near base of leaves covered with profuse squarrose hairs; terminal lobes of leaves short and broad, usually deltoid or deltoid-ovate, abruptly narrowed in cusp; ligules of peripheral flowers white ... . 14. **A. pannonica.**

+ More or less pubescent but green plants; stem in lower part more or less appressed-hairy; ligules of peripheral flowers white, less often pink .... 26.

26. Plant 15–60 cm high; basal and lower cauline leaves five to eight times as long as wide; terminal lobes of leaves usually lanceolate-ovate, quite abruptly narrowed in cusp, usually glabrous above; involucre 2.8–4.2 mm long; involucral bracts with light colored or scarcely brownish membranous border; heads in rather dense corymbs, small, ligules of peripheral flowers 1.2–2.2 mm long. 2n = 36... .... 15. **A. collina.**

+ Plants 20–100 cm high; basal and lower cauline leaves 7–15 times as long as wide; terminal lobes of leaves usually oblong-lanceolate, more gradually narrowed in cusp, usually more or less hairy above; involucre 3.0–4.6 mm long; involucral bracts with light brown membranous border; heads larger and in more lax corymbs; ligules of peripheral flowers 1.5–2.6 mm long. 2n = 54 .... 16. **A. millefolium.**

Section 1. Filipendulinae (DC.) Afan. 1961, Fl. SSSR, **26:** 90, p. p.—*Achillea *Filipendulinae* DC. 1837, Prodr. **6:** 27, s. str.

Involucre 3.3–4.0 mm long and 2–3 mm wide; receptacle narrow- and long-conical, usually even somewhat rising above involucre;

involucral bracts gradually merging with scales, ligulate flowers yellow; achenes 1.3–1.6 mm long; cauline leaves oblong or oblong-lanceolate, with intermediate teeth and lobules.

Type: *A. filipendulina* Lam.

1. **A. filipendulina** Lam. 1783, Encycl. Méth. Bot. **1**: 27; Afan. 1961, Fl. SSSR, **26**: 90; Fodor, 1974, Fl. Zakarp.: 137.—*A. filicifolia* Bieb. 1808, Fl. Taur.-Cauc. **2**: 338; id. 1819, ibid. **3**: 585.

Type: Southwestern Asia.

West (Carpathians: reported as an introduced plant); *Crimea* (lone specimen of doubtful origin).—In forest glades and edges, gardens and parks.—*General distribution*: Mediterranean (east), Asia Minor.—2n = 18.

Section 2. Micranthae Klok. et Krytzka, 1984, Tysyachelistniki: 171.—*Achillea* L. sect. *Filipendulinae* (DC.) Afan. 1961, Fl. SSSR, **26**: 90, p. p.

Involucre 2.5–5.0 mm long and 1.5–3.0 mm wide; receptacle convex; involucral bracts distinct from scales; ligulate flowers yellow; achenes 1.1–1.4 mm long; cauline leaves linear, usually without intermediate lobules.

Type: *A. micrantha* Willd.

2. **A. coarctata** Poir. 1810, in Lam. Encycl. Méth. Bot. Suppl. **1**: 94; Afan. 1961, Fl. SSSR, **26**: 91; Richards. 1976, Fl. Europ. **4**: 165.—*A. compacta* Willd. 1804, Sp. Pl. **3**: 2206, nom Lam. 1783; Ledeb. 1844, Fl. Ross. **2**, 2: 536.—*A. glomerata* Bieb. 1819, Fl. Taur.-Cauc. **3**: 585.—*A. tomentosa* auct. non L.: Stank. 1949, in Stank. and Taliev, Opred. Rast. Evrop. Chasti SSSR: 625.

Type: Described from specimens of obscure origin.

West (Moldavia; Black Sea: west).—On steppe slopes, in forest glades and edges, scrublands.—*General distribution*: Central Europe (southeast), Mediterranean (Balkans), Asia Minor.—2n = 18.

3. **A. biebersteinii** Afan. 1959, Bot. Mat. (Leningrad), **19**: 361; id. 1961, Fl. SSSR, **26**: 97; Sagalaev, 1988, Byull. Mosk. Obshch. Isp. Prir., Otd. Biol. **93**, 3: 112.—*A. micrantha* auct. non Willd. 1789; Willd. 1804, Sp. Pl. **3**, 3: 2209.

Type: Asia Minor ("in Cappadocia").

East (Lower Don).—On open clayey, stony and sandy slopes, gravel-beds, sometimes by roadsides, in open fields.—*General distribution*: Caucasus; Mediterranean (east), Asia Minor, Iran.

Note. This species is reported for the north of Manych bordering with Caucasus. In recent times, V.A. Sagaev reports it for Kalach-on-Don and the settlement of Golubinskaya.

4. **A. micrantha** Willd. 1789, Tract. de Achilleis: 33, non Willd. 1804; Afan. 1961, Fl. SSSR, **26**: 96; Gusev, 1971, Bot. Zhurn. **56,** 3: 357; Richards. 1976, Fl. Europ. **4**: 165; Mikheev, 1984, Bot. Zhurn. **69,** 5: 693; Klok. and Krytzka, 1984, Tysyachelistniki: 207; Fatare and Gavr. 1985, Fl. Rast. Latv. SSR (Vost.-Latv. Geob. R.-n.): 143.—*A. gerberi* Willd. 1804, Sp. Pl. **3,** 3: 2196, nom. illeg.; Ledeb. 1844, Fl. Ross. **2,** 2: 536.—*A. cancrinii* Grun. 1869, Bull. Soc. Nat. Moscou, **16,** 4: 417.

Type: Siberia ("in Sibiria"), apparently basin of the Don River.

North (Karelia-Murmansk: ecdemic near the station of Kondopog); *Baltic* (ecdemic in Riga and Daugavpils); *Center* (Ladoga-Ilmen; Upper Dnieper; Upper Volga: ecdemic by roadsides; Volga-Don: southeast, ecdemic northwards); *West* (Dnieper: south and east, ecdemic northward; Black Sea); *East*; *Crimea* (Kerch Peninsula, Arabat spit, as also ecdemic of roadsides).—On sands and terraces above floodplain, in sandy steppes and semideserts, sometimes on chalk outcrops, ecdemic of roadsides.—*General distribution*: Caucasus (Ciscaucasia), Western Siberia (south), Russian Central Asia.—2n = 36.

122 *Note.* Hybridizes with species from the section *Achillea*. The most common hybrids are: *A. micrantha* × *A. setacea* (=*A.* × *submicrantha*) Tzvel. sp. hybr. nova (Flores ligulati ochroleuci, folia latiora. Typus: "in viciniis opp. Sarepta, sabuletum in declivitate montium Ergeni, 14.VII.1913, M. Tomin"—LE) with yellowish-white flowers and leaves that are broader than in *A. micrantha*. The hybrid of *A. micrantha* × *A. collina* is possibly *A.* × *illiczevskyi* Tzvel. sp. hybr. nova (Flores ligulati atropurpurei. Typus: in viciniis urb Poltava in sabuletis. 1925. S. Illiczevsky"— LE) with dark-purple ligulate flowers, known from the vicinity of Poltava and Kirvograd.

○ 5. **A. birjuczensis** Klok. 1962, Fl. URSR, **11: 554**: id. 1950, Vizn. Rosl. URSR: 544, descr. ucrain; Klok. and Krytzka, 1984, Tysyachelistiniki: 204.—*A. micrantha* auct. non Willd.: Afan. 1961, Fl. SSSR, **26**: 96, p. p.; Richards, 1976, Fl. Europ. **4**: 165, p. min. p.

Type: Kherson Region, Biryuchi Island ("RSS Ucr., insula maeotica Birjuczij ostriv dicta, steppa sabulosa litoralis").

West (Black Sea: Azov seacoast); *Crimea* (Arabat Spit).—On marine sands and shell sands.—Endemic.—2n = 36.

Note. Possibly of hybrid origin: *A. micrantha* × *A. taurica*.

○ 6. **A. leptophylla** Bieb. 1808, Fl. Taur.-Cauc. **2:** 335; Afan. 1961, Fl. SSSR, **26:** 94, p. p.; Richards. 1976, Fl. Europ. **4:** 164, p. p.

Type: Crimea and southern Ukraine ("in campestribus Tauriae, tum ad Borysthenem circa urbem Cherson").

West (Moldavia: vicinity of Bendery; Black Sea); *East* (Lower Don).—On granite, limestone, and chalk outcrops.—2n = 18.

Note. Hybrid with *A. setacea*—*A.* × *leptophylloides* Tzvel. sp. hybr. nova (Flores ligulati ochroleuci; calathia sat magna. Typus: "Prov Lugansk, distr. Melovskoi pag. Ezhaczii, in sabuletis, 2.VII.1958, O. Dubovik"—LE) has yellowish-white ligulate flowers that are larger than in *A.* × *submicrantha*.

7. **A. taurica** Bieb. 1808, Fl. Taur.-Cauc. **2:** 334; Kondr. 1962, Fl. URSR, **11:** 259; Klok. and Krytzka, 1984, Tysyachelistniki: 209.—*A. leptophylla* Bieb. var. *taurica* (Bieb.) Bieb. 1819, Fl. Taur.-Cauc. **3:** 564.—*A. leptophylla* auct. non Bieb.: Afan. 1961, Fl. SSSR, **26:** 94: 94, p. p.; Richards. 1976, Fl. Europ. **4:** 164, p. p.

Type: Crimea ("in campis apricis Tauricae").

West (Black Sea); *East* (Lower Don: south; Trans-Volga: west; Lower Volga); *Crimea* (north, northwest and Kerch Peninsula).— On steppe slopes, outcrops of limestones, in wormwood steppes, sometimes by roadsides.—*General distribution*: Russian Central Asia (northwest); Central Europe (Dobruja).—2n = 18.

Note. Hybrids of *A. taurica* × *A. setacea* = *A.* × *subtaurica* Tzvel. sp. hybr. nova (Flores ochroleuci; pubescentia laxiuscula. Typus: "In viciniis lac. Elton. 14.V.1914, J. Janischevsky"—LE) are very similar to *A.* × *leptophylloides*, but have a more sparse pubescence on the whole plant.

○ 8. **A. glaberrima** Klok. 1925, Index Sem. Hort. Bot. Charjkov: 6; id. 1926, Ukr. Bot. Zhurn. **3:** 20; Afan. 1961, Fl. SSSR, **26:** 95; Richards. 1976, Fl. Europ. **4:** 164.—(Plate XIV, 2).

Type: Azov Area ("Donetzk Region, Volodarsky District, Kamennye Mogily Preserve").

---

123   Plate XIV.

1—*Achillea millefolium* L., 1a—head, 1b—ligulate flower, 1c—tubular flower, 1d—achene; 2—*A. glaberrima* Klok., 2a—head, 2b—ligulate flower, 2c—tubular flower, 2d—achene; 3—*Matricaria recutita* L., 3a—ligulate flower, 3b—tubular flower, 3c—achene; 4—*Lepidotheca suaveolens* (Pursh) Nutt., 4a—tubular flower, 4b—achene.

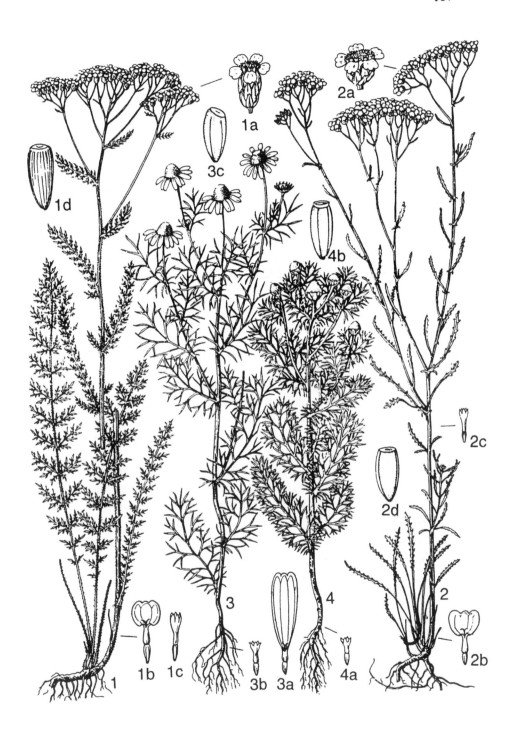

West (Black Sea: Kamennye Mogily Preserve and on the Miuss River").—On granite outcrops.—Endemic.—2n = 18.

Section 3. Achillea.

Involucre 2.7–5.0 mm long and 2–3 mm wide; receptacle convex; involucral bracts distinct from scales; ligulate flowers white or pink, rarely yellowish-white; achenes 1.3–2.2 mm long; cauline leaves oblong to linear, without intermediate lobes, less often with intermediate lobes.

Type: type species.

9. **A. stricta** Schleich. ex Gremli, 1881, Excurs.-Fl. Schweiz, ed. 4: 236; Schleich. 1821, Cat. Pl. Helv.: 5, nom. nud.; Afan, 1961, Fl. SSSR, **26**: 88; Richards. 1976, Fl. Europ. **4**: 162.—*A. tanacetifolia* All. var. *angustifolia* Weihe, 1823, Flora, **6**: 749.—*A. distans* Waldst. et Kit. ex Willd. subsp. *stricta* (Schleich. ex Gremli) Janchen, 1942, Österr. Bot. Zeitschr. **91**: 292.

Type: Switzerland ("Helvetia").

West (Carpathians).—In deciduous and mixed forests, forest glades, predominantly in middle mountain zone.—*General distribution*: Central Europe (mountains).—2n = 54.

10. **A. subtanacetifolia** Tzvel. nom. nov.—*A. tanacetifolia* All. 1785, Fl. Pedem. **1**: 183, non Mill. 1768; Fodor. 1974, Fl. Zakarp.: 137; Dostál, 1982, Seznam Cevn. Rostl. Kvet. Ceskoslov.: 264.—*A. distans* Waldst. et Kit. ex Willd. subsp. *tanacetifolia* (All.) Janchen, 1942, Österr. Bot. Zeitschr. **91**: 292; Richards, 1976, Fl. Europ. **4**: 162.

Type: Italy ("in pratis alpium").

West (Carpathians: Chernogora Range, Pozhizhevskaya mountain pasture).—In forest glades and edges, thinned-out forests, in middle mountain zone.—*General distribution*: Central Europe (mountains).

11. **A. distans** Waldst. et Kit. ex Willd. 1803, Sp. Pl.: **3**, 3: 2207; Afan. 1961, Fl. SSSR, **26**: 87; Richards. 1976, Fl. Europ. **4**: 162, p. max. p.—*A. dentifera* DC. 1815, Fl. Fr. **5**: 485.—*A. asplenifolia* auct. non Vent.: Kondr. 1962, Fl. URSR, **11**: 248.

Type: Hungary and Italy ("in sylvis Banatus, Italiae").

West (Carpathians; Dnieper: southwest; Black Sea: west).—In forest glades and edges, meadows, on stony slopes, in scrublands.—*General distribution*: Central Europe (southeast).—2n = 54.

12. **A. euxina** Klok. 1954, Bot. Mat. (Leningrad), **16**: 359; id. 1950, Vizn. Rosl. URSR: 544, descr. ucrain; Kondr. 1962, Fl. URSR, **11**: 250.—*A. asplenifolia* auct. non Vent.: Kondr. 1962, ibid.: 250.

Type: Kherson Region ("Dzharylgach Island, near the Sinei warf, on edges of depressions").

West (Black Sea); *Crimea* (Kerch Peninsula).—On coastal sands and shell sands, in steppe depressions, on solonetzs.—*General distribution*: Central Europe (southeast).—2n = 36.

Note. Very close to the southern European species *A. asplenifolia* Vent. (1803, *Descr. Pl. Nouv. Jard. Cels*.: tab. 95) but differs from it by being more pubescent, always with white ligulate flowers, and 2n = 36 (and not 18). Apparently, evolved as a result of assimilation of eastern populations of *A. asplenifolia* by *A. setacea* during hybridization.

13. **A. inundata** Kondr. 1962, Fl. URSR, **11**: 553; Richards, 1976, Fl. Europ. **4**: 163; Klok. and Krytzka, 1984, Tysyachelistniki: 232.—*A. millefolium* var. *macrophylla* Serg. 1949, in Kryl. Fl. Zap. Sib. **11**: 2722.—*A. millefolium* auct. non L.: Afan. 1961, Fl. SSSR, **26**: 78. p. p.

Type: Vicinity of Kiev ("Rss. Ucr., insula prope Kioviam, in pratis sabulosis ad ostium Desnae fl.").

Center (Volga-Kama: south and east; Volga-Don: south and east); *West* (Dnieper; Black Sea); *East* (Lower Don; Trans-Volga); *Crimea* (foothills and mountains).—In inundated, less often dry meadows, forest glades and edges, scrublands, thinned-out forests.—*General distribution*: Western Siberia (south), Russian Central Asia (north), Eastern Siberia (ecdemic in the vicinity of Norilsk).—2n = 54.

Note. Resembles larger plants of *A. millefolium* and some populations of *A. distans*, but apparently evolved through hybridization of more ancient species *A. asplenifolia* Vent. (2n = 18) and *A. collina* (2n = 36).

14. **A. pannonica** Scheele, 1845, Linnaea, **18**: 471; Afan. 1961, Fl. SSSR, **26**: 82; Richards. 1976, Fl. Europ. **4**: 163; Klok. and Krytzka, 1984, Tysyachelistniki: 234.—*A. millefolium* L. subsp. *pannonica* (Scheele) Hayek, 1929, in Hegi, Ill. Fl. Mitteleur. **6**, 2: 571.

Type: Hungary, vicinity of Budapest ("in Pannonia prope Pesthinum").

Baltic (south and islands of Estonia); *Center* (Ladoga-Ilmen: vicinity of Mozhaiskaya station, in the vicinity of Leningrad [St. Petersburg]; Volga-Don: rarely); *West* (Dnieper; Moldavia; Black Sea); *East* (Lower Don: near the coast of Azov Sea; Trans-Volga); *Crimea* (foothills and lower mountains).—On dry sandy and stony slopes, outcrops of chalk and limestone, in forest

170

glades and edges, sometimes by roadsides, and on edges of fields.—*General distribution*: Central Europe, Mediterranean (Balkans).—2n = 72.

Note. Differs from the preceding species by, on the average, smaller dimensions of the whole plant and heads, more abundant but lax pubescence of the stem and leaves, and probably evolved from it through hybridization with *A. setacea* (2n = 18).

15. **A. collina** J. Becker ex Reichenb. 1832, Fl. Germ. Excurs. **3**: 850; Richards. 1976, Fl. Europ. **4**: 163; Klok. and Krytzka, 1984, Tysyachelistniki: 227; Ruseikina, 1984, Izv. Akad. Nauk Mold. SSR, Ser. Biol. Khim. Nauk, **3**: 65; Tzvel. 1985, Novosti Sist. Vyssh. Rast. **22**: 273.—*A. millefolium* L. subsp. *collina* (J. Becker ex Reichenb.) Weiss. 1902, in Koch, Syn. Fl. Germ., ed. 3, **2**: 1404; Fodor, 1974, Fl. Zakarp.: 137.—*A. submillefolium* Klok. et Krytzka, 1984, Tysyachelistniki: 220, p. max. p.

Type: Central Europe (without mention of locality).

Baltic (south); *Center* (Ladoga-Ilmen: rarely and predominantly in the south; Upper Dnieper; Upper Volga: south; Volga-Kama: south; Volga-Don); *West*; *East* (Lower Don; Trans-Volga).—In dry meadows, forest glades and edges, on outcrops of chalk and limestones, sands, often by roadsides and on edges of fields.—*General distribution*: Central Europe, Mediterranean, Asia Minor (?).—2n = 36.

Note. I relate to this species a large part of the southern populations and forms of *A. submillefolium*, although the type of this name ("Moscow, Romashkovo, Serebryannyi pine forest, 2.IX.1966, M. Kolokov and others") as well as many other specimens belong to the next species. Notwithstanding different chromosome numbers, the plants transitional between *A. collina* and *A. millefolium* are not that uncommon.

16. **A. millefolium** L. 1753, Sp. Pl.: 899; Afan. 1961, Fl. SSSR, **26**: 78; Richards. 1976, Fl. Europ. **4**: 162.—*A. submillefolium* Klok. et Krytzka, 1984, Tysyachelistniki: 220, s. str.—(Plate XIV, 1).

Type: Europe ("in Europae pascuis pratisque").

Arctic (Arctic Europe: ecdemic); *North*; *Baltic*; *Center* (Carpathians; Dnieper: west and north); *East* (Lower Don: north; Trans-Volga: southern Urals).—In meadows, forest glades and edges, on sands and gravel-beds, in fields, by roadsides, in habitations, thinned-out forests.—*General distribution*: Western Siberia, Far East (ecdemic); Scandinavia, Atlantic and Central Europe, Mediterranean, Japan-China (ecdemic); North America (ecdemic).—2n = 54.

Note. Such features of this species as, on the average, more abundant pubescence of the whole plant, more gradually pointed terminal lobes of leaves, and larger heads distinguish it from *A. collina*, as also frequent occurrence of plants with pink flowers (var. *rosea* Desf.) make it possible to assume the origin of *A. millefolium* through ancient hybridization of the Europn *A. collina* with the asiatic *A. asiatica*.

17. **A. carpatica** Błocki ex Dubovik, 1974, Novosti Sist. Vyssh. Rast. (Kiev), 1974: 93; Klok. and Krytzka, 1984, Tysyachelistniki: 247.—*A. sudetica* auct. non Opiz.: Afan. 1961, Fl. SSSR, **26:** 80.

Type: Eastern Carpathians, Pozhizhev mountain pasture ("in prato subalpino "Pozyzewska" prope alpen Howerla Carpatorum orientalium Galiciae frequens").

West (Carpathians).—In meadows and on stony slopes, forest glades and edges, in upper mountain zone.—*General distribution*: Central Europe (eastern Carpathians).—2n = 54.

Note. Apparently, evolved through hybridization of *A. distans* × *A. sudetica* Opiiz. The latter species occurring in alpine Carpathians and the Sudates mountains and differing from *A. carpathica* by narrower terminal lobes of leaves, has so far not been found in Russia and bordering regions.

18. **A. apiculata** Orlova, 1966, Fl. Murm. Obl. **5:** 426; Tzvel. 1987, Arkt. Fl. SSSR, **10:** 111.—*A. millefolium* L. subsp. *sudetica* auct. non (Opiz) Weiss.: Kurtto, 1984, Retkeilyakasivo: 354.

Type: Murmansk Region ("West of the Kola Inlet, Malaya Volokhovka Inlet, in coastal meadows").

Arctic (Arctic Europe); *North* (Karelia-Murmansk: in the south up to vicinity of Belomorsk and Solovetskie Islands; Dvina-Pechora: north, in the south to the village of Ust-Tsilma on the Pechora River).—On coastal cultivated meadows, sands and gravel-beds, coastal buffs, in habitations.—*General distribution*: Scandinavia.—2n = 54.

Note. Differs from the Carpathian-Sudetes species *A. sudetica* Opiz, with which it is often clubbed, and from *A. nigrescens* by having considerably wider terminal lobes of leaves abruptly narrowing into the cusp.

19. **A. asiatica** Serg. 1946, Sist. Zam. Gerb. Tomsk. Univ. 1(72): 6; id. 1949, Fl. Zap. Sib. **11:** 2723; Afan. 1961, Fl. SSSR, **26:** 85.—*A. millefolium* L. subsp. *asiatica* (Serg.) Andrejev, 1974, Opred. Vyssh. Rast. Yakut.: 465, comb. invalid.—*A. setacea* Waldst.

172

et Kit. subsp. *asiatica* (Serg.) Worosch. 1985, Flor. Issled. i Razn. R-nakh SSSR: 195.

Type: Tomsk Region, valley of the Chulym River ("Prov. Tomsk, in valle flum. Czulym prope pag. Czerdatskoe, in margine agro").

North (Dvina-Pechora: to the east of the North Dvina River); *Center* (Volga-Kama: east); *East* (Trans-Volga: south).—In meadows, forest glades and edges, coastal buffs, sands and gravel-beds, by roadsides.—*General distribution*: Western and Eastern Siberia, Far East.

20. **A. nigrescens** (E. Mey.) Rydb. 1916, North Amer. Fl. **34**, 3: 221; Tzvel. 1987, Arkt. Fl. SSSR, **10**: 110.—*A. millefolium* L. *nigrescens* E. Mey. 1830, Fl. Labrad.: 65.—*A. asiatica* Serg. var. *alpina* Serg. 1946, Sist. Zam. Gerb. Tomsk. Univ. 1(72): 6.

Type: Labrador ("Labrador").

Arctic (Arctic Europe: lower reaches of the Pechora River and to its east); *North* (Dvina-Pechora: northeast).—In cultivated meadows, on gravel-beds, coastal buffs, stony slopes and rocks, floodplain terraces, sometimes by roadsides.—*General distribution*: Western and Eastern Siberia (north and mountains), Far East; Scandinavia (Iceland); North America.

Note. An arcto-alpine derivative of *A. asiatica*, also very close to the Carpathian-Sudetes species *A. sudetica* Opiz.

21. **A. stepposa** Klok. et Krytzka, 1984, Tysyachelistniki: 240.— *A. setacea* auct. non Waldst. et Kit.: Afan. 1961, Fl. SSSR, **26**: 83, p. p.; Richards, 1976, Fl. Europ. **4**: 163, p. p.

Type: Crimea ("Crimean Region, Kirov District, village of Vladislavovka, steppe slopes near the rail-road").

Center (Volga-Don: south and east); *West* (Dnieper; Black Sea); *East*; *Crimea*.—On steppe slopes, in forest glades and edges, outcrops of chalk and limestone, coastal sands and gravel-beds, sometimes by roadsides.—*General distribution*: Western Siberia (south), Russian Central Asia (north).—2n = 36.

Note. Possibly an autopolyploid of *A. setacea* (2n = 18). However, the hybrid origin of *A. stepposa* from *A. setacea* × *A. collina* is not ruled out.

22. **A. setacea** Waldst. et Kit. 1802, Pl. Rar. Hung. **1**: 82, tab. 80; Afan. 1961, Fl. SSSR. **26**: 83, p. p.; Richards. 1976, Fl. Europ. **4**: 163, p. p.; Klok. and Krytzka, 1984, Tysyachelistniki: 235.—*A. millefolium* L. subsp. *setacea* (Waldst. et Kit.) Weiss. 1902, in Koch, Syn. Fl. Germ., ed. 3, **2**: 1404.

Type: Hungary ("in clivis arenosis per planitiem guae inde a Danubio per Comitatum Pesthiensem ad Tibiscum extenditur").

Center (Volga-Don: south and east); *West* (Dnieper: west and south; Moldavia; Black Sea); *East* (Lower Don; Trans-Volga: west); *Crimea*.—In steppes, dry meadows, forest glades and edges, on sands and gravel-beds, by roadsides.—*General distribution*: Atlantic and Central Europe, Mediterranean, Asia Minor.—2n = 18.

Section 4. Nobilia Klok. et Krytzka, 1984, Tysyachelistniki: 174.

Involucre 2.0–3.5 mm long and 1.5–2.5 mm wide; receptacle convex; involucral bracts distinct from scales; ligulate flowers white, rarely yellowish-white; achenes 0.8–1.1 mm long; cauline leaves oblong or lanceolate-ovate with intermediate lobes or teeth.

127 Type: *A. nobilis* L.

23. **A. nobilis** L. 1753, Sp. Pl.: 899; Afan. 1961, Fl. SSSR, **26:** 76; Richards. 1976, Fl. Europ. **4:** 164.

Type: Europe ("in Helvetia, Misnia, Bohemia, G. Narbonensi, Tataria").

North (ecdemic by railroads); *Baltic* (ecdemic); *Center* (Ladoga-Ilmen: ecdemic; Upper Dnieper: south, ecdemic northward); Upper Don: south and east, ecdemic in the northwest; Volga-Kama: south and east, ecdemic in the northwest; Volga-Don; *West*; *East*; *Crimea*.—In dry meadows, on sands, in forest glades, and edges, steppes, by roadsides, in scrublands).—*General distribution*: Caucasus, Western Siberia, Russian Central Asia; Central and Atlantic Europe, Mediterranean, Asia Minor.—2n = 18.

24. **A. neilreichii** A. Kerner, 1871, Österr. Bot. Zeitschr. **21:** 141; Afan. 1961, Fl. SSSR, **26:** 77; Geideman, 1975, Opred. Vyssh. Rast. MoldSSR, ed. 2: 491; Klok. and Krytzka, 1984, Tysyachelistniki: 217.—*A. nobilis* L. subsp. *neilreichii* (A. Kerner) Velen. 1891, Fl. Bulg.: 263; Richards. 1976, Fl. Europ. **4:** 164.

Type: Hungary ("in mittelung Berglande in der Matra bei Parád und auf dem Sárhegy bei Gyöngyös").

West (Moldavia; Black Sea: southwest); *Crimea*.—On steppe slopes, in forest glades and edges, on sands, limestone outcrops.—*General distribution*: Central Europe, Mediterranean (Balkans).—2n = 45.

O 25. **A. micranthoides** Klok. 1954, Bot. Mat. (Leningrad), **16:** 360; id. 1950, Vizn. Rosl. URSR: 544, descr. ucrain.; Afan. 1961, Fl. SSSR, **26:** 100.

174

Type: Kherson Region ("Aksania Nova, virgin steppes, edge of heath bed").

West (Black Sea).—On steppe bottoms, solonetzs, gulley bottoms.—Endemic.—2n = 36.

Note. Possibly evolved through hybridization of *A. nobilis* (2n = 18) × *A. taurica* (2n = 18) and hence arbitrarily referred to section *Nobilia*.

Section 5. Ochroleucae Klok. et Krytzka, 1984, Tysyachelistniki: 174.

Involucre 3.0–3.5 mm long and as much wide; receptacle convex; involucral bracts distinct from scales; ligulate flowers light yellow or yellowish-white; achenes 0.9–1.2 mm long. Plants with elongate vegetative shoots with distant leaves; leaves linear, pinnatipartite.

Type: *A. ochroleuca* Ehrh.

26. **A. ochroleuca** Ehrh. 1792, Beitr. Naturk. **7**: 166, non Willd. 1804; Afan. 1961, Fl. SSSR, **26**: 93; Richards. 1976, Fl. Europ. **4**: 164.—*A. pectinata* Willd. 1804, Sp. Pl., **3**, 3: 2197, non Lam. 1783; Klok. 1950, Vizn. Rosl. URSR: 544.—*A. kitaibeliana* Soó, 1941, Acta Geobot. Hung. **4**: 193; Kondr. 1962, Fl. URSR, **11**: 263.

Type: Hungary ("Hungaria").

West (Moldavia; Black Sea: west).—On steppe slopes, outcrops of granites and other rocks, sands, in forest glades and edges.— *General distribution*: Central Europe (southeast), Mediterranean (Balkans).

GENUS 58. *SANTOLINA* L.
1753, Sp. Pl.: 842; id. 1754, Gen. Pl., ed. 5: 365

Heads homogamous, 5–12 mm in dia, at apices of stem and its branches, usually numerous. Involucre broadly cup-shaped, 3–6 mm long; involucral bracts imbricate. Receptacle almost hemispherical, with scalelike bracts. All flowers in head bisexual, regular, tubular; corolla yellow, five-toothed. Achenes 1.5–2.5 mm long, and 0.4–0.8 mm wide, conical-prismatic, with three to five raised veins, without crown, with obtuse upper margin somewhat raised on one side in larger peripheral achenes. Perennial herbaceous plants or semishrubs 15–100 cm high; stems usually numerous, densely leafy, more or less erect or ascending; leaves alternate, with very narrow pinnatipartite or pinnatilobate blades.

Type: *S. chamaecyparissus* L.

About 30 species in southern Europe and northern Africa; some of them cultivated as ornamental plants in many other countries, sometimes found as an escape.

1. Plant grayish-tomentose; leaves 2–5 mm wide ....
.... 1. **S. chamaecyparissus.**
+ Plant green, subglabrous; leaves 0.8–3.0 mm wide........
.... 2. **S. virens.**

1. **S. chamaecyparissus** L. 1753, Sp. Pl.: 842; Tzvel. 1961, Fl. SSSR, **26:** 126; Guinea and Tutin, 1976, Fl. Europ. 4: 145.

Type: Southern Europe ("in Europa australi").

West (Moldavia; Black Sea); *Crimea* (south).—Cultivated as ornamental plant and occurs as an escape in habitations, gardens and parks, by roadsides.—*General distribution*: Caucasus; Mediterranean, Asia Minor; cultivated in many other countries.— 2n = 18.

2. **S. virens** Mill. 1768, Gard. Dict., ed. 8, No. 4, Czer. 1981, Sosud. Rast. SSSR: 92.—*S. viridis* Willd. 1803, Sp. Pl. **3**, 3: 1798; Tzvel. 1961, Fl. SSSR, **26:** 127.—*S. chamaecyparissus* L. subsp. *viridis* (Willd.) Rouy, 1903, Fl. Fr. **8:** 224.—*S. chamaecyparissus* L. subsp. *squarrosa* auct. non (DC.) Nym.: Guinea and Tutin, 1976, Fl. Europe. **4:** 145, p. p.

Type: Described from garden specimen, probably originating from the western Mediterranean.

West (Black Sea: Melitopol); *Crimea*.—Cultivated as ornamental plant and found as an escape in habitations, parks, seacoasts.— *General distribution*: Mediterranean (west); cultivated in other countries.

Note. Possibly, only a glabrous variant of the preceding species.

GENUS 59. *ANACYCLUS* L.
1753, Sp. Pl.: 892; id. 1754, Gen. Pl., ed. 5: 381

Heads heterogamous, rarely homogamous (lacking ligulate flowers), 20–35 mm in dia, at apices of stem and its branches. Involucre cup-shaped, 5–8 mm long; involucral bracts imbricate. Receptacle hemispherical with scaly bracts. Peripheral flowers ligulate, pistillate, white, yellow, less often pink, sometimes absent; disk flowers bisexual, regular, tubular, their corolla yellow, five-toothed. Achenes strongly flattened, 2.5–3.5 mm long and

2.5–3.0 mm wide, with strongly raised, in peripheral achenes broadly winged, lateral ribs, with uneven crown at apex, 0.3–1.0 mm long, of irregular teeth or lobes, in peripheral achenes with ends of lateral wings rising above apex. Annual plants, 20–50 cm high, with usually erect, leafy stems and alternate leaves with more or less incised blades.

Lectotype: *A. valentinus* L.

The genus includes 13 species in the countries of the Mediterranean (predominantly in northwestern Africa). Three of these are cultivated as medicinal or ornamental plants in many other countries.

Literature: Ehrendorfer, F., D. Schweizer, H. Grezer, and C. Humpries, 1977. Chromosome banding and synthetic systematics in *Anacyclus* (Asteraceae—Anthemideae). *Taxon*, **26**, 4: 387–394.

1. Plant grayish-green from dense pubescence; stems usually strongly branched .... 1. **A. clavatus.**

129 + Plant scatteredly hairy to subglabrous; stems usually simple .... 2. **A. officinarum.**

1. **A. clavatus** (Desf.) Pers. 1807, Syn. Pl. **2:** 465; Afan. 1961, Fl. SSSR, **26:** 70, in adnot.; Dobrocz. 1962, Fl. URSR, **11:** 230.— *Anthemis clavata* Desf. 1800, Fl. Atl. **2:** 287.—*Ancyclus tomentosus* DC. 1815, Fl. Fr. **5:** 481; Schmalh. 1897, Fl. Sr. Yuzhn. Ross. **2:** 63.

Type: Northern Africa ("in arvis Barbariae").

Crimea (vicinity of Sevastopol).—As ecdemic of roadsides, in fields, habitations.—*General distribution*: Mediterranean.— 2n = 18.

2. **A. officinarum** Hayne, 1825, Getreua Darstell. Gew. **9:** 46; Afan. 1961, Fl. SSSR, **26:** 69; Dobrocz. 1987, Opred. Vyssh. Rast. Ukr.: 334.

Type: Described from cultivated plant.

West (Carpathians; Dnieper; Black Sea).—Occasionally cultivated as medicinal or aromatic plant, and as an escape, found in habitations, gardens and parks.—*General distribution*: Central Europe, Mediterranean.—2n = 18.

Note. Apparently evolved in cultivation through hybridization of *A. pyrethrum* (L.) Cass. × *A. radiatus* Lois. *A. officinarum* and *A. clavatus* have white ligulate flowers and the stalks of heads are expanded in the upper part. We may also have *A. radiatus* Lois. (1807, *Fl. Gall.*: 585) with larger yellow, less often pink, ligulate flowers can be found as an ornamental plant.

GENUS **60.** *MATRICARIA* L.
1753, Sp. Pl.: 890; id. 1754, Gen. Pl., ed. 5: 380

Heads heterogamous, 8–25 mm in dia, at apices of stem and its branches. Involucre cup-shaped, 1.8–4.0 mm long; involucral bracts imbricate. Receptacle obtusely conical, without bracts. Peripheral flowers ligulate, pistillate, white, usually bent down; disk flowers bisexual, regular, tubular, with yellow, five-toothed corolla. Achenes 0.7–1.3 mm long and 0.2–0.3 mm wide, conical-cylindrical, somewhat curved, with five lighter ribs, apically with somewhat truncate subobtuse margin, without crown. Annual plants, 8–30(40) cm high, with usually erect leafy stems and alternate pinnati- or bipinnatipartite leaves having linear terminal lobes.

Type: *M. recutita* L.

Two closely related species in warm temperate regions of the Northern Hemisphere; ecdemic in the extratropical regions of the Southern Hemisphere.

1. Plant 10–40 cm high, usually branched in upper half, subglabrous; heads 10–25 mm in dia; ligules of peripheral flowers 4–9 mm long; achenes 0.8–1.3 mm long .... 1. **M. recutita.**
+ Plant 8–20 cm high, usually branched from base, scatteredly hairy (especially near base of cauline leaves); heads 8–12 mm in dia; ligules of peripheral flowers 2.5–4.0 mm long; achenes 0.7–0.9 mm long .... 2. **M. tzvelevii.**

1. **M. recutita** L. 1753, Sp. Pl.: 891; Pobed. 1961, Fl. SSSR, **26:** 148.—*M. chamomilla* L. 1763, Sp. Pl.: 1256, nom. illeg.—*Chamomilla recutita* (L.) Rausch, 1974, Folia Geobot. Phytotax. (Praha), **9:** 255; Kay, 1976, Fl. Europ. **4:** 167.—(Plate XIV, 3).

Type: Europe ("in Europa").

Arctic (ecdemic in Murmansk); *North* (enough, rare; ecdemic in the north); *Baltic*; *Center*; *West*; *East*; *Crimea.*—In habitations, on edges of fields, by roadsides, sometimes cultivated as a medicinal plant.—*General distribution:* Caucasus, Western and Eastern 130 Siberia (south), Far East (south), Russian Central Asia; Scandinavia (south), Central and Atlantic Europe, Mediterranean, Asia Minor, Iran, Mongolia, Japan-China; North America.—2n = 18.

2. **M. tzvelevii** Pobed. 1961, Fl. SSSR, **26:** 871.—*M. chamomilla* L. subsp. *tzvelevii* (Pobed.) Soó, 1969, Acta Bot. Acad. Sci. Hung. **15,** 3–4: 344.—*Chamomilla tzvelevii* (Pobed.) Rausch, 1974, Folia Geobot. Phytotax. (Praha), **9:** 155; Kay, 1976, Fl. Europ. **4:** 167.

Type: Crimea ("vicinity of Sudak, on slope of Mt. Bolvan near the village of Uyutnoe").

Crimea (south).—On open stony and clayey slopes, gravel-beds, in habitations, by roadsides.—*General distribution*: Caucasus; Mediterranean (east), Asia Minor.

GENUS **61.** *LEPIDOTHECA* Nutt.

1841, Trans. Amer. Philos. Soc., nov. ser. 7: (454).—*Matricaria* L. sect. *Anactidea* DC. 1838, Prodr. **6**: 50.—*Lepidanthus* Nutt. 1841, op. cit.: 396, non Nees, 1830

Heads homogamous, 4–12 mm in dia, at apices of stem and its branches. Involucre cup-shaped, 2.5–6.0 mm long; involucral bracts imbricate. Receptacle obtusely conical, without bracts. All flowers bisexual, regular, tubular, with greenish-yellow, four-toothed corolla. Achenes 1.0–1.7 mm long and 0.4–0.5 mm wide, conical-cylindrical, somewhat flattened, with five veins of which two convergent on ventral side, with acute, up to 0.1 mm long margin at apex. Annual plants, 5–30(40) cm high, with usually erect leafy stems and alternate, usually bi- or tri-pinnatisect leaves with linear terminal lobes.

Type: *L. suaveolens* (Pursh) Nutt.

Of the three species, one is distributed in the countries of the Mediterranean, another in the western part of America, while the third is amphipacific, dispersed anthropogenically in almost all extratropical countries of both hemispheres.

1. **L. suaveolens** (Pursh) Nutt. 1841, Trans. Amer. Philos. Soc., nov. ser. 7: (454); Kovalevsk. 1962, Fl. Uzbek. **6**: 124; Tzvel. 1987, Arkt. Fl. SSSR, **10**: 132.—*Santolina suaveolens* Pursh, 1814, Fl. Amer. Sept. **2**: 520.—*Artemisia matricarioides* Less. 1831, Linnaea, **6**: 210.—*Matricaria discoidea* DC. 1838, Prodr. **6**: 50.—*M. matricarioides* (Less.) Porter, 1884, Mem. Torrey Bot. Club, **5**: 341; Pobed. 1961, Fl. SSSR, **26**: 150.—*M. suaveolens* (Pursh) Buchenau, 1894, Fl. Nordwest. Tiefebene: 496, non L. 1755.—*Chamomilla suaveolens* (Pursh) Rydb. 1916, North Amer. Fl. **34**, 3: 232; Kay, 1976, Fl. Europ. **4**: 167.—(Plate XIV, 4).

Terminal lobes of leaves 0.2–0.3 mm wide, gradually narrowed, terminating in spine.

Type: North America ("On the banks of the Kooskoosky").

Arctic (Arctic Europe: ecdemic); *North*; *Baltic*; *Center*; *West*; *East*; *Crimea* (rarely).—In habitations, by roadsides, in meadows, on coastal sands and gravel-beds.—*General distribution*: Caucasus, Western and Eastern Siberia, Far East, Russian Central Asia; Scandinavia, Central and Atlantic Europe, Mediterranean, Asia Minor, Iran, Dzhungaria-Kashgaria, Japan-China; North America, South America (ecdemic), Australia (ecdemic).—2n = 18.

GENUS 62. *TRIPLEUROSPERMUM* Sch. Bip.

1844, Tanaceteen: 31.—*Matricaria* auct. non L.: Rausch, 1974, Feddes Repert. **85**, 9–10: 652

Heads heterogamous, 12–50 mm in dia, at apices of stem and its branches, very rarely homogamous. Involucre narrow saucer-shaped to cup-shaped, 2.5–8.0 mm long; involucral bracts imbricate. Receptacle more or less convex to obtusely conical, without bracts. Peripheral flowers ligulate, pistillate, white, sometimes absent; disk flowers bisexual, regular, tubular, with yellow, five-toothed corolla. Achenes 0.8–2.7 mm long and 0.4–1.5 mm wide, triquetrous, with three strongly raised whitish ribs, dorsally with
131 two reddish-brown, roundish or oval glands in upper part, on remaining surface usually brownish or dark brown; often transversely rugose, with entire or three-lobed crown 0.1–1.0 mm long at apex. Annual, biennial, or perennial (but usually not long perennating) plants, 8–60 mm high, with usually erect, leafy stems, alternate bi- or tri-pinnatisect leaves and lanceolate terminal lobes.

Type: *T. inodorum* (L.) Sch. Bip.

About 30 species in the extratropical regions of the Northern Hemisphere, but predominantly in Eurasia.

Literature: Pobedimova, E.G. 1961. Zametki po sistematike nekotorykh rodov iz semeistva Compositae (triba Anthemideae) [Notes on the systematics of some genera of the family Compositae (Tribe Anthemideae)]. *Bot. Mat.* (*Leningrad*), **2**: 343–358.—Hämete Ahti, L. 1976. *Tripleurospermum* (Compositae) in the Northern parts of Scandinavia, Finland and Russia. *Acta Bot. Fenn.* **75**: 3–17.—Rauschert, S. 1974. Nomenklatural probleme in der Gattung *Matricaria. Folia Geobot. Phytotax.* (*Praha*), **9**, 3: 249–260.—Jeffrey, C. 1979. Notes on the lectotypification of the names *Cacalia* L., *Matricaria* L., and *Gnaphalium* L. *Taxon*, **28**, 4: 349–351.

1. Involucral bracts with rather broad (0.5–2.0 mm), dark brown membranous border. Perennials or biennials, north of the European part of Russia .... 2.

+ Involucral bracts with narrower (to 0.5 mm), light brown or whitish, membranous border .... 3.

2. Involucral bracts highly unequal, outermost almost half as long as innermost; heads 15–35 mm in dia; membranous border of involucral bracts 0.8–2.0 mm wide ....
.... 3. **T. hookeri.**

+ Involucral bracts slightly unequal, outermost more than two-thirds as long as innermost; heads 30–50 mm in dia; membranous border of involucral bracts 0.5–1.0 mm wide
.... 4. **T. subpolare.**

3. Heads 12–25 mm in dia; ligules of peripheral flowers 3–8 mm long. Annual plants of southeast of the European part of Russia, 8–25 cm high; achenes, excluding crown, 0.8–1.6 mm long, with strongly raised ventral rib, smooth between ribs, crown 0.8–1.0 mm long or absent ....
.... 5. **T. parviflorum.**

+ Heads 20–45 mm in dia; ligules of peripheral flowers 8–16 mm long. Plants 10–60 cm high; achenes 2.0–2.7 mm long, with less prominent ventral rib, transversely rugose between ribs, crown 0.1–0.3 mm long .... 4.

4. Perennial or biennial littoral plant, usually multicauline from base; terminal lobes of leaves somewhat fleshy, to 1 mm wide; involucral bracts with rather narrow, light brown border; ligules of peripheral flowers 15–20 mm long .... 1. **T. maritimum.**

+ Biennial or annual, predominantly weed plant, but extending to seacoasts; stem usually solitary but often branched above base; terminal lobes of leaves not fleshy, to 0.7 mm wide; involucral bracts with whitish or light brown, usually broader (0.3–0.5 mm), membranous border; ligules of peripheral flowers usually 10–15 mm long ....
.... 2. **T. inodorum.**

Section 1. Tripleurospermum.—*Tripleurospermum* Sch. Bip. sect. *Phaeocephala* Pobed. 1960, Bot. Mat. (Leningrad), **21:** 146.

Achenes, excluding crown, 1.8–2.5 mm long, finely transversely rugose between ribs, with finer ventral rib; heads 20–45 mm in dia; receptacle convex, often hemispherical.

Type: type species.

1. **T. maritimum** (L.) Koch, 1845, Syn. Fl. Germ., ed. 2, **3:** 1026; Pobed. 1961, Fl. SSSR, **26:** 177.—*Matricaria maritima* L. 1753, Sp. Pl.: 891; Kay, 1976, Fl. Europ. **4:** 166, excl. subsp.

Type: Northern Europe ("in Europae septentrionalis littoribus maris").

132   *Baltic* (near the coast, rarely); *Center* (Ladoga-Ilmen: on the coast of the Gulf of Finland and Chudskoe [Peipus] Lake.—In coastal cultivated meadows, on sands and gravel-beds.—*General distribution*: Scandinavia (south), Atlantic and Central Europe; ecdemic in North America.—2n = 18 + 0–48.

2. **T. inodorum** (L.) Sch. Bip. 1844, Tanaceteen: 31; Pobed. 1961, Fl. SSSR, **26:** 175.—*Matricaria inodora* L. 1755, Fl. Suec., ed. 2: 765.—*M. chamomilla* L. 1753, Sp. Pl.: 891, nom. ambig.— *Chrysanthemum inodorum* (L.) L. 1763, Sp. Pl., ed. 2: 1253.— *Matricaria perforata* Mérat, 1812, Nouv. Fl. Env. Paris.: 332; Key, 1976, Fl. Europ. **4:** 16.—*Pyrethrum conicum* Less. 1834, Linnaea, **9:** 189.—(Plate XV, 1).

Type: Sweden ("Suecia").

Arctic (Arctic Europe: ecdemic); *North*; *Baltic*; *Center*; *West*; *East*: Crimea.—In fields, more or less weedy meadows and forest glades, on coastal sands and in gravel-beds, in habitations, by roadsides.—*General distribution*: Caucasus, Western and Eastern Siberia, Far East (south), Russian Central Asia; Scandinavia, Central and Atlantic Europe, Mediterranean, Asia Minor, Iran, Dzhungaria-Kashgaria, Japan-China; North America, South America (ecdemic in south), Australia (ecdemic), Africa (north, ecdemic in south).—2n = 18 + 0–18, 36.

3. **T. hookeri** Sch. Bip. 1853, Bonplandia, **16:** 151; Orlova, 1966, Fl. Murm. Obl. **5:** 218, p. p.; Tzvel. 1987, **Arkt.** Fl. SSSR, **10:** 127, map. 44.—*Chrysanthemum grandiflorum* Hook. 1825, in Parry's 2nd Voy.: 398, non Brouss. 1804, nec Willd. 1809.— *Matricaria inodora* L. var. *phaeocephala* Rupr. 1845, Beitr. Pflanzenk. Russ. Reich. **2:** 42.—*M. phaeocephala* (Rupr.) Stefanss, 1924, Fl. Isl., ed. 2: 223.—*M. hookeri* (Sch. Bip.) Hutch. 1934, N. Rime-Ring. Sun: 252; Czer. 1981, Sosud. Rast. SSSR: 86.— *Tripleurospermum phaeocephalum* (Rupr.) Pobed. 1961, Bot. Mat. (Leningrad), **21:** 347; id. 1961, Fl. SSSR, **26:** 171.—*T. maritimum* (L.) Koch subsp. *phaeocephalum* (Rupr.) Hämet. Ahti, 1967, Acta Bot. Fenn. **75:** 9.—*Matricaria maritima* L. subsp. *phaeocephala* (Rupr.) Rausch, 1974, Folia Geobot. Phytotax. (Praha), **9:** 257; Kay, 1976, Fl. Europ. **4:** 168.

Type: Canada, coast of Hudson Bay ("von der York Factory").

Arctic (Novaya Zemlya; Arctic Europe); *North* (Karelia-Murmansk: north; Dvina-Pechora: northwest).—On rocks, stony and clayey slopes facing sea, coastal sands and gravel-beds.— *General distribution*: Eastern Siberia (Mt. Putoran, on the Yana and Kolyma rivers), Arctic; North America (in the south to southeastern coasts of the Hudson Bay).—2n = 18.

4. **T. subpolare** Pobed. 1961, Bot. Mat. (Leningrad), **21**: 347; id. 1961, Fl. SSSR, **26**: 172, map. 43.—*T. maritimum* L. var. *boreale* Hartm. 1849, Handb. Skand. Fl., ed. 5: 2.—*T. maritimum* L. subsp. *subpolare* (Pobed.) Hámet-Ahti, 1967, Acta Bot. Fenn. **75**: 5.— *Matricaria maritima* L. subsp. *subpolaris* (Pobed.) Rausch. 1974, Folia Geobot. Phytotax. (Praha), **9**: 257; Kay, 1976, Fl. Europ. **4**: 166.—*M. maritima* L. subsp. *borealis* (Hartm.) A. et D. Löve, 1976, Bot. Not. (Lund), **128**: 521.—*M. subpolaris* (Pobed.) Holub, 1977, Folia Geobot. Phytotax. (Praha), **12, 3**: 308.

Type: Solovetskie Islands ("Solovetskie Islands, Bolshoi Zayatski Island, on the seacoast").

Arctic (Arctic Europe); *North* (Karelia-Murmansk; Dvina-Pechora: north); *Center* (Ladoga-Ilmen: shores of Ladoga and Onega lakes).—On coastal, less often lacustrine sands and gravel-beds, on riverbanks and depressions in them, sometimes as ecdemic of roadsides, in habitations.—*General distribution*: Western Siberia (lower reaches of the Ob' River), Eastern Siberia (north), Far East (on the northern coast of the Sea of Okhotsk); Scandinavia; North America (south of Greenland).—2n = 18.

Section 2. Gastrosolum (Sch. Bip.) Tzvel. comb. nova.— *Gastrosolum* Sch. Bip. 1844, Tanaceteen: 29.—*Gastrosolum* sect. *Eugastrosolum* Sch. Bip. 1844, ibid.: 29, nom. illeg.— *Tripleurospermum* Sch. Bip. sect. *Eugastrosolum* (Sch. Bip.) Pobed. 1961, Fl. SSSR, **26**: 183, nom. illeg.

Achenes, excluding crown, 0.8–1.6 mm long, smooth between ribs, with strongly raised ventral rib; heads 12–25 mm in dia; receptacle obtusely conical.

133

Plate XV.

1—*Tripleurospermum inodorum* (L.) Sch. Bip., 1a—ligulate flower, 1b— tubular flower, 1c and 1d—achene from different sides; 2—*Leucanthemum vulgare* Lam. subsp. *ircutianum* (Turcz. ex DC.) Tzvel., 2a—ligulate flower, 2b—tubular flower, 2c—achene of peripheral flowers, 2d—achene of tubular flowers; 3—*Leucanthemella serotina* (L.) Tzvel., 3a—ligulate flower, 3b—tubular flower, 3c—achene; 4—*Pyrethrum corymbosum* (L.) Willd.: 4a—lower leaf, 4b—ligulate flower, 4c—tubular flower, 4d—achene.

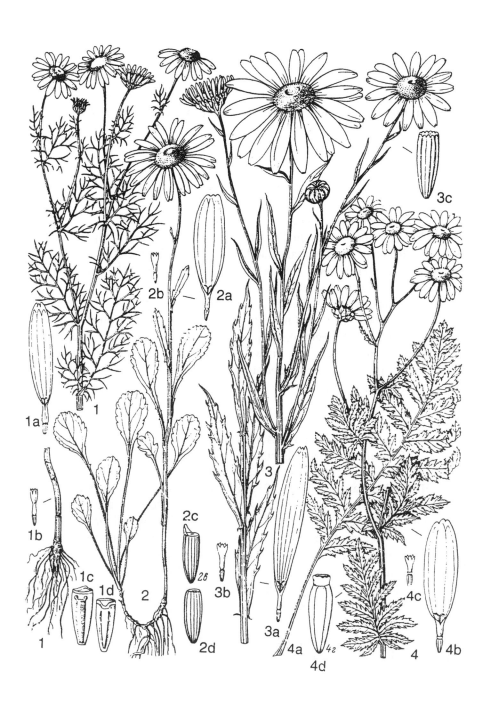

1a

1b

1c 1d

1

2b 2a

2c

2ß

3b

2d

3

3a

4a 4ß

4d

4

4c

4b

3c

134   Type: *T. praecox* (Bieb.) Bornm. (=*T. parviflorum* (Willd.) Pobed.).

5. **T. parviflorum** (Willd.) Pobed, 1961, Bot. Mat. (Leningrad), **21**: 354; id. 1961, Fl. SSSR, **26**: 183, p. p.—*Pyrethrum parviflorum* Willd. 1803, Sp. Pl.: **3**, 3: 2158.—*P. praecox* Bieb. 1808, Fl. Taur.-Cauc. **2**: 324.—*Matricaria parviflora* (Willd.) Poir. 1814, in Lam. Encycl. Méth. Bot. Suppl. 3: 608; Kay, 1976, Fl. Europ. **4**: 167.—*M. praecox* (Bieb.) DC. 1817, Prodr. **6**: 52; Stev. 1856, Bull. Soc. Nat. Moscou, **29**, 3: 383.—*Tripleurospermum praecox* (Bieb.) Bornm. 1940, Beih. Bot. Centralbl. **60**: 192.

Type: Locality not mentioned, possibly lower reaches of the Volga.

East (Trans-Volga: on the Ural River; Lower Volga).

Crimea (reported for the Sevastopol coast).—In sandy steppes and semideserts, on river sands and gravel-beds, by roadsides.— *General distribution*: Caucasus, Russian Central Asia; Mediterranean (east), Asia Minor, Iran.

Note. Besides the typical populations of this species with a well-developed crown (it is more or less three-lobed, one-third to half as long as the remaining part of the achene), there are populations in which the crown of the achenes is almost nonexistent (to 0.1 mm long) and usually with smaller (12–15 mm in dia), but more numerous heads. Their taxonomic status is as yet not clear.

GENUS 63. *ARGYRANTHEMUM* Sch. Bip.
1844, in Webb. and Berth. Phytogr. Canar. **2**: 258

Heads heterogamous, 25–50 mm in dia, solitary at apices of stem and its branches, but usually numerous. Involucre saucer-shaped, 10–18 mm wide and 4–6 mm long; involucral bracts imbricate. Receptacle strongly convex, without bracts. Peripheral flowers pistillate, ligulate, white; disk flowers bisexual, tubular, with yellow, five-toothed corolla. Achenes heteromorphic; in peripheral flowers broadly cuneate, 3.0–4.5 mm long and 2.5–3.5 mm wide, with three, strongly raised, winged ribs, strongly truncate in upper part, without corona; in disk flowers achenes narrow-cuneate, 1.8–2.5 mm long and 0.8–1.2 mm wide, more or less flattened from sides, with five to eight longitudinal ribs, of which one ventral strongly raised, apically with truncate corona to 0.4 mm long. Glabrous or subglabrous semishrubs, with erect, usually strongly branched stems, and alternate distant leaves with more or less incised or lobate blades.

Type: *A. frutescens* (L.) Sch. Bip.

The genus includes 12–15 species, distributed mainly in the Canary Islands, of these one species is widely cultivated as a medicinal plant and occasionally found as an escape.

1. **A. frutescens** (L.) Sch. Bip. 1844, in Webb. and Berth, Phytogr. Canar. **2:** 264, tab. 91; Tzvel. 1961, Fl. SSSR, **26:** 147.— *Chrysanthemum frutescens* L. 1753, Sp. Pl.: 887.

Plant 40–100 cm high; leaves to 10 cm long and 5 cm wide, ovate to oblong, pinnatisect or bipinnatisect.

Type: Canary Islands ("in Canariis insulis").

West (Black Sea); *Crimea*.—Cultivated as an ornamental plant and occasionally found as an escape, found in habitations, by roadsides, in gardens and parks.—*General distribution*: Mediterranean (Canary Islands); cultivated and escape in many other countries.—2n = 18, 27.

GENUS 64. *CHRYSANTHEMUM* L.
1753, Sp. Pl.: 887; id. 1754, Gen. Pl., ed. 5: 379

Heads heterogamous, 25–70(80) mm in dia, solitary at apices of stem and its branches. Involucre saucer-shaped or broadly cup-shaped, 5–10 mm long; involucral bracts imbricate. Receptacle almost hemispherical. Peripheral flowers pistillate, ligulate, yellow 135 (rarely white); disk flowers bisexual, tubular, with yellow, less often red, five-toothed corollas. Achenes heteromorphic, lacking crown, 2.0–4.5 mm long and 1–4 mm wide, in peripheral flowers broadly cuneate, more or less flattened with three or two, strongly raised, winged ribs and two to six veins between them; in disk flowers with 6–12 more or less evenly distributed ribs, of which only one or two (ventral and dorsal) strongly raised. Glabrous or weakly hairy annual plants, with erect stem and alternate distant leaves more or less incised or lobed, less often undivided.

Type: *C. coronarium* L.

Five species, predominantly in the countries of the Mediterranean, of these three are extensively cultivated as ornamental plants and often occur as escape while one apparently evolved in cultivation, found in eastern Asia.

1. Middle and lower cauline leaves usually undivided, irregularly coarsely toothed, less often shallow-pinnatilobate; ligules of peripheral flowers 8–15 mm long, with two-winged lateral ribs .... 1. **C. segetum.**

+ Middle and lower cauline leaves pinnatisect or bipinnatisect; ligules of peripheral flowers 15–25 mm long; achenes of peripheral flowers with three-winged ribs (ventral and laterals) .... 2.

2. Corolla of tubular flowers yellow, of peripheral flowers usually unicolorous, yellow, rarely white; achenes of disk flowers weakly flattened from sides, with one or two strongly raised ribs (ventral and lateral) and five to eight finer ribs .... 2. **C. coronarium.**

+ Corolla of tubular flowers dark red, of peripheral flowers bi- or tri-colorous (yellow, white, and reddish); achenes of disk flowers strongly flattened from sides, with two raised ribs (ventral and dorsal), other ribs scarcely visible .... 3. **C. carinatum.**

1. **C. segetum** L. 1753, Sp. Pl.: 889; Tzvel. 1961, Fl. SSSR, **26:** 136; Schultz, 1976, Bot. Zhurn., **61:** 10: 1452; Heyw. 1976, Fl. Europ. **4:** 168; Liacavicius, 1982, Bot. Zhurn. **67,** 2: 233.

Type: Europe ("in Scaniae, Germaniae, Belgii, Angliae, Galliae agris").

Baltic; *Center* (Ladoga-Ilmen; Upper Dnieper; Volga-Don); *West*; *East* (Lower Don).—Sometimes cultivated as an ornamental plant, as also ecdemic or escaped by roadsides, in fields, habitations.—*General distribution*: Caucasus; Scandinavia (south), Central and Atlantic Europe, Mediterranean, Asia Minor, ecdemic in other countries.—2n = 18.

2. **C. coronarium** L. 1753, Sp. Pl.: 890; Tzvel. 1961, Fl. SSSR, **26:** 134; Gusev, 1968, Bot. Zhurn. **53,** 2: 269; Heyw. 1976, Fl. Europ. **4:** 169.

Type: Crete and Sicily ("in Creta, Sicilia").

Baltic; *Center*; *West*; *East*; *Crimea*.—Cultivated as ornamental plant, sometimes as escape found by roadsides, in habitations.— *General distribution*: Mediterranean; cultivated in many other countries, often escaped.—2n = 18.

3. **C. carinatum** Schousb. 1800, Jagttag. Vextrig. Marokko: 198, tab. 6; Tzvel. 1961, Fl. SSSR, **26:** 133; Heyw. 1976, Fl. Europ. **4:** 168.

Type: Morocco ("Marokko").

Baltic; *Center*; *West*; *Crimea*.—Cultivated as ornamental plant, sometimes as escape in southern regions, found by roadsides, in habitations.—*General distribution*: Africa (north); cultivated in many other extratropical countries.

136

GENUS **65.** *COLEOSTEPHUS* Cass.
1826, Dict. Sic. Nat. **41:** 43

Heads heterogamous, 20–50 mm in dia, solitary at apices of stem and its branches. Involucre narrow-saucer-shaped, 4–6 mm long and 8–20 mm wide; involucral bracts imbricate. Receptacle strongly convex, obtusely conical, without bracts. Peripheral flowers ligulate, pistillate, but often sterile, yellow; disk flowers bisexual, tubular, with yellowish, five-toothed corolla and tube strongly flattened at base. Achenes all alike (but in peripheral flowers often not developed), 1.8–2.0 mm long and about 0.7 mm wide, conical-cylindrical with eight or nine fine ribs and very loosely fitting testa, with more or less truncate, 1.2–1.8 mm long crown at apex. Annual, glabrous or weakly hairy plants with erect stems; leaves alternate, distant, undivided, but more or less serrate-dentate, elliptical or oblong.

Type: *C. myconis* (L.) Cass.

Five species in the countries of the Mediterranean and northern Africa, one of them ecdemic in many other countries.

1. **C. myconis** (L.) Reichenb. f. 1853, Icon. Fl. Germ. **16:** 49; Tzvel. 1961, Fl. SSSR, **26:** 147; Heyw. 1976, Fl. Europ. **4:** 174.— *Chrysanthemum myconis* L. 1763, Sp. Pl., ed. 2: 1254.—*Pyrethrum myconis* (L.) Moench. 1802, Meth. Pl. Suppl. 247; Schmalh. 1897, Fl. Sredn. Yuzhn. Ross. **2:** 70.

Ligules of peripheral flowers 6–15 mm long. Plant 10–40 cm high.

Type: Spain ("in Hispania").

Crimea (vicinity of Sevastopol, ecdemic).—By roadsides, in habitations, plantations.—*General distribution*: Central Europe (ecdemic), Mediterranean, Asia Minor.—2n = 18.

GENUS **66.** *LEUCANTHEMUM* Mill.
1754, Gard. Dict., ed. 4; DC. 1838, Prodr. **6:** 45

Heads heterogamous, 25–80 mm in dia, solitary at apices of stem and its branches. Involucre saucer-shaped, 4–8 mm long and 8–22 mm in dia; involucral bracts imbricate. Receptacle somewhat convex, without bracts. Peripheral flowers ligulate, pistillate, white; disk flowers bisexual, tubular with yellow, five-toothed corolla. Achenes heteromorphic, 1.8–3.2 mm long and 0.6–0.8 mm wide, more less terete, but narrowed toward base, with 8–12 rather uniformly disposed ribs with rows of muciferous cells on them, between ribs with longitudinal secretory canals; peripheral achenes somewhat more flattened, with ribs closer on one side, often with unilateral, to 1 mm-long corona, inner (in disk flowers) not flattened, without or with crown as five subobtuse teeth. Perennial plants, with erect leafy stem and alternate, undivided but more or less toothed or more or less pinnatilobate leaves.

Type: *L. vulgare* Lam.

The genus includes 15–25 species, distributed predominantly in the mountainous regions of the Mediterranean and southern Europe. Some of them are cultivated as ornamental plants and one species is widely distributed in extratropical Eurasia and is ecdemic in other countries.

Literature: Przywara, L. 1974. Karyological studies in *Leucanthemum vulgare* s. 1. from Poland. *Acta Biol. Cracov.*, *Ser. Bot.*, **17**, 1: 55–73.—Przywara, L. 1974. Biosystematic studies on the collective species *Leucanthemum vulgare* from Poland. *Fragm. Flor. Geobot.* **20**, 4: 413–471.—Prager, L., H. Schuwerk, and R. Schuwerk. 1985. Zur Verbreitung der *Leucanthemum*-Arten im Altmuhljura und den benachbarten Gebieten. *Berichte bayer Bot. Ges.* **56**: 231–233.

137   1. Involucral bracts with whitish or brownish membranous border .... 2.
       + Involucral bracts with broader dark brown membranous border. Mountainous plants of the Carpathians .... 3.
       2. All cauline leaves more or less uniformly serrate-dentate and relatively gradually narrowed into acute tip rather large (often more than 10 cm long), usually numerous; one or two upper leaves strongly reduced. Cultivated plants 35–150 cm high; heads 50–80 mm in dia; achenes of peripheral flowers with unilateral crown 0.4–0.8 mm long .... 2. **L. maximum.**
       + Cauline leaves irregularly toothed, often partly pinnatilobate, rather abruptly narrowed into subacute or subobtuse tip, not so large (almost always to 10 cm long) and less numerous, usually more than two upper

leaves strongly reduced. Predominantly wild plants 20–80 cm high; heads 25–60 mm in dia; achenes of peripheral flowers with or without unilateral, to 0.5 mm long crown .... 3. **L. vulgare.**

3. Blades of at least some basal leaves more or less orbicular, abruptly narrowed into petiole, middle and upper cauline leaves uniformly serrate-dentate, gradually narrowed into acute tip, strongly and rather gradually narrowed toward base, without auriculate expansions; achenes of peripheral flowers with distinct but unilateral more or less toothed, crown 0.4–0.8 mm long; achenes of disk flowers apically with teeth as many as ribs. Forest plants, 25–80 cm high .... 1. **L. waldsteinii.**

+ Blades of basal leaves obovate and elliptical to suborbicular, and then cuneate, petiolate; middle and upper cauline leaves not uniformly toothed, rather abruptly narrowed into usually auriculately expanded base; achenes of disk flowers without teeth at apex. Alpine plants, 5–40 cm high .... 4.

4. Achenes of peripheral flowers lacking crown; usually one or two upper cauline leaves strongly reduced .... 4. **L. subalpinum.**

+ Achenes of peripheral flowers with unilateral 0.4–0.8 mm long crown; usually three to six upper cauline leaves strongly reduced .... 5. **L. margaritae.**

1. **L. waldsteinii** (Sch. Bip.) Pouzar, 1975, Preslia, **47,** 2: 158; Heyw. 1976, Fl. Europ. **4:** 177.—*Tanacetum waldsteinii* Sch. Bip. 1844, Tanaceteen.: 35.—*Chrysanthemum rotundifolium* Waldst. et Kit. 1803, in Willd. Sp. Pl. **3,** 3: 2144.—*Leucanthemum rotundifolium* (Waldst. et Kit.) DC. 1837, Prodr. **6:** 46, non Opiz. 1825.

Type: Carpathians ("in alpibus Hungariae").
West (Carpathians).—In coniferous and mixed forests, forest glades and edges, in middle mountain zone.—*General distribution*: Central Europe (Carpathians).

2. **L. maximum** (Ramond) DC. 1836, Prodr. **6:** 46; Tzvel. 1961, Fl. SSSR, **26:** 145, in adnot; Heyw. 1976, Fl. Europ. **4:** 175.— *Chrysanthemum maximum* Ramond, 1800, Bull. Soc. Philom. Paris, **2:** 140.

Type: Pyrenees ("in pratis Pyreneorum circa Bareges...").
Baltic; Center; West.—Quite extensively cultivated as ornamental plant and sometimes found as an escape in habitations,

190

gardens and parks.—*General distribution*: Atlantic Europe (Pyrenees); cultivated in many other extratropical countries.— 2n = 36.

3. **L. vulgare** Lam. 1779, Fl. Fr. **2**: 137; Tzvel. 1961, Fl. SSSR, **26**: 143; Heyw. 1976, Fl. Europ. **4**: 175.—*Chrysanthemum leucanthemum* L. 1753, Sp. Pl.: 888.

Type: Europe ("in pratis Europae").

138   a. Subsp. **vulgare.**—Stem usually glabrous, 10–40 cm high; cauline leaves usually deeply toothed to pinnatilobate, middle and upper rather narrow, with almost parallel margins, with long teeth or lobes near base, forming auricles; peripheral achenes usually without crown.

Arctic (Arctic Europe: ecdemic); *North*; *Baltic*; *Center* (Ladoga-Ilmen; Upper Dnieper; Upper Volga; Volga-Kama: predominantly in the Urals; Volga-Don: very rarely); *West* (Carpathians; Dnieper; Moldavia; rarely); *East* (Trans-Volga: southern Urals); *Crimea*.—In meadows, forest glades and edges, on river and coastal sands and gravel-beds, by roadsides.— *General distribution*: Caucasus; Scandinavia, Central and Atlantic Europe, Mediterranean; ecdemic in other countries.—2n = 18.

b. Subsp. **ircutianum** (Turcz. ex DC.) Tzvel. comb. nova.— *Leucanthemum ircutianum* Turcz. ex DC. 1838, Prodr. **6**: 47.— *Chrysanthemum ircutianum* (Turcz. ex DC.) Ledeb. 1845, Fl. Ross. **2**, 6: 543.—*Leucanthemum vulgare* var. *ircutianum* (Turcz. ex DC.) Kryl. 1904, Fl. Alt. i Tomsk. Obl.: 618.—*Chrysanthemum leucanthemum* subsp. *ircutianum* (Turcz. ex DC.) Sukacz. 1918, Izv. Ross. Akad. Nauk, 1918: 965.—Stems more or less hairy, usually throughout or only in lower part, 20–80 cm high; cauline leaves usually toothed, middle and upper on the average broader and with more convex lateral margins, with or without short teeth near base; peripheral achenes usually with crown.—(Plate XV, 2).

Type: Irkutsk ("in pratis Sibiriae, circa Irkutsk").

North; *Baltic*; *Center*; *West*; *East* (Lower Don; Trans-Volga; Lower Volga: very rarely); *Crimea*.—In meadows, forest glades and edges, thinned-out forests, by roadsides, in fields and habitations, on river sands and gravel-beds; often cultivated as an ornamental plant.—*General distribution*: Caucasus, Western and Eastern Siberia, Far East (south), Russian Central Asia (rarely); Scandinavia, Central and Atlantic Europe, Mediterranean, Asia Minor, Iran, Dzhungaria-Kashgaria, Mongolia, Japan-China; North America; ecdemic in other extratropical countries.—2n = 36.

Note. Both the subspecies of *L. vulgare* occur in the territory of the Flora almost throughout but subsp. *vulgare* is distinctly less common and predominantly in the western regions. At present, it is not clear if *L. vulgare* be treated s. str. Some authors consider it as a tetraploid subspecies and for the diploid subspecies use the epithet *L. praecox* (Horvatič) Horvatič (1963, *Acta Bot. Croat.* **22:** 211; Dmitrieva and Parfenov, 1991, *Kariologiya flory kak osnova tsitogeneticheskogo monitoringa* [Karyology of Flora as the Basis of Cytogenetic Monitoring]: 1–230).

4. **L. subalpinum** (Schur) Tzvel. 1961, Fl. SSSR, **26:** 143; Fodor, 1974, Fl. Zakarp.: 138.—*Chrysanthemum leucanthemum* L. var. *subalpina* Schur, 1859, Verh. Siebenb. Ver. Naturw. **10:** 137.— ?*Leucanthemum rotundifolium* Opiz, 1825, Naturalientausch, **9:** 122.—*Tanacetum subalpinum* (Schur) Simonk. 1886, Enum. Fl. Transsilv.: 313.—*Leucanthemum raciborskii* M. Pop. et Chrsh. 1949, in M. Pop. Opred. Rastit. Fl. Karpat.: 248; Dobrocz. 1962, Fl. URSR, **11:** 273; Fodor, 1974, op. cit.: 138.—*L. vulgare* L. subsp. *subalpinum* (Schur) Soó, 1972, Feddes Repert. **83,** 3: 154.

Type: Eastern Carpathians ("Siebenbürgen").

West (Carpathians).—In cultivated meadows, on stony slopes and rocks in upper mountain zone.—*General distribution*: Central Europe (mountains).—2n = 18.

Note. There is much confusion in the nomenclature of the alpine European races of *L. vulgare* s. 1.—group. It is quite likely that *L. rotundifolium* Opiz (loc. cit.) relates to this species and is its priority name.

5. **L. margaritae** (Gáyer ex Jav.) Soó, 1972, Feddes Repert. **83,** 3: 154; Soják. 1983, Sborn. Nar. Muz. Praze, **39,** 1: 53.— *Chrysanthemum leucanthemum* L. var. *margaritae* Gáyer ex Jav. 1925, Magyar Fl.: 1128.—*Leucanthemum vulgare* Lam. subsp. *margaritae* (Gayer ex Jav.) Soó, 1971, Acta Bot. Hung. 17: 125.— *L. adustum* (Koch) Gremli subsp. *margaritae* (Gayer ex Jav.) Holub. 1974, Folia Geobot. Phytotax. (Praha), **9:** 273.

139   Type: Hungary ("A. Balatonnál").

West (Carpathians).—On limestone rocks and stony slopes in upper mountain zone.—*General distribution*: Central Europe (Carpathians).—2n = 54.

Note. Closer to the alpine species *L. adustum* and replaces it in the alpine Carpathians.

GENUS **67.** *LEUCANTHEMELLA* Tzvel.
1961, Fl. SSSR, **26:** 137

Heads heterogamous, 35–60 mm in dia, solitary at apieces of stem and its branches. Involucre 4–7 mm long and 10–20 mm wide; involucral bracts imbricate. Receptacle rather strongly convex, without bracts. Peripheral flowers ligulate, sterile, white; disk flowers bisexual, tubular, with yellow, five-toothed corolla. Achenes 2–3 mm long and 0.7–0.8 mm wide, more or less terete but narrowed toward base, with (8)10(12) raised, and almost uniformly disposed, ribs terminating in subobtuse teeth to 0.3 mm long, without secretory canals and muciferous cells. Perennial plants, 30–150 cm high, with erect leafy and alternate undivided or three- to five-parted leaves.

Type: *L. serotina* (L.) Tzvel.

Two species, sporadically distributed in the countries of southeastern Europe and eastern Asia.

1. **L. serotina** (L.) Tzvel. 1961, Fl. SSSR, **26:** 139; Heyw. 1976, Fl. Europ. **4:** 171.—*Chrysanthemum serotinum* L. 1753, Sp. Pl.: 888.—*Leucanthemum serotinum* (L.) Stank. 1949, in Stank. and Taliev, Opred. Vyssh. Rast. Evrop. Chasti SSSR: 630.—(Plate XV, 3).

Cauline leaves sessile, undivided, oblong-lanceolate, serrate-dentate.

Type: Garden grown plant, originating from southeastern Europe, although North America mentioned in error as native place ("in America septentrionali").

West (Dnieper: valley of the Uzh River; Moldavia: lower reaches of the Danube River).—In swamps and swampy meadows.—*General distribution*: Central Europe (southeast).

GENUS **68.** *PYRETHRUM* Zinn.
1757, Cat. Pl.: 414; Scop. 1772, Fl. Carniol., ed. 2, **2:** 148

Heads heterogamous, 8–50 mm in dia, rarely homogamous and then 7–10 mm in dia, solitary at apices of stem and its branches or more numerous (to 100) and then clustered in corymbose or corymbose-paniculate inflorescences. Involucre saucer-shaped to cup-shaped, 4–7 mm long; involucral bracts imbricate. Receptacle more or less convex, without bracts. Peripheral flowers usually ligulate, pistillate, white, rarely pink, sometimes absent; disk flowers bisexual, tubular, with yellow, five-toothed corolla. Achenes 1.0–3.5 mm long and 0.3–1.0 mm wide, ribbed-terete, but narrowed toward base, with 5–10 more or less raised ribs, in peripheral flowers often more or less shifted to one side of achene, apically with more or less toothed or lobed, crown 0.1–1.5 mm long, usually lacking secretory canals and muciferous cells. Perennial plants sometimes somewhat woody at base, 10–150 cm high, with erect leafy stem and alternate more or less incised leaves.

Type: *P. corymbosum* (L.) Scop.

About 100 species in the extratropical regions of Eurasia (in the east to Baikal) and in northern Africa. Very similar to genus *Tanacetum* L. and often clubbed with it.

140

1. Heads, 8–14 mm in dia, always with ligulate flowers, 20–100, clustered in dense corymbose inflorescences; involucre cup-shaped; ligules of peripheral flowers rotund or reniform, 2–4 mm long; achenes 1.5–2.0 mm long, with crown 0.2–0.3 mm long; leaves pinnately parted or pinnatilobate .... 4. **P. macrophyllum.**

+ Heads with ligulate flowers 12–45 mm in dia, rarely without ligulate flowers and then smaller, solitary at apices of stem and its branches or clustered in corymbose infloresence, and then less numerous (to 30); involucre saucer-shaped; ligulate or peripheral flowers broadly elliptical to oblong-linear, 3–18 mm long .... 2.

2. Plant 15–45 cm high, silver-grayish from dense, appressed pubescence; stems rather weakly leafy, predominantly in lower part; leaves pinnatisect or bipinnatisect, with very narrow (linear to oblong) terminal lobes; ligules 8–16 mm long; achenes 2.5–3.2 mm long .... 3. **P. cinerariifolium.**

+ Plant green, less often somewhat grayish from rather profuse, somewhat lax, pubescence; stems with numerous leaves; terminal lobes and segments of leaves broader, usually more or less toothed .... 3.

3. Plants 25–100 cm high; leaves without punctate glands; ligulate flowers always present, their ligules 10–18 mm long .... 4.

+ Plants 10–70 cm high; leaves with numerous punctate glands; ligules of ligulate flowers (sometimes absent) 3–10 mm long .... 5.

4. Involucral bracts with light, colored or brownish membranous border; receptacle two to three times as wide as high .... 1. **P. corymbosum.**

+ Involucral bracts with dark brown, usually broader, membranous border; receptacle usually one and one-half times as wide as high ....2. **P. clusii.**

5. Predominantly wild, green plant; ligulate flowers always present, their ligules (6)7–12(15) mm long......................
.... 5. **P. parthenifolium.**

+ Predominantly cultivated, yellowish-green plants; ligules of ligulate flowers (often absent), 3–7 mm long ....
.... 6. **P. parthenium.**

Section 1. Pyrethrum.

Heads fewer, relatively large; achenes 2.0–3.5 mm long, without punctate glands and secretory canals, with crown 0.5–0.7 mm long; leaves without punctate glands.

Type: type species.

1. **P. corymbosum** (L.) Willd. 1803, Sp. Pl. **3**, 3: 2155; Tzvel. 1961, Fl. SSSR, **26**: 232.—*Chrysanthemum corymbosum* L. 1753, Sp. Pl: 89.—*Tanacetum corymbosum* (L.) Sch. Bip. 1844, Tanaceteen: 57; Heyw. 1976, Fl. Europ.: 170.—*Pyrethrum tauricum* Zelenetzky, 1906, Mat. Fl. Kryma: 296.—(Plate XV, 4).

Type: Europe and Siberia ("in Thuringia, Bohemia, Helvetia, Sibiria").

North (Dvina-Pechora: vicinity of Vytegra); *Baltic* (south); *Center* (Ladoga-Ilmen: northeast; Upper Dnieper: south; Upper Volga: on the Oka River; Volga-Kama: south and east; Volga-Don); *West*; *East*; *Crimea.*—In deciduous and mixed forests, forest glades and edges, scrublands.—*General distribution*: Caucasus, Western Siberia (south); Atlantic Europe (south), Central Europe, Mediterranean.—2n = 36.

Note. Populations, sometimes found in Zhiguly and Southern Urals, with more deeply incised leaves and more gradually pointed lobes, have been identified in herbaria in 1987 as "*Tanacetum corymbosum* subsp. *grusinum* Voith-Drescheret et Ehrendorfer" and possibly actually belong to the separate Caucasian race of *P. corymbosum* s. 1. The majority of populations from Crimea (*P. tauricum* Zelenetzky) have somewhat smaller, very numerous heads and, possibly, also deserve to be included in a separate subspecies.

2. **P. clusii** Fisch. ex Reichenb. 1830, Fl. Germ. Excurs. **2**: 231; Tzvel. 1961, Fl. SSSR, **26**: 234.—*Chrysanthemum subcorymbosum* Schur, 1859, Verh. Siebenb. Ver. Naturw. **10**: 146.—*Pyrethrum subcorymbosum* (Schur) Schur, 1866, Enum. Pl. Tranassilv.: 337; M. Pop. 1948, Ocherk Rastit. Fl. Karp.: 250.—*Tanacetum clusii*

(Fisch. ex Reichenb.) Soják, 1971, Acta Mus. Nat. Prag. **27**, 2: 50.—*T. corymbosum* (L.) Willd. subsp. *clusii* (Fisch. ex Reichenb.) Heyw. 1976, Journ. Linn. Soc. London (Bot.), **71**: 272; id. 1976, Fl. Europ. **4**: 171.

Type: The Alps and Carpathians ("in Bergwaldern im südlichen Gebiete, schon von Clusius in Ungarn und Oestereich wohl unterschieden").

West (Carpathians).—In meadows, forest glades, and edges, in upper and middle mountain zones.—*General distribution*: Central Europe (southern Alps and Carpathians), Mediterranean (Balkans).—2n = 18.

Section 2. Cinerariifolia (Heyw.) Tzvel. 1961, Fl. SSSR, **26**: 231.—*Tanacetum* L. subsect. *Cinerariifolia* Heyw. 1953, Anal. Inst. Bot. Cavanilles, **12**, 2: 325.

Heads relatively large and less numerous; achenes 2.5–3.2 mm long, with secretory canals and punctate glands, crown 0.5–0.8 mm long; leaves with punctate glands.

Type: *P. cinerariifolium* Trev.

3. **P. cinerariifolium** Trev. 1820, Index Sem. Hort. Vratisl. App. 2: 2; id. 1826, Nova Acta Acad. Leop.-Carol. **13**: 204; Tzvel. 1961, Fl. SSSR, **26**: 213.—*Tanacetum cinerariifolium* (Trev.) Sch. Bip. 1844, Tanaceteen: 58; Heyw. 1976, Fl. Europ. **4**: 171.

Type: Yugoslavia ("in collibus saxosis Dalmatiae").

West (Black Sea); *East* (Lower Don); *Crimea*.—Cultivated as insecticidal plant and occasionally found as an escape in fields, by roadsides in habitations.—*General distribution*: Mediterranean (Balkans); in cultivation—Central Europe, Mediterranean, where occasionally occurs as an escape.—2n = 18.

Section 3. Gymnocline (Cass.) DC. 1838, Prodr. **6**: 57, p. p.—*Gymnocline* Cass. 1816, Bull. Soc. Philom. Paris, 1816: 199, s. str.

Heads small and numerous (20–100), with cup-shaped involucre; achenes 1.5–2.0 mm long, without secretory canals, but with punctate glands, crown 0.2–0.3 mm long; leaves with punctate glands.

Type: *P. macrophyllum* (Waldst. et Kit.) Willd.

4. **P. macrophyllum** (Waldst. et Kit.) Willd. 1803, Sp. Pl.: **3**, 3: 2154; Tzvel. 1961, Fl. SSSR, **26**: 194; Dobrocz. 1962, Fl. URSR, **11**: 285.—*Chrysanthemum macrophyllum* Waldst. et Kit. 1801, Descr. Icon. Pl. Rar. Hung. **1**: 97, tab. 94.—*Tanacetum macrophyllum* (Waldst. et Kit.) Sch. Bip. 1844, Tanaceteen: 53; Heyw. 1976, Fl. Europ. **4**: 171.

Type: Hungary ("in sylvis croaticis ad viam carolinam, silavonicis montis Papuk, et banaticis ad thermas Herculis atque ad limites Vallachiae").

West (Dnieper: in "Alexandria" Park near the city of Belaya Tserkov).—As cultivated and escaped plant in gardens and parks.—*General distribution*: Caucasus; Central Europe (southeast), Mediterranean (Balkans), Asia Minor; occasionally cultivated in other regions of Europe and at places naturalized.

Section 4. Parthenium (Briq.) Willk. 1870, Prodr. Fl. Hisp. **2**: 99; Tzvel. 1961, Fl. SSSR, **26**: 201.—*Tanacetum* L. subsect. *Parthenium* Briq. 1916, in Burnat, Fl. Alp. Marit. **6**: 119.

Heads medium-sized, often rather numerous, sometimes without ligulate flowers; achenes 1.0–1.8 mm long, without secretory canals, but with punctate glands, crown 0.1–0.5 mm long; leaves with punctate glands.

Type: *P. parthenium* (L.) Smith.

5. **P. parthenifolium** Willd. 1803, Sp. Pl.: **3**, 3: 2156; Tzvel. 1961, Fl. SSSR, **26**: 203.—*Tanacetum parthenifolium* (Willd.) Sch. Bip. 1844, Tanaceteen: 56; Heyw. 1976, Fl. Europ. **4**:171.

Type: Garden plant of obscure origin, possibly from the Caucasus.

142 *Crimea* (mountains).—On more or less shady, stony slopes and rocks, in forest on scrublands.—*General distribution*: Caucasus, Russian Central Asia; Mediterranean (east), Asia Minor, Iran, Dzhungaria-Kashgaria.—2n = 18.

6. **P. parthenium** (L.) Smith, 1800, Fl. Brit. 2: 900; Tzvel. 1961, Fl. SSSR, **26**: 204.—*Matricaria parthenium* L. 1753, Sp. Pl.: 890.—*Tanacetum parthenium* (L.) Sch. Bip. 1844, Tanaceteen: 55; Heyw. 1976, Fl. Europ. **4**: 171.

Type: Europe ("in Europae cultis, ruderatis").

North (Karelia-Murmansk: south; Dvina-Pechora: south); *Baltic*; *Center*; *West*; *East*; *Crimea*.—Cultivated as ornamental plant (usually as a border) and often as an escape in gardens and parks, in habitations.—*General distribution*: Caucasus, Western and Eastern Siberia (south), Russian Central Asia; Scandinavia (south), Central and Atlantic Europe, Mediterranean, Asia Minor; apparently evolved in cultivated.—2n = 18.

Note. Apparently, known throughout as cultivated, sometimes escape or ecdemic plant. Several varieties are known; of these must be mentioned the extensively distributed 'border' form with yellowish-green leaves and 'velutinous' form with numerous ligulate flowers in heads. The ligulate flowers are often strongly reduced and are often absent (var. *flosculosum* DC.).

GENUS **69.** *BALSAMITA* Mill.
1754, Gard. 1754, Gard. Dict. Abridg., ed. 4

Heads heterogamous, 14–22 mm in dia, or homogamous (lacking peripheral ligulate flowers) and then 6–8 mm in dia, usually clustered in corymbose inflorescence, rather numerous. Involucre narrow-saucer-shaped, 3–5 mm long; involucral bracts imbricate. Receptacle weakly convex without bracts. Peripheral flowers ligulate, pistillate, white, often absent; disk flowers bisexual, tubular, with yellow five-toothed corolla. Achenes 1.2–2.0 mm long, 0.4–0.7 mm wide, ribbed-terete but narrowed toward base, with five to eight ribs and 0.1–0.3 mm long crown, without secretory canals and muciferous cells, but with punctate glands between ribs. Perennial plants, 30–120 cm high, with erect leafy stem and alternate, undivided, more or less toothed leaves.

Type: *B. major* Desf.

The genus includes four closely related species in southwestern Asia, and the Mediterranean; one of them is cultivated and found as an escape in many extratropical countries. Very close to the preceding genus, the last two sections of which, probably deserve to be related to this genus.

1. **B. major** Desf. 1792, Acta Soc. Hist. Nat. Paris, **1:** 3; Heyw. 1976, Fl. Europ. **4:** 171.—*Tanacetum balsamita* L. 1753, Sp. Pl.: 845.—*Pyrethrum tanacetum* DC. 1838, Prodr. **6:** 63.—*P. majus* (Desf.) Tzvel. 1961, Fl. SSSR, **26:** 198.

Heads without ligulate flowers, clustered in dense inflorescence.

Type: Mediterranean ("in Hetruria, Narbona").

Center (Upper Dnieper: Belorussia); *West* (Black Sea); *East* (Lower Don); *Crimea.*—Cultivated as an aromatic vegetable plant and sometimes found as an escape in habitations, by roadsides, in parks.—*General distribution*: Caucasus, Russian Central Asia; Central and Atlantic Europe, Mediterranean, Asia Minor, Iran.— 2n = 18, 54.

Note. Apparently evolved in cultivation. Its probable ancestor, *B. balsamitoides* (Sch. Bip.) Tzvel. comb. nova (=*Tanacetum balsamitoides* Sch. Bip. 1844, *Tanaceteen*: 51) with white ligulate flowers and less dense inflorescence, may be found in botanical gardens and parks in the south of the "Flora" territory.

198

GENUS 70. *TANACETUM* L.
1753, Sp. Pl.: 843; id. 1754, Gen. Pl., ed. 5: 366

Heads heterogamous, 4–15(30) mm in dia, often clustered in corymbose inflorescence, less often solitary or fewer at apices of
143 stem and its branches. Involucre usually cup-shaped, 3–8 mm long; involucral bracts imbricate. Receptacle weakly convex or flat, without bracts. Peripheral flowers pistillate, yellow, ligulate (but usually with short ligules), indistinctly ligulate, or tubular; disk flowers bisexual, tubular, with yellow, five-toothed corolla. Achenes 1.2–3.5 mm long, and 0.4–0.9 mm wide, ribbed-terete, but narrowed toward base, with 5–10(15) more or less raised ribs and 0.1–0.7 mm long crown, without secretory canals and muciferous cells, but usually with punctate glands. Perennial plants, 10–180 cm high, with erect, leafy stem and alternate, pinnatisect or bipinnatisect, less often pinnatilobate leaves.

Type: *T. vulgare* L.
About 50 species, distributed in countries of the Northern Hemisphere but predominantly in northeastern Europe, Russian Central Asia and southwestern Asia.
Literature: Muradyan, L.T. 1976. Sravnital'naya anatomiya obolochek semyanok predstavitelei roda *Tanacetum* s. 1. [Comparative anatomy of the testa of achenes of the members of the genus *Tanacetum* s. 1.]. *Biol. Zhurn. Armenii*, **29**, 8: 38–43.

1. Plant (30)40–150(170) cm high, usually without short vegetative shoots; cauline leaves numerous; heads 4–8 mm in dia, 10–100, clustered in rather dense corymbose inflorescences; peripheral flowers pistillate but tubular; achenes 1.2–2.0 mm long .... 2. **T. vulgare.**
+ Plants (6)10–40(50) cm high, usually with short vegetative shoots; cauline leaves less numerous; heads 4–25(30) mm in dia, 1–10(15), solitary or in corymbose or corymbose-paniculate inflorescences; peripheral flowers ligulate but often with very short ligules .... 2.
2. Involucre 8–18 mm in dia; involucral bracts with dark brown or brown, membranous border; ligules of peripheral flowers 3–7 mm long; achenes 2.4–3.5 mm long. Northern plant, covered with rather long simple hairs ....
.... 1. **T. bipinnatum.**
+ Involucre 4–12 mm in dia, involucral bracts with whitish, less often light brown, membranous border; ligules of peripheral flowers 1.5–3.0(4.0) mm long; achenes 1.5–2.8

mm long. Steppe or forest-steppe plants, more or less covered with short, bifurcate (but often mixed with simple) hairs to subglabrous .... 3.

3. All leaves to 6 mm wide, their primary lobes to 3 mm long; terminal lobes short, fleshy with inconspicuous apical spinule; involucre 4–7 mm in dia; ligules of peripheral flowers to 2 mm long, usually inconspicuous ....
.... 4. **T. santolina.**

+ At least lower leaves more than 7 mm wide, their primary lobes more than 5 mm long; terminal lobes longer (linear to lanceolate), more or less flat with distinct apical spinule; ligules of peripheral flowers 1–4 mm long .... 4.

4. Outer involucral bracts lanceolate, acute, with or without very narrow membranous border; involucre (5)7–10(12) mm in dia, more or less hairy to subglabrous plants ...
.... 3. **T. kittaryanum.**

+ Outer involucral bracts ovate or broadly lanceolate, subobtuse, with broad membranous border, fragile with fruits .... 5.

5. Leaves green, weakly hairy, often subglabrous; involucre 5–9 mm in dia; ligules of peripheral flowers 1–2 mm long. Plants of Crimea .... 6. **T. paczoskii.**

+ Leaves more or less grayish from rather dense pubescence
.... 6.

144   6. Involucre 4–7 mm in dia; heads (1)3–12(15) on longer stalk; ligules of peripheral flowers 1–2 mm long; basal leaves to 1.5 cm wide .... 5. **T. achilleifolium.**

+ Involucre 6–12 mm in dia; heads (1)2–10(12); ligules of peripheral flowers 1.8–3.0 mm long; basal leaves to 2 cm wide .... 7.

7. Stem and leaves appressed-hairy; heads in lax corymbose inflorescence on stalk 5(8) cm long ... 7. **T. millefolium.**

+ Stem and leaves more densely and partly squarrose-hairy; heads in dense corymbose inflorescence on stalk 2(4) cm long .... 8. **T. odessanum.**

Section 1. Omalotes (DC.) Tzvel. 1961, Fl. SSSR, **26:** 324.—
Omalotes DC. 1838, Prodr. **6:** 83.

Heads usually few, with peripheral ligulate flowers; receptacle strongly convex; achenes 2.4–3.5 mm long, with crown 0.3–0.8 mm long. Plants more or less covered with long simple hairs, 6–40 cm high, with short vegetative shoots and relatively fewer cauline leaves.

Type: *T. camphoratum* Less.

1. **T. bipinnatum** (L.) Sch. Bip. 1844, Tanaceteen: 48; Tzvel. 1961, Fl. SSSR, **26:** 324; Heyw. 1976, Fl. Europ. **4:** 170.—*Chrysanthemum bipinnatum* L. 1753, Sp. Pl.: 890.—*Pyrethrum bipinnatum* (L.) Willd. 1803, Sp. Pl. **3**, 3: 2160.

Type: Siberia ("in Sibiria").

Arctic (Arctic Europe); *North* (Karelia-Murmansk: southeast of Kola Peninsula; Dvina-Pechora: near the mouth of the North Dvina River; basins of the Mezen and Pechora rivers, the Urals); *Center* (Volga-Kama: central Urals).—On coastal sands and gravel-beds, coastal buffs, slopes of river terraces, inundated meadows, in lower reaches and alder groves.—*General distribution*: Western and Eastern Siberia (north), Arctic, Far East (Kolyma Upland); North America (Alaska).—2n = 54, 56.

Section 2. Tanacetum.

Heads usually numerous, small, with peripheral tubular but pistillate flowers; receptacle weakly convex; achenes 1.2–2.0 mm long, with crown 0.1–0.4 mm long. Plants more or less covered with short simple and bifurcate hairs, 30–170 cm high, usually without short vegetative shoots and with numerous cauline leaves.

Type: type species of genus.

2. **T. vulgare** L. 1753, Sp. Pl.: 844; Tzvel. 1961, Fl. SSSR, **26:** 326; Heyw. 1976, Fl. Europ. **4:** 170.—(Plate XVI, 1).

Type: Europe ("in Europae aggeribus").

Arctic (Arctic Europe: ecdemic); *North*; *Baltic*; *Center*; *West*; *East*; *Crimea*.—In inundated, less often dry meadows, forest glades and edges, on coastal sands and gravel-beds, coastal buffs, in scrublands, by roadsides.—*General distribution*: Caucasus, Western and Eastern Siberia, Far East (ecdemic), Russian Central Asia; Scandinavia, Central and Atlantic Europe, Mediterranean, Asia Minor, Iran, Dzhungaria-Kashgaria, Mongolia, Japan-China; North America.—2n = 18.

Note. In cultivation there is a variety: var. *crispum* L. (loc. cit.: 845) with crispate leaves. The populations from the basin of

145   Plate XVI.

1—*Tanacetum vulgare* L.: 1a—peripheral flower, 1b—inner flower, 1c—achene; 2—*T. millefolium* (L.) Tzvel., 2a—lower part of plant, 2b—head, 2c—ligulate flower, 2d—tubular flower; 3—*Dendranthema zawadskii* (Herbich) Tzvel., 3a—ligulate flower, 3b—tubular flower, 3c—achene; 4—*Arctanthemum arcticum* (L.) Tzvel. subsp. *polare* (Hult.) Tzvel., 4a—ligulate flower, 4b—tubular flower, 4c—achene.

202

the North Sosva River (Urals) have a darker membranous border on the involucral bracts, which brings them closer to the Siberian species *T. boreale* Fisch. ex DC.

Section 3. Xanthoglossa (DC.) Sch. Bip. 1844, Tanaceteen: 48; Tzvel. 1961, Fl. SSSR, **26:** 330.—*Pyrethrum* sect. *Xanthoglossa* DC. 1838, Prodr. **6:** 60.

Heads relatively less numerous, usually small, with peripheral ligulate, but often very small, flowers; receptacle weakly, less often strongly convex; achenes 1.5–3.5 mm long, with crown 0.1–0.7 mm long. Plants more or less covered with bifurcate hairs 146 (often mixed with simple hairs), 6–40 cm high, usually with short vegetative shoots and less numerous cauline leaves.

Lectotype: *T. millefolium* (L.) Tzvel.

3. **T. kittaryanum** (C.A. Mey.) Tzvel. 1961, Fl. SSSR, **26:** 345.—*T. millefolium* auct. non (L.) Tzvel.: Heyw. 1976, Fl. Europ. **4:** 170, p. p.

Type: Orenburg Region and Bashkiria ("in collibus prov. Orenburg, ab urbe Ufa occidentem versus prope pag, Kilimova").

a. Subsp. **kittaryanum.**—Leaves grayish-green from rather abundant long-persisting pubescence, not stiff; heads one to six (eight) on rather long (usually 5–10 cm) stalk, forming lax and not always regular corymbose inflorescence.

Center (Volga-Kama: southeast; Volga-Don: east); *East* (Trans-Volga).—In steppes, on steppe slopes, in steppefied forest meadows.—*General distribution*: Western Siberia (south), Russian Central Asia (north).

b. Subsp. **uralense** (Krasch.) Tzvel. comb. nova.—*Pyrethrum uralense* Krasch., 1946, Bot. Mat. (Leningrad), **9:** 162, fig. 4.—*Chrysanthemum uralense* Krasch. 1936, Fl. Yugo-Vost. Evrop. Chasti SSSR, **6:** 347, descr. ross.—*Tanacetum uralense* (Krasch.) Tzvel. 1961, Fl. SSSR, **26:** 346; Heyw. 1976, Fl. Europ. **4:** 170, in adnot.—Leaves initially more or less grayish-green, later green, scatteredly hairy to subglabrous, relatively stiff; heads one to eight on shorter (usually 2–7 cm long) stalk, forming more regular corymbose inflorescences.

Type: Southern Urals ("the village of Andreevka, Orenburg District, stony peaks and slopes of mountains").

Center (Volga-Kama: southeast; Volga-Don: east); *East* (Trans-Volga).—On stony slopes and in stony steppes.—*General distribution*: Western Siberia (south), Russian Central Asia (northwest).

O c. Subsp. **sclerophyllum** (Krasch.) Tzvel. comb. nova.—*Pyrethrum sclerophyllum* Krasch. 1946, Bot. Mat. (Leningrad),

9: 164, fig. 5.—*Tanacetum sclerophyllum* (Krasch.) Tzvel. 1961, Fl. SSSR, **26**: 347; Heyw. 1976. Fl. Europ. **4**: 170, in adnot.— Leaves initially more or less pubescent, grayish-green, later green, often subglabrous, stiff; heads (one)two to four(six) on 3–10 cm long stalk, clustered in somewhat lax corymbose inflorescence, on the average larger (9–12 mm in dia).

Type: Saratov Region ("vicinity of Khvalynsk, southern slopes of chalk outcrops of Mt. Kalancha").

Center (Volga-Don: east); *East* (Lower Don: east; Trans-Volga: west).—On chalk and limestone outcrops.—Endemic.

4. **T. santolina** Winkl. 1891, Tr. Peterb. Bot. Sada, **11**, 12: 375; Tzvel. 1961, Fl. SSSR, **26**: 343; Heyw. 1976, Fl. Europ. **4**: 170.

Type: Kazakhstan, Kyzyl-Orda Region ("in valle fluminis Syrdaria non procul a pago Karmatschu").

East (Lower Volga: south and east).—On solonetzes in sandy and clayey semideserts.—*General distribution*: Western Siberia (south), Russian Central Asia; Dzhungaria-Kashgaria.

5. **T. achilleifolium** (Bieb.) Sch. Bip. 1844, Tanaceteen: 47; Tzvel. 1961, Fl. SSSR, **26**: 351; Heyw. 1976, Fl. Europ. **4**: 170.— *Pyrethrum achilleifolium* Bieb. 1808, Fl. Taur.-Cauc. **2**: 327; id. 1819, ibid. **3**: 580.

Type: Basin of the Don and lower reaches of the Volga River ("Ex campis apricis ad Tanain et Wolgam inferiorem").

Center (Volga-Don: southeast); *West* (Black Sea); *East* (Lower Don; Trans-Volga: south; Lower Volga); *Crimea*.—In alkaline steppes and semideserts, on solonetzes.—*General distribution*: Caucasus (Ciscaucasia), Western Siberia (south), Russian Central Asia (north); reported for Central Europe (southeast).

O 6. **T. paczoskii** (Zefir.) Tzvel. 1961, Fl. SSSR, **26**: 349; Heyw. 1976, Fl. Europ. **4**: 170.—*Pyrethrum paczoskii* Zefir. 1957, Bot. Mat. (Leningrad), **18**: 255.

147 Type: Crimea ("Aibary, Kamennaya steppe").

Crimea (west and northern foothills of yailas).—On stony and rubbly slopes, in steppes.—Endemic.

7. **H. millefolium** (L.) Tzvel. 1961, Fl. SSSR, **26**: 348; Heyw. 1976, Fl. Europ. **4**: 170, excl. syn.—*Anthemis millefolia* L. 1753, Sp. Pl.: 896.—*Chrysanthemum millefolium* L. 1767, Syst. Veg., ed. 12: 563.—*Pyrethrum millefoliatium* (L.) Willd. 1803, Sp. Pl., **3**, 3: 2160.—*Tanacetum tauricum* Sch. Bip. 1844, Tanaceteen: 48.— *Pyrethrum baumanii* Stev. 1856, Bull. Soc. Nat. Moscou,

204

29, 4: 383, p. p.; Tzvel. 1964, Novost Sist. Vyssh. Rast. 1964: 312.—(Plate XVI, 2).

Type: Garden plants of obscure origin, probably from the south of the European part of Russia.

Center (Volga-Don: east); *West* (Dnieper: southeast; Moldavia: southeast; Black Sea); *East* (Lower Don; Trans-Volga: south; Lower Volga: west and north); *Crimea.*—In steppes, steppefied forest glades, on outcrops of chalk and limestone.—*General distribution*: Caucasus (north), Western Siberia (southwest), Russian Central Asia (northwest).

○ 8. **T. odessanum** (Klok.) Tzvel. 1961, Fl. SSSR, **26**: 348; Heyw. 1976, Fl. Europ. **4**: 170, in adnot.—*Pyrethrum odessanum* Klok. 1926, Index Sem. Hort. Bot. Charkow: 7; id. 1950, Vizn. Rosl. URSR: 540.

Type: Odessa ("Odessa, herb. Ledebour").

West (Moldavia: southeast; Black Sea).—In stony steppes, on outcrops of limestone.—Endemic.

GENUS 71. *LEUCANTHEMOPSIS* (Giroux) Heyw.
1975, Anal. Inst. Bot. Cavanilles, **32**, 2: 182; id. 1976, Fl. Europ. **4**: 172.—*Tanacetum* L. subsect. *Leucanthemopsis* Giroux, 1933, Bull. Soc. Hist. Nat. Afr. Nord. 24: 54

Heads heterogamous, 20–40 mm in dia, solitary. Involucre saucer-shaped, 3.5–6.0 mm long; involucral bracts imbricate, with wide blackish-brown membranous border. Receptacle weakly convex, without bracts. Peripheral flowers ligulate, pistillate, white, rarely yellow or pinkish; disk flowers bisexual, tubular with yellow, five-toothed corolla. Achenes 2–3 mm long and 0.7–1.0 mm wide, cuneate-terete, with 3–10 weak ribs and crown 0.5–1.0 mm long, without secretory canals, but with rows of muciferous cells along ribs. Perennial plants, 5–20 cm high, with usually erect stems and alternate, predominantly basal, pinnatilobate or pinnatipartite, leaves on rather long petiole.

Type: *L. alpina* (L.) Heyw.

The genus includes 5–10 species, distributed in the mountainous regions of Europe, predominantly on the Iberian Peninsula.

1. **L. alpina** (L.) Heyw. 1975, Anal. Hist. Bot. Cavanilles, **32**, 2: 182; id. 1976, Fl. Europ. **4**: 172.—*Chrysanthemum alpinum* L. 1753, Sp. Pl.: 889.—*Pyrethrum alpinum* (L.) Schrank, 1792, Prim.

Fl. Salisb.: 215; Tzvel. 1961, Fl. SSSR, **26**: 228; Dobrocz. 1987, Opred. Vyssh. Rast. Ukr.: 338.

Ligules of peripheral flowers white, 7–15 mm long; basal and lower cauline leaves more or less ovate, cuneately narrowed in petiole, pinnatilobate, with two to five lobes on each side.

Type: Switzerland ("in alpibus Helvetiae ad Thermas Piperinas").

West (Carpathians).—On stony slopes, in cultivated meadows and on rocks in upper mountain zone.—*General distribution*: Central Europe, Mediterranean.—2n = 18, 36.

Note. This species is reported for the Carpathians within the Soviet Union. However, I have not seen its plants from this area. The Carpathian plants of *L. alpina* s. l., apparently relate to a separate subspecies, subsp. *tatrae* (Vierh.) Tzvel. or to *L. tatrae* (Vierh.) Holub.

148 GENUS **72. *ARCTANTHEMUM*** (Tzvel.) Tzvel. 1985, Novosti Sist. Vyssh. Rast. **22**: 274.—*Dendranthema* (DC.) Des Moul. sect. *Arctanthemum* Tzvel. 1961, Fl. SSSR, **26**: 879, 384

Heads heterogamous, 28–60 mm in dia, solitary at apices of stem and its branches. Involucre saucer-shaped, 4–8 mm long; involucral bracts imbricate. Receptacle somewhat convex, without bracts. Peripheral flowers ligulate, pistillate, white; disk flowers bisexual, tubular, with yellow, five-toothed corolla. Apical appendages of anthers oblong, broadly rounded at apex. Achenes 1.8–2.6 mm long and about 0.5 mm wide, more or less terete, but narrowed toward base, with five to eight more or less distinct ribs, without crown and secretory canals as also muciferous cells. Perennial plants, 5–30 cm high, with erect or ascending stems and alternate leaves approximate in lower part of shoots; leaves more or less deeply pinnatilobate, less often toothed, on rather long petiole.

Type: *A. arcticum* (L.) Tzvel.

A monotypic genus.

1. **A. arcticum** (L.) Tzvel. 1985, Novosti Sist. Vyssh. Rast. **22**: 274.—*Chrysanthemum arcticum* L. 1753, Sp. Pl.: 889.—*Leucanthemum arcticum* (L.) DC. 1838, Prodr. **6**: 45, quoad. nom.—*Tanacetum arcticum* (L.) Sch. Bip. 1844, Tanaceteen: 35.—*Pyrethrum arcticum* (L.) Stank. 1949, in Stank. and Taliev, Opred. Rast. Evrop. Chasti SSSR: 630.—*Dendranthema arcticum* (L.) Tzvel. 1961, Fl. SSSR, **26**: 386; Heyw. 1976, Fl. Europ. **4**: 169.

Type: Kamchatka and North America ("in Kamtschatka, America Septentrionali").

a. Subsp. **polare** (Hult.) Tzvel. 1987, Arkt. Fl. SSSR, **10**: 116, map 39.—*Chrysanthemum arcticum* L. subsp. *polaris* Hult. 1949, Svensk. Bot. Tidskr. **43**: 776.—*Leucanthemum hultenii* A. et D. Löve, 1961, Bot. Not. (Lund), 1961: 44.—*Dendranthema hultenii* (A. et D. Löve) Tzvel. 1961, Fl. SSSR, **26**: 387.—*D. arcticum* (L.) Tzvel. subsp. *polare* (Hult.) Heyw. 1975, Journ. Linn. Soc. London (Bot.) **71**, 4: 272; id. 1976, Fl. Europ. **4**: 169.—*Arctanthemum hultenii* (A. et D. Löve) Tzvel. 1985, Novosti Sist. Vyssh. Rast. **22**: 274.—Short (5–20 cm high) plant, with less incised (usually three-toothed or shallowly three-lobed leaves).—(Plate XVI, 4).

Type: Northeastern Europe.

Arctic (Arctic Europe); *North* (Karelia-Murmansk: reported for vicinity of Kandalaksha).—In solonetz meadows, on sands and gravel-beds of the seacoast and in river mouths.—*General distribution*: Arctic, Far East (Okhotsk seacoast); North America (coast of Hudson Bay).—2n = 18.

Note. The type subspecies—subsp. *arcticum*—is distributed in the northern part of the Pacific Coast of Asia and North America.

GENUS 73. *DENDRANTHEMA* (DC.) Des Moul.

1855, Actes Soc. Linn. Bordeaux, **20**: 561, s. str.; Tzvel. 1961, Fl. SSSR, **26**: 364.—*Pyrethrum* Zinn. sect. *Dendranthema* DC. 1838, Prodr. **6**: 62

Heads heterogamous, 25–150 mm in dia, solitary at apices of stem and its branches. Involucre saucer-shaped, 4.5–20.0 mm long; involucral bracts imbricate. Receptacle somewhat convex, without bracts. Peripheral flowers (in "double" garden forms) ligulate, pistillate, yellow, pink, or white, of most diverse colors in cultivated varieties; disk flowers bisexual, tubular, with yellow, five-toothed corolla. Upper appendages of anthers lanceolate-ovate, subobtuse. Achenes 1.8–3.0 mm long and 0.5–0.6 mm wide, more or less terete but narrowed toward base, with indistinct ribs, without crown and secretory canals but with rows of 149 muciferous cells, becoming slippery on wetting. Perennial herbaceous plants or semishrubs, with more or less erect stems and alternate pinnatipartite, pinnatilobate or more or less toothed leaves on rather long petiole.

Type: *D. indicum* (L.) Des Moul.

More than 50 species, predominantly in eastern Asia, one species reaches the west to the Carpathians and two others are extensively cultivated as ornamental plants.

1. Wild herbaceous plant, 15–50 cm high; involucre 10–20 mm in dia; ligulate flowers pink or white, usually in one whorl .... 1. **D. zawadskii.**

+ Cultivated and occasionally escape plant, 25–150 cm high; involucre 8–40 mm in dia; ligulate flowers of various colors, usually in several whorls, often entirely replacing tubular disk flowers (heads "double") .... 2.

2. Heads 25–50 mm in dia, usually numerous; involucre 8–20 mm in dia; ligulate flowers often yellow ....

.... 2. **D. indicum.**

+ Heads 50–150 mm in dia, less numerous; involucre 20–40 mm in dia; ligulate flowers of various colors less often yellow .... 3. **D. morifolium.**

1. **D. zawadskii** (Herbich) Tzvel. 1961, Fl. SSSR, **26**: 376; Heyw. 1976, Fl. Europ. **4**: 69; Dobrocz. 1987, Opred. Vyssh. Rast. Ukr.: 339.—*Chrysanthemum zawadskii* Herbich, 1831, Addit. Fl. Galic.: 44, tab. 1.—*Tanacetum alaunicum* K.-Pol. 1930, Delect. Sem. Hort. Bot. Univ. Voronez. **2**: 30.—*Chrysanthemum arcticum* L. subsp. *alaunicum* (K.-Pol.) K.-Pol. 1931, 25-let Nauch.-pedag. i Obshch. Deyat. B.A. Keldera: 318.—*Tanacetum zawadskii* (Herbich) B. Pawl. 1934, Ochron. Przyr.: 14.—*Chrysanthemum kozo-poljanskii* Golitz, 1949, Byull. Obshch. Estestvoisp. Voronezh. Univ. **5**: 21, descr. ross.—*Leucanthemum alaunicum* (K.-Pol.) Worosch. 1953, Spisok. Sem. Glavn. Bot. Sada Akad. Nauk SSSR, **8**: 6.—*L. sibiricum* DC. subsp. *alaunicum* (K.-Pol.) Golitz, 1954, in Majevskii, Fl. Sredn. Pol. Evrop. Chasti SSSR, ed. 8: 581, nom. illeg.—*L. sibiricum* subsp. *kozo-poljanskii* Golitz, 1954, op. cit.: 581, descr. ross.—(Plate XVI, 3).

Type: Carpathians ("in Galiciae cacuminibus montium Pienninorum").

North (Dvina-Pechora: basin of the Pinega and Ural rivers); *Center* (Volga-Kama; Volga-Don: on the Don and Oskola rivers); *West* (Carpathians: reported); *East* (Trans-Volga: southern Urals).— On outcrops of limestone and chalk in pine and deciduous forests, in forest glades and edges.—*General distribution*: Western and Eastern Siberia, Far East; Central Europe (Carpathians), Mongolia (north), Japan-China (north),—2n = 54.

208

Note. Separation of chalk populations of this species from Kursk, Belgorod, and Voronezh regions in separate taxa is not fully justified.

2. **D. indicum** (L.) Des Moul. 1855, Actes Soc. Linn. Bordeaux, **20**: 561; Tzvel. 1961, Fl. SSSR, **26**: 371; Heyw. 1976, Fl. Europ. **4**: 169.—*Chrysanthemum indicum* L. 1753, Sp. Pl.: 889.

Type: India ("in India").

West (Moldavia; Black Sea); *East* (Lower Don); northward as green house and indoor plant, often planted in gardens and parks in summer.—Extensively cultivated as ornamental plant.— *General distribution*: Japan-China; cultivated and as an escape in many other countries.—2n = 36, 54.

3. **D. morifolium** (Ramat.) Tzvel. 1961, Fl. SSSR, **26**: 373; Heyw. 1976, Fl. Europ. **4**: 169.—*Chrysanthemum morifolium* Ramat. 1792, Journ. Hist. Nat. (Paris), **2**: 240.—*C. sinense* Sabine, 1823, Trans. Linn. Soc. **14**: 142.—*Dendranthema sinense* (Sabine) Des Moul. 1855, Actes Soc. Linn. Bordeaux, **20**: 562.

Type: Cultivated plant, originating from China or Japan.

Same as the preceding species.—Widely cultivated as an ornamental plant.—*General distribution*: Evolved in cultivation 150 as a result of hybridization of several species and prolonged selection; cultivated in many other countries and occasionally found as an escape.—2n = 42, 54.

GENUS **74.** *ARTEMISIA* L.[1]
1753, Sp. Pl.: 845; id. 1754, Gen. Pl., ed. 5: 357

Heads numerous, small, cylindrical and narrow-ovate to globose and saucer-shaped, in paniculate, racemose, less often spicate or subcapitate inflorescences. Involucral bracts herbaceous or coriaceous-herbaceous, imbricate in two to six irregular rows of almost similar length or outer considerably shorter than inner. Receptacle conical or hemispherical, less often almost flat, punctate-alveolate, glabrous or more or less hairy, without bracts. Flowers tubular, bisexual or peripheral pistillate in one whorl; disk flowers more numerous, bisexual or staminate with two- to five-toothed corolla, almost colorless or yellow, rarely reddish-violet. Apical appendage of anthers lanceolate or oblong-lanceolate. Achenes small, obovate or oblong-ovate, subcylindrical or more

---

[1]Treatment by T.G. Leonova.

or less flattened, glabrous with few longitudinal veins, without crown, very rarely with rudimentary crown. Perennial, biennial, and annual herbs, semishrubs, and semishrublets. Leaves alternate, more or less incised, less often undivided and entire.

Lectotype: *A. vulgaris* L.

About 400 species, distributed predominantly in warm temperate zone of the Northern Hemisphere but some entering the Arctic and subtropics.

Literature: Krascheninnikov, I.M. 1946. Opyt filogeneticheskogo analiza nekotorykh evraziatskikh grupp roda *Artemisia* L. v svyazi s osobennostyami paleogeografii Evrazii [An attempt at phylogenetic analysis of some Eurasian groups of the genus *Artemisia* L. in connection with special features of paleogeography of Eurasia]. *Mat. Ist. Fl. Rastit. SSSR*, **2**: 87–196.—Monoszon, M.Kh. 1950. Opisanie pyltsy vidov polynei proizrastayuschikh na territorii SSSR [Description of pollen of the wormwood species growing in the territory of the USSR]. *Tr. Inst. Geogr. Akad. Nauk SSSR*, **46**: 271–360.—Krascheninnikov, I.M. 1958. Rol' i znachenie Angarskogo floristicheskogo tsentra v filogeneticheskom razvitii osnovnykh evraziatskikh grupp polynei podroda *Euartemisia* [The role and significance of the Angara floristic center in the phylogenetic development of the main Eurasian groups of wormwoods of the subgenus *Euartemisia*]. *Mat. Ist. Fl. Rastit. SSSR*, **3**: 62–128.—Poljarkov, P.P. 1961. Materialy k sistematike roda polyn—*Artemisia* L. [Materials on the systematics of the genus of wormwoods—*Artemisia* L.]. *Tr. Inst. Bot. Akad. Nauk KazSSR*, **11**: 134–177.—Ehrendorfer, F. 1964. Notizen zur Cytotaxonomie und Evolution der Gattung *Artemisia*. *Österr. Bot. Zeitschr.* **111**, 1: 84–142.—Singh, G. and R.D. Joshi, 1969. Pollen morphology of some Eurasian species of *Artemisia*. *Grana Palynol.* **9**, 1–3: 50–62.—Persson, K.1974. Biosystematic studies in the *Artemisia maritima* complex in Europe. *Opera Bot. (Lund)*, **35**: 1–188.—Leonova, T.G. 1987. Konspekt roda *Artemisia* L. (Asteraceae) flory evropeiskoi chasti SSSR [Conspectus of the genus *Artemisia* (Asteraceae) in the flora of the European part of the USSR]. *Novosti Sist. Vyssh. Rast.*, **24**: 177–201.—Leonova, T.G. 1988. Klyuch dlya opredeleniya vidov r. *Artemisia* L. (Asteraceae) evropeiskoi chasti SSSR [Key to the identification of species of the genus *Artemisia* L. (Asteraceae) in the European part of the USSR]. *Novosti Sist. Vyssh. Rast.*, **25**: 137–142.—Mosyakin, S.L. 1990. New and noteworthy alien species of *Artemisia* L. (Asteraceae) in the Ukrainian SSR. *Ukr. Bot. Zhurn.*, **47**: 10–13.

210

1. All leaves entire, linear to lanceolate and oblong, entire, rarely lower and middle more or less deeply three-lobed in upper part .... 2.
+ Lower and usually middle cauline leaves more or less pinnatisect, pinnatipartite or ternate and palmately divided ... 3.
2. Plant green, more or less pubescent with stellate hairs, later glabrous or subglabrous; heads 2–4 mm wide ......
.... 28. **A. dracunculus.**
+ Plant grayish from dense tomentum of short stellate hairs; heads 1.5–2.5 mm wide ....29. **A. glauca.**
3. Leaves bicolorous, green and glabrous or sparsely pubescent above, whitish or grayish arachnoid-tomentose beneath .... 4.
+ Leaves more or less concolorous, green on both sides, glabrous or sparsely hairy or grayish, silvery, less often whitish from dense pubescence .... 13.
4. Semishrublets with stems woody in lower part, densely leafy from middle and above; leaves with narrow lateral segments and small intermediate. undivided lobes between them; heads nutant, hemispherical or saucer-like .... 5.
+ Herbaceous perennials with rhizome; stem more or less densely leafy throughout; leaves with broad lateral segments, without intermediate lobes; heads erect, oblong to hemispherical .... 6.
5. Leaves 4–8 cm long, 2–7 cm wide, with distinct intermediate lobes, glabrous or subglabrous above, white-tomentose beneath, terminal lobes lanceolate, 1.0–1.5 mm wide, scarcely pointed; heads 4–8 mm wide, in narrow and lax raceme or panicle branched only in lower part ....
.... 11. **A. macrantha.**
+ Leaves 1.5–5.0 cm long, 0.9–3.0 cm wide, with inconspicuous intermediate lobes, weakly grayish-pubescent above, densely grayish-tomentose beneath, terminal lobes filiform-linear, 0.5–0.9 mm wide, acute; heads 2.5–4.5 mm wide, in more or less broad and dense panicle ... 10. **A. pontica.**
6. Heads oblong or narrow-campanulate, 1.5–3.0 mm wide, loosely and uniformly borne on all branches of panicle; entire plant weakly hairy; leaves with or without auricles at base of petiole .... 7.
+ Heads hemispherical or broadly campanulate, 4–7(8) mm wide, densely borne on short branches of panicle or in raceme, initially subcapitate; entire plant, particularly

151

young, with dense whitish pubescence; leaves without auricles .... 12.

7. Leaves bi- or tri-pinnatisect with wide, primary segments always incised, 2.5–9.0 cm wide; auricles at base of petiole distinct, with one to five pairs of lobes. Widely distributed plant, with more or less many-headed caudex at base, rarely with short stolons; heads on 0.5–1.0 mm long stalk .... 1. **A. vulgaris.**

+ Usually only lower leaves bipinnatisect, middle cauline leaves pinnatisect with undivided segments, rarely few of them bipinnatisect, upper leaves often undivided. Rather rare adventive plant, with long horizontal rhizome, usually producing large groups of distant shoots .... 8.

8. Leaf segments with short and acute teeth .... 3. **A. selengensis.**

+ Leaf segments entire, less often with fewer distant teeth .... 9.

9. Leaves glabrous and green above, somewhat lustrous, rarely weakly pubescent in beginning of leaf growth; involucre moderately pubescent to subglabrous; cauline leaves basally with auricles of one to three pairs of lobes ..... 10.

+ Leaves pubescent above at least in the beginning, dull green to grayish-green; involucre rather densely hairy; cauline leaves without auricles or with auricles of one(two) pairs of small lobes .... 11.

10. Heads 3–5 mm long and 2.5–4.5 mm wide, broadly campanulate; general inflorescence compact and relatively narrow .... 2. **A. verlotiorum.**

+ Heads 2–3(3.5) mm long and 1.5–2.0 mm wide, campanulate, clustered in broadly paniculate inflorescence .... 4. **A. rubripes.**

11. Segments of middle cauline leaves linear to linear-lanceolate; heads 3.0–4.5(5.0) mm long and 2.5–3.5(4.0) mm wide .... 5. **A. umbrosa.**

152

+ Segments of middle cauline leaves lanceolate to lanceolate-ovate; heads 2.5–3.5 mm long, and 2.0–2.5 mm wide ... . 6. **A. argyi.**

12. Stem 10–35(50) cm high; leaves narrow- or broadly elliptical, 2–7(11) cm long, 1–3(4) cm wide, pinnatisect, rarely weakly and distantly bipinnatisect, with more or less long cuneate base, upper and middle cauline leaves sessile, lower on petiole about 2.5 cm long; primary segments of leaves

212

undivided and entire, linear or linear-lanceolate, sessile; terminal lobules (if present) acute .... 7. **A. tilesii.**

+ Stem 35–65 cm high; leaves broadly ovate, suborbicular, less often elliptical, 5–8 cm long and 4–7 cm wide, bi- or tri-pinnatisect with round base, others petiolate, petiole 3–5 mm long; primary segments of leaves incised, broadly rhombic, petiolate, terminal lobe of lower leaves orbicular, others short-acuminate .... 8. **A. leucophylla.**

13(3). Heads hemispherical or saucer-shaped, many-flowered (20–85), less often broadly campanulate, mostly nutant, in lax inflorescence .... 14.

+ Heads ovate or obovate, less often broadly campanulate, few-flowered (1–10, less often 15, flowers), erect, less often deflected, in dense inflorescence........................ 30.

14. Plant white-tomentose or silver-grayish throughout vegetative period from dense appressed sericeous hairs; receptacle hairy... 15.

+ Plant glabrous at flowering, green or with sericeous pubescence; receptacle glabrous, less often hairy .... 20.

15. Plants whitish tomentose, stems more or less woody at base, 15–35 cm high, sparsely leafy; short vegetative shoots numerous, strongly clustered; leaves bipinnatisect or pinnatisect, 2.5–6.0 cm long, 2.0–2.5 cm wide, without auricles at base of petiole, terminal lobes oblong, 2–7 mm long, 1 mm wide, obtuse; heads 3.0–4.5 mm wide, in lax and spreading panicle; branchlets glabrous....................
.... 26. **A. hololeuca.**

+ Plants silver-grayish, more or less densely leafy; leaves bi- or tri-pinnatisect or ternate .... 16.

16. Plants pulvinate, with less numerous stems, 6–27 cm high; leaves pinnatisect, bipinnatisect or tripinnatisect, 1–2(4) cm long, lower with 0.1–0.8(1.6) cm long petiole; cauline leaves sessile, auriculate; terminal lobes oblong to linear, 3–13 mm long and 0.7–1.5 mm wide; heads 4–6 mm wide, in narrow and sparse raceme; corolla in upper part lanate-tomentose .... 22. **A. caucasica.**

+ Plants not pulvinate; heads in branched, more or less spreading panicle, less often raceme .... 17.

17. Herbs with ribbed, nonwoody stems; leaves bi- or tri-pinnatisect, 1.5–15.0 cm long, 2–12 cm wide, lateral segments not incised to midrib, terminal lobes oblong-ovate or oblong, 1–10 mm long .... 18.

+ Semishrublets; stems woody in lower part, cylindrical; leaves bipinnatisect or ternate, 2.5–3.5 cm long and 2.0–3.5 cm wide; lateral segments incised to midrib; terminal lobes linear to oblong-elliptical, (1.5)4.0–23.0 mm long .... 19.

18. Perennials, 40–125 cm high, with relatively thick root; leaves broadly ovate, without intermediate lobes; middle cauline leaves without auricles; heads 2.5–4.0 mm wide, in dense panicle .... 20. **A. absinthium.**

153 + Biennials, less often annuals, 30–100 cm high, with slender root; leaves broadly deltoid, sometimes with smaller lobes in upper part; middle and lower cauline leaves with linear auricles at base; heads 4–6(7) mm wide, in lax panicles .... 21. **A. sieversiana.**

19. Plant grayish-sericeous, with numerous branches and stems, 15–35 cm high; leaves thin, bipinnatisect, 0.6–2.2 cm long, 0.3–1.2(2.0) cm wide, orbicular-broadly ovate or orbicular-broadly rhombic, with petiole 0.5–1.0 cm long; terminal lobes 1.5–5.0 mm long, 0.3–1.0 mm wide; heads 2–4(6) mm wide, on stalk 1–5 mm long; corolla glabrous .... 23. **A. frigida.**

+ Plant grayish-sericeous, with isolated or fewer branches and stems, 20–80 cm high; leaves subcoriaceous, biternate or pinnatisect, less often ternate, 2.5–3.5 cm long, 2.0–3.5(6.0) cm wide, broadly obovate to suborbicular, less often elliptical; cauline leaves mostly sessile; terminal lobes 4–23 mm long, 1–3 mm wide; heads 6–8 mm wide, on stalk 2–20 mm long; corolla densely hairy................ .... 25. **A. sericea.**

20(14). Semishrubs, with stem woody to considerable height or semishrublets with stems woody in lower part .... 21..

+ Herbaceous perennials (often with thick woody root), less often annuals or biennials.................................... 22.

21. Stems 50–150 cm high, to 15 mm thick, woody to considerable height; leaves bi- or tri-pinnatisect, 4–8 cm long, 3–6 cm wide, without intermediate lobes; petioles 1.5–4.5 cm long, terminal lobes filiform-linear, (3)5–22 mm long; heads broadly campanulate or hemispherical, 2–3 mm wide, in dense panicles; corolla in upper part not densely hairy... 9. **A. abrotanum.**

+ Stems (7) 10–50 (80) cm high, 1.5–6.0 mm thick, woody only in lower part; leaves tripinnatisect (some leaves bipinnatisect), 2–5 cm long, 1.5–2.5 cm wide, with intermediate lobes in upper half; petioles 1.0–2.5 cm long; terminal lobes lanceolate, 1–4 mm long; heads hemispherical, 3–5(6) mm wide, in narrow racemes or panicle; corolla glabrous .... 12. **A. santolinifolia.**

22. Leaves ovate to obovate and broadly elliptical, 1.3–5.0 cm long, 0.7–2.0 cm wide, lower leaves bi- or tri-, middle and upper cristately pinnatisect, along petiole (less often also in upper part of leaf) usually with small undivided lobes [auricles] 2–4 mm long; petioles 0.5–1.0 mm long; terminal lobes elliptical to linear, 1–4(8) mm long; heads hemispherical, 6–8 mm wide, on stalk 1–2(5) mm long, in racemes or panicle branched only in lower part; involucral bracts with long-ciliate, blackish, membranous margin; receptacle densely hairy ....27. **A. rupestris.**

+ Leaves with auricles, not cristate; involucral bracts without cilia on margin; receptacle glabrous ....23.

23. Heads usually three to eight, in spreading or capitate raceme or panicle branched only in lower part; leaves 2–10 cm long, 0.5–3.5 cm wide, with one to three lateral segments on each side, without reduced lobes; vegetative shoots numerous, forming sod......................................24.

+ Heads numerous, in more or less dense panicle; leaves 4–25 cm long, 2–8 cm wide, with four to eight sometimes petiolulate, lateral segments on each side, often with intermediate lobes in upper part; vegetative shoots less numerous, not forming sod ....25.

154   24. Heads three to eight, saucer-shaped, green, as also stems, 7–15 mm wide on nutant stalk (1) 4–65 mm long, in lax racemes; involucral bracts with blackish membranous margin; disk flowers bisexual; leaves pinnatisect ....
.... 17. **A. norvegica.**

+ Heads in large numbers, subglobose, often violet, as also stem, 3.5–6.0 mm wide, sessile or on stalk 1–5 mm long, erect in capitate or sparse racemes or panicle branching only in lower part; involucral bracts with light-colored membranous margin; disk flowers staminate; leaves ternate or pinnatisect, middle leaves sometimes undivided, linear ....38. **A. borealis.**

25. Annual or biennial, glabrous, densely leafy plant; heads in dense panicle ....26.

+ Perennial, mostly pubescent, sparsely leafy plant; heads in racemes or panicle branched only in lower part ...27.

26. Leaves elliptical-oblong, 7–16 cm long and 4–6 cm wide, lateral segments five to seven on each side, with petioles 2–5 mm long wingless or weakly winged; terminal lobes gradually pointed; heads broadly ovate, about 2 mm wide, on upward directed stalk 1 mm long, in dense and narrow, sparsely leafy panicle ....
.... 19. **A. tournefortiana.**

+ Leaves broadly ovate, (1.5)2.5–11.0 cm long, (0.7)2.5–4.0 cm wide; lateral segments two or three(five) on each side, with broadly winged petiole 1–2 mm long; terminal lobes abruptly narrowed; heads spherical, 2–4 mm wide, on stalk 1–3 mm long, nutant, in broad, lax, densely leafy panicle .... 18. **A. annua.**

27. Leaves 5–37 cm long, 2–8 cm wide, more or less densely (particularly beneath) pubescent, with petiole 3–20 cm long; primary segments incised to lateral veins, petiolulate, often with reduced intermediate lobes along petiolule; terminal lobes one to three, asymmetric, 1–5 mm long, 0.5–1.0 mm wide, oblong to linear, with long cartilaginous cusp .... 16. **A. laciniata.**

+ Primary segments of leaves deeply incised but not to lateral veins, decurrent on midrib, without petiolule; terminal lobes (2)4–10, symmetric, 1.0–1.5 mm long, lanceolate .... 28.

28. Leaves glabrous or subglabrous, 4–22(23) cm long, (2.0)3.5–11.0(16.0) cm wide, with winged midrib and petiole 4–15 mm long; heads 3–5 mm wide, on stalk 1–3 mm long; involucral bracts with broad colorless membranous margin, glabrous, as also corolla .... 14. **A. latifolia.**

+ Leaves more or less densely pubescent, particularly beneath, with wingless, less often weakly winged midrib; heads on stalk (2)3–30 mm long; involucral bracts and corolla hairy .... 29.

29. Plants (particularly upper part of stems and leaves beneath) densely grayish-sericeous from long appressed hairs; leaves 10–20(36) cm long, 3–8(10) cm wide, with petiole 5–12(24) cm long; heads 3–7 mm wide; involucral bracts with colorless membranous margin ....
.... 15. **A. armeniaca.**

+ Plants not densely pubescent, sometimes more or less glabrous; leaves 7–21(30) cm long, 2–6(10) cm wide, with petiole 1.5–10.0(14.0) cm long; heads (5)7–9 mm wide;

216

involucral bracts with brownish membranous margin ...
.... 13. **A. tanacetifolia.**

30(13). Annual or biennial plants with solitary (less often two or three), reddish stem and slender root; terminal lobes of cauline leaves linear-filiform; heads ovate or broadly oval, 1.0–1.5 mm wide, in dense panicle; involucral bracts glabrous.................................... 39. **A. scoparia.**

155   + Perennial plants with thick woody root or rhizome; stems often woody at base; heads larger .... 31.

31. Plant with long creeping rhizome, grayish from dense sericeous pubescence; corolla densely hairy at apex ...
.... 24. **A. austriaca.**

    + Plant with more or less thick, usually many-headed root; corolla glabrous or scatteredly hairy .... 32.

32. Involucre of two or three rows of bracts little different in size; peripheral flowers pistillate; disk flowers staminate (but with abortive pistil). Plants glabrous or with appressed silver-grayish pubescence, sometimes with stellate hairs .... 33.

    + Involucre imbricate; outer bracts considerably shorter than inner; all flowers bisexual. Plant whitish- or grayish-tomentose .... 41.

33. Plants mostly grayish from dense sericeous pubescence, less often somewhat glabrescent; heads clustered on short lateral branches; involucral bracts more or less densely hairy; disk flowers four to eight, apically (particularly in buds) hairy .... 34.

    + Plants green, initially pubescent, later glabrous or subglabrous; heads more or less distant; involucral bracts glabrous, less often scatteredly hairy; disk flowers about 20, glabrous .... 35.

34. Heads 2–3 mm long, 1.5–2.0 mm wide, narrow-elliptical or narrow-ovate, sessile or subsessile; stems light-colored .... 34. **A. marschalliana** (var. *marschalliana*).

    + Heads 3–4 mm long, 2–3 mm wide, elliptical, on stalk 1–3 mm long; stems brownish-violet .... 35. **A. bottnica.**

35. Stems virgate, often strongly branched almost from base, woody in lower part .... 36.

    + Stems herbaceous, weakly branched only in upper part
.... 39.

36. Plants 17–35 cm high, with pubescence of stellate hairs, later glabrescent; leaves ternate or pinnatisect (some leaves bipinnatisect), 2.5–4.5 cm long, with winged petiole

without auricles; terminal lobes linear, 0.7–2.5 cm long, 1.0–1.5 cm wide; heads elliptical, 4–5 mm long, on stalk 1–4 mm long erect, in racemes or panicle ....30. **A. salsoloides.**

+ Plants (30)50–100 cm high, with simple hairs or glabrous; leaves bipinnatisect, with wingless petiole; heads broadly ovate or ovate, (2)3–4 mm long, sessile or on stalk 1–3 mm long, in panicles..37.

37. Leaves more or less coriaceous, terminal lobes spatulate-broadly linear or oblong-spatulate, 0.7–4.0 cm long, 1.5–2.0 mm wide, orbicular; heads broadly ovate, 3–4 mm long, on stalk 1.0–2.5 mm long, nutant, glabrous or pubescent; involucral bracts fleshy, pubescent or glabrous ....36. **A. trautvetteriana.**

+ Leaves thin, terminal lobes narrow-linear or oblong, 0.5–3.5 cm long, 0.5–1.0 mm wide, acute; heads 2–3 mm long; involucral bracts glabrous, less often with isolated hairs ....38.

38. Heads oblong, distant, on nutant, glabrous, stalk 2–3 mm long; involucral bracts thin, straw-yellow, lustrous .... 37. **A. arenaria.**

+ Heads ovate or oval, more or less clustered, subsessile; involucral bracts fleshy, green, matte............................... ....34. **A. marschalliana** (var. *tschernieviana*).

156 39. Heads broadly ovate, 4–5 mm long, 3–5 mm wide, on erect, stiff, stalk 1–6 mm long, in narrow, dense panicle; auricles at base of petiole small, undivided.................... ....33. **A. bargusinensis.**

+ Heads ovate, (1.5)3.0–4.0 mm long, 1.5–2.0 mm wide, subsessile or on nutant, less often erect, stalk 1–2(4) mm long, in more or less lax and spreading panicle; auricles at base of petiole long, pinnately lobed......................40.

40. Stems 35–70 cm high, 2–8 mm thick, brownish-violet, assurgent, often many; lower leaves 2–8(11) cm long, 2–4 cm wide, with petiole 1–5(7) cm long; terminal lobes 3–10(20) mm long; heads 2–3 mm long, in spreading panicles ....31. **A. campestris.**

+ Stems 15–60 cm high, 1–4 mm thick, light green or straw-green, erect, mostly solitary or three or four; lower leaves (2)7–16 cm long, 1.5–8.0 cm wide, with broad petioles 2.0–8.5 cm long; terminal lobes (3)10–30 mm long; heads 1.5–4.0 mm long, in narrow panicles .... 32. **A. commutata.**

41(32). Leaves pinnati- or bipinnatisect .... 42.

+ Leaves bi- or tri-pinnatisect on same stem .... 44.

42. Leaves bipinnatisect; terminal lobes oblong to oblong-obovate, 1–4(5) mm long, 0.2 mm wide, acuminate. Plants pulvinate, vegetative shoots woody to considerable height; stems 5–15(25) cm high, 0.5–1.5 mm thick, during flowering leafy; heads narrow-obovate, 2.5–3.0 mm long .... 49. **A. gracilescens.**

+ Leaves pinnatisect, less often also bipinnatisect; terminal lobes linear or linear-filiform, acute. Plants more or less densely caespitose; vegetative shoots woody only at base; stems leafless or with few leaves at flowering; heads oblong-obovate, (3.0)3.5–4.5 mm long .... 43.

43. Stems 15–30(35) cm high, 9–20(50), densely tomentose throughout vegetative period; leaves 1.0–2.0(2.5) cm long, 0.5–1.5 cm wide; terminal lobes 1.2–7.0 mm long; panicles lax, broadly pyramidal, with almost horizontal or deflected branches .... 50. **A. terrae-albae.**

+ Stems 10–30(40) cm high, 4–11, glabrous or subglabrous at flowering; leaves 2.5–6.0 cm long, 1.0–3.5 cm wide; terminal lobes 7–25 mm long; panicle narrow, dense with its branches obliquely upward directed or adpressed to stem .... 51. **A. lessingiana.**

44. Heads clustered at apices of stem and short, more or less closeset, branches, sessile, broadly campanulate, (6)7–11-flowered; lower leaves 1–10 cm long, 1.0–2.5(5.5) cm wide, with petiole 1.2–2.5 cm long; terminal lobes elliptical and oblong to linear (1.5)4.0–9.0(15.0) mm long. Plant whitish, 9–40 cm high, rather densely caespitose ....
.... 47. **A. compacta.**

+ Heads more or less uniformly distributed often stalked, narrow, (1)3–5(6–9)-flowered; lower leaves smaller .... 45.

45. Plants white, whitish or yellowish from dense arachnoid or lanate-tomentum, more or less persisting till end of vegetation; vegetative shoots mostly numerous, approximate .... 46.

+ Plants grayish or grayish-green from more or less dense tomentum, at flowering glabrous, subglabrous or more or less pubescent; vegetative shoots not many or absent ... 49.

157 46. Plants, particularly young, yellowish, lanate-to-mentose, densely leafy till end of vegetation, (27)35–55 cm high; terminal lobes of leaves obovate to almost oblong,

expanded in upper part, 0.7–1.2 mm wide, orbicular; panicles (7)10–30 cm long, (4)7–40 cm wide; heads broadly obovoid, (4)5–6 mm long, 3–4 mm wide ....
.... 43. **A. dzevanovskyi.**

+ Plants white or whitish, arachnoid-tomentose by end of flowering, with somewhat sparser leaves; heads narrow-obovate or elliptical, 3–4(5) mm long, 2.0–2.5(3.0) mm wide .... 47.

47. Heads distant, mostly nutant, narrow-campanulate, 3.5–5.0 mm long, 2.0–2.5 mm wide, in lax and spreading panicles with almost horizontal branches; terminal lobes of leaves oblong or oblong-obovate, 1–5 mm long, 0.7–1.5 mm wide
.... 42. **A. nutans.**

+ Heads densely arranged, erect, in more or less dense panicles with obliquely upward directed branches; terminal lobes of leaves 0.2–0.5(1.0) mm wide .... 48.

48. Plants with fibrous root system and few assurgent vegetative shoots; terminal lobes of leaves oblong-obovate or oblong, 3–7(10) mm long, 0.5–1.0 mm wide; heads in panicles, 6–13 cm long and (1)2–4(6) cm wide ....
.... 40. **A. maritima.**

+ Plants with tap root and mostly numerous approximate, short, erect vegetative shoots, somewhat caespitose; terminal lobes of leaves linear or oblong, 1–5(7) mm long, 0.2–0.5 mm wide; heads in panicles, 4–25 cm long, 4–15 cm wide .... 41. **A. lerchiana.**

49(45). Terminal lobes of leaves linear to linear filiform; heads sessile .... 50.

+ Terminal lobes of leaves oblong to oblong-obovate, less often partly linear; heads stalked .... 51.

50. Subtending leaves considerably exceeding heads; leaves persisting till end of vegetation, their terminal lobes 4–6(12) mm long, obtuse; stems 6(10)–45(95); heads elliptical, 2.5–3.0 mm long, 2–3 mm wide .... 44. **A. taurica.**

+ Subtending leaves shorter than heads; lower leaves falling by flowering; terminal lobes of leaves 2–9(11) mm long, acute; stems one to eight; heads oblong or obovate, 3.0–4.5 mm long, 1.5–2.0 mm wide .... 46. **A. nitrosa.**

51. Plants with fibrous root system, often soboliferous, flowering stems 1–10(15), 15–75(100) cm high. Vegetative shoots 3–30 cm long, one to three; terminal lobes of leaves 3–11(16) mm long, acute or obtuse; heads more or

220

less nutant stalk on 0.5–3.0 mm long; in spreading panicles
.... 45. **A. santonica.**
+ Plants with tap root, flowering stems numerous, 10–30(40)
cm high, vegetative shoots 0.5–5.0 cm high, rather densely
caespitose; terminal lobes of leaves 1.0–3.5(4.0) mm long,
acute, with rather long cusp; heads upward directed, on
1–2 mm long stalk, in dense, narrow panicles................
.... 48. **A. pauciflora.**

SUBGENUS 1. *ARTEMISIA*

Peripheral flowers in head pistillate, with two-toothed, narrow-tubular corolla, fertile; disk flowers bisexual, with four- or five-toothed corolla, fertile. Involucre of two or three rows of bracts little different in size. Receptacle conspicuous, glabrous or hairy.

Type: lectotype of genus.
Section 1. Artemisia.
Short- or long-rhizomatous, herbaceous perennials; leaves mostly bicolorous green and glabrous or weakly hairy above, white or grayish arachnoid-tomentose beneath, less often glabrous on both sides, pinnati or bipinnatisect without intermediate lobes; heads narrow to broadly campanulate and subglobose; receptacle and flowers glabrous.

Type: lectotype of genus.
1. **A. vulgaris** L. 1753. Sp. Pl. 848; Poljak. 1961. Fl. SSSR. **26:** 438; Tutin, 1976. Fl. Europ. **4:** 180; Leonova 1987. Novosti Sist. Vyssh. Rast. **24:** 178.

Type: Europe ("in Europae cultis ruderatis").
North; *Baltic*; *Center*; *West*; *East*; *Crimea*. In thinned-out forests, forest glades and edges, meadows, scrublands, on banks of water bodies, by roadsides, in habitations, on edges of fields.— *General distribution*: Caucasus, Western and Eastern Siberia, Far East, Russian Central Asia; Scandinavia, Central and Atlantic Europe, Mediterranean, Asia Minor, Iran; North America (in cultivation), Africa (north).—2n = 16.
Note. A highly polymorphic species. Three variants of the species can be identified based on the width and degree of incision of the primary segments of leaves: a) var. *vulgaris*— most extensively distributed, with moderately incised primary segments and usually oblong-elliptical, 6–12 mm wide, lateral segments; b) var. *latiloba* (Ledeb.) Filat. [1964. *Bot. Mat. (Alma-*

Ata), **2:** 64 (=*A. vulgaris* γ. *latiloba* Ledeb. 1833, *Fl. Alt.* **4:** 83]—differs from the preceding variety by subcoriaceous leaves that are whitish-pubescent beneath, with weakly incised primary segments and wide, mostly elliptical, undivided, 10–20 mm wide lateral segments, occurs rarely; c) var. *coarctata* (Forsell.) Afzel. ex Fries [1819, *Fl. Halland*: 132 (=*A. coarctata* Forsell. 1807, *Linn. Inst. Skriff.* **1:** 12. = *A. vulgaris* subsp. *coarctata* (Forsell.) Ameljcz. 1980. *Fl. Krasnoyarsk. Kraya,* **10:** 41]—characterized by strongly incised primary segments with narrow (usually 2–4 mm wide), linear-lanceolate to almost linear lateral segments, found occasionally throughout the range of the species but predominantly in the northern regions and near the seacoast.

Lellep (1978, *Eesti NSV Fl.* **6:** 255) reports for Estonia a relatively closer North American species, *A. gnaphalodes* Nutt., collected only once by Rammel in 1957 on the berm of the railroad near the city of Pyarna. The lower and middle cauline leaves of this species are undivided, unevenly toothed above middle, and lack auricles. Galinis (1980, *Lietuvos TSR Fl.* **6:** 118) mentions that, in Vilnius, there is in cultivation another North American species, *A. purshiana* Bess., which is sometimes found as an escape.

2. **A. verlotiorum** Lamotte, 1877, Assoc. Fr. Avancem. Sci. Compt. Rend. (Clern.-Ferr.), **5:** 513; Gams, 1929, in Hegi, Ill. Fl. Mitteleur. **6,** 2: 631; Tutin, 1976, Fl. Europ. **4:** 180, p. p.; Mosyakin, 1990, Ukr. Bot. Zhurn. **47,** 4: 10.

Type: France, vicinity of Clermont-Ferrand ("a 6 kilometres environ de Clermont").

Crimea (south).—Endemic in Nikitskii-Botanical Garden.—*General distribution*: Japan-China; ecdemic in many other countries.—2n = 16, 18.

Note. In the Soviet Union, this species is known only from Crimea (Mosyakin, loc. cit.). All other references refer to *A. umbrosa* (Bess.) Pamp.—a species more adapted to the climatic conditions of the warm temperate zone.

3. **A. selengensis** Turcz. ex Bess. 1834, Nouv. Mém. Soc. Nat. Moscou, **3:** 50; Poljak. 1961, Fl. SSSR, **26:** 454; Vynaev and Tretyakov, 1978, Botanika, **20:** 104; Puzyrev, 1986, Bot. Zhurn. **71,** 2: 256; Mosyakin, 1990, Ukr. Bot. Zhurn. **47,** 4: 10.

Type: Buryat Autonomous Republic, village of Selenginsk ("in insulis Selengae ad Selenginsk").

Center (Upper Dnieper: Minsk and Brest; Volga-Kama: Udmurtia, railroad station of Kez and Izhevsk); *West* (Dnieper: west).—By roadsides, in habitations; only as ecdemic.—*General distribution*: Eastern Siberia (southeast), Far East; Mongolia, Japan-China (northwestern China).—2n = 16.

Note. For Belorussia (Upper Dnieper), a more southern Siberian species, *A. integrifolia* L. (1753, *Sp. Pl.*: 848), is known as an adventive plant. It has entire, but deeply toothed, leaves and the heads are 3–4 (and not 2.0–2.5) mm in dia.

159 4. **A. rubripes** Nakai, 1917, Bot. Mag. Tokyo, **31**: 112; Poljak. 1961, Fl. SSSR, **26**: 443; Mosyakin, 1990, Ukr. Bot. Zhurn. **47**, 4: 11.

Type: Korean Peninsula ("Korea sept.: inter Pu-tyong-po et Po-tyai-dong ... in fruticetis Pu-nyong ... Korea media: circa Shin-po-ri").

Center (Volga-Kama); *West* (Dnieper: Kiev).—Ecdemic in habitations, by roadsides.—*General distribution*: Far East; Japan-China; ecdemic in other countries.

5. **A. umbrosa** (Bess.) Pamp. 1930, Nuovo Giorn. Bot. Ital., nov. ser. **36**, 4: 448; Poljak. 1961, Fl. SSSR, **26**: 435; Mosyakin, 1990, Ukr. Bot. Zhurn. **47**, 4: 11.—*A. vulgaris* L. var. *umbrosa* Bess. 1834, Nouv. Mém. Soc. Nat. Moscou, **3**: 52.—*A. verlotiorum* auct. non Lamotte: Rasinš , 1960, Latv. PSR Veg. **3**: 113; Tutin, 1976, Fl. Europ. **4**: 180, p. p.; Vynaev and Tretyakov, 1978, Botanika, **20**: 104; Gusev, 1980, Bot. Zhurn. **65**, 2: 250; Ignatov and others, 1983, Byull. Glavn. Bot. Sada Akad. Nauk SSSR, **129**: 47; Puzyrev, 1985, Bot. Zhurn. **70**, 2: 268; Fatere and Gavrilova, 1985, Fl. Rastit. Latv. SSSR: 145; Voloskova, 1986, Nauchn. Dokl. Vyssh. Shkoly, Biol. Nauki, **8**: 74.

Type: Baikal Region, the Selenga River ("Talis ad Selengam in umbrosis Turtschan").

Baltic (Latvia and Lithuania); *Center* (Ladoga-Ilmen; Upper Dnieper; Upper Volga; Volga-Kama); *West* (Dnieper: vicinity of Kiev).—Endemic of roadsides, in habitations.—*General distribution*: Siberia (southeast), Far East; Mongolia, Japan-China; ecdemic in other extratropical countries.

6. **A. argyi** Lévl. et Vaniot, 1910, Feddes Repert. **8**: 138; Poljak. 1961, Fl. SSSR, **26**: 451; Mosyakin, 1990, Ukr. Bot. Zhurn. **47**, 4: 11.

Type: North China ("Kiang-Sou: Ka-ngai-Bon").

Center (Upper Volga; Volga-Kama); *West* (Dnieper: vicinity of Kiev).—Ecdemic of roadsides, in habitations.—*General distribution*: Far East (south); Japan-China.—2n = 36.

7. **A. tilesii** Ledeb. 1815, Mém. Acad. Sci. Pétersb. **5**: 568; Poljak. 1961, Fl. SSSR, **26**: 443; Tutin, 1976, Fl. Europ. **4**: 180.

Type: Kamchatka ("in Kamtschatka, Tilesius").

Arctic (Novaya Zemlya; Arctic Europe: Kanin and to the east from it); *North* (Dvina-Pechora: basin of the Mezen River and northeast).—In inundated meadows, scrublands, on river sands and gravel-beds, slopes of river and coastal terraces, taluses and stony slopes.—*General distribution*: Western and Eastern Siberia (north), Arctic, Far East (Kamchatka); North America (Arctic).—2n = 18.

8. **A. leucophylla** (Turcz. ex Bess.) Turcz. ex Clarke, 1876, Comp. Ind.: 162; Leonova, 1971, Novosti Sist. Vyssh. Rast. **8**: 234; id. 1987, ibid., 24: 179; Korobkov, 1987, Arkt. Fl. SSSR, **10**: 143.—*A. vulgaris leucophylla* Turcz. ex Bess. 1834, Nouv. Mém. Soc. Nat. Moscou, **3**: 54 (in textu).

Type: Northern Mongolia, Khubsagul Lake ("In glareosis ad lacum Kossogol Ircut").

Arctic (Arctic Europe: northeast); *North* (Karelia-Murmansk: Karelia; Dvina-Pechora: north and northeast).—On sandy stony and rubbly slopes, coastal sands and gravel-beds, inundated meadows, in shrub thickets.—*General distribution*: Western Siberia (Altai), Eastern Siberia, Arctic, Far East (north); Mongolia, Tibet, Japan-China (northwestern China).—2n = 16, 18.

Section 2. Abrotanum Bess. 1829, Bull. Soc. Nat. Moscou, **1**: 222.—*Abrotanum* Neck. 1790, Elem. Bot. **1**: 98, nom. illeg.—*Artemisia* subgen. *Abrotanum* (Bess.) Rydb. 1916, North Amer. Fl. **34**, 3: 247.—*Artemisia* sect. *Annua* Ameljcz. 1980, Fl. Krasnoyarsk. Kraya, **10**: 53.

Semishrubs or herbaceous perennials, less often biennials or annuals; leaves glabrous or pubescent beneath, bi- or tri-pinnatisect, often with intermediate lobes; heads hemispherical, broadly campanulate or broadly ovate, less often saucer-shaped; receptacle glabrous.

Lectotype: *A. abrotanum* L.

9. **A. abrotanum** L. 1753, Sp. Pl.: 845; Poljak. 1961, Fl. SSSR, **26**: 456; Tutin, 1976, Fl. Europ. **4**: 183; Galinis, 1980, Lietuv. TSR Fl. **6**: 119.—*A. procera* Willd. 1803, Sp. Pl. **3**, 3: 1818.—*A. elatior* Klok. 1962, Fl. URSR, **11**: 552, 319.—*A. paniculata* auct. non Lam.: Bess. 1845, Mém. Acad. Sci. Pétersb. **4**: 471.

Lectotype: Syria, Turkey ("Syriae, Galatiae, Cappadociae montibus apricis").

North (Dvina-Pechora: south); *Baltic* (south; cultivated and escaped); *Center* (Upper Volga; Volga-Kama; Volga-Don; in Ladoga-Ilmen and Upper Dnieper, cultivated and as escape); *West* (Dnieper; Black Sea; in the Carpathians and Moldavia, cultivated); *East.*— On banks of water bodies, in inundated meadows, forest glades and edges, on outcrops of chalk and limestone, in scrublands, often cultivated and found as an escape or ecdemic of roadsides; in gardens).—*General distribution*: Caucasus, Western Siberia, Russian Central Asia (north); Central Europe, Mediterranean, Asia Minor, Iran (west).—2n = 18.

10. **P. pontica** L. 1753, Sp. Pl.: 847; Poljak. 1961, Fl. SSSR, **26:** 461; Tutin, 1976, Fl. Europ. **4:** 183; Galinis, 1980, Lietuv. TSR Fl. **6:** 120; Ignatov and others, 1990, Flor. Issl. Mosk. Obl.: 94.

Type: Hungary, northern Greece, Turkey ("in Hungaria interiore, Pannonia, Thracia, Mysia").

Baltic (sometimes cultivated in Latvia and Lithuania); *Center* (Upper Volga: ecdemic; Volga-Kama; Volga Don); *West* (excluding the Carpathians); *East* (Lower Don; Trans-Volga; Lower Volga; north and west); *Crimea.*—In dry meadows, forest glades and edges, steppes and scrublands, sometimes as weed in fields.— *General distribution*: Caucasus (Ciscaucasia), Western Siberia, Russian Central Asia (north); Central Europe, Mediterranean, Asia Minor; North America (in cultivation).—2n = 18.

11. **A. macrantha** Ledeb. 1815, Mém. Acad. Sci. Pétersb. **5:** 573; Poljak. 1961, Fl. SSSR, **26:** 462; Tutin, 1976, Fl. Europ. **4:** 183; Puzyrev, 1985, Bot. Zhurn. **70,** 20: 268.

Type: Siberia ("in Sibiria ad lacum Tschumasof").

Center (Volga-Kama: south and east); *East* (Lower Don: vicinity of Volgograd; Trans-Volga).—In steppefied and herbage meadows, forest glades and edges, on stony steppe slopes, in scrublands, thinned-out pine and birch forests.—*General distribution*: Western and Eastern Siberia, Russian Central Asia (north).

12. **A. santolinifolia** Turcz. ex Bess. 1834, Nouv. Mém. Soc. Nat. Moscou, **3:** 87, in textu; Poljak. 1961, Fl. SSSR, **26:** 465; Tutin, 1976, Fl. Europ. **4:** 183; Leonova, 1987, Novosti Sist. Vyssh. Rast. **24:** 181.

Type: Eastern Siberia, Trans-Baikal ("In campis ad fluvium Onam, ubi cum fl. Uda se jungit").

East (Trans-Volga: southern Urals).—On stony and clayey-rubbly slopes.—*General distribution*: Western and Eastern Siberia (south), Russian Central Asia (southeast); Mongolia.—2n = 18.

13. **A. tanacetifolia** L. 1753, Sp. Pl.: 848; Poljak. 1961, Fl. SSSR, **26:** 468; Leonova, 1978, Bot. Zhurn. **63,** 1: 81; id. 1987, Novosti Sist. Vyssh. Rast. **24:** 181.

Type: Siberia ("In Sibiria").

North (Dvina-Pechora: basins of the Pinega, Soyana, and Mezen rivers); *Center* (Volga-Kama: northern and central Urals).— In thinned-out larch and pine forests, forest glades and edges, on outcrops of limestone and allied rocks.—*General distribution*: Eastern Siberia, Far East; Mongolia; North America (Alaska).

Note. The localities of this East Siberian taiga species in the European part of Russia are probably relicts of the last glaciation.

14. **A. latifolia** Ledeb. 1815, Mém. Acad. Sci. Pétersb. **5:** 569; Poljak. 1961, Fl. SSSR, **26:** 472; Tutin, 1976, Fl. Europ. **4:** 182; Leonova, 1978, Bot. Zhurn. **63,** 1: 80; id. 1987, Novosti Sist. Vyssh. Rast. **24:** 181; Ignatov and others, 1983, Byull. Glavn. Bot. Sada Akad. Nauk SSSR, **129:** 47; Voloskova, 1986, Nauchn. Dokl. Vyssh. Shkoly. Biol. Nauk, **8:** 74.

Type: Siberia ("In Sibiria").

Center (Upper Dnieper and Upper Volga—ecdemic; Volga–Kama: south and east; Volga-Don: excluding northwest); *East* 161 (Lower Don: vicinity of Kamyshin; Trans-Volga)—In thinned-out forests, forest glades, and edges, steppes, scrublands, on outcrops of chalk and limestone, in alkaline meadows.—*General distribution*: Western Siberia (south), Eastern Siberia, Far East (south), Russian Central Asia (north).—2n = 36.

Note. Besides var. *latifolia* with glabrous involucral bracts and corolla of disk flowers, *A. latifolia* var. *pilosiuscula* (Krasch.) Leonova (1987, op. cit.: 182.—*A. latifolia*, f. *pilosiuscula* Krasch. 1949, in Kryl. *Fl. Zap. Sib.* **11:** 2810) with weakly hairy corollas or involucral bracts (less often both) is occasionally found. Possibly, this variant is a hybrid of *A. latifolia* × *A. armeniaca*. It is recorded in Bashkiria, Orenburg, Chelyabinsk and Tambov regions, and in the vicinity of the city of Kazan.

15. **A. armeniaca** Lam. 1789, Encycl. Méth. **1:** 263; Poljak. 1961, Fl. SSSR, **26:** 471; Klok. 1962, Fl. URSR, **11:** 317; Tutin, 1976, Fl. Europ. **4:** 182; Leonova, 1978, Bot. Zhurn. **63,** 1: 83.

Type: Armenia ("l'Armenie").

Center (Volga-Kama; Volga-Don); *West* (Dnieper: Dnepropetrovsk Region; Black Sea: Kherson Region); *East* (Lower Don; Trans-Volga; Lower Volga: north).—In scrublands, thinned-out pine,

226

birch, and oak forests, herbage and sheep's fescue-feather grass steppes, inundated meadows, on outcrops of limestone and allied rocks.—*General distribution*: Caucasus, Western Siberia (south); Asia Minor (east), Iran (west).

Note. Two varieties are recorded: a) var. *armeniaca*—with dense pubescence of leaves on both sides and usually also of stem (sometimes subglabrous), found in dry localities, mostly in the steppe and forest steppe regions; b) var. *excelsa* Filat. (1964, *Bot. Mat. (Alma-Ata)*, **2**: 64)—with leaves not so densely pubescent beneath and scatteredly hairy or glabrous above, confined to more humid and shady localities, predominantly in the forest zone.

16. **A. laciniata** Willd. 1803, Sp. Pl. **3**, 3: 1843 (excl. var.); Poljak. 1961, Fl. SSSR, **26**: 473; Tutin, 1976, Fl. Europ. **4**: 182; Leonova, 1987, Novosti Sist. Vyssh. Rast. **24**: 182.

Lectotype: Siberia ("Sibiria").

Center (Ladoga-Ilmen: ecdemic in Vyborg; Volga-Don: Voronezh Region, vicinity of Bobrov); *East* (Trans-Volga: south and east).— On steppefied and solonetz meadows, meadow slopes, solonetz steppes, less often in scrublands, thinned-out deciduous forests, and forest-steppe zone.—*General distribution*: Scandinavia (Norway), Atlantic Europe (northwestern Scotland).—2n = 18.

17. **A. norvegica** Fries, 1817, Nov. Fl. Suec. ed. 1: 56, tab. 3; Poljak. 1961, Fl. SSSR, **26**: 489; Tutin, 1976, Fl. Europ. **4**: 183; Leonova, 1987, Novosti Sist. Vyssh. Rast. **24**: 182.

Type: Norway ("E rupibus Dovrefjäll Norvegiae adeoque extra fines Florae nostrae").

Arctic (Arctic Europe: Polar Urals, Karsk tundra); *North* (Dvina-Pechora: Cispolar and Northern Urals); *Center* (Volga-Kama: Urals).—In mountain lichen, moss and shrub-tundra, on stony slopes, taluses, gravel-beds and sandy banks of rivers in alpine zone; larch-birch sparse forest at upper limit of forest.—*General distribution*: Scandinavia (Norway), Atlantic Europe (northwestern Scotland).—2n = 18.

Note. The species is represented by two varieties a) var. *norvegica* (=*A. norvegica* var. *villosula* Trautv. ex Korsh)—Whole plant densely pubescent, less often leaves glabrescent; involucral bracts and corolla of middle flowers densely hairy; b) var. *uralensis* Rupr. (1850, *Beitr. Planzenk. Russ. Reich.* **7**: 64; id. 1856, *Fl. bor. ural*: 37 (=*A. ruprechti* Poljak. 1955, *Bot. Mat. (Leningrad)*, **17**: 409))—whole plant, including corollas and involucral bracts, glabrous. The latter variety is found quite rarely.

18. **A. annua** L. 1753, Sp. Pl.: 847; Poljak. 1961, Fl. SSSR, **26:** 489; Schultz, 1976, Bot. Zhurn. **61,** 10: 1452; Tutin, 1976, Fl. Europ. **4:** 185; Leonova, 1978, Bot. Zhurn. **63,** 1: 80; Efimova and others, 1981, Bot. Zhurn. **66,** 7: 1049; Ignatov and Makarov, 1984, Byull. Glavn. Bot. Sada Akad. Nauk SSSR, **132:** 50; Bosek, 1986, Bot. Zhurn., **71,** 1: 100.

162   Lectotype: Siberia ("Sibiriae montosis").

Baltic (ecdemic); *Center* (Ladoga-Ilmen; Upper Dnieper; Upper Volga; Volga-Kama; Volga-Don; ecdemic everywhere); *West*; *East* (Lower Don; Trans-Volga); *Crimea.*—On river sands and gravel-beds, by roadsides, in habitations, sometimes in weedy meadows, in more northern regions only as a rare ecdemic plant, southward more common.—*General distribution*: Caucasus, Western and Eastern Siberia, Far East, Russian Central Asia; Central Europe, Mediterranean, Asia Minor, Iran, Mongolia, Japan-China; North America (ecdemic).—2n = 18.

19. **A. tournefortiana** Reichenb. 1823, Icon. ex Cent. **1:** 6, tab. 5; Poljak. 1961, Fl. SSSR, **26:** 488; Schultz, 1977, Bot. Zhurn. **62,** 10: 1516; Tzvel. 1982, Novosti Sist. Vyssh. Rast. **19:** 230; Leonova, 1987, Novosti Sist. Vyssh. Rast. **24:** 183; Mosyakin, 1990, Ukr. Bot. Zhurn. **47,** 4: 10.

Lectotype: East ("Orient").

Baltic; *Center* (Upper Dnieper; Upper Volga); *West* (Dnieper; Black Sea: east).—In habitations, by roadsides, only as ecdemic weed.—*General distribution*: Caucasus (Transcaucasia), Russian Central Asia; Central Europe (ecdemic), Asia Minor, Iran, Dzhungaria-Kashgaria, Mongolia, Tibet, Himalayas.—2n = 18.

Note. Another very closely related North American species, *A. biennis* Willd., has been reported for Vyborg (Suominen, 1979, *Acta Bot. Fennica*, **111:** 79).

Section 3. Absinthium (Lam.) DC. 1805, in Lam. and DC. Fl. Fr., ed. 3, **4:** 189.—*Absinthium* Lam. 1778, Fl. Fr. **2:** 45.—*Artemisia* subgen. *Absinthium* (Lam.) Rydb. 1916, North Amer. Fl. **34,** 3: 247.—*Artemisia* sect. *Artemisia* auct.: Tutin, 1976, Fl. Europ. **4:** 180, p. p.

Herbaceous perennials, less often annuals or biennials or semishrubs, usually rather densely hairy or tomentose, more or less grayish-silvery from dense, appressed or squarrose hairs, less often subglabrous, green; pubescence of simple and bifid hairs. Leaves similarly pubescent on both sides, bi- or tri-pinnatisect, less often bi- or tri-ternate or undivided, without

228

short intermediate lobes. Heads hemispherical, large, less often saucer-shaped or campanulate. Receptacle pilose, rarely scatteredly hairy or glabrous.

Lectotype: *A. absinthium* DC.

20. **A. absinthium** L. 1753, Sp. Pl.: 848; Poljak. 1961, Fl. SSSR, **26**: 515; Tutin, 1976, Fl. Europ. **4**: 180.

Type: Europe ("Europae ruderatis aridis").

Arctic (Arctic Europe: vicinity of Vorkuta); *North*; *Baltic*; *Center*; *West*; *East*; *Crimea.*—In meadows, forest glades and edges, scrublands, on banks of water bodies, by roadsides.— *General distribution*: Caucasus, Western Siberia (south), Eastern Siberia (southwest), Russian Central Asia; Scandinavia, Central and Atlantic Europe, Mediterranean, Asia Minor, Iran, Himalayas; North America (ecdemic), Africa (north).—2n = 18.

21. **A. sieversiana** Willd. 1803, Sp. Pl. **3**, 3: 1845; Poljak. 1961, Fl. SSSR, **26**: 517; Worosch. and others, 1966, Opred. Rast. Mosk. Obl.: 329; Gusev, 1975, Bot. Zhurn. **60**, 3: 380; Tutin, 1976, Fl. Europ. **4**: 180; Lellep, 1978, Eesti NSV Fl. **6**: 239; Bosek, 1979, Bot. Zhurn. **64**, 2: 244; Leonova, 1979, Bot. Zhurn. **64**, 3: 434; Galinis, 1980, Lietuv. TSR Fl. **6**: 124; Tzvel. 1982, Novosti Sist. Vyssh. Rast. **19**: 230; Mosyakin, 1990, Ukr. Bot. Zhurn. **47**, 4: 12.

Type: Siberia ("In Sibiria").

North (south of Karelia-Murmansk and Dvina-Pechora— ecdemic); *Baltic* (ecdemic); *Center* (Ladoga-Ilmen; Upper Dnieper; Upper Volga; Volga-Don—ecdemic; Volga-Kama); *West* (Dnieper— ecdemic).—Usually as ecdemic of roadsides, in habitations, fields, less often in meadows, forest glades and edges, on coastal buffs, sands and gravel-beds.—*General distribution*: Western and Eastern Siberia, Far East, Russian Central Asia; Mongolia, Tibet (west), Himalayas, Japan-China.—2n = 18.

Note. One more closely related East Siberian species, *A. jacutica* Drob., has been reported for the former Vyatka Governance 163 as an adventive plant (1914, *Tr. Bot. Muz. Akad. Nauk SSSR*, **12**: 108), with shorter (20–40 cm high) cylindrical (and not ribbed) stem and smaller leaves with narrow-linear lobes.

22. **A. caucasica** Willd. 1803, Sp. Pl.: **3**, 3: 1823; Poljak. 1961, Fl. SSSR, **26**: 492; Kotov, 1965, Vizn. Rosl. Ukr., ed. **2**: 687; Leonova, 1969, in E. Wulf, Fl. Kryma, **3**, 3: 214.—*A. argentata* Klok. 1954, Bot. Mat. (Leningrad), **16**: 366; id. 1962, Fl. URSR,

11: 315; id. 1987, Opred. Vyssh. Rast. Ukr.: 340.—*A. lanulosa*
Klok. 1962, op. cit.: 554, 313, sine descr. lat.; id. 1987, op. cit.:
340.—*A. pedemontana* auct. non Balb.: Tutin, 1976, Fl. Europ. **4:**
184, p.p.

Lectotype: Caucasus ("In Caucaso").

West (Black Sea: Donetzk Region); *Crimea* (east).—On stony
slopes and rocks, predominantly limestone.—*General distribution*:
Caucasus; Central Europe (Romania), Mediterranean (Bulgaria),
Asia Minor (north), Iran (northwest).—2n = 16, 18(?).

Note. Crimean populations of this species have been described
as *A. lanulosa* Klok. while the populations from Donetzk Region
(basin of the Kalmius River) as *A. argentata* Klok.

23. **A. frigida** Willd. 1803, Sp. Pl. **3**, 3: 1838; Krasch. 1936, Fl.
Yugo-Vost. Evrop. Chasti SSSR, **6:** 361; Poljak. 1961, Fl. SSSR,
26: 494; Tutin, 1976, Fl. Europ. **4:** 183; Leonova, 1987, Novosti
Sist. Vyssh. Rast. **24:** 185.

Type: Eastern Siberia, Dauria ("in aridis frigidis Dahuriae").

Center (Volga-Kama: east and south); *East* (Trans-Volga:
east).—On stony and rubbly (predominantly limestone) slopes,
in thinned-out pine forests and its edges, steppes, less often on
abandoned lands.—*General distribution*: Western Siberia (south),
Eastern Siberia, Far East, Russian Central Asia (north); Mongolia;
North America.—2n = 18.

24. **A. austriaca** Jacq. 1773, Fl. Austr. **1:** 61; Poljak. 1961, Fl.
SSSR, **26:** 498; Florovskaya, 1965, Fl. Leningr. Obl. **4:** 248; Worosch.
1966, Opred. Vyssh. Rast. Mosk. Obl.: 330; Tutin, 1976, Fl. Europ.
4: 183; Lellep, 1978, Eesti NSV Fl. **6:** 231; Galinis, 1980, Lietuv.
TSR Fl. **6:** 121.—*A. repens* Pall. ex Willd. 1803, Sp. Pl. **3**, 3: 1840;
Klok. 1962, Fl. URSR, **11:** 318.

Type: Austria ("In apricis siccis tenuigramine vestitis et asperis
locis").

North (ecdemic); *Baltic* (ecdemic); *Center*; *West* (excluding
Carpathians); *East*; *Crimea*.—In dry, often more or less saline
meadows, steppes, forest glades and edges, on stony slopes, by
roadsides, in habitations, in northern and northwestern regions
as ecdemic, but often naturalized; forest steppe and steppe zone,
very common.—*General distribution*: Caucasus, Western Siberia
(south), Far East (south; ecdemic), Russian Central Asia (north
and east); Central Europe, Mediterranean, Asia Minor (north),
Iran (north and northwest).—2n = 16, 36.

25. **A. sericea** Web. ex Sechm. 1775, Dissert. Artem.: **16;** Poljak. 1961, Fl. SSSR, **26:** 501; Worosch. 1966, Opred. Rast. Mosk. Obl.: 329; Tutin, 1976, Fl. Europ. **4:** 184.

Type: Siberia, from the Urals to Angara River ("In campis siccis et locis montosis, a jaico fluvio orientem versus ad Angaram usque").

Center (Upper Volga: Moscow—ecdemic; Volga-Kama: south and east; Volga-Don); East (Trans-Volga).—In herbage, cereal, and scrub steppes, in glades of pine and birch forests, on outcrops of limestone and allied rocks, sometimes on railroad dumps and in open fields.—*General distribution*: Western Siberia (south), Eastern Siberia, Russian Central Asia; Mongolia (north).—2n = 18.

O 26. **A. hololeuca** Bieb. ex Bess. 1834, Nouv. Mém. Soc. Nat. Moscou, **3:** 46; Poljak. 1961, Fl. SSSR, **26:** 507; Klok. 1962, Fl. URSR, **11:** 325; Tutin, 1976, Fl. Europ. **4:** 185; Leonova, 1983, Spisok Rast. Gerb. Fl. SSSR, **24:** 19.

Type: Voronezh and Kharkov regions ("Colles cretacei ad fluvium Oskol et Donez in gub. Woronez et distr. Kupensk gub. Chark.").

Center (Volga-Don: Voronezh and Belgorod regions); *West* (Dnieper: Kharkov and Lugansk regions; Black Sea: Donetzk Region); *East* (Lower Don).—On chalk outcrops, less often on sands of river terraces.—Endemic.—2n = 18.

164 *Note.* This rare species is found only in the basins of the North Donets, Don, and Mius rivers (the Krynka River).

27. **A. rupestris** L. 1753, Sp. Pl.: 847; Poljak. 1961, Fl. SSSR, **26:** 508; Tutin, 1976, Fl. Europ. **4:** 185; Lellep, 1978, Eesti NSV Fl. **6:** 233; Ramensk. and Andreeva, 1982, Opred. Vyssh. Rast. Murm. Obl. i Karel.: 392; Puzyrev, 1985, Bot. Zhurn. **70,** 2: 268.

Type: Siberia and Sweden, Åland Islands ("In Sibiria, Oelandiae rupibus calcareis").

North (Karelia-Murmansk: Karelia); *Baltic* (northwest of Estonia); *Center* (Volga-Kama: central Urals, Udmurtia—ecdemic); *East* (Trans-Volga: southern Urals).—On stony, predominantly limestone, slopes, in saline meadows and solonetzes, cereal-herbage and meadow steppes, forest glades and edges.—*General distribution*: Western Siberia (south), Eastern Siberia, Russian Central Asia (mountains); Scandinavia (Åland and Gotland islands), Central Europe (basin of Saale River), Mongolia.—2n = 18.

SUBGENUS **2. *DRACUNCULUS*** (Bess.) Peterm.
1848, Deutschl. Fl.: 294.—*Artemisia* sect. *Dracunculus* Bess. 1835,
Bull. Soc. Nat. Moscou, **8:** 16.—*Oligosporus* Cass. 1817, Bull.
Soc. Philom. Paris, 1817: 33; Poljak. 1961, Tr. Inst. Bot. Akad.
Nauk KazSSR, **11:** 166.—*Artemisia* subgen. *Oligosporus* (Cass.)
Galinis, 1980, Lietuv. TSR Fl. **6:** 124, sine auct. comb.

Peripheral flowers pistillate, with two- or three-toothed, tubular
or narrow-conical corolla, often at base somewhat inflated, fertile;
disk flowers staminate but with rudimentary pistil and five-toothed
corolla. Receptacle flat or somewhat convex, glabrous.
Lectotype: *A. dracunculus* L.
Section 4. Dracunculus Bess. 1835, Bull. Soc. Nat. Moscou,
8: 16.
Herbaceous perennials, with usually horizontal rhizome;
pubescence of stellate, less often simple or bifid, hairs; leaves
simple, undivided or linear to lanceolate, without auricles, lower and
middle sometimes three-lobed near apex; heads subglobose, small.

Type: type of subgenus.

28. **A. dracunculus** L. 1753, Sp. Pl.: 849; Poljak. 1961, Fl.
SSSR, **26:** 529; Klok. 1962, Fl. URSR, **11:** 327; Tutin, 1976, Fl.
Europ. **4:** 185; Oktyabreva and others, 1978, Nauchn. Dokl. Vyssh.
Shkoly, Biol. Nauki, **18:** 93; Lellep, 1978, Eesti NSV Fl. **6:** 242;
Galinis, 1980, Lietuv. TSR Fl. **6:** 125; 1 Leonova, 1987, Novosti
Sist. Vyssh. Rast. **24:** 187; Kotov, 1987, Opred. Vyssh. Rast. Ukr.:
340.—*Oligosporus dracunculus* (L.) Poljak. 1961, Tr. Inst. Bot.
Akad. Nauk KazSSR, **11:** 166.

Type: Siberia and Tataria ("in Sibiria, Tataria").
North (Dvina-Pechora: Kotlas—ecdemic); *Baltic* (ecdemic and
in cultivation); *Center* (Ladoga-Ilmen; Upper Dnieper and Upper
Volga:–ecdemic; Volga-Kama: south; Volga-Don); *West* (Dnieper;
Moldavia; Black Sea); *East*; *Crimea*.—In saline meadows, on
coastal buffs, river sands and gravel-beds, forest glades and
edges, in meadows, sheep's fescue-feather grass and feather
grass steppes, scrublands, inundated forests, by roadsides, in
habitations; cultivated as vegetable and aromatic plant, often
escaped.—*General distribution*: Caucasus, Western and Eastern
Siberia (south), Far East (south), Russian Central Asia; Central
Europe, Mediterranean, Asia Minor, Mongolia, Japan-China (north);
North America.—2n = 18, 36, 54, 72.

Note. In the European part of the former USSR, two varieties of this species have been recorded: var. *dracunculus*—leaves and stems entirely glabrous, and var. *pilosus* Krasch.—leaves and stems more or less densely pilose, later scatteredly hairy.

29. **A. glauca** Pall. ex Willd. 1803, Sp. Pl. **3**, 3: 1831; Poljak. 1961, Fl. SSSR, **26**: 535; Tutin, 1976, Fl. Europ. **4**: 185; Oktyabreva and others, 1978, Nauchn. Dokl. Vyssh. Shkoly, Biol. Nauki, **12**: 94; Galinis, 1980, Lietuv. TSR Fl. **6**: 125; Voloskova, 1986, Nauchn. Dokl. Vyssh. Shkoly, Biol. Nauki, **8**: 74.—*Oligosporus glaucus* (Pall. ex Willd.) Poljak. 1961, Tr. Inst. Bot. Akad. Nauk KazSSR, **11**: 166.

Type: Siberia ("In Sibiria").

North (Karelia-Murmansk: Kandalaksha—ecdemic; Dvina-Pechora—ecdemic); *Baltic* (Lithuania—ecdemic); *Center* (Upper Volga—ecdemic; Volga-Kama: east); *West* (Black Sea: east); *East* (Trans-Volga: east).—On solonetzes and saline meadows, coastal buffs, river sands and gravel-beds, in sheep's fescue-feather grass and herbage steppes, birch forests, by roadsides.—*General distribution*: Western and Eastern Siberia (south), Russian Central Asia (north); Mongolia; North America.—2n = 18, 36.

Section 5. Salsoloides Leonova, 1988, Novosti Sist. Vyssh. Rast. **25**: 144.

Semishrublets with stems more or less woody in lower part; pubescence of stellate hairs; leaves ternate or pinnatisect (partly bipinnatisect) with narrowly winged petioles without auricles; heads elliptical.

Type: *A. salsoloides* Willd.

30. **A. salsoloides** Willd. 1803, Sp. Pl., **3**, 3: 1832; Poljak. 1961, Fl. SSSR, **26**: 538; Tutin, 1976, Fl. Europ. **4**: 185; Kotov, 1987, Opred. Vyssh. Rast. Ukr.: 340.—*A. tanaitica* Klok. 1954, Bot. Mat. (Leningrad), **16**: 362, 364; id. 1962, Fl. URSR, **11**: 337.—*Oligosporus salsoloides* (Willd.) Poljak. 1961, Tr. Inst. Bot. Akad. Nauk SSSR, **11**: 168.—(Plate XVII, 1).

Type: The Don Region ("In montibus cretaceis Tanais").

Center (Volga-Kama: south; Don: south and east); *West* (Dnieper: east; Black Sea: east); *East* (Lower Don; Trans-Volga).—On outcrops of limestone, chalk and marl, in stony steppes.—*General distribution*: Caucasus (Ciscaucasia), Western Siberia (southwest).

Section 6. Campestres Korobkov, 1981, Polyny Sev.-Vost. SSSR: 112; Amelczenko, 1986, Novoe o Flore Sib.: 241.—*Artemisia* sect. *Scopariae* Krasch. ex Ameljcz. 1986, Novoe o Fl. Sib.: 241.

Semishrublets, herbaceous perennials, less often annuals and biennials; pubescence of simple hairs or hairs branched at base. Leaves with ternate or pinnati- to tri-pinnatisect leaves, predominantly with auricles. Heads ovate and ellipsoidal to subglobose.

Lectotype: *A. campestris* L.

Note. Besides the species described below, in Moscow Region there is one more Far Eastern species, *A. desertorum* Spreng. (Ignatov and others, 1990, in *Flor. Issl. Mosk. Obl.* [Floristic Studies in Moscow Region]: 94).

31. **A. campestris** L. 1753, Sp. Pl.: 846; Poljak. 1961, Fl. SSSR, **26**: 553; Tutin, 1976, Fl. Europ. **4**: 185.—*Oligosporus campestris* (L.) Cass. 1826, Dict. Sci. Nat. **36**: 25; Poljak. 1961, Tr. Inst. Bot. Akad. Nauk KazSSR, **11**: 167.—*A. dniproica* Klok. 1962, Fl. URSR, **11**: 556, 330, p. p.; Klok. 1987, Opred. Vyssh. Rast. Ukr.: 340.—*A. campestris* subsp. *campestris* Tutin, 1976, op. cit.: **4**: 186, p. p.

Type: Europe ("In Europae campis apricis, aridis").

North (Karelia-Murmansk; Dvina-Pechora: Arkhangelsk and Vologoda regions); *Baltic*; *Center*; *West* (Dnieper: north).—In meadows, on railroad dumps, in forest glades and edges, pine forests, by roadsides, on edges of fields, in habitations, predominantly in sandy and sandy-loam soils as also on outcrops of limestone.—*General distribution*: Scandinavia, Central and Atlantic Europe; North America (ecdemic).—2n = 36.

32. **A. commutata** Bess. 1835, Bull. Soc. Nat. Moscou, **8**: 70; Poljak. 1961, Fl. SSSR, **26**: 551; Tutin, 1976, Fl. Europ. **4**: 185.—*A. commutata* var. (β.) *gebleriana* Bess. 1835, op. cit.: 72; Krasch. 1949, in Kryl. Fl. Zap. Sib. **11**: 2447.—*A. commutata* var. *uralensis* Krasch. 1976, Fl. Yugo-Vost. Evrop. Chasti SSSR, **6**: 352.—*Oligosporus commutatus* (Bess.) Poljak. 1961, Tr. Inst. Bot. Akad. Nauk KazSSR, **11**: 167.

Lectotype: Bashkiria ("In Baschkiria").

Center (Volga-Kama: south and east; Volga-Don: Zhiguli); *East* (Trans-Volga; Lower Volga: east).—On outcrops of limestone 166 and allied rocks, clayey slopes, saline meadows, forest glades and edges, river sands and gravel-beds, in steppes, by roadsides, on edges of fields.—*General distribution*: Western and Eastern Siberia, Far East (south); Mongolia.—2n = 36.

33. **A. bargusinensis** Spreng. 1826, Syst. Veg. **3**: 493; Poljak. 1961, Fl. SSSR, **26**: 555; Tutin, 1976, Fl. Europ. **4**: 185.—*A. commutata* Bess. var. (α) *helmiana* Bess. 1835, Bull. Soc. Nat. Moscou,

234

8: 71.—*Oligosporus bargusinensis* (Spreng.) Poljak. 1961, Tr. Inst. Bot. Akad. Nauk KazSSR, **11**: 167.

Type: Eastern Siberia, the Barguzin River ("Deserta Bargusin ultra lacum Baical").

Center (Volga-Kama: central Urals); *East* (Trans-Volga: southern Urals).—On steppe, stony and rubbly slopes, river sands and gravel-beds, in forest glades and edges, scrublands.—*General distribution*: Western and Eastern Siberia (south), Far East.

34. **A. marschalliana** Spreng. 1826, Syst. Veg. **3**: 496; Krasch. 1949, in Kryl. Fl. Zap. Sib. **11**: 2773; Klok. 1962, Fl. URSR, **11**: 328; Leonova, 1987, Novosti Sist. Vyssh. Rast. **24**: 189.—*A. inodora* Bieb. 1808, Fl. Taur.-Cauc. **2**: 295, non Mill. 1768; Krasch. 1936, Fl. Yugo.-Vost. Evrop. Chasti SSSR, **6**: 351.—*A. campestris* L. var. (β.) *sericea* Fries, 1819, Fl. Halland. Diss.: 131.—*A. sericophylla* Rupr. 1845, Beitr. Pflanzenk. Russ. Reich. **2**: 41, in textu.—*A. campestris* f. *sericea* (Fries) Korsh. 1898, Tent. Fl. Ross. Or.: 216.—*A. sosnovskyi* Krasch. et Novopokr. 1949, Bot. Mat. (Leningrad), **11**: 178, p. p.—*A. campestris* var. *marschalliana* (Spreng.) Poljak. 1961, Fl. SSSR, **26**:. 554.—*A. campestris* var. *sericophylla* (Rupr.) Poljak. 1961, op. cit.: 554.—*A. campestris* subsp. *sericea* (Fries) Lemke et Rothm. 1976, in Rothm. Excursionsfl. Deutschl.: 548; Galinis, 1980, Lietuv. TSR Fl. **6**: 127, 316.—*A. campestris* var. *sericea* (Fries) Lellep, 1978, Eesti NSV Fl. **6**: 246.—*A. campestris* subsp. *campestris* auct.: Tutin, 1976, Fl. Europ. **4**: 186, p. p.—*A. arenaria* auct. non Bess.: Kotov, 1987, Opred. Vyssh. Rast. Ukr.: 340.

Lectotype: Northwestern Ukraine ("Volhynia, Besser. Hb. M.B.").

North (Karelia-Murmansk: Shore of Lake Ladoga; Dvina-Pechora: mouth of North Dvina, vicinity of Arkhangelsk, Plesetsk Dist. Arkhangelsk Region); *Baltic* (Saarema Island, Baltic Sea coast, the Viliya River near Kaunas); *Center* (Ladoga-Ilmen; Upper Volga; Volga-Kama: south; Volga-Don); *West*; *East*; *Crimea*.—In steppes, pine forests, forest glades and edges, in sandy and sandy-loam meadows, sands of terraces above floodplain, outcrops of limestone, chalk and other rocks, by roadsides.—*General distribution*: Caucasus, Western and Eastern Siberia (south), Russian Central Asia; Mediterranean (east), Asia Minor (north).

Note. Represented by varieties: a) var. *marschalliana*—whole plant, excluding involucral bracts and corolla, densely hairy; b) var. *tschernieviana* (Bess.) Leonova (1987, *Novosti Sist. Vyssh. Rast.* **24**: 191.—*A. tschernieviana* Bess. 1835, *Bull. Soc. Nat. Moscou*, **8**: 31; Tutin, 1976, *Fl. Europ.* **4**: 185, p. p.; Kotov, 1987,

Opred. Vyssh. Rast. Ukr.: 340.—*A. sosnovskyi* Krasch. et Novopokr. 1949, *Bot. Mat.* (*Leningrad*), **11**: 178.—*A. dniproica* Klok. 1962, *Fl. URSR*, **11**: 556, 350, p. p.—*A. arenaria* var. *tschernieviana* (Bess.) Grossh. 1934, *Fl. Kavk.* **4**: 141.—*A. arenaria* auct. non DC.: Krasch. 1936, op. cit.: 352, p. p.; Poljak. 1961, op. cit.: 540; Klok. 1962, op. cit.: 333, p. p.)—whole plant glabrous, less often involucral bracts and corolla with isolated hairs. The latter variety was described from Astrakhan Region between the villages of Krasnyi Yar and Selitrenoe (lectotype: "Krasnojar ad Selitrenoi (herb. Fischer)") but occurs through the range of the species.

35. **A. bottnica** Lundstr. ex Kindb. 1877, Svensk. Fl.: 301; Nym. 1899, Consp. Fl. Europ. Suppl. **2**: 170; Leonova, 1978, Bot. Zhurn. **63**, 1: 85.—*A. campestris* L. var. (γ.) *bottnica* (Lundstr. ex Kindb.) Hartm. 1879, Handb. Scand. Fl., ed. 11: 7.—*A. commutata* Bess. var. *bottnica* (Lundstr. ex Kindb.) Krasch. 1946, Nat. Ist. Fl. Rastit. SSSR, **2**: 152.—*A. borealis* Pall. subsp. *bottnica* (Lundstr. ex Kindb.) Hult. 1950, Atlas: map 1721.—*A. campestris* L. subsp. *bottnica* (Landstr. ex Kindb.) Tutin, 1976, Fl. Europ. **4**: 186.—*A. campestris* auct. non L.: Poljak. 1961, Fl. SSSR, **26**: 553, p. p.

Type: Northern Sweden, mouth of the Pite-Elb River ("Norrl. Vb. vid. utloppet af Piteelf").

North (Karelia-Murmansk; Dvina-Pechora); *Center* (Ladoga-Ilmen).—Only on coastal sands of Lake Onega.—*General distribution*: Scandinavia (coast of the Gulf of Bothnia).—2n = 36.

168   O 36. **A. trautvetteriana** Bess. 1845, Mém. Pres. Acad. Sci. Pétersb. Div. Sav. **4**: 464, excl. var. *erecta*; Poljak. 1961, Fl. SSSR, **26**: 537, Tutin, 1976, Fl. Europ. **4**: 185; Leonova, 1987, Novosti Sist. Vyssh. Rast. **24**: 191.—*A. pauciflora* Bieb. var. (β.) *cernua* Bess. 1835, Bull. Soc. Nat. Moscou, **8**: 19.—*A. arenaria* DC. var. (β.) *cernua* (Bess.) DC. 1838, Prodr. **6**: 94.—*Oligosporus trautvetterianus* (Bess.) Poljak. 1961, Tr. Inst. Bot. Akad. Nauk KazSSR, **11**: 168.—*Artemisia arenaria* auct. non DC. 1838: Klok. 1961, Fl. URSR, **11**: 336, p. p.

Type: Coast of the Black Sea between Odessa and the mouth of the Dniester River ("In sabulosis maritimis Odessam Inter Ostia Turae").

West (Moldavia: south; Black Sea: coast of the Black and Azov seas).—On coastal sands.—Endemic.

37. **A. arenaria** DC. 1838, Prodr. **6**: 94. excl. var. *cernua*; Poljak. 1961, Fl. SSSR, **26**: 540, p. p.; Leonova, 1987, Novosti Sist. Vyssh. Rast. **24**: 192.—*Oligosporus arenaria* (DC.) Poljak.

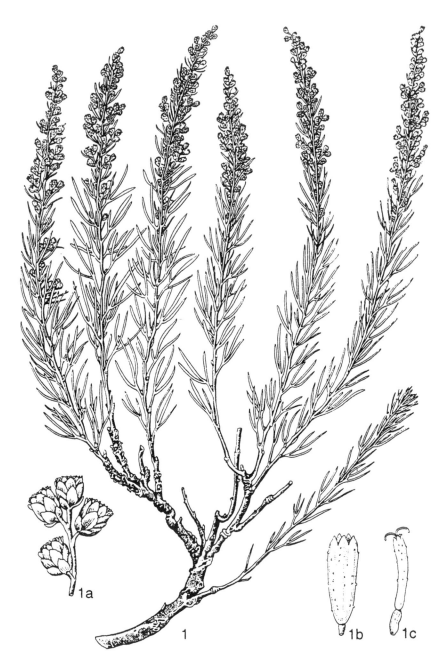

Plate XVII.
1—*Artemisia salsoloides* Willd., 1a—part of inflorescence, 1b—disk flower,
1c—peripheral flower.

1961, Tr. Inst. Bot. Akad. Nauk KazSSR, **11**: 168, p. p.—*Artemisia pauciflora* auct. non Web. 1775: Bieb. 1808, Fl. Taur.-Cauc. **2**: 290.—*A. albicerata* auct. non Krasch.: Filat. 1966, Fl. Kazakhst. **9**: 110, p. p.—*A. tschernieviana* auct. non Bess.: Tutin, 1976, Fl. Europ. **4**: 185, p. p.

Lectotype: The Volga River near Astrakhan ("Ex areni mobilis ad Wolgam circa Astrachan").

West (Black Sea: Dnepropetrovsk and Kherson regions); *East* (Lower Don; Lower Volga).—On shifting and more or less fixed, sands.—*General distribution*: Caucasus (Ciscaucasia and eastern Transcaucasia), Russian Central Asia (north).

38. **A. borealis** Pall. 1776, Reise, 3, Anhang.: 755; Poljak. 1961, Fl. SSSR, **26**: 557.—*A. campestris* L. subsp. *borealis* (Pall.) H.M. Hall et Clements, 1923, Publ. Carnegie Inst. Wash.: 326, 122; Tutin, 1976, Fl. Europ. **4**: 186.—*Oligosporus borealis* (Pall.) Poljak. 1961, Tr. Inst. Bot. Akad. Nauk KazSSR, **11**: 167.

Type: Western Siberia, lower reaches of the Ob' River ("In rupestribus arcticae plagae, circa Obum flavium").

Arctic (Novaya Zemlya; Arctic Europe).—In stony, rubbly and lichen tundras, on coastal sands and gravel-beds, stony slopes.—*General distribution*: Western Siberia (Altai), Eastern Siberia, Arctic, Far East (Sakhalin, Kamchatka); Scandinavia; North America.—$2n = 18, 36$.

Note. There are two varieties in the circumscription of this species: a) var. *borealis*—whole plant, excluding involucral bracts, densely hairy; b) var. *adamsii* (Bess.) Leonova (1987, *Novosti Sist. Vyssh. Rast.* **24**: 194.—*A. borealis* β. *adamsii* Bess. 1835, *Bull. Soc. Nat. Moscou*, **8**: 83).—whole plant glabrous, less often involucral bracts and corollas with isolated hairs.

39. **A. scoparia** Waldst. et Kit. 1802, Descr. Icon. Pl. Rar. Hung. **1**: 66, tab. 65; Poljak. 1961, Fl. SSSR, **26**: 560; Tutin, 1976, Fl. Europ. **4**: 186; Lellep, 1978, Eesti NSV Fl. **6**: 247; Galinis, 1980, Lietuv. TSR Fl. **6**: 127; Leo nova, 1987, Novosti Sist. Vyssh. Rast. **24**: 194.—*Oligosporus scoparius* (Waldst. et Kit.) Less. 1834, Linnaea, **9**: 191; Poljak. 1961, Tr. Inst. Bot. Akad. Nauk KazSSR, **11**:. 167.

Type: Hungary ("Crescit per totam Hungarian inde a limitibus Vallachiae et Serviae usque in Austriam, ubi eam ad Leytam fluvium legimus").

North (Karelia-Murmansk: ecdemic); *Baltic* (ecdemic); *Center*; *West*; *East*; *Crimea*.—In more or less weedy meadows, on coastal sands and gravel-beds, coastal buffs, in forest glades and edges, on steppe slopes, alkali lands, outcrops of limestones and chalk, sands of terraces above floodplain, edges of fields, in habitations.— *General distribution*: Caucasus, Western and Eastern Siberia (south), Far East, Russian Central Asia; Central and Atlantic Europe, Mediterranean, Asia Minor, Iran.—2n = 16, 36.

170   SUBGENUS **3. *SERIPHIDIUM*** (Bess.) Peterm. 1848, Deutschl. Fl.: 294.—*Artemisia* sect. *Seriphidium* Bess. 1829, Bull. Soc. Nat. Moscou, **1**, 8: 222; Poljak. 1961, Fl. SSSR, **26:** 562.—*Seriphidium* (Bess.) Poljak. 1961, Tr. Inst. Bot. Akad. Nauk KazSSR, **11:** 171

Heads obovate, ellipsoidal, or subcylindrical; flowers (1) 3–7(11) all bisexual; outer involucral bracts (one-fourth) one-third to one-half as long as inner; corolla and receptacle glabrous.

Lectotype: *A. maritima* L.

Section 7. Seriphidium Bess. 1829, Bull. Soc. Nat. Moscou, **2**, 8: 222; Poljak. 1961, Fl. SSSR, **26:** 573.—*Artemisia* sect. *Sclerophyllum* Filat. subsect. *Kazachstanicae* Filat. 1986, Novosti Sist. Vyssh. Rast. **23:** 227.—*Artemisia* sect. *Halophilum* Filat. 1986, ibid.

Semishrublets, more or less densely pubescent with simple hairs, less often subglabrous, with perennial shoots more or less woody near base; leaves small, one- to three-pinnate or palmate, terminal lobes almost filiform and linear to oblong and elliptical; heads in panicle, less often in conusoid panicle.

Type: lectotype of subgenus.

40. **A. maritima** L. 1753, Sp. Pl.: 846; Poljak. 1961, Fl. SSSR,· **26:** 573, p. p.; Leonova, 1971, Novosti Sist. Vyssh. Rast. **7:** 282; Tutin, 1976, Fl. Europ. **4:** 181; Filat. 1984, Novosti Sist. Vyssh. Rast. **21:** 177.—*Seriphidium maritimum* (L.) Poljak. 1961, Tr. Inst. Bot. Akad. Nauk KazSSR, **11:** 172.—(Plate XVIII, 1).

Lectotype: Northern part of Western Europe ("In Europae septentrionalioris littoribus maris").

a. Subsp. **humifusa** (Fries ex C. Hartm.) K. Persson, 1974, Opera Bot. (Lund), **35:** 150; Tutin and K. Persson, 1976, Fl. Europ. **4:** 181; Leonova, 1987, Novosti Sist. Vyssh. Rast. **24:** 195.—*A. maritima* var. *humifusa* Fries ex C. Hartm. 1879, Handb. Skand. Fl., ed. 11: 8.—*A. maritima* var. *osiliensis* Lellep, 1963, Uch. Zap. Tartusk. Univ. 136, Tr. Bot. **6:** 42; id. 1978, Eesti NSV

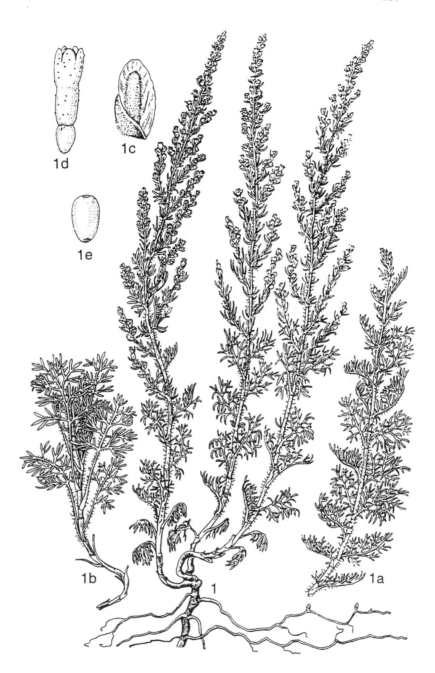

Plate XVIII.

1—*Artemisia maritima* L., 1a—part of young flowering shoot, 1b—vegetative shoot, 1c—involucral bract, 1d—flower, 1e—achene.

Fl. **6**: 251.—*A. maritima* var. *osiliensis* f. *subsalina* Lellep, 1963, op. cit.: 42; id. 1978, op. cit.: 252.

Sterile shoots rather numerous, 2–9 cm high; stems 5–25(40) cm high, densely lanate-tomentose, somewhat woody at base; lower cauline leaves 2.0–5.0(5.5) cm long, 1.7–3.0 cm wide; heads 4–5 mm long, 2–3 mm wide, sessile or on stalk 1.0–1.5 mm long, in panicles 6–13 cm long, 1–4(6) cm wide, with branches 1–6(15) cm long divergent at acute angle, yellowish, less often reddish-violet in upper part.

Lectotype: Sweden, Gotland Island ("Gotland Boxarfve").

North (Karelia-Murmansk: Petrozavodsk—ecdemic); *Baltic* (Estonia: western part of Saarema Island).—On stony coasts, in coastal swamps and meadows.—*General distribution*: Scandinavia (Åland and Gotland islands).—2n = 54.

Note. Apparently, subsp. *maritima*, widely distributed on the coasts of northern and western Europe, is not found in the former USSR. It has been reported for Leningrad [St. Petersburg] Region (Florovskaya, 1965, *Fl. Leningr. Obl.* **4**: 251, etc.), but has not been confirmed in recent times.

41. **A. lerchiana** Web. ex Stechm. 1775, Dissert. Artem: 24, 25, s. str. (excl. var. β.); Willd. 1803, Sp. Pl. **3**, 3: 1838; Poljak. 1961, Fl. SSSR, **26**: 579; Leonova, 1971, Novosti Sist. Vyssh. Rast. **7**: 283; Tutin and K. Persson, 1976, Fl. Europ. **4**: 181; Filat. 1984, Novosti Sist. Vyssh. Rast. **21**: 168, p. p.—*A. fragrans* Willd. 1803, op. cit.: 1835.—*A. astrachanica* Poljak. 1954, Bot. Mat. (Leningrad), **16**: 422.—*Seriphidium lerchianum* (Web. ex Stechm.) Poljak. 1961, Tr. Inst. Bot. Akad. Nauk KazSSR, **11**: 172.—*S. fragrans* (Willd.) Poljak. 1961, ibid.: 173.—*Artemisia pseudofragrans* Klok. 1962, Fl. URSR, **11**: 342, 557.

Lectotype: Astrakhan Region, village of Chernyi Yar ("Astrachaniae ut at ad ripam latum nigram (Tschornoi Jar) Wolgae fluvii").

171 *West* (Black Sea: Kherson Region—coast of Sivash); *East* (Lower Don; Trans-Volga: south; Lower Volga); *Crimea.*—In saline and stony steppes, on banks of water bodies, depressions in sand, rarely on limestones.—*General distribution*: Caucasus, Western Siberia (southwest), Russian Central Asia (northwest); Central Europe (Romania, Dobruja), Asia Minor, Iran.—2n = 18, 36.

42. **A. nutans** Willd. 1803, Sp. Pl. **3**, 3: 1831; Leonova, 1971, Novosti Sist. Vyssh. Rast. **7**: 284; id. 1987, ibid. **24**: 197; Tutin and K. Persson, 1976, Fl. Europ. **4**: 182; Kotov, 1987, Opred.

Vyssh. Rast. Ukr.: 340.—*A. cretacea* Kotov, 1962, Fl. URSR, **11:**
571, 347.—*A. monogyna* auct. non Waldst. et Kit.: Poljak. 1961,
Fl. SSSR, **26:** 574, p. p.—*A. santonica* auct. non L.: Filat. 1984,
Novosti Sist. Vyssh. Rast. **21:** 176, p. p.

Lectotype: Ukraine, chalk slopes facing the Severnyi Donets
River ("E montibus cretacais ad fluvium Donez").

West (Volga-Don: south and east); *West* (Dnieper: east); *East*
(Lower Don: northeast; Trans-Volga: south; Lower Volga: vicinity
of Uralsk).—On chalk outcrops.—*General distribution*: Western
Siberia (southwest), Russian Central Asia (northwest).

O 43. **A. dzevanovskyi** Leonova, 1969, in E. Wulf, Fl. Kryma,
3, 3: 222; id. 1970, Novosti Sist. Vyssh. Rast. **6:** 198; Tutin and
K. Persson, 1976, Fl. Europ. **4:** 181.—*A. lerchiana* auct. non Web.
et Stechm.: Filat. 1984, Novosti Sist. Vyssh. Rast. **21:** 168, p. p.

Type: Crimea, southwestern part of Tarkhankut Peninsula
("Tarkhankut Peninsula, 3–4 km southeast of the village of Olenevka
near the fish factory*, on limestone buffs facing the sea").

Crimea (Tarkhankut Peninsula).—On limestone buffs facing
the sea).—Endemic.

44. **A. taurica** Willd. 1803, Sp. Pl. **3,** 3: 1837; Poljak. 1961, Fl.
SSSR, **26:** 578; Leonova, 1969, in E. Wulf Fl. Kryma, **3,** 3: 22;
Tutin and K. Persson, 1976, Fl. Europ. **4:** 182; Filat. 1984, Novosti
Sist. Vyssh. Rast. **21:** 175.—*A. caspia* B. Keller et N. Kom. 1936,
Fl. Yuzo-Vost. Evrop. Chasti SSSR, **6:** 356, p. p.—*Seriphidium
tauricum* (Willd.) Poljak. 1961, Tr. Inst. Bot. Akad. KazSSR, **11:**
173.—*Artemisia graveolens* Minat. 1965, Ukr. Bot. Zhurn. **22,** 3:
30.—*A. taurica* subsp. *graveolens* (Minat.) Vlas. 1968, Tr.
Vsesoyuzn. Inst. Lek. Arom. Rast. **13:** 489.—(Plate XIX, 1).

Type: Crimea ("In Tauria").

Center (Volga-Kama: Udmurtia, Uva station—ecdemic); *West*
(Black Sea: near Sivash); *East* (Lower Don; Lower Volga); *Crimea*.—
In solonetz and saline steppes, on stony slopes, in lower and
middle mountain zones, by roadsides, in habitations.—*General
distribution*: Caucasus; Asia Minor.—2n = 18, 36.

45. **A. santonica** L. 1753, Sp. Pl.: 845; Leonova, 1969, in E.
Wulf, Fl. Kryma, **3,** 3: 218, 360; id. 1971, Novosti Sist. Vyssh.
Rast. **7:** 285; Tutin and K. Persson, 1976, Fl. Europ. **4:** 182; Filat.
1984, Novosti Sist. Vyssh. Rast. **21:** 176.—*A. monogyna* Waldst.
et Kit. 1802, Descr. et Icon. Pl. Rar. Hung. **1:** 77, tab. 75; Poljak.

*where tinned fish food is made—General Editor.

1961, Fl. SSSR, **26**: 574.—*A. maritima* var. (μ.) *boschniakiana*
Bess. 1834, Bull. Soc. Nat. Moscou, **7**, 1: 39.—*A. boschniakiana*
(Bess.) DC. 1838, Prodr. **6**: 104.—*A. maritima* subsp. *monogyna*
(Waldst. et Kit.) Gams, 1928, Ill. Fl. Mitteleur. **6**, 2: 664.—*A.
stepposa* B. Keller, 1936, Fl. Yugo-Vost. Evrop. Chasti SSSR, **6**:
355.—*A. praticola* Klok. 1962, Fl. URSR, **11**: 557, 345.—*Seriphidium
monogynum* (Waldst. et Kit.) Poljak. 1961, Tr. Inst. Bot. Akad.
Nauk KazSSR, **11**: 172.—*A. kumykorum* Minat. 1965, Ukr. Bot.
Zhurn. **27**, 3: 32.—*A. salina* auct. non Willd.: B. Keller, 1936, Fl.
Yugo-Vost. Evrop. Chasti SSSR, **6**: 354.

Lectotype: Vicinity of Astrakhan ("In Tataria, Persia, etc.").

Center (Volga-Kama—ecdemic, in the vicinity of Kazan; Volga-
Don: south and east); *West*; *East*; *Crimea*.—On solonetzes and
solonchaks, in saline meadows, steppe depressions, on banks of
water bodies, saline slopes in lower mountain zone.—*General
distribution*: Caucasus, Russian Central Asia (northwest); Central
Europe (southeast).

46. **A. nitrosa** Web. ex Stechm. 1775, Dissert. Artem.: 24;
Poljak. 1961, Fl. SSSR, **26**: 580; Leonova, 1971, Novosti Sist.
Vyssh. Rast. **7**: 286; Tutin and K. Persson, 1976, Fl. Europ. **4**:
181; Filat. 1984, Novosti Sist. Vyssh. Rast. **21**: 166; Puzyrev, 1986,
Bot. Zhurn. **71**, 12: 256.—*A. maritima* L. f. *citriodora* Kazak. ex
Poljak. 1961, Fl. SSSR, **26**: 574.—*Seriphidium nitrosum* (Web. ex
Stechm.) Poljak. 1961, Tr. Inst. Bot. Akad. Nauk KazSSR, **11**: 172.

Type: Western Siberia ("In montosis lacus salsi Utschjum
Krasnojarensis tractus sub finem Augusti adhuc florentem Juneni").

Center (Volga-Kama: Udmurtia, Uva station and votkinsk—
ecdemic); *East* (Trans-Volga: center and south; Lower Volga: delta
and left bank of the Volga River).—In saline meadows and alkaline
soils, on banks of salt lakes and by roadsides.—*General distribution*:
Western and Eastern Siberia (south), Russian Central Asia (north);
Dzhungaria-Kashgaria (Dzhungaria), Mongolia (north).

47. **A. compacta** Fisch. ex DC. 1838, Prodr. **6**: 102; Poljak.
1961, Fl. SSSR, **26**: 585; Leonova, 1971, Novosti Sist. Vyssh.
Rast. **7**: 288; Filat. 1984, Novosti Sist. Vyssh. Rast. **21**: 170.—
Seriphidium compactum (Fisch. ex DC.) Poljak. 1961, Tr. Inst.
Bot. Akad. Nauk KazSSR, **11**: 175.

Type: Altai, the Chui River ("ad Tschujam").

East (Trans-Volga: vicinity of Novouzensk; Lower Volga: coast
of Lakes Elton and Baskunchak).—On dry alkaline and rubbly-
clayey slopes, sandy banks of salt lakes.—*General distribution*:

Western Siberia (south), Eastern Siberia (southwest), Russian Central Asia (eastern part of the Kazak "melkosopochnik" [low-hilly area], basin of Lake Zaisan and Alakul), Dzhungaria-Tarbagatai, Mongolia (northwest).

48. **A. pauciflora** Web. 1775, in Stechm. Dissert. Artem.: 26; Poljak. 1961, Fl. SSSR, **26**: 589; Leonova, 1971, Novosti Sist. Vyssh. Rast. **7**: 289; Tutin and K. Persson, 1976, Fl. Europ. **4**: 182; Filat. 1984, Novosti Sist. Vyssh. Rast. **21**: 166.—*Seriphidium pauciflorum* (Web. ex Stechm.) Poljak. 1961, Tr. Inst. Bot. Akad. Nauk KazSSR, **11**: 175.

Type: Banks of the Volga River in the vicinity of Volgograd ("In ripa clata nigra Wolgae fluvii ut et Zarizinae ad Wolgam fluvium").

East.—On solonetzes and solonchaks, banks of saline lakes.— *General distribution*: Western Siberia (south), Russian Central Asia (north).—2n = 18.

49. **A. gracilescens** Krasch. et Iljin, 1949, Sist. Zam. Gerb. Tomsk. Univ. **1–2**: 2; Poljak. 1961, Fl. SSSR, **26**: 591; Leonova, 1971, Novosti Sist. Vyssh. Rast. **7**: 291; Tutin and K. Persson, 1976, Fl. Europ. **4**: 182; Filat. 1984, Novosti Sist. Vyssh. Rast. **21**: 167.—*Seriphidium gracilescens* (Krasch. et Iljin) Poljak. 1961, Tr. Inst. Bot. Akad. Nauk KazSSR, **11**: 175.

Type: Altai ("Altai Territory, Kuznetskii Steppe, saline lakes in pine groves, on solonetzes").

East (Lower Volga: delta of the Volga and on the Ural River).— On solonetzes and solonchaks, usually on sandy, sandy loam, and stony soils, on banks of saline lakes.—*General distribution*: Western Siberia (southwest), Russian Central Asia (north and northeast); Dzhungaria-Kashgaria, Mongolia (southwest).

50. **A. terrae-albae** Krasch. 1930, Otchet o Rab. Pochv.—Bot. Otd. Kazakh. Eksp. Akad. Nauk SSSR, **4**, 2: 269; Poljak. 1961, Fl. SSSR, **26**: 592; Leonova, 1971, Novosti Sist. Vyssh. Rast. **7**: 292; Filat. 1984, Novosti Sist. Vyssh. Rast. **21**: 169.—*A. maritima* L. subsp. *terrae-albae* (Krasch.) Krasch. 1935, Sorn. Rast. SSSR, **4**: 255; id. 1921, Bot. Mat. (Leningrad), **2**: 187, nom. nud.—*A. caspica* B. Keller et N. Kom. 1936, Fl. Yuvo-Vost. Evrop. Chasti SSSR, **6**: 356, p. p.—*Seriphidium terrae-albae* (Krasch.) Poljak. 1961, Tr. Inst. Bot. Akad. Nauk KazSSR, **11**: 175.

Lectotype: Kazakhstan, Sarysu River ("Kazakhstan, Turgai, Sarysu River in the lower reaches of vicinity of Kizil-Dzhangil area, near Kutala-Sai wormwood steppe").

East (Lower Volga: on banks of the Ural River).—On solonetzes and solonchaks with clayey, rubbly and rubbly-sandy soils.—*General distribution*: Russian Central Asia; Dzhungaria-Kashgaria, Mongolia (southwest).

51. **A. lessingiana** Bess. 1841, Linnaea, **13**: 90, 103; Poljak. 1961, Fl. SSSR, **26**: 594; Leonova, 1971, Novosti Sist. Vyssh. Rast. **7**: 292; Tutin and K. Persson, 1976, Fl. Europ. **4**: 182; Filat. 174 1984, Novosti Sist. Vyssh. Rast. **21**: 171.—*Seriphidium lessingianum* (Bess.) Poljak. 1961, Tr. Inst. Bot. Akad. Nauk KazSSR, **11**: 173.

Type: Orenburg Region, Ilek District, dry hills "Mertvye-soli" (E Regio Uralensis Marswa").

Center (Volga-Kama: Bashkiria, Karatau Range); *East* (Trans-Volga: southern spurs of Obshchii Syrt in the Urals).—In steppes, usually stony or rubbly slopes, steppe depressions.—*General distribution*: Western Siberia (south), Russian Central Asia (Mugodzhary, basin of the Irgiz River).

TRIBE 6. **ASTEREAE** Cass.[1]

Heads heterogamous or homogamous. Peripheral flowers ligulate; inner (disk) flowers tubular. Receptacle without scales. Anthers usually obtuse at base, without appendages. Style branches flattened, acute to subobtuse, with stigmatic surface of branches peripheral and not reaching their tips. Anthers helianthoid type. Achenes with pappus of scabrous bristles at apex, rarely without pappus. Leaves alternate, usually undivided.

Type: *Aster* L.

GENUS **73**. *GRINDELIA* Willd.
1807, Ges. Nutrf. Freunde Berlin Mag. **1**: 259

Heads heterogamous, 20–35 mm in dia, solitary at apices of stem and its branches, but often numerous. Involucre narrow, saucer-shaped, 11–23 mm in dia and 7–10 mm long; involucral bracts imbricate, linear-lanceolate or linear, without membranous border, rather stiff, usually deflexed, sticky. Receptacle somewhat convex, without scales. Peripheral flowers pistillate, ligulate, yellow,

---

[1]Treatment by N.N. Tzvelev.

Plate XIX.
1—*Artemisia taurica* Willd., 1a—middle leaf, 1b—part of inflorescence,
1c—involucral bracts.

sometimes absent; disk flowers tubular, bisexual or sterile, with yellow, five-toothed corolla. Achenes 2.5–4.3 mm long and 0.7–1.2 mm wide, more or less flattened from sides, cuneate-prismatic, glabrous, with five to nine veins, of which only two(four) more or less raised as ribs; pappus of 2–8 stiff, subulate, more or less scabrous, readily detaching bristles. More or less hairy (to subglabrous) perennials, biennials, or annual plants with erect, often strongly branched, leafy stems and alternate, undivided, usually more or less toothed leaves.

Type: *G. inuloides* Willd.

About 50 species in North and South America; some of them are ecdemic in other countries or are cultivated as medicinal or oil-producing plants.

Literature: Protopopova, V.V. and V.S. Tkachenko, 1979. Istoriya ta prognoz poshirennya *Grindelia squarrosa* (Pursh) Dunal. [The history and forecast of dispersal of *Grindelia squarrosa* (Pursh) Dunal]. *Ukr. Bot. Zhurn.*, **36**, 5: 457–461.

1. **G. squarrosa** (Pursh.) Dunal, 1819, Mém. Muz. Hist. Nat. (Paris), **5**: 50; Tamamsch. 1959, Fl. SSSR, **25**: 30; Schultz, 1976, Bot. Zhurn. **61**, 10: 1451; Kozhevnikov and Makhaeva, 1976, Bot. Zhurn. **61**, 4: 566; A. Hansen, 1976, Fl. Europ. **4**: 109; Borotnyjk, 1978, Ukr. Bot. Zhurn. **35**, 2: 128; Puzyrev, 1985, Bot. Zhurn. **70**, 2: 269; Myrza and others, 1987, Ukr. Bot. Zhurn. **44**, 6: 42; Tretyakov, 1990, Bot. Zhurn. **75**, 2: 261; Ignatov and others, 1990, Flor. Issled. Mosk. Obl.: 86.—*Donia squarrosa* Pursh, 1814, Fl. Am. Septentr. **2**: 559.—(Plate XX, 1).

Leaves sessile, linear-oblong or oblong-lanceolate, usually more or less toothed, less often partly entire.

Type: USA, the Missouri River ("In open prairies on the banks of the Missouri").

176 *Baltic* (vicinity of Riga and Kaliningrad); *Center* (Upper Dnieper: vicinity of Vitebsk; Upper Volga: Moscow Region); Volga-Kama: Udmurtia; Volga-Don: south); *West* (Dnieper; Moldavia; Black Sea); *Crimea* (vicinity of the village of Nikita).—By roadsides, in habitations, forest glades and edges, weedy meadows and pastures, on steppe slopes, in forest belts, sometimes cultivated in experimental fields and often as escape; entirely naturalized in many regions of Ukraine.—*General distribution*: Caucasus (ecdemic near Anapa), Far East (ecdemic in south); North America; sometimes cultivated and ecdemic in other countries.— $2n = 12, 24$.

175

Plate XX.

1—*Grindelia squarrosa* (Pursh) Dunal, 1a—achene; 2—*Solidago virgaurea*
L. subsp. *virgaurea*, 2a—achene; 3—*Bellis perennis* L., 3a—achene; 4—*Aster
amellus* L. subsp. *bessarabicus* (Bernh. ex Reichenb.) Soó, 4a—achene.

GENUS 76. *SOLIDAGO* L.
1753, Sp. Pl.: 878; id. 1754, Gen. Pl., ed. 5: 374

Heads heterogamous, 2.5–20.0 mm in dia, usually numerous, clustered in paniculate, less often corymbose or racemose inflorescences. Involucre 2–8 mm long and 2–10 mm in dia, bowl-shaped, less often cup-shaped; involucral bracts imbricate, ovate to linear. Receptacle flat or somewhat convex, without scales. Peripheral flowers pistillate, ligulate, yellow, sometimes very small and almost not exserted from involucre; disk flowers tubular, bisexual, with yellow, five-toothed corolla. Achenes 0.7–4.5 mm long and 0.2–0.8 mm wide, more or less terete, with 5–12 more or less distinct veins, short-hairy or less often glabrous; pappus of numerous scabrous bristles, usually longer than achene. More or less hairy (to subglabrous) perennial plants, with erect leafy stem and alternate, entire but usually more or less toothed leaves.

Lectotype: *S. virgaurea* L.

Nearly 120 species in the extratropical regions of the Northern Hemisphere, predominantly in North America.

Literature: Evtyukhova, M.A. 1959. Geograficheskie rasy zolotoi rozgi v klimaticheskikh usloviyakh Moskvy [Geographical races of European goldenrod in the climatic conditions of Moscow]. *Byull. Glavn. Bot. Sada Akad. Nauk SSSR*, **34**: 39.—Schultz, G.E. 1973. Severoevropeiskoe taksony is rodstva *Solidago virgaurea* L. s. l. (sem. Asteraceae) [North European taxa in the *Solidago virgaurea* L. s. l.—group (Family Asteraceae)]. *Novosti Sist. Vyssh. Rast.*, **10**: 248–257.—Zaitseva, T.A. 1976. Izmenchivost' morfologicheskih priznakov i ritmov razvitiya *Solidago virgaurea* L. v svyazi s geograficheskoi shirotoi i vertikal'noi poyasnost'yu [Variation of morphological characters and development rhythms of *Solidago virgaurea* L. in connection with geographic latitude and vertical zonality]. In *Okhrana Sredy i Rastsional'noe Ispolzovanie Rastitel'nykh Resursov* ["Conservation of Environment and Rational Exploitation of Plant Resources"]. Moscow: 36–37.—Orlova, T.G. 1986. O vidovoi samostoyatel'nosti *Solidago lapponica* Wither [Species independence of *Solidago lapponica* Wither]. In *Sovremennye Problemy Filogenii Rastenii* [Current Problems in Plant Phylogeny]. Moscow: 97–99.—Guzikowa, M. and P.F. Maycock. 1986. The invasion and expansion of three North American species of goldenrod [*Solidago canadensis* L. sensu lato, *S. gigantea* Ait. and *S. graminifolia* (L.) Salisb.] in Poland. *Acta Soc. Bot. Polon.* **55**, 3: 384.

1. Leaf blades with a few pairs of almost similar lateral veins; heads clustered in narrow-paniculate or racemose inflorescence; involucre 5–8 mm long; ligules of peripheral flowers 4–8 mm long. Plants 10–100 cm high, with tap root .... 1. **S. virgaurea** s. 1.

+ Leaf blades with one pair of distinct lateral veins arising from lower part of midrib and running almost parallel to it to leaf apex; heads clustered in broadly paniculate or corymbose inflorescence; involucre 2.0–4.5 mm long; ligules of peripheral flowers, 0.6–2.0 mm long. Plants 50–150 cm high, with long rhizome .... 2.

2. Heads sessile or on very short (to 0.7 mm) peduncles, clustered in fasicles at apices of branches of corymbose inflorescence; ligules of peripheral flowers 0.6–1.0 mm long, scarcely exserted from bowl-shaped involucre; cauline leaves linear or linear-lanceolate, scabrous from short hairs .... 5. **S. graminifolia.**

+ Heads on more than 1 mm long peduncle, on one side of branches of rather broad paniculate inflorescence; ligules of peripheral flowers 1–2 mm long, exserted from cup-shaped involucre; cauline leaves lanceolate, oblong- or linear-lanceolate .... 3.

177 3. Stems glabrous throughout or with few short hairs (usually near base of leaves); leaves rather stiff, obliquely upward directed, glabrous beneath or with few hairs on veins...
   .... 2. **S. serotinoides.**

+ Stems short-hairy throughout or atleast in upper half; leaves less stiff and usually strongly deflected sideways, more or less short-hairy beneath .... 4.

4. Leaves scatteredly hairy (often only on veins) beneath, as also stems, usually glabrous above; involucre 2–3 mm long; corolla of tubular flowers 2.4–2.8 mm long ....
   .... 3. **S. canadensis.**

+ Leaves densely short-hairy beneath, as also stem, usually also short-hairy above; involucre 3.0–4.5 mm long; corolla of tubular flowers 3–4 mm long .... 4. **S. altissima.**

Section 1. Solidago.

Heads relatively large, clustered in narrow-paniculate or racemose inflorescences; involucre cup-shaped, 5–8 mm long.

Type: type species.

1. **S. virgaurea** L. 1753, Sp. Pl.: 880; Juz. 1959, Fl. SSSR, **25:** 34; McNeill, 1976, Fl. Europ. **4:** 110, p. max. p.

Type: Europe ("in Europae pascuis siccis").

a. Subsp. **virgaurea**.—Inflorescence usually branched and rather lax; involucre 5.0–6.5 mm long; ligules of peripheral flowers 4.5–6.0 mm long; outermost involucral bracts two-fifths to half as long as the longest [inner]. Plant 25–100 cm high, with usually short-hairy stem; cauline leaves 10–16, gradually and relatively weakly reducing in upper part, their blades more or less elliptical, more or less hairy on margin and veins.—(Plate XX, 2).

Arctic (Arctic Europe: south of Kanin, lower reaches of the Pechora River); *North*; *Baltic*; *Center*; *West*; *East* (Lower Don; Trans-Volga).—In forest glades and edges, dry meadows and thinned-out forests, scrublands.—*General distribution*: Caucasus, Western- Siberia; Scandinavia, Central and Atlantic Europe, Mediterranean, Asia Minor.—2n = 18.

b. Subsp. **taurica** (Juz.) Tzvel. comb. nova.—*S. taurica* Juz. 1959, Fl. SSSR, **25:** 575, 36; Dobrocz. 1962, Fl. URSR, **11:** 27.— Like the preceding subspecies, but stems usually subglabrous and cauline leaves numerous (14–20) and rather abruptly and strongly reduced above, approximate.

Type: Crimea ("on sandy loam from the Ai-Petri mountains between Pendikyul and Uchansu near the highway").

Baltic (islands of Estonia); *Center* (Ladoga-Ilmen: Mozhaiskaya station and vicinity of St. Petersburg); *West* (Carpathians: south; Moldavia; Black Sea); *East* (Lower Don; Trans-Volga: southern Urals); *Crimea* (mountains).—In thinned-out forests, forest glades and edges, on stony slopes, usually on outcrops of limestones and other rocks.—*General distribution*: Central Europe (southeast), Mediterranean.

Note. Apparently, a more southern, mountain and calciphilic race.

c. Subsp. **stenophylla** (G.E. Schultz) Tzvel. comb. nova.—*S. lapponica* with subsp. *stenophylla* G.E. Schultz, 1973, Novosti Sist. Vyssh. Rast. **10:** 249.—*S. virgaurea* L. var. *angustifolia* Meinsh. 1878, Fl. Ingr.: 155, non Koch, 1837.—*S. stenophylla* (G.E. Schultz) Tzvel. 1993, Byull. Mosk. Obshch. Ispyt. Prir. Otd. Biol. **98,** 6: 99.—Inflorescence relatively shorter, usually lax; involucre 6–7 mm long. Plants 20–70 cm high, with usually sub-glabrous stem; cauline leaves 8–12, lower narrow-lanceolate, others usually oblong-linear; other characters as in subsp. *virgaurea*.

Type: Leningrad Region, vicinity of Luga ("Herb Fl. Ingr. N. 3056. In pratis siccioribus vel arenosis hinc inde copiose").

178   *North* (Karelia-Murmansk: south); *Baltic*; *Center* (Ladoga-Ilmen; Upper Dnieper: north; Upper Volga: rarely and predominantly in the western parts; Volga-Don: northwest).—In pine forests, heaths, on stony ridges and hills.—*General distribution*: Scandinavia (southeast).

Note. Apparently, related to sandy forms of topography of glacial origin, particularly with eskers. Unlike subsp. *virgaurea* that flowers in August–September, this subspecies usually flowers in July and has a lesser number of ligulate flowers in heads.

d. Subsp. **lapponica** (With.) Tzvel. comb. nova.—*S. lapponica* With. 1779, Arrang. Brit. Pl., ed. 3: 728; Juz. 1959, Fl. SSSR, **25**: 41; Rebristaya, 1987, Arkt. Fl. SSSR, **10**: 65, map. 15.—?*S. minuta* L. 1763, Sp. Pl., ed. 2: 1235.—?*S. virguarea* L. subsp. *minuta* (L.) Arcang. 1882, Comp. Pl. Ital.: 339; McNeill, 1976, Fl. Europ. **4**: 110, p. max. p.—General inflorescence dense, often racemose; involucre 6.5–8.0 mm long; ligules of peripheral flowers 6.0–7.5 mm long; outermost involucral bracts two-fifths to half as long as the longest. Plant 10–50 cm high, with usually subglabrous stem; cauline leaves 6–10, somewhat reducing in upper part, more or less elliptical, lower leaves usually broadly elliptical, more or less hairy on margin and veins.

Type: Fennoscandia ("in Lapland plant").

Arctic (Arctic Europe); *North* (Karelia-Murmansk: north and along the coast of the White Sea, also on Segozero Lake; Dvina-Pechora: predominantly north and basin of the Pechora River); *Center* (Volga-Kama: the Urals); *East* (Trans-Volga: southern Urals high mountains).—In various tundras, on meadow slopes of coastal terraces, in scrublands, forest glades and edges.—*General distribution*: Scandinavia.

Note. Possibly, this arctic and subarctic race of the species is not distinguishable from the alpine one, and then it should be called subsp. *minuta* (L.) Arcang. However, I am not fully convinced about this.

O e. Subsp. **jailarum** (Juz.) Tzvel. comb. nova.—*S. jailarum* Juz. 1959, Fl. SSSR, **25**: 575, 39; Dobrocz. 1962, Fl. URSR, **11**: 27.—Like the preceding subspecies, but with 8–12 cauline leaves, more dense inflorescence and with outermost involucral bracts one-third to two-fifths as long as the longest (involucral bracts in general more numerous).

Type: Crimea, Babugan-yaila ("slope of Babugan-yaila Bolshaya Polyana in beech elfin forests).

Crimea (yaila and in the upper part of the forest zone).—In cultivated meadows and on stony slopes, in glades and edges near upper boundary of forest.—Endemic.

f. Subsp. **alpestris** (Waldst. et Kit. ex Willd.) Reichenb. 1831, Fl. Germ. Excurs. **2**; 246.—*S. alpestris* Waldst. et Kit. ex Willd. 1803, Sp. Pl. **3**, 3: 2065; Juz. 1959, Fl. SSSR, **25**: 39.—*S. virgaurea* L. subsp. *minuta* auct. non (L.) Arcang.: McNeill, 1976, Fl. Europ. **4**: 110, p. p.—Like two preceding subspecies, but with more profuse pubescence (stems and leaves short-hairy beneath, and often also above), and the outermost involucral bracts not less than half as long as the longest.

Type: Carpathians ("in alpibus Carpathicis, Austriacis, Bohemicis").

West (Carpathians).—In alpine meadows and on stony slopes, in forest glades and edges, at upper boundary of forest, in sparse forest.—*General distribution*: Central Europe (southeast), Mediterranean (Balkans).—2n = 18.

Section 2. Unilateralis G. Don, 1830, in London, Hort. Brit.: 348.

Rather small heads on short stalk on one side of branches of broadly paniculate inflorescence; involucre cup-shaped, 2.0–4.5 mm long.

Lectotype: *S. canadensis* L.

2. **S. serotinoides** A. and D. Löve, 1982, Taxon, **31**, 2: 358.—*S. serotina* Ait. 1789, Hort. Kew. **3**: 211, non Retz, 1781: Juz. 1959, Fl. SSSR, **25**: 48; Dobrocz. 1987, Opred. Vyssh. Rast. Ukr.: 321.—*Aster latissimefolius* Kuntze var. *serotinus* (Ait.) Kuntze, 1891, Revis. Gen. Pl. **1**: 314.—*S. gigantea* Ait. subsp. *serotina* (Kuntze) McNeill, 1973, Bot. Journ. Linn. Soc. (London), **67**: 280; id. 1976, Fl. Europ. **4**: 110.—*S. gigantea* auct. non Ait.: Czer. 179 1973, Svod. Dopoln. Izm. "Fl. SSSR": 591; Oktyabreva and Tikhomir., 1987, Fl. Meshchery, **2**: 98; Puzyrev, 1989, Bot. Zhurn., **74**, 5: 763.

Type: North America ("America borealis").

Baltic; *Center*; *West* (Carpathians; Dnieper; Black Sea).—Cultivated as an ornamental plant in gardens and parks and often found as an escape.—*General distribution*: North America; cultivated and escaped in other extratropical countries.—2n = 36.

Note. The closely related American species, *S. gigantea* Ait. with 2n = 18, has not been found in the former USSR so far. It differs from *S. serotinoides* by glabrous or subglabrous (and not short-hairy) achenes and the leaves rather profusely hairy (and not glabrous or subglabrous) beneath, along veins.

3. **S. canadensis** L. 1753, Sp. Pl.: 879; Juz. 1959, Fl. SSSR, **25:** 48; McNeill, 1976, Fl. Europ. **4:** 110; Dobrocz. 1987, Opred. Vyssh. Rast. Ukr.: 32; Oktyabreva and Tikhomir., 1987, Fl. Meshchery, 2: 98.

Type: North America ("in Virginia, Canada").

Center (Upper Volga; Upper Don).—Cultivated as an ornamental plant in gardens and parks, sometimes found as an escape.—*General distribution*: North America; cultivated and escaped in other extratropical countries.

Section 3. Euthamia (Nutt.) DC. 1836, Prodr. **5:** 341.—*Solidago* L. subgen. *Euthamia* Nutt. 1818, Gen. North Amer. Fl. **2:** 162; Juz. 1959, Fl. SSSR, **25:** 50.

Heads small, in fasicles at apices of branches of corymbose inflorescence, subsessile; involucre bow-shaped, 3.5–4.2 mm long.

Type: *S. graminifolia* (L.) Salisb.

5. **S. graminifolia** (L.) Salisb. 1796, Prodr.: 199; Juz. 1959, Fl. SSSR, **25:** 50; McNeill, 1976, Fl. Europ. **4:** 111.—*Chrysocoma graminifolia* L. 1753, Sp. Pl.: 841.

Type: Canada ("in Canada").

West.—Cultivated as an ornamental plant in gardens and parks, sometimes as escaped.—*General distribution*: North America; cultivated and as escape in other extratropical countries.—2n = 18.

GENUS 77. *BELLIS* L.
1753, Sp. Pl.: 886; id. 1754, Gen. Pl., ed. 5: 378

Heads heterogamous, 10–30(50) mm in dia, solitary at apices of more or less long (2–25 cm high), leafless stems—scapes. Involucre 3–8 mm long, 5–12 mm in dia, saucer-shaped; involucral bracts biseriate, oblong-elliptical or elliptical, all almost equal, less often outer longer. Receptacle subobtuse-conical, less often weakly convex, without scales. Peripheral flowers pistillate, ligulate, white or pink, less often red; disk flowers tubular, bisexual, with yellow, four(five)-toothed corolla. Achenes, 1.2–1.5 mm long and 0.6–0.8 mm wide, strongly flattened from sides, obovate or oblong-obovate, short-hairy, with thick lateral ribs, bordering entire achenes, without pappus, less often with very short pappus. More or less hairy perennials, less of annual plants, with scape-like stem and rosulate, alternate, undivided but often toothed leaves.

Type: *B. perennis* L.

180   The genus includes seven species, distributed in Europe, southwestern Asia, and northwestern Africa as also in the Canaries and Azores islands, one of them is cultivated as ornamental plant in many other extratropical countries and often found as an escape.

1. Involucre 3–6 mm long; involucral bracts obtuse, all almost equal; leaves obovate, oblong- or orbicular-obovate, gradually narrowed into rather long (often longer than blade) petiole; root fibers of various thickness (0.2–1.0 mm) .... 1. **B. perennis.**
+ Involucre 6.5–8.0 mm long; involucral bracts subobtuse, outer distinctly exceeding inner; leaves oblong, gradually narrowed in shorter (always shorter than blade) petiole; root fibers numerous, almost all about 1 mm thick ....
....2. **B. sylvestris.**

1. **B. perennis** L. 1753, Sp. Pl.: 856; Tamamsch. 1959, Fl. SSSR, **25**: 54, Plate II, Fig. 2; Tutin, 1976, Fl. Europ. **4**: 111.— (Plate XX, 3).

Type: Europe ("in Europae apricis pascuis").

North (Karelia-Murmansk: south; Dvina-Pechora: southwest, vicinity of Arkhangelsk); *Baltic*; *Center* (Ladoga-Ilmen; Upper Dnieper; Upper Volga; Volga-Kama: vicinity of Izhevsk; Volga-Don); *West*; *East* (Lower Don); *Crimea*.—In meadows, forest glades and edges, thinned-out forests, scrublands, often cultivated as an ornamental plant and found as an escape.—*General distribution*: Caucasus, Far East (south, escape), Russian Central Asia (escape); Scandinavia, Atlantic and Central Europe, Mediterranean, Asia Minor, Japan-China (escape); North America (escaped); cultivated and occasionally escaped in other extratropical countries.—2n = 18.

Note. The "double" varieties are common in cultivation, in which all or almost all tubular flowers modified into ligulate flowers.

2. **B. sylvestris** Cyr. 1792, Pl. Rar. Neap. **2**: 22, tab. 4; Dobrocz. 1962, Fl. URSR, **11**: 35, in adnot; Ryndina and Kossykh, 1965, Novosti Sist. Vyssh. Rast. 1965: 301; Tutin, 1976, Fl. Europ. **4**: 112.

Type: Italy, vicinity of Naples ("Neapoli").

Crimea (south).—In juniper and hornbean forests, forest glades and edges, scrublands.—*General distribution*: Mediterranean.— 2n = 36, 54.

Note. In the south of Crimea also could be found the Mediterranean annual species *B. annua* L. with weakly developed root system having root fibers up to 0.4 mm thick, often somewhat separated inner leaves of the rosette, and smaller heads. Collections of *B. perennis* from Belogorsk District (12.V.1955, M. Popova and D. Dobroczaeva) are closer to this species.

GENUS **78.** *CALLISTEPHUS* Cass.
1825, Dict. Sci. Nat. **37:** 491

Heads heterogamous, 30–130 mm in dia, solitary at apices of stem and its branches. Involucre 20–50 mm long and 15–80 mm in dia, saucer-shaped or broadly cup-shaped, in many rows; outer involucral bracts larger than the ones following them, herbaceous, green, ciliate on margin, subobtuse or acute. Receptacle somewhat convex, without scales. Peripheral flowers (and often all flowers in a head) pistillate, ligulate, of various colors (usually pink, white, red, violet); disk flowers, if present, tubular, bisexual, with yellow five-toothed corolla. Achenes 3–5 mm long, more or less flattened, short-hairy, cuneate-oblong, with two to four ribs; pappus 2.8–6.0 mm long, of numerous scabrous bristles of almost same length. More or less crisped-hairy annual plants, 10–60 cm high, with erect leafy stems and alternate, undivided but usually toothed, less often lobate leaves.

Type: *C. chinensis* (L.) Neas.
A monotypic genus.

181   1. **C. chinensis** (L.) Nees. 1832, Gen. Sp. Aster.: 222; Golubkova, 1959, Fl. SSSR, **25:** 74.—*Aster chinensis* L. 1753, Sp. Pl.: 1877.

Type: China ("in China").
North; *Baltic*; *Center*; *West*; *East*; *Crimea.*—Cultivated as an ornamental plant in habitations, gardens and parks, in southernmost regions sometimes found as an escape.—*General distribution.* Far East (south); Japan-China; extensively cultivated in extratropical countries and occasionally found as an escape.—2n = 18.

Note. Numerous varieties of this species may be separated into two groups: large-headed, with usually solitary heads on stem, over 70 mm in dia, and small headed (or dwarf) with usually numerous but smaller (30–70 mm in dia) heads. The more ornamental varieties are "double" headed and all tubular flowers in them are modified into ligulate flowers.

GENUS **79. *BELLIDASTRUM*** Scop.
1760, Fl. Carniol.: 376

Heads heterogamous, 20–32 mm in dia, solitary at apices of more or less long (6–30 cm high) scapes. Involucre 7–9 mm long and 7–10 mm in dia, broadly cup-shaped; involucral bracts biseriate, almost equal, linear-lanceolate, acute, more or less scatteredly hairy. Receptacle obtusely conical, without scales. Peripheral flowers pistillate, ligulate, white or pink; disk flowers tubular, bisexual, with yellow, five-toothed corolla. Achenes 1.2–1.6 mm long and 0.3–0.5 mm wide, flattened, short-hairy, predominantly in upper half, oblong, with two ribs; pappus 3.5–4.0 mm long, of numerous scabrous bristles. More or less crisped-hairy perennial plants 5–30 cm high, with scapes and rosettes of alternate, undivided but usually crenate-toothed leaves on more or less long petiole.

Type: *B. michelii* Cass.
A monotypic genus.

1. **B. michelii** Cass. 1817, Dict. Sci. Nat. **4**, Suppl. 7: 11; Dostál, 1982, Seznam Cévn. Rostl. Květ Československ.: 254.—*Doronicum bellidiastrum* L. 1753, Sp. Pl.: 886.—*Aster bellidiastrum* (L.) Scop. 1769, Annus Hist.-Nat. **2**: 64; Fodov, 1974, Fl. Zakarp.: 132; Merxm. and Schreib. 1976, Fl. Europ. **4**: 115.

Type: The Alps ("in Alpibus Helveticis, Italicis, Tyrolensibus").
West (reported for the Carpathians).—In cultivated meadows and on stony slopes, on banks of streams and rivulets in upper mountain zone.—*General distribution*: Central Europe (mountains), Mediterranean (mountains).—2n = 18.

GENUS **80. *ASTER*** L.
1753, Sp. Pl.: 872; id. 1754, Gen. Pl., ed. 5: 373

Heads heterogamous, rarely homogamous, 10–45 mm in dia, solitary to numerous clustered in paniculate inflorescence. Involucre 4–13 mm long and 6–18 mm in dia, cup-shaped to broadly saucer-shaped; involucral bracts imbricate, in many rows, usually unequal, less often almost equal, lanceolate-ovate to linear. Receptacle flat or weakly convex, without scales, but with irregularly projected margins of alveoli. Peripheral flowers pistillate, ligulate, of various colors but not yellow, sometimes small or even entirely reduced; disk flowers tubular, bisexual, with yellow, less often pink, five-toothed corolla. Achenes 1.3–4.0 mm long and 0.5–1.5

mm wide, usually strongly flattened, short-hairy, rarely subglabrous, oblong or cuneate-obovate, with two to four more or less prominent veins; pappus of numerous scabrous bristles usually longer than achene and almost equal. More or less hairy, less often glabrous perennial plants, 6–150 cm high, with usually erect, leafy stems and alternate, undivided, but often more or less toothed, leaves.

182

Lectotype: *A. amellus* L.

About 250 species in extratropical countries of both hemispheres, but predominantly in North America and eastern Asia, as also in mountainous regions of the tropics.

1. Inflorescence with rather numerous heads 20–30 mm in dia, on peduncles covered with dense glandular hairs; involucral bracts dorsally glandular hairy. Cultivated, but occasionally escaped, 30–150 cm high plant, with undivided and entire leaves short-hairy on both sides, distinctly auriculate at base ....5. **A. novae-angliae.**
+ Stalks of heads more or less with short simple hairs; involucral bracts dorsally glabrous or with short simple hairs .... 2.
2. Wild plant, 10–70 cm high, with short rhizome and often with rosulate basal leaves; stems throughout and leaves beneath (and often also above) short-hairy; heads 1–10(15) on each stem, 25–50 mm in dia, rarely without ligulate flowers, and then 10–15 mm in dia .... 3.
+ Cultivated, but often escaped plant 30–140 cm high, with long rhizome, without rosulate basal leaves; stems glabrous or with longitudinal stripes of short hairs; leaves glabrous or subglabrous, but on margin more or less scabrous from fine spines; heads (10)15–50(80) on each stem, 12–35 mm in dia, always with ligulate flowers .... 5.
3. Cauline leaves serrate-dentate, upper often entire, scatteredly hairy to subglabrous above, with several pairs of weakly developed lateral veins; heads usually one to three on each stem; involucral bracts acute, outermost usually half as long as the longest inner bracts ....
.... 6. **A. sibiricus.**
+ Cauline leaves usually entire, rarely the lowermost and rosulate basal leaves coarsely and scatteredly toothed, rather densely hairy on both sides, usually with one pair of distinct lateral veins arising from lower part of midrib; involucral bracts obtuse or subacute, outermost usually two-thirds and more as long as the longest inner bracts .... 4.

4. Plant 20–70 cm high, with rather large cauline leaves, usually not exceeding rosulate basal leaves if present, and almost always with more or less branched stems, bearing several heads; outermost involucral bracts two-thirds as long as the inner bracts .... 7. **A. amellus.**

+ Plant 6–35 cm high, with rather small and usually less numerous (up to 10) cauline leaves, almost always smaller than rosulate basal leaves; stem always simple with one head; outermost involucral bracts only slightly shorter than the longest inner bracts .... 8. **A. alpinus.**

5. Most cauline leaves linear-lanceolate or narrow-lanceolate, 6–10 mm wide, entire, less often weakly toothed, semiamplexicaul; ligulate flowers usually white; outer involucral bracts light green, two-fifths to half as long as the longest inner bracts .... 4. **A. lanceolatus.**

+ Most cauline leaves broadly lanceolate or lanceolate-ovate, 10–25 mm wide, all or at least some with distinct teeth; ligulate flowers pink, bluish-violet, or lilac, very rarely white .... 6.

6. Stalks of heads with rather numerous (usually more than four), approximate, small, subtending leaves; outermost involucral bracts nearly half as long as the longest inner, short-acuminate, and usually with recurved tips ....
.... 3. **A. versicolor.**

+ Stalks of heads with fewer (one to three), distant, small subtending leaves; outermost involucral bracts two-thirds and more as long as the longest inner, often as long or even longer, usually without recurved tips .... 7.

7. Outermost involucral bracts herbaceous, green, short-acuminate, usually not shorter than inner bracts or even longer, often more or less divergent from them ....
.... 1. **A. novi-belgii.**

+ Outermost involucral bracts light green, partly more or less membranous, gradually pointed, nearly two-thirds as long as innermost bracts and more or less adpressed to them ....2. **A. salignus.**

Section 1. Genuini Nees, 1832, Gen. Sp. Aster.: 52, s. str.

Heads numerous, 12–35 in dia; achenes 1.3–2.5 mm long. Plants 30–150 cm high with long rhizome, without vegetative shoots with rosulate leaves.

Lectotype: *A. novi-belgii* L.

1. **A. novi-belgii** L. 1753, Sp. Pl.: 877; Tamamsch. 1959, Fl. SSSR, **25:** 83; Dobrocz. 1962, Fl. URSR, **11:** 38; Yeo, 1976, Fl. Europ. **4:** 114.

Type: United States of America, states of Virginia and Pennsylvania ("in Virginia, Pensylvania").

North (Dvina-Pechora: south); *Baltic*; *Center*; *West.*— Cultivated as an ornamental plant in habitations, gardens, and parks and escaped, also as ecdemic in other extratropical countries and often escaped.—2n = 18, 48, 54.

2. **A. salignus** Willd. 1803, Sp. Pl. **3,** 3: 2040; Tamamsch. 1959, Fl. SSSR, **25:** 84; Resinš, 1960, Latv. PSR Veg. **3:** 115; Yeo, 1976, Fl. Europ. **4:** 114.—*A. salicifolius* Scholl. 1787, Fl. Barb. Supp.: 328, non Lam. 1783.—*A. praecox* Meinsh. 1878, Fl. Ingr.: 156.

Type: Central Europe ("in Germania ad ripas Albis, et in Hungaria").

North (Karelia-Murmansk: south; Dvina-Pechora: west and south); *Baltic*; *Center*; *West.*—Cultivated as ornamental plant and often found as an escape or ecdemic, often wholly naturalized plants in habitations, by roadsides, on banks of water bodies, in scrublands.—*General distribution*: Western Siberia (south); Scandinavia, Central and Atlantic Europe; North America; cultivated in other extratropical countries and often escaped.—2n = 18.

Note. Often considered as a hybrid of *A. novibelgii* L. × *A. lanceolatus* Willd. and, possibly, actually has hybrid origin. But presently, it is a fully stabilized species.

3. **A. versicolor** Willd. 1803, Sp. Pl. **3,** 3: 2045; Yeo, 1976, Fl. Europ. **4:** 114; Ignatov and Makarov, 1985, Bot. Zhurn. **70,** 6: 853.

Type: Described from a garden plant of obscure origin.

Baltic; *Center* (Ladoga-Ilmen; Upper Volga; Volga-Don); *West* (Carpathians; Dnieper; Black Sea).—Like the preceding species, but more rare.—*General distribution*: Central and Atlantic Europe, Mediterranean; North America.

Note. Usually considered as a hybrid of *A. novibelgii* × *A. laevis* L., possibly, evolved in Europe. Presently it is a fully stabilized species. In gardens and parks, there is a closely related species *A. laevis* L. with glaucous-green amplexicaul leaves (Ignatov and others, 1990, *Fl. Issl. Mosk. Obl.*: 87).

4. **A. lanceolatus** Willd. Sp. Pl., **3,** 3: 2050; Fodor, 1974, Fl. Zakarp.: 132; Yeo, 1976, Fl. Europ. **4:** 114; Ignatov and Makarov, 1985, Bot. Zhurn. **70,** 6: 853.

184   Type: North America ("in America boreali").

Baltic; *Center*; *West* (Carpathians; Dnieper; Moldavia).—
Cultivated as an ornamental plant and sometimes escaped, found
in habitations, by roadsides, in parks.—*General distribution*:
North America; cultivated and escaped in Scandinavia (south),
Central and Atlantic Europe.—2n = 18.

Note. For Transcarpathia reported as wild plants, two more
closely related species with more numerous but small heads: *A.
tradescantii* L. (=*A. parviflorus* Nees) and *A. dumosus* L. (Fodor,
1974, *Fl. Zakarp.*: 132). He also reports for Transcarpathia one
more species, *A. ericoides* L. with numerous very small leaves on
branches of the inflorescence and spinescent, more or less
attenuate, outer involucral bracts. For Moscow Region, *A. puniceus*
L. is reported as cultivated and occasionally escaped species
(Ignatov, and others, 1990. *Flor. Issl. Mosk. Obl.*: 87).

5. **A. novae-angliae** L. 1753, Sp. Pl.: 875; Tamamsch. 1959, Fl.
SSSR, **25**: 86; Fodor, 1974, Fl. Zakarp.: 132; Yeo, 1976, Flk. Europ.
4: 114; Dobrocz. 1987, Opred. Vyssh. Rast. Ukr.: 321.

Type: United States of America ("in Nova Anglia").

West (Carpathians; Dnieper; Moldavia).—Cultivated as an
ornamental plant and often escaped, found in habitations, parks
and gardens, by roadsides.—2n = 10.

Section 2. Aster.

Heads 1–10(15), 25–40 mm in dia; achenes 2.5–4.0 mm long.
Plant 15–70 cm high with short rhizome, often with reduced
vegetative shoots with rosulate leaves.

Type: lectotype of genus.

6. **A. sibiricus** L. 1753, Sp. Pl.: 872; Tamamsch. 1959, Fl.
SSSR, **25**: 94; Merxm. and Schreib. 1976, Fl. Europ. **4**: 115, p. p.;
Petrovskii, 1987, Arkt. Fl. SSSR, **10**: 67, map. 16, p. p.

Type: Siberia ("in Sibiria").

a. Subsp. **sibiricus.**—Outermost involucral bracts half to two-
thirds as long as the longest inner; strongly reduced subtending
leaves usually shifted from base of head; all leaves serrate-dentate;
stems 15–40 cm high, usually with two to five heads.

North (Dvina-Pechora: except southwest); *Center* (Volga-Kama:
northwest).—In meadows, forest glades and edges, river sands
and gravel-beds, on stony slopes, in thinned-out forests.—*General
distribution*: Western and Eastern Siberia, Far East; Japan-China.—
2n = 18.

b. Subsp. **subintegerrimus** (Trautv.) A. et D. Löve, 1976, Bot. Not. (Lund), **128:** 521.—*A. sibiricus* L. var. *subintegerrimus* Trauṭv. 1847, in Middend. Reise Sibir. **1:** 161.—*A. arcticus* Eastw. 1902, Bot. Gaz. (London), **23:** 295.—*A. subintegerrimus* (Trautv.) Ostenf. et Resvoll, 1916, Nyt. Mag. Naturvid. **54:** 163; Tamamsch. 1959, Fl. SSSR, **25:** 95.—*A. sibiricus* auct. non L.: Merxm. and Schreib. 1976, Fl. Europ. **4:** 115, p. p.—Outermost involucral bracts more than two-thirds as long as the longest inner; strongly reduced subtending leaves usually approximate to base of heads; usually many leaves weakly toothed, partly almost entire; stem 10–30 cm high, usually with one(two or three) heads.

Type: Eastern Siberia ("Ad fl. Boganidam").

Arctic (Arctic Europe); *North* (Karelia-Murmansk: Kola Peninsula; Dvina-Pechora: basin of the Pechora River).—On river sands and gravel-beds, on meadow slopes of river and coastal terraces, in stony tundras, on outcrops of limestone and allied rocks, in scrublands, sometimes in larch forests.—*General distribution*: Eastern Siberia, Far East (north); Scandinavia (Norway).—2n = 18.

7. **A. amellus** L. 1753, Sp. Pl.: 873; Tamamsch. 1959, Fl. SSSR, **25:** 86; Merxm. and Schreib. 1976, Fl. Europ. **4:** 115, p. p: Efimova and others, 1981, Bot. Zhurn. **66,** 7: 1048.

185   Type: Southern Europe ("in Europae australis asperis collibus").

a. Subsp. **amellus.**—Middle and inner involucral bracts obtuse, usually with more or less membranous tips; pubescence of entire plant and involucre usually less dense; leaves green. Stem usually with 5–10 distant heads.

Center (Upper Dnieper; Upper Volga: along the Oka River; Volga-Kama: east; Volga-Don); *West* (Carpathians; Dnieper: north; Moldavia); *East* (Lower Don: north and east; Trans-Volga: southern Urals).—In meadow steppes, forest glades and edges, pine forests, scrublands.—*General distribution*: Western Siberia (southwest), Central and Atlantic Europe, Mediterranean.—2n = 18, 36, 54.

b. Subsp. **bessarabicus** (Bernh. ex Reichenb.) Soó, 1968, Acta Bot. Acad. Sci. Hung. **12,** 3–4: 366.—*A. bessarabicus* Bernh. ex Reichenb. 1831, Fl. Germ. Excurs. **2:** 246.—*A. amelloides* Bess. 1822, Enum. Pl. Volhn.: 33, non Hoffm. 1801; Tamamsch. 1959, Fl. SSSR, **25:** 87.—*A. scepusiensis* Kit. ex Kanitz, 1863, Linnaea, **32:** 373.—*A. amellus* auct. non L.: Merxm. and Schreib. 1976, Fl. Europ. **4:** 115, p. p.—Middle and inner involucral bracts acute or

262

subacute, often without membranous border near tip; pubescence of entire plant and involucre, on the average, more dense; leaves green or somewhat grayish; stems usually with 10–15(20) distant heads.—(Plate XX, 4).

Type: Western Ukraine ("in dumetis Volhyniae").

Center (Volga-Kama: southeast; Volga-Don: south and east); *West* (Dnieper; Moldavia; Black Sea); *East* (Lower Don; Trans-Volga); *Crimea* (mountains).—In steppes, steppefied glades and forest edges, on outcrops of chalk and limestone, in scrublands.— *General distribution*: Caucasus, Russian Central Asia (western Kopetdag); Central Europe (southeast), Mediterranean (Balkans).— $2n = 36, 54$.

Note. In regions with mixed occurrence with the preceding subspecies, it is usually found on more arid steppe slopes, particularly on limestone and chalk outcrops. In southern Urals, hybrids have been reported with *A. alpinus* L. subsp. *parviceps* Novopokr.—*A.* × *alpino-amellus* Novopokr. sp. hyb. nova[1] with solitary heads, but with involucral bracts of different length and more numerous cauline leaves. In the herbarium of the Komarov Botanical Institute, there are original specimens collected from the vicinity of Chernogovka and Kiev, from Perm Region (village of Ilinskoe) and from the southern coast of Crimea (collection of Steven), which have smaller (17–25 mm in dia) heads with ligules of peripheral flowers 4–8 mm long as also with almost unequal and narrower involucral bracts. Apparently, this is the result of hybridization of *Aster ammellus* L. s. l. × *Erigeron acris* L. s. l. = × *Asterigeron ucrainicus* Tzvel. sp. hybr. nova.[2] The commonly used name for this subspecies—*A. ammelloides* Bess.—cannot be considered valid because there is a prior name *A. amelloides* Hoffm. 1801, in Roem. *Arch. Bot.* **2**, 2: 297 (described from Pyrenees). *A. bessarabicus* Bernh. ex Reichenb. is the new name for *A. amelloides* Bess. and is based on the very same type.

c. Subsp. **ibericus** (Bieb.) Avetis. 1972, Biol. Zhurn. Armen. **25**, 10: 63.—*A. ibericus* Bieb. 1808, Fl. Taur.-Cauc. **2**: 311; Tamamsch. 1959, Fl. SSSR, **25**: 87.—*A. amellus* L. f. *ibericus* (Stev.) Nees,

---

[1]Capitula vulgo solitaria. Involucri phylla sat valde inequalia. Folia caulina sat numerosa. Typus: "Prov. Orenburg, distr. Troitzk, declivitas occidentalis montis Shartynka, in steppa, 1.VII.1916, V. Krascheninnikova" [LE].

[2]Capitula 17–25 mm in diam. Ligulae 4–8 mm lg., quam in *Aster amellus* s. l. angustiores. Involucri phylla angustiora, lanceolata, minus inaequalia. Typus: "Prov. Tschernigow, XII.1969, leg. Regel" [LE].

1831, Gen. Sp. Aster.: 45.—*A. jailicola* Juz. in herb.—Middle and inner involucral bracts acute or subacute, usually without membranous border at tip, shortest outer bracts two-thirds as long as the longest inner; pubescence of whole plant more profuse and long; stems usually almost villous from about 1 mm long hairs; leaves grayish-green; stems usually with (1)2–8(15) rather densely crowded heads.

Type: Georgia ("in Iberia").

Crimea (yaila in the upper part of forest zone).—In cultivated meadows and on stony slopes, in forest glades and edges in upper and middle mountain zones.—*General distribution*: Caucasus; Asia Minor.

186   *Note.* I am not fully convinced that the Crimean plants are completely identical with the Caucasian plants. It is quite likely that they deserve a separate subspecific epithat and were identically named by S.V. Juzepczuk in herbarium as "*A. jailicola* Juz."

Section 3. Alpigeni Nees, 1833, Gen. Sp. Aster.: 24, s. str.—*Aster* sect. *Alpinaster* Tamamsch. 1959, Fl. SSSR, **25**: 104.

Heads solitary, usually 25–45 mm in dia, sometimes without ligulate peripheral flowers, and then smaller outer involucral bracts more than two-thirds as long as the longest inner; achenes 2.4–3.0 mm long. Plant (5)10–25(35) cm high, with short but strongly branched rhizome and usually shorter vegetative shoots bearing rosulate leaves.

Lectotype: *A. alpinus* L.

8. **A. alpinus** L. 1753, Sp. Pl.: 872; Tamamsch. 1959, Fl. SSSR, **25**: 105; Merxm. and Schreib. 1976, Fl. Europ. **4**: 115, p. p.; Petrovsky, 1987, Arkt. Fl. SSSR, **10**: 69, map. 17.—*A. alpinus* var. *grandiflorus* Korsh. 1898, Tent. Fl. Ross. Or.: 202.—*A. alpibus* subsp. *grandiflorus* (Korsh.) Novopokr. 1936, Fl. Yugo-Vost. Evrop. Chasti SSSR, **6**: 308.—*A. korshinskyi* Tamamsch. 1959, Fl. SSSR, **25**: 578, 109.

Type: Mountains of Europe ("in Austria, Vallesia, Helvetia, Pyreneis").

a. Subsp. **alpinus.**—Heads 30–45 mm in dia, always with ligulate flowers; lowermost cauline and rosulate leaves usually strongly deflected and considerably larger than strongly reduced and relatively fewer (3–10) cauline leaves.—(Plate XXI, 1).

North (Dvina-Pechora: basin of the Pinega River, Timan Ridge, basin of the Pechora River, Urals); *Center* (Volga-Kama: Urals; Volga-Don: Zhiguli); *West* (Carpathians); *East* (Trans-Volga:

southern Urals).—On outcrops of limestone and allied rocks, in forest glades and edges, scrublands.—*General distribution*: Caucasus, Western and Eastern Siberia, Arctic, Far East, Russian Central Asia (mountains); Central Europe (mountains), Atlantic Europe (Pyrenees), Mediterranean (mountains), Asia Minor (mountains); Dzhungaria-Kashgaria, Mongolia, Tibet; North America.—2n = 18, 27, 36.

Note. Despite high polymorphism, the majority of populations of this species from the European part of the former Soviet Union do not differ significantly from the typical populations from Austria and Switzerland. However, the very large-headed plants from the vicinity of Krasnoufimsk (Mt. Sokolov Kamen) were described as *A. korshinskyi*. Specimens collected from Mt. Ivdel, have white ligulate flowers (usually they are pink or pinkish-blue). For the Carpathians, a variety was reported with very sparse pubescence.— var. *glabratus* (Herb.) Voloszcz. (Fodor, 1974, *Fl. Zakarp.*: 132). The Zhiguli populations are closer to the next subspecies.

b. Subsp. **parviceps** Novopokr. 1936, Fl. Yugo-Vost. Evrop. Chasti SSSR, **6**: 307.—*A. alpinus* var. *minor* Ledeb. 1845, Fl. Ross. **2**, 2: 472.—? *A. scapigerum* Ledeb. 1845, ibid.: 472.—*A. alpinus* var. *parviflorus* Korsch. 1898, Tent. Fl. Ross. Or.: 202.— *A. serpentimontanus* Tamamsch. 1959, Fl. SSSR, **25**: 108, descr. ross.—*A. alpinus* subsp. *serpentimontanus* (Tamamsch.) A. et D. Löve, 1976, Bot. Not. (Lund), **128**: 521, comb. illeg.— Heads 20–30 mm in dia, always with ligulate flowers; lowermost cauline and rosulate leaves usually more or less upward directed and only slightly larger than often more numerous (up to 15) cauline leaves.

Lectotype: Altai ("Sibiria altaica").

Center (Volga-Kama: southeast); *East* (Trans-Volga: southern Urals and adjoining areas).—On stony and rubbly slopes, in steppes.—*General distribution*: Western and Eastern Siberia (south), Russian Central Asia (east); Dzhungaria-Kashgaria, Mongolia.

Note. Apparently, this is the most xerophilous plains and low-mountain race of the species. Its correct nomenclature is difficult to decide because of the extremely confusing synonymy in the circumscription of *A. alpinus* s. 1.

c. Subsp. **tolmatschevii** (Tamamsch.) A. et D. Löve, 1976, Bot. Not. (Lund), **128**: 521.—*A. tolmatschevii* Tamamsch. 1959, Fl. SSSR, **25**: 107.—*A. chrysocomoides* Turcz. ex DC. 1838, Prodr. **7**, 1: 272, non Desf. 1815.—*A. alpinus* var. *discoideus* Ledeb.

187

Plate XXI.
1—*Aster alpinus* L. subsp. *alpinus*, 1a—achene; 2—*Tripolium vulgare* Nees,
2a—achene; 3—*Galatella rossica* Novopokr., 3a—achene; 4—*G. villosa*
(L.) Reichenb. fil., 4a—achene.

266

1845, Fl. Ross. **2,** 2: 472.—Heads 9–15 mm in dia, without ligulate
flowers, or with ligulate flowers only slightly exceeding involucre;
lowermost cauline and rosulate leaves more or less upward
188 directed and somewhat longer than fewer (usually to 10) cauline
leaves.

Type: Trans-Baikal ("in apricis Dahuriae prope Norinhoroi
seu Norin-choroiskoi Karaal").

North (Dvina-Pechora: northern Urals); *Center* (Volga-Kama:
central Urals).—In cultivated meadows, on stony slopes and
rocks in alpine tundra belt.—*General distribution*: Western Siberia
(Altai), Eastern Siberia, Arctic, Russian Central Asia (east).

Note. This is an alpine and subalpine tundra race of the
species, which is identical in habit with the preceding subspecies,
but with lesser number of cauline leaves and more or less reduced
ligulate flowers.

GENUS 81. *TRIPOLIUM* Nees
1832, Gen. Sp. Aster.: 152.—*Aster* L. sect. *Tripolium* (Nees) Benth.
1873, in Benth. and Hook. fil. Gen. Pl. **2:** 273.

Heads heterogamous, 15–35 mm in dia, solitary at apices of
stem and its branches, but often numerous. Involucre 5–8 mm
long and 7–12 mm in dia, broadly cup-shaped; involucral bracts
imbricate, oblong or linear-oblong, obtuse or subacute, membranous on margin, outer bracts one-third to half as long as the
longest inner. Receptacle somewhat convex, without scales, but
with irregularly raised margin of alveoli. Peripheral flowers pistillate,
ligulate, bluish, pink or whitish, sometimes reduced; disk flowers
tubular, bisexual, with yellow, five-toothed corolla. Achenes 1.6–
4.0 mm long and 0.6–0.8 mm wide, oblong or oblong-ovate, rather
strongly flattened, with two slightly raised veins, more or less
hairy; pappus 6–10 mm long, of numerous scabrous bristles of
almost same length. Glabrous or subglabrous annual or biennial
plants, 10–50 cm high, with erect or ascending leafy stems and
alternate, somewhat fleshy, undivided and usually entire leaves.

Type: *T. vulgare* Nees.
Two closely related halophytic species in the extratropical
countries of the Northern Hemisphere.

1. Achenes 2.6–4.0 mm long, little different in size within a
single head; pappus 4–6 mm long; stems usually branched
in upper part .... 1. **T. vulgare.**

+ Achenes of peripheral flowers in head 1.6–2.5 mm long with 6–9 mm long pappus, inner much different in size, 2.8–4.0 mm, less hairy, with pappus 7–10 mm long; stems often branched almost from base; leaves on the average less fleshy .... 2. **T. pannonicum.**

1. **T. vulgare** Nees, 1832, Gen. Sp. Aster.: 152; Tamamsch. 1959, Fl. SSSR, **25:** 184, p. p.—*Aster tripolium* L. 1753, Sp. Pl.: 872; Merxm. and Schreib. 1976, Fl. Europ. **4:** 115.—*Tripolium pannonicum* (Jacq.) Dobrocz. subsp. *maritimum* Holub, 1973, Folia Geobot. Phytotax. (Praha), **8,** 2: 177.—(Plate XXI, 2).

Type: Europe and Siberia ("in Europae littoribus maritimis et ad Sibiriae lacus salsos"), lectotype in Europe.
Arctic (Arctic Europe: in the east to the mouth of the Pechora River); *North* (Karelia-Murmansk and Dvina-Pechora: seacoast); *Baltic*; *Center* (Ladoga-Ilmen: seacoast, ecdemic in Leningrad and Pskov).—In more or less saline meadows, on sands and gravel-beds, seacoasts, sometimes in saline places by roadsides.—*General distribution*: Scandinavia, Atlantic and Central Europe; North America.—2n = 18.

2. **T. pannonicum** (Jacq.) Dobrocz. 1962, Fl. URSR, **11:** 63; Tzvel. 1990, Novosti Sist. Vyssh. Rast. 27: 145.—*Aster pannonicum* Jacq. 1770, Hort. Bot. Vindob. **1:** 3, tab. 8.—*A. tripolium* L. subsp. *pannonicum* (Jacq.) Soó, 1925, Bot. Közlem. **22:** 64; Merxm. and Schreib. 1976, Fl. Europ. **4:** 115.—*Tripolium vulgare* auct. non Nees: Tamamsch. 1959, Fl. SSSR, **25:** 184, p. p.

Type: Hungary ("in paludosis Hungariae").
Center (Volga-Kama: southeast; Volga-Don: south and east); *West* (Dnieper; Moldavia; Black Sea); *East*; *Crimea*.—In alkaline meadows and solonchaks, on banks of saline water bodies, sometimes as ecdemic of roadsides, in habitations.—*General distribution*: Caucasus, Western and Eastern Siberia (south), Far East (south), Russian Central Asia; Central Europe, Mediterranean, Asia Minor, Iran, Dzhungaria-Kashgaria, Mongolia, Japan-China.—2n = 18.

GENUS **82.** *GALATELLA* Cass.
1825, Dict. Sci. Nat. **37:** 463, 448.—*Aster* L. sect. *Galatella* (Cass.) Benth. et Hook. fil.: 1876, Gen. Pl. **2:** 273

Heads heterogamous, 5–35 mm in dia, clustered in corymbose or corymbose-paniculate inflorescences. Involucre 3–8 mm long and 4–9 mm in dia, saucer-shaped or cup-shaped; involucral bracts imbricate, lanceolate-ovate to narrow-linear, acute or obtuse, outer one-fourth to five-sixths as long as the longest inner bracts. Receptacle somewhat convex, without scales, but with irregularly raised margins of alveoli. Peripheral flowers ligulate, sterile (with rudiment of style entire and usually scarcely exserted from throat), pink or bluish, rarely white, often absent; disk flowers tubular, bisexual, with yellow, rarely somewhat pinkish, five-toothed corolla. Achenes 2.0–5.3 mm long and 0.4–1.2 mm wide, narrow-oblong, more or less flattened, with two prominent and one or two inconspicuous veins, short-hairy; pappus 4–8 mm long, of numerous scabrous bristles of almost equal length. More or less hairy (to subglabrous) perennial plants 10–130 (150) cm high, with erect, leafy stems and alternate, entire and undivided, linear, oblong or narrow-elliptical leaves often with punctate glands.

Lectotype: *G. punctata* (Waldst. et Kit.) Nees.

The genus includes 40–50 species in central and southern Europe, as also in a large part of extratropical Asia (excluding its northern regions).

1. Outer and middle involucral bracts narrow-lanceolate or linear, outermost more than half (usually two-thirds) as long as the longest inner bracts; heads without ligulate flowers; tubular flowers 15–40; leaves narrow-linear, 1–3 mm wide, with one vein, glabrous or subglabrous, more or less scabrous on margin. Plant 10–50 cm high ....
.... 12. **G. linosyris.**

+ Outer and middle involucral bracts lanceolate or lanceolate-ovate, outermost less than half (usually one-fourth to two-fifths) as long as the longest inner bracts .... 2.

2. Plants 10–40(50) cm high; all leaves with one vein or only lower leaves with three veins; stalk of heads more or less arachnoid-tomentose .... 3.

+ Plants 25–130(150) cm high; lower and middle leaves with three distinct veins, upper often with one vein; stalk of heads with very short papilliform hairs .... 7.

3. Heads always with 5–10(15) blue or bluish-violet ligulate flowers; lowermost leaves, usually withering by flowering, linear-lanceolate, 2–4 mm wide, three-veined, rest numerous, narrow-linear, 1–2 mm wide, one-veined, glabrous or subglabrous, more or less scabrous on margin from spines
.... 7. **G. angustissima.**

190   + Heads without ligulate flowers, rarely with one to five ligulate flowers, and then middle and upper cauline leaves

2.5–4.0 mm wide or ligulate flowers pale, weakly developed
.... 4.

4. Entire plant grayish-tomentose; leaves oblong-linear or oblong, (3)4–10 mm wide, almost not reducing in upper part; heads in rather dense corymbose inflorescence...
.... 11. **G. villosa.**

+ Plant green or somewhat grayish from rather dense pubescence, but not tomentose .... 5.

5. Lower, and usually also middle cauline, leaves oblong-linear or elliptical-linear, 4–8 mm wide, three-veined, rather abruptly changing to narrow, one-veined upper leaves; stalk of heads with small subtending leaves; inner involucral bracts rather gradually pointed; ligulate flowers present or absent...................................... 8. **G. divaricata.**

+ All leaves more or less identical in width, linear, 1.5–4.0 mm wide, usually one-veined; inflorescence rather dense, usually corymbose; stalk of heads to 30 mm long; inner involucral bracts obtuse .... 6.

6. Outer and middle involucral bracts acute or subacute, lanceolate; stalk of heads often arcuately bent, with small subtending leaves; heads with 10–20 flowers ....
.... 9. **G. crinitoides.**

+ Only outermost involucral bracts subacute, lanceolate-ovate or ovate, others obtuse; stalk of heads more or less erect, with solitary subtending leaf; heads with 5–8(10) flowers .... 10. **G. tatarica.**

7(2). Heads without ligulate flowers, very rarely with one to three ligulate flowers; leaves rather gradually pointed ...... 8.

+ Heads with (3)4–10 light pink, pink, or lilac ligulate flowers
.... 9.

8. Stems throughout or almost throughout (except lower part) covered with short papilliform hairs; leaves usually on both surfaces with scattered papillate hairs; stalk of heads densely covered with papilliform hairs, usually more or less erect with fewer subtending leaves; general inflorescence usually dense .... 5. **G. biflora.**

+ Stems in lower and middle part glabrous or subglabrous, in upper part and on stalk of heads with very short (inconspicuous even at high magnification) and not very dense papillae; leaves glabrous, on margin scabrous from spines; stalk of heads usually arcuately bent and with large number of small subtending leaves; general inflorescence lax .... 6. **G. trinervifolia.**

270

9. General inflorescence very lax; stalk of heads usually arcuately bent and with numerous subtending leaves; leaves relatively short-acuminate; heads with 14–20 flowers. Plant of southeastern European part of the former USSR .... 4. **G. pastuchovii.**
+ General inflorescence more dense; stalk of heads usually more or less erect and with fewer subtending leaves ...... 10.
10. Leaves abruptly short-acuminate (particularly lower and middle cauline), often subacute; heads with 6–15 flowers; involucral bracts always one-veined ....
.... 3. **G. dracunculoides.**
+ Leaves gradually long-acuminate; heads with 14–25 flowers; involucral bracts often three-veined .... 11.
11. Stems covered only with papillate hairs; outer involucral bracts lanceolate .... 1. **G. rossica.**
+ Stems at least in upper part near base of leaves with thin arachnoid tomentum (besides papilliform hairs); outer involucral bracts lanceolate-ovate. Plants of southwest European part of Russia .... 2. **G. punctata.**

191

Section 1. Galatella.

Plant 25–130(150) cm high; almost all leaves three-veined; heads with or without ligulate flowers, their stalks with very short papilliform hairs; outer involucral bracts two-sevenths to half as long as the longest inner.

Type: lectotype of genus.

1. **G. rossica** Novopokr. 1948, Tr. Bot. Inst. Akad. Nauk SSSR, ser. 1, **7**: 122; id. 1949, Bot. Mat. (Leningrad) **11**: 212; Dobrocz. 1962, Fl. URSR, **11**: 53.—?*G. strigosa* Weinm. 1850, Bull. Soc. Nat. Moscou, **23**: 546.—*G. ledebouriana* Novopokr. 1948, op. cit.: 122; id. 1949, op. cit.: 214.—*G. punctata* (Waldst. et Kit.) Nees subsp. *rossica* (Novopokr.) Novopokr. 1949, op. cit.: 212, in syn.—*Aster punctatus* Waldst. et Kit. subsp. *rossicus* (Novopokr.) Soó, 1966, Acta Bot. Acad. Sci. Hung. **12**, 3–4: 366.—*Galatella punctata* auct. non (Waldst. et Kit.) Nees: Tzvel. 1959, Fl. SSSR, **25**: 154, p. max. p.; Tabaka and others, 1988, Fl. Sosud. Rast. LatvSSR: 135.—*Aster sedifolius* auct. non L.: Merxm. and Schreib. 1976, Fl. Europ. **4**: 116, p. p.—(Plate XXI, 3).

Type: Bashkiria, vicinity of Ufa ("in forest meadows between Kamennyi Perevoz on the Ufa River and the village of Kariovka").

North (Dvina-Pechora: near the mouth of the Pinega River and vicinity of the village of Kholmogory); *Baltic* (in the town of Yurmala, possibly introduced); *Center* (Upper Dnieper: southeast; Upper Volga: south and east; Volga-Kama: south and east; Volga-Don); *West* (Dnieper; Moldavia; Black Sea); *East* (Lower Don: north; Trans-Volga; Lower Volga: north).—In inundated, less often dry meadows, forest glades and edges, thinned-out deciduous forests, scrublands.—*General distribution*: Western Siberia (south), Russian Central Asia (east).

Note. More xerophilic populations with shorter stems and narrow leaves from Trans-Volga and southern Urals were described as a separate species.—*G. ledebouriana* Novopokr.

2. **G. punctata** (Waldst. et Kit.) Nees, 1833, Gen. Sp. Aster: 161; Tzvel. 1959, Fl. SSSR, **25**: 154, p. min. p.; Dobrocz. 1962, Fl. URSR, **11**: 51.—*Aster punctata* Waldst. et Kit. 1802, Descr. Icon. Pl. Rar. Hung. **2**: 113, tab. 109; Willd. 1803, Sp. Pl., **3**, 3: 2022.—*A. sedifolius* auct. non L.: Merxm. and Schreib. 1976, Fl. Europ. **4**: 116, p. p.

Type: Hungary ("in salsis Hungariae").

West (Moldavia: south; Black Sea: southwest).—In forest glades and edges, on steppe slopes, in inundated meadows, scrublands.—*General distribution*: Central Europe (southeast).

3. **G. dracunculoides** (Lam.) Nees, 1832, Gen. Sp. Aster: 164; Tzvel. 1959, Fl. SSSR, **25**: 162.—*Aster dracunculoides* Lam. 1780, Encycl. Méth. Bot. **1**: 303.—*A. sedifolius* L. subsp. *dracunculoides* (Lam.) Merxm. 1974, Bot. Journ. Linn. Soc. (London), **68**: 279, p. p.; Merxm. and Schreib. 1976, Fl. Europ. **4**: 116, p. p.

Type: Garden plant, originating from the Caucasus or Ukraine.
Center (Volga-Don: southeast); *West* (Dnieper and Black Sea: left bank of the Dnieper River); *East* (Lower Don); *Crimea*.—In forest glades and edges, deciduous forests, on steppe slopes, in scrublands.—*General distribution*: Caucasus; Asia Minor.

4. **G. pastuchovii** (Kem.-Nat.) Tzvel. 1959, Fl. SSSR, **25**: 159.— *G. dracunculoides* (Lam.) Nees var. *pastuchovii* Kem.-Nat. 1927, Vestn. Tifl. Bot. Sada, 3–4: 132.

Type: Azerbaidzhan, vicinity of Kuba ("Prov. Baku, distr. Kuba, pag. Divitchi, leg. Pastuchov").

192 *East* (Lower Don: southeast; Lower Volga: south).—In saline meadows, on slopes of river and coastal terraces, as also bare knolls, in scrublands.—*General distribution*: Caucasus (east).

5. **G. biflora** (L.) Nees, 1832, Gen. Sp. Aster.: 159; Tzvel. 1959, Fl. SSSR, **25**: 151; Ignatov and others, 1990, Fl. Issl. Mosk. Obl.: 87.—*Chrysocoma biflora* L. 1753, Sp. Pl.: 841.—*G. biflora* subsp.

krascheninnikovii Novopokr. 1948, Tr. Bot. Inst. Akad. Nank SSSR, ser. 1. **7**: 120.—*G. krascheninnikovii* Novopokr. 1948, ibid.: 120, nom. altern.—*G. novopokrovskii* Zefir. 1957, Bot. Mat. (Leningrad), **18**: 249; Dobrocz. 1962, Fl. URSR, **11**: 50.—*Aster sedifolius* L. subsp. *dracunculoides* (Lam.) Merxm. 1974, Bot. Journ. Linn. Soc. (London), **68**: 279, p. p.; Merxm. and Schreib. 1976, Fl. Europ. **4**: 16, p. p.

Type: Siberia ("in Sibiria").

Center (Upper Volga: ecdemic in Moscow; Volga-Kama: southeast; Volga-Don: south and east); *West* (Dnieper: left bank of the Dnieper River; Moldavia: Kordy; Black Sea); *East* (Lower Don; Trans-Volga; Lower Volga: north); *Crimea*.—On solonetzes, steppe slopes, clayey precipes, chalk and limestone outcrops, in steppefied forest glades.—*General distribution*: Western and Eastern Siberia (south), Russian Central Asia (north); Dzhungaria-Kashgaria.

Note. Populations with lesser number of flowers in head (usually 7–15) than in typical populations from southern Siberia (usually 12–20) predominate in the European part of the former USSR and undoubtedly this species deserves to be split into two subspecies: predominantly Siberian but extending into Southern Urals, and predominantly European described from Kerch Peninsula by the name of *G. novopokrovskii*. Smaller and narrow-leaved populations from more arid localities were described as subsp. *krascheninnikovii* Novopokr. from Kazakhstan.

6. **G. trinervifolia** (Less.) Novopokr. 1948, Tr. Bot. Inst. Akad. Nauk SSSR, ser, 1, **7**: 121; Tzvel. 1959, Fl. SSSR, **25**: 152.—*Aster trinervifolius* Less. 1835, Linnaea, **9**: 183.—*Galatella glabra* Novopokr. 1936, Fl. Yugo-Vost. Evrop. Chasti SSSR, **6**: 314, descr. ross.—*G. subglabra* Novopokr. 1948, op. cit.: 120.

Type: Vicinity of Orenburg ("In graminosis procul ab Orenburg").

East (Trans-Volga: south; Lower Volga: northeast).—On solonetzes and saline meadows, clayey semideserts, on outcrops of limestone.—*General distribution*: Western Siberia (southwest), Russian Central Asia (north).

Section 2. Fastigiatae Novopokr. 1918, Izv. Ross. Akad. Nauk, ser. 6, **12**: 2280, 2282; Tzvel. 1959, Fl. SSSR, **25**: 165.

Plants 10–40(50) cm high; lower leaves three-veined, others one-veined; heads with, less often without, ligulate flowers, their stalks covered with thin arachnoid tomentum; outer involucral bracts one-fourth to two-fifths as long as the longest inner bracts.

Lectotype: *G. hauptii* (Ledeb.) Lindel.

7. **G. angustissima** (Tausch) Novopokr. 1948, Tr. Bot. Inst. Akad. Nauk SSSR, ser. 1, **7**: 136; Tzvel. 1959, Fl. SSSR, **25**: 167.— *Aster angustissimus* Tausch, 1828, Flora, **11**: 487.—*Gelasia desertorum* Less. 1835, Linnaea, **9**: 185.—*Galatella desertorum* (Less.) Kar. et Kir. 1842, Enum. Pl. Song.: 107.—?*G. strigosa* Weinm. 1850, Bull. Soc. Nat. Moscou, **23**: 546.—*Aster sedifolius* L. subsp. *angustissimus* (Tausch) Merxm. 1974, Bot. Journ. Linn. Soc. (London), **68**: 279; Merxm. and Schreib. 1976, Fl. Europ. **4**: 116.

Type: Siberia ("in Sibiria").

Center (Volga-Kama: southeast; Volga-Don); *East* (Lower Don: north and east; Trans-Volga; Lower Volga: north).—On outcrops of chalk and limestone, in forest glades and edges, thinned-out pine groves, sandy steppes.—*General distribution*: Western and Eastern Siberia (south), Russian Central Asia (north); Mongolia (northwest).

8. **G. divaricata** (Fisch. ex Bieb.) Novopokr. 1918, Izv. Ross. Akad. Nauk, ser. 6, **12**: 2274; Tzvel. 1959, Fl. SSSR, **25**: 171; Khmelev and Kunaeva, 1985, Bot. Zhurn. **70**, 10: 1415.— *Chrysocoma divaricata* Fisch. ex Bieb. 1818, Fl. Taur.-Cauc. **3**: 563.—*Aster divaricatus* (Fisch. ex Bieb.) Schmalh. 1898, Fl. Sr. 193 Yuzhn. Ross. **2**: 43, non L. 1753.—*A. kirghisorum* Korsh. 1898, Tent. Fl. Ross. Or.: 205; Merxm. and Schreib. 1976, Fl. Europ. **4**: 116.

Type: Vicinity of Volgograd ("Sarepta").

Center (Volga-Don: Kalach and Vorob'evka districts, Voronezh Region); *East* (Lower Don: northeast; Trans-Volga; Lower Volga: north).—In alkaline and saline meadows, on clayey and stony slopes, in steppes and semideserts.—*General distribution*: Western Siberia (south), Russian Central Asia (north).

Note. Of the two variants of this species, the more common is var. *divaricata* lacking ligulate flowers and the more rare var. *radiata* (Trautv.) Tzvel. comb. nova [=*Linosyris divaricata* (Fisch. ex Bieb.) DC. var. *radiata* Trautv. 1866, *Bull. Soc. Nat. Moscou*, **39**: 341] with one to seven pinkish-violet ligulate flowers.

9. **G. crinitoides** Novopokr. 1948, Tr. Bot. Inst. Akad. Nauk SSSR, ser. 1, **7**: 137; Tzvel. 1959, Fl. SSSR **25**: 166.

Type: Kazakhstan ("in declivibus stepposis Kasachstaniae septenetrionali-orientalis: Karkaralinsk, Pavlodar").

Center (Volga-Don: east and southeast); *West* (Lower Don: east; Trans-Volga).—On steppe slopes, outcrops of chalk and limestone.—*General distribution*: Western Siberia (south), Russian Central Asia (north).

Note. Evolved through hybridization of *Galatella angustissima* × *G. villosa*, but possibly, in certain areas already became a stable hybrid. In Zhiguli one more hybrid has been noted: *G. biflora* × *G. villosa*, which is similar to *G. critinoides* but is taller (usually 30–60 cm high) and always lacks ligulate flowers (in *G. critinoides* usually there are a few weakly developed and pale colored ligulate flowers).

Section 3. Chrysocomella Novopokr. sect. nova[1]; Novopokr. 1936, Fl. Yugo-Vost. Evrop. Chasi SSSR, **6**: 316, descr. ross.

Plants 10–35 cm high; all leaves one-veined; heads without ligulate flowers, their stalks finely arachnoid-tomentose; outer involucral bracts one-fifth to half as long as the longest inner.

Type: *G. villosa* (L.) Reichenb. fil.

10. **G. tatarica** (Less.) Novopokr. 1918, Izv. Ross. Akad. Nauk, ser. 6, **12**: 2275.—*Chrysocoma tatarica* Less. 1835, Linnaea, **9**: 186.—*Linosyris glabrata* Lindl. ex DC. 1837, Prodr. **5**: 353.—*L. tatarica* (Less.) C.A. Mey. 1841, in Bong. and C.A. Mey. Verz. Saisang.—Nor. Pfl.: 38; Tzvel. 1959, Fl. SSSR, **25**: 179.—*L. tarbagatensis* C. Koch, 1850, Linnaea, **23**: 703.—*Aster glabratus* (Lindl.) Korsh. 1898, Tent. Fl. Ross. Or.: 205, non Kuntze, 1891.— *A. tarbagatensis* (C. Koch) Merxm. 1974, Bot. Journ. Linn. Soc. (London), **68**: 280; Merxm. and Schreib. 1976, Fl. Europ. **4**: 116.— *Crinitaria tatarica* (Less.) Czer. 1981, Sosud. Rast. SSSR: 61.

Type: Southern Urals, vicinity of Orsk and Sol-Iletsk ("in Mertwii ssol pr. Ilezkaja Saschtschita et ad fl. Or. supra Orsk").

Center (Volga-Don: vicinity of Saratov); *West* (Dnieper: vicinity of the village of Prishib on the Seversky Donetz River); *East* (Lower Don: east; Trans-Volga; Lower Volga: north).—On solonetzes and saline meadows, alkaline clayey slopes, in steppe depressions, on outcrops of chalk and limestone.—*General distribution*: Western Siberia (south), Russian Central Asia; Dzhungaria-Kashgaria.

Note. From the vicinity of Saratov is known a hybrid of *G. tatarica* × *G. villosa* = *G.* × *subtatarica* Tzvel. sp. hybr. nova (*G. villosae* similis, sed foliis angustioribus et minus villosis. Typus: "In viciniis urb. Saratov. 9.VI.1913, D. Janischevsky"—

---

[1]Plantae 10–35 cm alt. Folia uninervia. Capitulae discoideae, pedunculis arachnoideo-tomentosis.—Typus: *G. villosa* (L.) Reichenb. fil.

LE) with narrower and less villous leaves than in *G. villosa*. *Crinita punctata* Moench (1794, *Meth.*: 578) the type species of *Criniaria* Cass. is apparently, a more recent synonym of *Galatella biflora* (L.) Nees, and hence *Crinitaria* should be considered a more recent synonym of the genus *Galatella* s. str.

11. **G. villosa** (L.) Reichenb. fil. 1853, Icon. Fl. Germ. **16**: 8; Novopokr. 1986, Fl. Yugo-Vost. Evrop. Chasti SSSR, **6**: 316.— *Chrysocoma villosa* L. 1753, Sp. Pl.: 841.—*Conyza oleifolia* Lam. 1786, Encyl. Méth. **2**: 86.—*Crinitaria villosa* (L.) Cass. 1825, Dict. Sci. Nat. **37**: 476; Czer. 1981, Sosud. Rast. SSSR: 61.—*Linosyris* 194 *villosa* (L.) DC. 1836, Prodr. **5**: 352; Tzvel. 1959, Fl. SSSR, **25**: 178.—*Aster villosus* (L.) Sch. Bip. 1855, in F. Schultz, Arch. Fl. **1**: 130, non Thunb. 1800.—*A. cinereus* Korsh. 1898, Tent. Fl. Ross.: 205.—*A. oleifolius* (Lam.) Wagenitz, 1964, Bot. Jahrb. 83: 329; Merxm. and Schreib. 1976, Fl. Europ. **4**: 116.—(Plate XXI, 4).

Type: Probably southeast of the European part of Russia ("in Sibiria, Tataria").

Center (Volga-Kama: southeast; Volga-Don: south, east and Mt. Galichi); *West* (Dnieper: left bank of the Dnieper River; Moldavia; Black Sea); *East*; *Crimea*.—In steppes, steppefied forest glades and edges, solonetzes, on outcrops of chalk and limestone.— *General distribution*: Caucasus, Western Siberia (south), Russian Central Asia (north and east); Central Europe.

Section 4. Linosyris (Reichenb. fil.) Tzvel. comb. nova.— *Galatella* subgen. *Linosyris* (Cass.) Reichenb. fil. 1853, Icon. Fl. Germ. **16**: 8.—*Linosyris* Cass. 1825, Dict. Sci. Nat. **37**: 460, 467, non Ludw. 1757.—*Aster* L. sect. *Linosyris* (Reichenb. fil.) Hoffm. 1894, in Engl. and Prantl, Nat. Pflanzenfam. **4**, 5: 163.

Plants 10–50 cm high; all leaves one-veined; heads without ligulate flowers, their stalks more or less arachnoid-tomentose and papillate; outer involucral bracts more than two-thirds as long as the longest inner.

Type: *G. linosyris* (L.) Reicheb. fil.

12. **G. linosyris** (L.) Reichenb. fil. 1853, Icon. Fl. Germ. **16**: 8; Novopokr. 1936, Fl. Yugo-Vost. Evrop. Chasti SSSR, **6**: 316.— *Chrysocoma linosyris* L. 1753, Sp. Pl.: 841.—*Aster linosyris* (L.) Bernh. 1800, Syst. Verz. Erfurt.: 151; Merxm. and Schreib. 1976, Fl. Europ. **4**: 116.—*Linosyris vulgaris* Cass. ex Less. 1832, Syn. Gen. Compos: 195, in syn.; Tzvel. 1959, Fl. SSSR, **25**: 175; V. Tikhomir, 1966, Opred. Rast. Mosk. Obl.: 321.—*Crinitaria linosyris* (L.) Less. 1832, op. cit.: 195; Czer. 1981, Sosud. Rast. SSSR: 61.

Type: Europe ("in Europa temperatiore").

Center (Upper Dnieper: south; Upper Volga: reported from the Oka River; Volga-Don); *West*; *East* (Lower Don); *Crimea*.—In forest glades and edges, on steppe slopes, in solonetzes, on outcrops of chalk and limestone, in scrublands.—*General distribution*: Caucasus; Scandinavia (south), Atlantic and Central Europe, Mediterranean, Asia Minor.—2n = 18, 36.

Note. Occasionally present are the hybrids of *G. linosyris* × *G. angustissima* = *G.* × *sublinosyris* Tzvel. sp. hybr. nova (Specie *G. linosyris* similis, sed phylla externa phyllis internis duplo breviora. Typus: "Prov. Saratov. pag. Razboitschina, in betuleto-querceto. 10.VII.1912. D. Janischewsky"—LE) and *G. linosyris* × *G. villosa* = *G.* × *subvillosa* Tzvel. sp. hybr. nova (A specio antecedente foliis latioribus et villosioribus differt. Typus: "Prov. Saratov, pag. Razboitschina. 29.VII.1913, D. Janischewsky"—LE).

GENUS 83. *CONYZANTHUS* Tamamsch.
1959, Fl. SSSR, **25**: 583, 185

Heads heterogamous, rather small, clustered in paniculate or racemose inflorescences. Involucre 5–7 mm long and 3–5 mm in dia, bowl-shaped; involucral bracts imbricate, lanceolate to linear, acute, glabrous or subglabrous, outermost one-fourth to one-third as long as the longest inner bracts. Receptacle flat, without scales, with raised margins of alveoli. Peripheral flowers pistillate, 15–25, with ligule very short (almost as long as style branches), two- or three-toothed, pinkish, light azure or pale violet; disk flowers tubular, bisexual, four to eight, with five-toothed pale yellow or lilac corolla. Achenes 1.6–2.0 mm long and about 0.4 mm wide, ellipsoidal-cylindrical, somewhat flattened, with four to six raised veins, scatteredly short-hairy; pappus 3.5–4.0 mm long, of one row of scabrous bristles of almost similar length. Glabrous or subglabrous, annual, biennial or perennial plants 20–100 cm high, with erect, leafy stems and alternate, undivided, entire or more or less toothed, linear or oblanceolate leaves.

195   Type: *C. graminifolius* (Spreng.) Tamamsch.

Three or four species in South America, two of them ecdemic and partly naturalized in many parts of the world, predominantly in the tropical and subtropical countries.

1. **C. graminifolius** (Spreng.) Tamamsch. 1959, Fl. SSSR, **25**: 186; Kozhevnikova and Makhaeva, 1976, Bot. Zhurn. **61**, 4: 566.—

Conyza graminifolia Spreng. 1826, Syst. Veg. **3**: 515.—*Aster squamatus* (Spreng.) Hieron. var. *graminifolius* (Spreng.) Cabrera, 1941, Rev. Mus. La Plata (Bot.), **5**, 4: 69.—*A. squamatus* auct. non (Spreng.) Hieron.: Yeo, 1976, Fl. Europ. **4**, 115, p. p.

Ligules of peripheral flowers lilac or pinkish, not exceeding pappus. Annuals or biennials.

Type: Uruguay ("Monte Video").

Crimea (ecdemic in the vicinity of the village of Nizhnyaya Massandra).—In coastal, gravelly and stony places, by roadsides, in habitations.—*General distribution*: South America, ecdemic in the Caucasus and the Mediterranean and many other countries.—2n = 20.

GENUS 84. *BRACHYACTIS* Ledeb.
1845, Fl. Ross. **2**, 2: 495

Heads heterogamous, rather small, clustered in paniculate or racemose inflorescences. Involucre 7–9 mm long and 4–7 mm in dia, bowl-shaped or cup-shaped; involucral bracts linear-lanceolate or linear, in three or four rows, of almost similar length, light green, scabrous on margin, one-veined, spinescent. Receptacle flat, without scales, with somewhat raised margins of alveoli. Peripheral flowers pistillate, in several rows, with short, obliquely truncate tube, with style far exserted from tube, pinkish or whitish; disk flowers shorter than pappus, tubular, bisexual, almost colorless, four- or five-toothed. Achenes 2.0–2.5 mm long and 0.4–0.5 mm wide, ellipsoidal-cylindrical but distinctly flatened, with short, semiappressed hairs and three to five veins, of which two or three raised as ribs; pappus 5.0–6.5 mm long, of nearly two rows of scabrous bristles of almost similar length. More or less hairy plants, with erect leafy stems and alternate, linear or linear-lanceolate entire leaves.

Type: *B. ciliata* Ledeb.

About five species in warm temperate regions of the Northern Hemisphere.

Literature: Dubina, D.V., V.V. Protopopova, and O.N. Dubovik. 1986. Novyi dlya flory URSR rid *Brachyactis* Ledeb. [The genus *Brachyctis* Ledeb.—new in the flora of the Ukraine]. *Ukr. Bot. Zhurn.*, 43, 2: 51–54.

1. **B. ciliata** (Ledeb.) Ledeb. 1845, Fl. Ross. **2**, 2: 495; Botsch. 1959, Fl. SSSR, **25**: 189; Tzvel. 1979, Novosti Sist. Vyssh. Rast.

16: 203; Dubina and others, 1986, Ukr. Bot. Zhurn. **43**, 2: 51.—
Erigeron ciliatus Ledeb. 1829, Icon. Pl. Fl. Ross. **1**, 1: 24, tab.
100; id. 1833, Fl. Alt. **4**: 92.

Leaves 1–3 mm wide, with rather large spines on margin,
usually one-veined.

Type: Plant raised in Tartu from seeds collected between
Barnaul and Lotkevsk.

West (Moldavia: vicinity of Kishinev; Black Sea); *East* (Trans-
Volga: Southern Urals).—In saline meadows, steppe depressions,
coastal sands and gravel-beds, by roadsides.—*General
distribution*: Western and Eastern Siberia, Far East (south), Russian
Central Asia; Central Europe (southeast), Dzhungaria-Kashgaria,
Mongolia, Japan-China.

196

GENUS 85. *HETEROPAPPUS* Less.
1832, Syn. Gen. Compos.: 189

Heads heterogamous, 1.5–5.0 cm in dia, clustered in corymbose
or corymbose-paniculate inflorescence. Involucre 5–10 mm long
and 8–17 mm in dia, saucer-shaped or cup-shaped; involucral bracts
imbricate, in many rows, of almost similar length, linear-lanceolate
or linear, all gradually pointed, outer often merging with subtending
leaves. Receptacle somewhat convex, without scales, but often
with somewhat raised margins of alveoli. Peripheral flowers ligulate,
pistillate, pink, lilac or light blue, rarely white; disk flowers tubular,
bisexual, with yellow, five-toothed corolla. Achenes 1.8–3.0 mm
long and 0.9–1.5 mm wide, oblong-ovate but rather strongly flattened,
with two(three) raised veins, rather densely short-hairy; pappus
3–4 mm long, of numerous, often brownish or reddish scabrous
bristles of almost similar or variable length. More or less hairy
biennial or perennial plants, 10–100 cm high, with erect or ascending,
leafy stems, and alternate leaves; lower leaves (fast withering)
often more or less toothed, others entire, lanceolate to linear, sessile.

Type: *H. hispidus* (Thunb.) Less.
About 20 species in extratropical regions of Asia.

1. **H. biennis** (Ledeb.) Tamamsch. ex Grub. 1972. Novosti Sist.
Vyssh. Rast. **9**: 281; Tamamsch. 1959, Fl. SSSR, **25**: 72, nom.
provis.—*Aster biennis* Ledeb. 1811, Delect. Sem. Hort. Bot. Dorpat.
Suppl. **1**: 1, in nota; Ledeb. 1833, Fl. Alt. **4**: 97, in nota, non Nutt.
1818.—*Calimeris tatarica* Lindl. 1836, in DC. Prodr. **5**: 259.—
Heteropappus tataricus (Lindl.) Tamamsch. 1959, Fl. SSSR, **25**:

71; Tuganaev and Puzyrev, 1988, Gemerofity Vytsko-Kam. mezhdurech. 97.

Cauline leaves linear, 1.5–6.0 cm long and 1–5 mm wide; heads 30–50 mm in dia, with 15–30 ligulate flowers; ligules 12–20 mm long; achenes 1.8–2.2 mm long; pappus 3.0–3.5 mm long.

Type: Siberia ("in Sibiria").

Center (Volga-Kama: ecdemic in the vicinity of the Agryz railroad Station of Tataria).—On railroad dumps.—*General distribution*: Western Siberia (Altai), Eastern Siberia, Far East; Mongolia, Japan-China.

GENUS **86.** *ERIGERON* L.
1753, Sp. Pl.: 863; id. 1754, Gen. Pl., ed. 5: 371

Heads heterogamous, 4–50 mm in dia, clustered in paniculate, corymbose-paniculate or racemose inflorescences, less often solitary. Involucre 5–8 mm long and 4–12 mm in dia, cup-shaped or saucer-shaped; involucral bracts linear-lanceolate or linear, in several rows, outermost usually more than half as long as innermost bracts, often almost as long, acute, less often subacute, one-veined. Receptacle flat or weakly convex, without scales and raised margins of alveoli. Peripheral flowers pistillate, in several rows, ligulate, but often with very short (almost as long as involucre) and narrow ligules, pink, lilac, light pink, bluish or whitish; disk flowers tubular, bisexual, with yellowish, whitish or pinkish, five-toothed corolla, often with intermediate flowers of third type: pistillate but tubular, with short obliquely truncate and often two- to four-toothed corolla tube and exserted style. Achenes 1.5–2.8 mm long and 0.4–0.6 mm wide, narrow-ellipsoidal, but somewhat flattened, short-hairy, two- or three-veined; pappus 3.5–7.0 mm long, of one or two rows of scabrous bristles, the
197 outermost of them often considerably smaller. More or less hairy (to subglabrous), perennial or biennial, rarely annual plants, with usually erect, more or less leafy stem and alternate, entire, rarely partly toothed crenate leaves.

Lectotype: *E. uniflorus* L.

Over 200 species in extratropical countries of both hemispheres.

Literature: Cronquist, A. 1947. Revision of North American species of *Erigeron*, north of Mexico. *Brittonia*, **6**, 2: 121–302.—Botschantzev, V.P. 1954. Zametki ob Astereae. 1 [Notes on the Astereae. 1]. *Bot. Mat. (Leningrad)*, **16**: 379–394.—Nesom, G.L. 1989. The separation of *Trimorpha* (Compositae: Astereae) from

Erigeron. Phytologia, **67**, 1: 61–66.—Nesom, G.L. 1989. Infrageneric taxonomy of New World *Erigeron* (Compositae: Astereae). *Phytologia*, **67**, 1: 67–93.—Tzvelev, N.N. 1991. Zametki o rodakh *Erigeron* L. s. 1. i. *Cirsium* Mill. (Asteraceae) v evropeiskoi chasti SSSR [Notes on the genera *Erigeron* L. s. 1. and *Cirsium* Mill. (Asteraceae) in the European part of the USSR]. *Novosti Sist. Vyssh. Rast.*, **28**: 147–152.

1. Heads 30–45 mm in dia, 1–10 on each stem; ligules of peripheral flowers pink or light pink, 0.4–0.6 mm wide, exceeding involucre by more than 10 mm. Cultivated and occasionally escaped perennial plant 25–80 cm, high with long rhizome .... 3. **E. speciosus.**

+ Heads to 25 mm in dia; ligules of peripheral flowers exceeding involucre by less than 10 mm. Wild plants ...... 2.

2. Ligules of peripheral flowers narrow-elliptical, flat and strongly deflected sideways, 1.0–1.6 mm wide, exceeding more or less saucer-shaped involucre by 6–10 mm, white or light pink. Perennial of alpine tundra belt to 20 cm high; stems always with one head, weakly leafy .... 3.

+ Ligules of peripheral flowers narrow-linear, usually with more or less involute margin and not deflected sideways, 0.2–0.5 mm wide, exceeding more or less cup-shaped involucre by 1–6 mm, whitish or pinkish .... 4.

3. Involucral bracts covered with rather long, erect or somewhat bent, hairs predominantly in lower half, lower part of outer bracts visible almost to base; stems relatively short-hairy .... 1. **E. silenifolius.**

+ Involucral bracts almost to tip (but predominantly near their margin) covered with long, flexuous hairs, lower part of outer bracts covered with dense villous pubescence; stems in upper part villous-hairy .... 2. **E. komarovii.**

4. Heads solitary on stems, rarely (some plants in a population) two or three on each stem; involucre near base, as also upper part of stem, often rather densely villous-hairy. Perennial arcto-alpine plants, 5–25(30) cm high ... 5.

+ Heads almost always more than three on stem, rarely two or three, and then stem and involucre weakly hairy to sub-glabrous; flowers in head of three types: pistillate with ligule, pistillate without ligule, and bisexual, five-toothed tubular .... 9.

5. Flower in head of three types: peripheral pistillate with very narrow ligules, followed by pistillate and tubular

with weakly developed indistinctly toothed corolla, and inner bisexual, tubular with five-toothed corolla .... 6.

+ Flowers in head of two types: peripheral pistillate with very narrow ligules and inner bisexual tubular with five-toothed corolla .... 7.

6. Pubescence of involucre of moderately long, bent, and always colorless hairs. Plant of the Carpathians, with less profuse pubescence .... 11. **E. alpinus.**

+ Pubescence of involucre of longer, flexuous hairs, at least some of them reddish or pinkish (in dry specimens, the pigment is seen as dark-colored septum between cells). Plant of the Arctic and adjoining regions with more profuse pubescence .... 12. **E. borealis.**

198

7. Pubescence of involucre of only colorless hairs, on the average, less dense and short. Plant of the Carpathians .... 13. **E. uniflorus.**

+ Pubescence of involucre of colored (reddish or pinkish) hairs. Plants of the Arctic, on the average, less densely long-pubescent .... 8.

8. Stalks of heads and stems in upper part covered with short simple, semiappressed hairs besides long simple hairs, lacking glandular hairs .... 14. **E. eriocalyx.**

+ Stalks of heads and stems in upper part covered with very short glandular hairs besides long simple hairs; pubescence of involucre usually more profuse and long .... 15. **E. eriocephalus.**

9 (4). Stalks of heads densely covered with rather long squarrose simple hairs; stems also usually throughout (and particularly in lower part) covered with hairs............. 10.

+ Stalks of heads covered with very short simple or glandular hairs, but often with mixture of long squarrose hairs; stems scatteredly hairy, often subglabrous .... 12.

10. Relatively high-mountain perennial plant of Crimea, with strongly developed, usually many-headed root; ligulate flowers light pink, exceeding involucre by 2.5–4.5 mm; inflorescence usually dense, of 3–10 heads................... .... 4. **E. orientalis.**

+ Plains and low-mountain biennial plants, with weakly developed tap root; ligulate flowers whitish or light pink, exceeding involucre by 1–3 mm .... 11.

11. Stems with numerous (15–30, excluding basal) approximate leaves; pappus of achenes with brownish or reddish tinge.

Plant of steppe zone, entering forest steppe ....
.... 5. **E. podolicus.**

+ Stems with 5–12 (excluding basal) distant and more sideways deflected leaves; pappus of achenes without brownish or reddish tinge. Plant of forest zone, entering forest-steppe .... 7. **E. acris.**

12. Cauline leaves 12–20, distant, up to inflorescence (to base of its first branches); basal and lower cauline leaves rather large, more or less abruptly changing to numerous smaller and less divergent middle and upper cauline leaves. Plant 26–60 cm high, with usually numerous heads ......
.... 6. **E. droebachiensis.**

+ Cauline leaves 5–10(12), distant, on unbranched part of stem; leaves rather gradually and relatively less reduced above and usually strongly divergent .... 13.

13. Stems 20–30 cm high, usually branched in upper fourth, less often third, with numerous heads on stalks to 30 mm long. Plant of forest zone and forest-steppe ....
.... 8. **E. uralensis.**

+ Stems 10–30(40) cm high, usually branched in upper half, less often third, with less numerous (usually 3–20) heads on longer (at least some longer than 30 mm) stalks. Northern or relatively high-mountain plants .... 14.

14. Involucral bracts usually rather densely long-hairy; basal leaves quite gradually pointed. Plant of the Carpathians
.... 9. **E. angulosus.**

+ Involucral bracts very short-hairy, with little mixture of long squarrose hairs or entirely without them; basal leaves more abruptly pointed, on the average wider. Plant of the north of the European part of Russia ....
.... 10. **E. politus.**

199   **Section 1. Fruticosus** G. Don, 1830, in Loud. Hort. Brit.: 343.—*Erigeron* sect. *Platyglossa* Botsch. 1954, Bot. Mat. (Leningrad), **16:** 388; id. 1959, Fl. SSSR, **25:** 200.

Ligules of peripheral flowers narrow-elliptical, 1.3–1.5 mm wide, bent, exceeding involucre by 6–10 mm; heads usually solitary. Littoral perennials, 5–25 cm high, with short rhizome and less numerous leaves; lower leaves persisting at fruiting.

Type: *E. glaucus* Ker-Gawl.

1. **E. silenifolius** (Turcz.) Botsch. 1954, Bot. Mat. (Leningrad), **16:** 392; id. 1959, Fl. SSSR, **25:** 212; Halliday, 1976, Fl. Europ. **4:**

120; Petrovskii, 1987, Arkt. Fl. SSSR, **10**: 74.—*Aster silenifolius* Turcz. 1836, in DC. Prodr. **5**: 227.

Type: Eastern Siberia ("inter Jacutiam et Ochotiam").

North (Dvina-Pechora: Cispolar Urals along the Kozhim River).—In riverine meadow plots and gravel-beds, limestone rocks and stony slopes.—*General distribution*: Arctic, Eastern Siberia.

2. **E. komarovii** Botsch. 1954, Bot. Mat. (Leningrad), **16**: 391; id. 1959, Fl. SSSR, **25**: 213.—*Aster consanguineus* Ledeb. 1845, Fl. Ross. **2**, 2: 473.—*Erigeron consanguineus* (Ledeb.) Novopokr. 1938, Bot. Mat. (Leningrad), **7**: 137, non Cabrera, 1937.—*E. muirii* auct. non A. Gray: Petrovskii, 1987, Arkt. Fl. SSSR, **10**: 75, map. 19.

Type: Karaginskii Island ("in insula Karjaginski").

Arctic (Arctic Europe: Polar Urals).—On stony slopes and rocks, riverine meadow plots and gravel-beds.—*General distribution*: Eastern Siberia, Arctic, Far East (Kamchatka).

Note. This species can hardly be merged with the Alaskan *E. muirii* A. Gray, which has much more profuse pubescence of the whole plant.

Section 2. Phoenactis Nutt. 1840, Trans. Amer. Philos. Soc., ser. 2, **7**: 310.

Ligules of peripheral flowers narrow-linear, 0.4–1.3 mm wide, exceeding involucre by 10–20 mm; stems with 1–10 heads. Plains and low-mountain perennials 25–100 cm high, with rather long rhizomes, usually without short vegetative shoots; cauline leaves numerous, lower withering by flowering.

Lectotype: *E. speciosus* (Lindl.) DC.

3. **E. speciosus** (Lindl.) DC. 1836, Prodr. **5**: 284; Bosek, 1975, Rast. Bryansk Obl.: 392; Halliday, 1976, Fl. Europ. **4**: 117.—*Stenactis speciosa* Lindl. 1833, Bot. Reg. 19, tab. 1577.

Type: Garden-raised plant from California ("California").

Baltic (south); *Center* (Upper Dnieper; Volga-Don); *West* (Carpathians; Dnieper).—Cultivated as an ornamental plant, sometimes as escape.—*General distribution*: North America; cultivated and as escape in many other extratropical countries.

Section 3. Trimorpha (Cass.) DC. 1836, Prodr. **5**: 290 ("Trimorphaea").—*Trimorpha* Cass. 1817, Bull. Soc. Philom. Paris, 1817: 137; id. 1825, Dict. Sci. Nat. **37**: 462, 482 ("Trimorphaea"); Nesom, 1989, Phytologia, **67**, 1: 61.

Ligules of peripheral flowers narrow-linear, 0.2–0.5 mm wide, involute, not deflected, exceeding involucre by less than 5 mm; besides central tubular bisexual flowers also with intermediate tubular pistillate flowers having weakly developed, indistinctly toothed corolla; stems with 1–50 heads. Biennials or perennials 5–60 cm high.

Type: *Trimorpha vulgaris* Cass. nom. illeg. (=*Erigeron acris* L.).

4. **E. orientalis** Boiss. 1856, Diagn. Pl. Or., ser. 2, **3**: 7; Botsch. 1959, Fl. SSSR, **25**: 261; Halliday, 1976, Fl. Europ. **4**: 118.

Type: Northeastern Turkey and northern Iran ("circa Erzeroum . . . in monte Savalan prov. Aderbidjan Pérsiae borealis . . . in herbidis Tauri Isaurici").

200　*Crimea* (mountains).—In meadow plots, on stony slopes and rocks, predominantly in yaila, partly also in the upper part of forest zone.—*General distribution*: Caucasus, Russian Central Asia (Kopetdag); Asia Minor, Iran.

5. **E. podolicus** Bess. 1822, Enum. Pl. Volhyn.: 76; Botsch. 1959, Fl. SSSR, **25**: 249.—*E. asteroides* Andrz. ex Bess. 1822, op. cit.: 33, non Roxb. 1814.—*E. acer* L. subsp. *macrophyllus* auct. non (Herb.) Guterm.: Halliday, 1976, Fl. Europ. **4**: 118, p. p.

Type: Podolia uplands ("in Podolia australis").
Center (Volga-Kama: southeast; Volga-Don); *West* (Carpathians: southwest; Dnieper; Black Sea); *East*; *Crimea* (lower mountains and vicinity of Bakhchisarai and Simferopol).—On steppe slopes, steppefied forest glades and edges, outcrops of chalk and limestone.—*General distribution*: Caucasus, Western Siberia (south), Russian Central Asia; Central Europe (southeast).

6. **E. droebachiensis** O.F. Muell. 1782, Fl. Dan. 0 **5**, 15: 4, tab. 874; Botsch. 1959, Fl. SSSR, **25**: 252, in adnot.; Tzvel. 1990, Novosti Sist. Vyssh. Rast. **27**: 146.—*E. acer* L. subsp. *droebachiensis* (O.F. Muell.) Arcang. 1882, Comp. Fl. Ital.: 340; Halliday, 1976, Fl. Europ. **4**: 118.

Type: Norway, vicinity of Oslo ("An den felsigten Ufern des Dröbacker Meerbusens").
Baltic; *Center* (Ladoga-Ilmen: rarely; Upper Dnieper; Upper Volga: vicinity of Melenki; Volga-Kama: southeast; Volga-Don); *West* (Dnieper); *East* (Lower Don: Archeda sands; Trans-Volga: Buzuluk pine forest).—In pine forests and their sandy glades, less often in sandy meadows, by roadsides.—*General distribution*: Scandinavia (south), Central Europe.

Note. Rather sporadic distribution of this species confirms its considerable antiquity.

7. **E. acris** L. 1753, Sp. Pl.: 863; Botsch. 1959, Fl. SSSR, **25:** 246; Halliday, 1976, Fl. Europ. **4:** 118, p. p.; Petrovskii, 1987, Arkt. Fl. SSSR, **10:** 81, p. p.—(Plate XXII, 1).

Type: Europe ("in Europae apricis, siccis").

North; *Baltic*; *Center*; *West*; *East* (Lower Don: north; Trans-Volga; Lower Volga: north); *Crimea* (*Ai*-Petri mountains, vicinity of Yalta, probably ecdemic).—In meadows, forest glades and edges, on steppe slopes, edges of fields, river sands and gravel-beds, in pine forests, by roadsides.—*General distribution*: Caucasus, Western Siberia, Eastern Siberia, Russian Central Asia; Scandinavia, Central Europe and Atlantic Europe, Mediterranean, Asia Minor, Iran, Dzhungaria-Kashgaria, Mongolia, Japan-China; North America.—2n = 18.

Note. The most common species of the section; in the north, it reaches up to Lake Imandra and limestone outcrops in the lower part of the Pechora Basin.

8. **E. uralensis** Less. 1834, Linnaea, **9:** 154; Botsch. 1959, Fl. SSSR, **25:** 248, in adnot.; Tzvel. 1990, Novosti Sist. Vyssh. Rast. **27:** 148.—*E. acer* L. subsp. *elongatiformis* Novopokr. 1936, Fl. Yugo-Vost. Evrop. Chasti SSSR, **6:** 318, descr. ross.—*E. brachycephalus* Lindb. fil. 1944, Sched. Pl. Finl. Exs. 21–42: 88 (No. 1372).—*E. decoloratus* Lindb. fil. 1944, ibid.: 88 (Nos. 1376, 1377).—*E. elongatiformis* (Novopokr.) Serg. 1949, in Kryl. Fl. Zap. Sib. **11:** 2687.—*E. acer* L. subsp. *brachycephalus* (Lindb. fil.) Hiit. 1971, Ann. Bot. Fenn. **8:** 78.—*E. acer* L. subsp. *decoloratus* Lindb. fil.) Hiit. 1971, ibid.: 77.—*E. politus* auct. non Fries.: Rasiņš, 1960, Latv. PSR Veg. **3:** 124.

Type: Urals, vicinity of Zbatoust ("In colle aprica quadam pr. Slatoust").

North (Karelia-Murmansk; Dvina-Pechora: west and south); *Baltic*; *Center*; *East* (Trans-Volga).—In pine forests and their sandy slopes, sandy meadows, river shoals and gravel-beds, on stony slopes and rocks.—*General distribution*: Western and Eastern Siberia, Far East; Scandinavia.

Note. Possibly, evolved through postglacial hybridization of *E. politus* × *E. acer*. Here I also include the probable result of iterative hybridization of *E. uralensis* × *E. politus* = *E.* × *decoloratus* Lindb. fil.; its plants are distinctly closer to *E. politus*. For the latter, whitish ligulate flowers have been reported.

286

202   9. **E. angulosus** Gaud. 1829, Fl. Helvet. **5:** 265; Fodor. 1974, Fl. Zakarp.: 133.—*E. acer* L. subsp. *angulosus* (Gaud.) Vacc. 1909, Cat. Pl. Vall. Aoste, **1:** 350; Halliday 1976, Fl. Europ. **4:** 118.

Type: Switzerland ("in Alpibus, supra Bagnes").

West (reported for the Carpathians).—In cultivated meadows and on stony slopes in upper mountain zone.—*General distribution*: Central Europe (mountains), Mediterranean (mountains).

10. **E. politus** Fries, 1843, Bot. Not. (Lund), 1843: 120; id. 1846, Summa Veg. Scand.: 184; Orlova, 1966, Fl. Murm. Obl. **5:**198, map 52.—*E. elongatus* Ledeb. 1829, Icon. Pl. Fl. Ross. **1:** 9, tab. 31, non Moench, 1802; Botsch, 1959, Fl. SSSR, **25:** 251.—*E. acer* L. var. *elongatus* (Ledeb.) Mela et Cajand. 1906, Suom. Kasv.: 566; Tokarevskikh, 1977, Fl. Ser.-Vost. Evrop. Chasti SSSR, **4:** 168.—*E. acer* L. subsp. *politus* (Fries) Lindb. fil. 1901, Enum. Pl. Fennoscand. Or.: 56; Halliday, 1976, Fl. Europ. **4:** 118; Petrovskii, 1987, Arkt. Fl. SSSR, **10:** 82.

Type: Norway ("Norvegia, Vaage").

Arctic (Arctic Europe: except islands); *North* (Karelia-Murmansk: north; Dvina-Pechora: except southwest).—In meadows, river sands and gravel-beds, on open slopes of river and coastal terraces, in pine forests and their glades.—*General distribution*: Western and Eastern Siberia, Far East; Scandinavia.—2n = 16.

Note. Differences of this species from the preceding species are insignificant and, possibly, not always persistent.

11. **E. alpinus** L. 1753, Sp. Pl.: 864; Botsch. 1959, Fl. SSSR, **25:** 252; Dobrocz. 1962, Fl. URSR, **11:** 74; Halliday, 1976, Fl. Europ. **4:** 118.

Type: Switzerland ("in Alpibus Helvetiae").

West (Carpathians: Chernogora Range and Mt. Pope Ivan of Marmarosh).—In cultivated meadows, on stony slopes and rocks in upper mountain zone.—*General distribution*: Caucasus; Central and Atlantic Europe (mountains); Mediterranean (mountains), Asia Minor.

12. **E. borealis** (Vierh.) Simm. 1913, Lunds Univ. Årsskr., nov. ser. 2, **9,** 19: 127; Botsch. 1959, Fl. SSSR, **25:** 253; Halliday, 1976, Fl. Europ. **4:** 119; Petrovskii, 1987, Arkt. Fl. SSSR, **10:** 80, map

201   Plate XXII.
1—*Erigeron acris* L., 1a—achene; 2—*E. borealis* (Vierh.) Simm.; 3—*Phalacroloma annuum* (L.) Dumort., 3a—achene of inner (tubular) flowers, 3b—achene of peripheral (ligulate) flowers; 4—*Conyza canadensis* (L.) Cronq., 4a—achene.

1a

1

2

3 3a 3b 4 4a

22.—*Trimorpha borealis* Vierh. 1906, Bieh. Bot. Centralbl. **19**, 2: 447.—(Plate XXII, 2).

Type: Northern Eurasia and southern Greenland ["Skandinavien, Lofoten, Schottland, Faröer, Island (Verbreitet), Grönland (nur im südlichen Teil)"].

Arctic (Arctic Europe); *North* (Karelia-Murmansk: Khibiny; Dvina-Pechora: vicinity of Ust-Pinega and Ust-Tsilym, the village of Adak on the Usa River).—In cultivated meadows, on stony slopes and rocks, riversands and gravel-beds.—*General distribution*: Western Siberia (north), Arctic; Scandinavia, Atlantic Europe (mountains of Scotland); North America.—2n = 18.

Section 4. Erigeron.

Ligules of peripheral flowers narrow-linear, 0.2–0.5 mm wide, involute, not deflected, exceeding involucre by less than 5 mm, intermediate tubular pistillate flowers absent; stem with one, very rarely with two or three heads. Perennials, to 25 cm high.

Type: type species.

13. **E. uniflorus** L. 1753, Sp. Pl.: 864; Botsch. 1959, Fl. SSSR, **25**: 227; Dobrocz. 1962, Fl. URSR, **11**: 68; Halliday, 1976, Fl. Europ. **4**: 119, p. p.

Type: Lapland and Switzerland ("in Alpibus Lapponiae, Helvetiae").

West (reported for the Carpathians).—In cultivated meadows, on stony slopes and rocks, in upper mountain zone.—*General distribution*: Caucasus; Scandinavia, Central Europe (mountains), Mediterranean (mountains), Asia Minor (mountains); North America (northeast).—2n = 18.

14. **E. eriocalyx** (Ledeb.) Vierh. 1906, Bieh. Bot. Centralbl. **19**, 2: 512; Botsch, 1959, Fl. SSSR, **25**: 228; Petrovskii, 1987, Arkt. Fl. SSSR, **10**: 78, map 19.—*E. alpinus* L. var. *eriocalyx* Ledeb. 1833, Fl. Alt. **4**: 91.—*E. uniflorus* L. subsp. *eriocalyx* (Ledeb.) A. and D. Löve, 1976, Bot. Not. (Lund), **128**: 521.—*E. uniflorus* auct. non L.: Halliday, 1976, Fl. Europ. **4**: 119, p. p.

Type: Altai ("in humidis alpinis et subalpinis non rarus").

203 *Arctic* (Novaya Zemlya: south; Arctic Europe: Bolshezemelsk and Karsk tundra, Pai-Khoi, Polar Urals); *North* (Dvina-Pechora: Cis-Polar Urals).—In cultivated meadows, gravel-beds, on stony slopes and rocks.—*General distribution*: Western and Eastern Siberia, Far East (southwest); Scandinavia, Dzhungaria-Kashgaria, Mongolia.

15. **E. eriocephalus** Vahl, 1840, in Hornem. Icon. Fl. Dan. **13,** 39: t, tab. 2299; Botsch. 1959, Fl. SSSR, **25:** 232; Petrovskii, 1987, Arkt. Fl. SSSR, **10:** 79, map 21.—*E. uniflorus* L. subsp. *eriocephalus* (Vahl) Cronq. 1947, Brittonia, **6,** 2: 236; Halliday, 1976, Fl. Europ. 4: 119.

Type: Greenland ("ad 200 pedes supra mare et prope rivulum ad Niakarnak Coloniae Umanak Grönlandiae").

Arctic (Novaya Zemlya; Arctic Europe: Vaigach, Bolshezemelsk and Karks tundra, Pai-Khoi).—In stony placës, on rocks, gravel-beds, in cultivated meadows, on river sands.—*General distribution*: Western and Eastern Siberia (north), Arctic; Scandinavia; North America.—2n = 18.

GENUS 87. *PHALACROLOMA* Cass.

1826, Dict. Sic. Nat. **39:** 404.—*Stenactis* Cass. 1825, ibid., **37:** 462, 485, p. p. (excl. typo!).—*Erigeron* L. sect. *Phalacroloma* (Cass.) Torr. et Gray, 1841, Fl. North Amer. **2:** 176; Botsch. 1959, Fl. SSSR, **25:** 243

Heads heteromorphic, 13–20 mm in dia, clustered in corymbose-paniculate or paniculate inflorescence, sometimes (in very weak plants) solitary. Involucre 3–5 mm long and 4–8 mm in dia, broadly cup-shaped or saucer-shaped; involucral bracts 30–60, linear-lanceolate or linear, acute, one-veined, in several rows, outermost more than two-thirds as long as longest inner bracts. Receptacle more or less convex, without scales and without raised margins of alveoli. Peripheral flowers pistillate, in two irregular rows, ligulate, with narrow-linear, white, light pink or light lilac ligules 4–8 mm long, considerably exceeding involucre; disk flowers bisexual, tubular, with yellow, five-toothed corolla. Achenes 0.8–1.2 mm long and 0.2–0.3 mm wide, somewhat flattened, scatteredly short-hairy, two-veined, pappus of peripheral flowers 0.2–0.3 mm long, of very short bristles connate in lower part, in disk flowers of outer row of bristles connate in lower part, 0.2–0.3 mm long, inner row of fragile scabrous bristles 1.6–3.0 mm long. More or less hairy, biennial or annual plants, with erect leafy stem and alternate, coarsely toothed or entire leaves.

Type: *P. acutifolium* Cass. [=*P. annuum* (L.) Dumort.].

Three closely related species, often considered as subspecies of a single species, distributed in the warm temperate regions of both hemispheres. Apparently native of North America. Often clubbed with *Erigeron* L., however, new data (Adema, 1984) speak in favor of its independent status.

Literature. Adema, F. 1984. De madelief-fijnstraal, *Phalacroloma annuum* (L.) Dumort. in Niderland. *Gorteria,* **12,** 3–4: 51–56.

1. Stem rather densely covered with short, appressed hairs; upper, and usually also middle, cauline leaves entire, lower often more or less toothed, usually subobtuse; ligulate flowers usually white .... 3. **P. strigosum.**

+ Stem, at least in lower part, covered with more or less squarrose hairs of variable length, often subglabrous; cauline leaves usually acute .... 2.

2. Cauline leaves, except smaller uppermost, distantly coarsely toothed; stem rather densely hairy; ligulate flowers usually light pink or light lilac, sometimes white ....

.... 1. **P. annuum.**

+ Upper, and usually also middle, cauline leaves entire, lower often distantly coarsely toothed; stem usually weakly hairy (with scattered hairs) to subglabrous; ligulate flowers usually white, sometimes light pink ....

.... 2. **P. septentrionale.**

1. **P. annuum** (L.) Dumort. 1827, Florula Belg.: 67; Adema, 1984, Gorteria, **12,** 3–4: 51, p. p.—*Aster annuus* L. 1753, Sp. Pl.: 875, s. str.—*Erigeron annuus* (L.) Pers. 1806, Syn. Pl. **2:** 43; Botsch. 1959. Fl. SSSR, **25:** 244, p. p.; Halliday, 1976, Fl. Europ. **4:** 117, p. p.—*Stenactis annuua* (L.) Cass. 1825, Dict. Sci. Nat. **39:** 405; Ledeb. 1845, Fl. Ross. **2,** 2: 491; Tzvel. 1990, Novosti Sist. Vyssh. Rast. **27:** 149.—(Plate XXII, 3).

Type: Canada ("in Canada").

Center (Ladoga-Ilmen: Leningrad; Upper Dnieper; Upper Volga; Volga-Kama); *West* (Carpathians; Dnieper: west).—As ecdemic in gardens and parks, by roadsides, in forest glades and edges.— *General distribution*: North America; ecdemic in Atlantic and Central Europe.—2n = 26, 27, 36.

2. **P. septentrionale** (Fern. et Wieg.) Tzvel. 1991, Novosti Sist. Vyssh. Rast. **28:** 148.—*Erigeron ramosus* (Walt.) Britt., Sterns et Pogg. var. *septentrionalis* Fern. et Wieg. 1913, Rhodora, **15:** 60.—*E. annuus* (L.) Pers. subsp. *septentrionalis* (Fern. et Wieg.) Wagenitz, 1965, in Hegi, Ill. Fl. Mitteleur., ed. 2, **6,** 3: 96; Halliday, 1976, Fl. Europ. **4:**117.—*Stenactis septentrionalis* (Fern. et Wieg.) Holub. 1974, Folia Geobot. Phytotax. (Praha), **9:** 273; Tzvel. 1990,

Novosti Sist. Vyssh. Rast. 27: 149.—*Phalacroloma annuum* (L.) Dumort. subsp. *septentrionale* (Fern. et Wieg.) Adema, 1984, Gorteria, 12, 3–4: 53.—*Erigeron annuus* auct. non (L.) Pers.: Botsch. 1959, Fl. SSSR, **25:** 244, p. max. p.—*Stenactis annua* auct. non (L.) Cass.: Dobrocz. 1962, Fl. URSR, **11:** 76, p. max. p.; Vorosch. 1966, Opred. Rast. Mosk. Obl.: 321.

Type: Canada, Newfoundland ("gravelly thickets along Harry's River, Newfoundland").

North (Karelia-Murmansk: near Lake Sandal); *Baltic*; *Center* (Ladoga-Ilmen: vicinity of Leningrad; Upper Dnieper; Upper Volga; Volga-Kama; Volga-Don); *West*; *East.*—In meadows, forest glades and edges, on banks of water bodies, by roadsides; in the north only as ecdemic of roadsides, in habitations, on edges of fields.— *General distribution*: Caucasus, Far East; Scandinavia (south), Central and Atlantic Europe, Mediterranean; North America.— 2n = 26, 27.

Note. This very species is widely distributed in the European part of the USSR while the previous species is known only from relatively small number of localities. In many areas in the south of the forest zone and in the forest-steppe zone it has been completely naturalized. All species of the genus are apomicts, which explains the frequent occurrence of 2n = 27 in them.

3. **P. strigosum** (Muchl. ex Willd.) Tzvel. 1991, Novosti Sist. Vyssh. Rast.: 148.—*Erigeron strigosus* Muehl. ex Willd. 1803, Sp. Pl. **3,** 3: 1956; Botsch. 1959, Fl. SSSR, **25:** 245.—*Stenactis strigosa* (Muehl. ex Willd.) DC. 1836, Prodr. **5:** 299; Fodor, 1974, Fl. Zakarp.: 133.—*Erigeron pseudoannuus* Makino, 1929, Journ. Jap. Bot. **6,** 1: 5.—*Stenactis pseudoannus* (Makino) Worosch. 1954, Spisok. Sem. Glavn. Bot. Sada Akad. Nauk SSSR, **9:** 65, comb. illeg.— *Erigeron annuus* (L.) Pers. subsp. *strigosus* (Muchl. ex Willd.) Wagenitz, 1965, Hegi in Ill. Fl. Mitteleur., ed. 2, **6:** 96; Halliday, 1976, Fl. Europ. **4:** 117.—*Phalacroloma annuum* (L.) Dumort. subsp. *strigosum* (Muchl. ex Willd.) Adema, 1984, Gorteria, **12,** 3–4: 53, 117.

Type: USA, state of Pennsylvania ("in Pensylvania").

Baltic (reported); *Center* (reported for Ladoga-Ilmen); *West* (reported for the Carpathians).—By roadsides, in habitations as ecdemic plant.—*General distribution*: Far East (south); Central and Atlantic Europe; North America.

GENUS **88.** *CONYZA* Less.

1832, Syn. Gen. Compos.: 203, nom. conserv., non L. 1753, nom. rej.—*Erigeron* L. sect. *Caenotus* Nutt. 1818, Gen. North Amer. Pl. **2:** 148; Botsch. 1959, Fl. SSSR, **25:** 239.

Heads heterogamous, 2.5–6.0 mm in dia, clustered in corymbose-paniculate or paniculate inflorescences, rarely (in very small plants) solitary. Involucre 3–6 mm long and 2.5–6.0 mm in dia, cup-shaped or goblet-shaped, involucral bracts in several rows, three-veined, outermost linear-lanceolate, usually one-third to two-fifths as long as the longest linear inner bracts. Receptacle almost flat, without scales, with slightly raised margins of alveoli. Peripheral flowers pistillate, numerous, ligulate; ligules very short, not exserted or slightly exserted from involucre, not deflected sideways, whitish but often somewhat lilac or pinkish by end of flowering, 0.5–1.5 mm long, apically with two teeth; disk flowers less numerous, bisexual, tubular, with pale yellow or whitish, three- to five-toothed corolla. Achenes 0.8–1.5 mm long and 0.2–0.3 mm wide, flattened, scatteredly short-hairy, with two raised and often also with one or two fine veins; pappus 2.0–3.5 mm long, of 20–30 scabrous bristles of more or less equal length. More or less hairy (to subglabrous) annual or biennial plants, with erect leafy stem and alternate entire or coarsely toothed leaves.

Type: *C. chilensis* Spreng.

About 50 species in tropical and warm temperate regions of both hemispheres, but predominantly in America.

1. **C. canadensis** (L.) Cronq. 1943, Bull. Torr. Bot. Club. **70:** 632; id. 1976, Fl. Europ. **4:** 120.—*Erigeron canadensis* L. 1753, Sp. Pl.: 863; Botsch. 1959, Fl. SSSR, **25:** 239.—(Plate XXII, 4).

Involucre 3–4 mm long and 2.5–4.0 mm in dia; ligules of peripheral flowers 0.5–1.0 mm long; corolla of tubular flowers three- or four-toothed.

Type: North America and Europe ("in Canada, Virginia, nune in Europa australi").

North; *Baltic*; *Center*; *West*; *East*; *Crimea*.—In fields, weedy meadows, coastal sands and gravel-beds, forest glades and edges, in habitations, by roadsides and pathways.—*General distribution*: Caucasus, Western and Eastern Siberia, Far East, Russian Central Asia; Scandinavia, Central and Atlantic Europe, Mediterranean,

Asia Minor, Iran, Dzhungaria-Kashgaria, Mongolia, Japan-China; North America; ecdemic in other countries.—2n = 18.

Note. In the southern regions, it is possible to find a closely related species *C. bonariensis* (L.) Cronq., which is distinguished by the densely hairy, involucre 4–6 mm long, five-toothed corolla of the tubular flowers, and more profuse pubescence of the whole plant.

GENUS **89.** *BACCHARIS* L.
1753, Sp. Pl.: 860; id. 1754, Gen. Pl., ed. 5: 370

Heads homogamous, dioecious, 4–7 mm in dia, clustered in paniculate, more or less leafy inflorescences. Involucre 3.5–4.5 mm long and 4–7 mm in dia, cup-shaped; involucral bracts imbricate, in several rows, almost coriaceous, outer ovate, one-fifth to one-third as long as the longest, lanceolate or oblong, inner. Receptacle convex, without scales, with irregularly toothed margin of alveoli. Pistillate flowers in female head tubular, with whitish or light yellow, two- to five-toothed corolla 2.5–3.5 mm long, considerably shorter than pappus; staminate flowers (usually with rudiment of pistil) also tubular, but with wider corolla tube and five larger deflected yellowish teeth. Achenes with 10 raised veins, glabrous; pappus biseriate, consisting of weakly scabrous bristles 6–12 mm long, in staminate flowers 3–4 mm long. Glabrous or subglabrous shrubs, 0.5–2.0 m high, with rather densely leafy branches and alternate, more or less toothed, less often entire, somewhat fleshy leaves.

Lectotype: *B. halimifolia* L.

206 About 500 species in tropical and warm temperate regions of America, one of them is cultivated and often as escape in the wild in many other countries.

1. **B. halimifolia** L. 1753, Sp. Pl.: 860; Botsch. 1959, Fl. SSSR, **25:** 289; Geideman, 1975, Opred. Vyssh. Rast. MoldSSR, ed. 2: 481; Tutin, 1976, Fl. Europ. **4:** 120.

Leaves light green, 1.2–8.5 cm and 0.5–6.0 cm wide, ovate to oblanceolate.

Type: USA, State of Virginia ("in Virginia").

West (Moldavia: town of Soroki).—Cultivated in gardens and parks as ornamental plant.—*General distribution*: North America; cultivated and escaped in the wild in the Caucasus, Russian Central Asia; Central and Atlantic Europe, Mediterranean.—2n = 18.

294

TRIBE 7. **EUPATORIEAE** Cass.[1]

Heads homogamous; all flowers tubular, bisexual, pink, lilac, light azure, or white. Receptacle without scales, glabrous. Anthers usually without basal appendages. Style branches obtuse, usually covered with short papillae. Pollen helianthoid type. Achenes with pappus of scabrous bristles, less often with a crown or few scale-like bristles. Leaves opposite, undivided or more or less incised.

Type: *Eupatorium* L.

GENUS **90.** *EUPATORIUM* L.
1753, Sp. Pl.: 836; id. 1754, Gen. Pl., ed. 5: 363

Heads small, homogamous, usually with five bisexual tubular, pinkish or lilac flowers clustered in dense corymbose inflorescence. Involucre cylindrical, with two or three rows of imbricate lanceolate bracts. Receptacle flat, glabrous. Stamens with basally roundish anthers and oblong-ovate connectives. Style branches terete, densely covered with papillae. Achenes black, conusoid, pentagonal, with pappus of scabrous bristles. Perennial herbaceous plants, with creeping rhizome, erect stems, and opposite or whorled leaves.

Lectotype: *E. cannabinum* L.

44 species in the warm temperate and subtropical regions of Eurasia and North America.

Literature: King, R.N. and H. Robinson. 1970. *Eupatorium*, a Compositae genus of arcto-tertiary distribution. *Taxon,* **19,** 5: 769–774.

1. **E. cannabinum** L. 1753, Sp. Pl.: 838; Tamamsch. 1959, Fl. SSSR, **25:** 19; Tutin, 1976, Fl. Europ. **4:** 109.

Short-hairy plant 50–150 cm high, with opposite leaves cut into three to five lanceolate irregularly coarsely-toothed lobes. Involucral bracts pinkish or lilac. Achenes about 3 mm long, with pappus 3–4 mm long.

Type: Europe ("Europa ad aquas").

Baltic; *Center*; *West*; *East*; *Crimea*.—On banks of water bodies, in swampy forests, and forest gullies, in swampy meadows and forest glades.—*General distribution*: Caucasus, Russian

---
[1]Treatment by G.Yu. Konechnaya.

Central Asia (southwest); Scandinavia, Central and Atlantic Europe, Mediterranean, Asia Minor, Iran.—2n = 20.

207 GENUS **91.** *AGERATUM* L.
1753, Sp. Pl.: 839; id. 1754, Gen. Pl., ed. 5: 363

Heads rather small, homogamous, many-flowered, with bisexual, tubular, violet, light azure, pinkish-lilac or whitish flowers, clustered in corymbose or corymbose-paniculate inflorescence. Involucre cup-shaped or hemispherical, with imbricate, linear, acuminate bracts in two or three rows. Receptacle flat, without scales. Stamens with basally rounded anthers and deltoid appendage of connective. Style branches long, more or less cylindrical. Achenes narrow-conical, somewhat curved, pentagonal, almost black; apically with a crown of more or less connate and free scales. Annual or perennial herbaceous plants, less often semishrubs, with erect stem and opposite, less often alternate, orbicular-ovate or lanceolate leaves.

Lectotype: *A. conyzoides* L.

About 40 species, predominantly in the south of North America and South America; some of them occur in other countries as ornamental or ecdemic plants, often naturalized.

Literature: Johnson, M.F. 1971. A monograph of the genus *Ageratum* L. (Compositae–Eupatorieae). *Ann. Missouri Bot. Gard.* **58,** 1: 6–88.

1. **A. houstonianum** Mill. 1768, Gard. Dict., ed. 8: 2; Tamamsch. 1959, Fl. SSSR, **25:** 17; Tutin, 1976, Fl. Europ. **4:** 109.

Entire plant rather densely covered with simple hairs. Leaves opposite, ovate, more or less cordate at base, toothed. Involucre 4–5 mm long, 7–10 mm in dia. Achenes 2–3 mm long.

Type: Mexico ("ad La Vera Cruz").

North; *Baltic*; *Center*; *West*; *East*; *Crimea*.—Cultivated as ornamental plant, occasionally as escape.—*General distribution*: Central America (south), South America (north), cultivated in many other countries in both hemispheres.—2n = 20.

TRIBE 8. **CARDUEAE** Cass.

Heads homogamous or heterogamous; flowers usually all tubular but peripheral often more or less large, tubular-infundibuliform. Receptacle with scaly or setaceous bracts, less often without them. Anthers sagittate at base. Styles below fork with crown of hairs or thickening. Pollen predominantly anthemoid type. Achenes apically with pappus of scabrous or plumose bristles, rarely without. Leaves alternate, with undivided or more or less incised, often spiny lamina.

Type: *Carduus* L.

GENUS **92.** *ECHINOPS* L.[1]
1753, Sp. Pl.: 814; id. 1754, Gen. Pl., ed. 5: 356

Heads homogamous, with one bisexual tubular flower, clustered in very dense glomerular general inflorescence of globose form. Involucre cylindrical; involucral bracts numerous, imbricate, in three to five rows, membranous or coriaceous, outermost setaceous. Receptacle convex, glabrous, alveolate. Corolla white to azure. Anthers with sagittate basal appendages. Achenes fusiform, densely appressed-hairy; pappus as crown, its bristles short, flat, connate in lower half or only at base. Perennial, less often annual herbaceous plant, usually with more or less incised, less often undivided spiny leaves.

Lectotype: *E. sphaerocephalus* L.

The genus includes 120 species, distributed in extratropical (excluding the Arctic and considerable part of forest zone) and subtropical regions of Eurasia and Africa (excluding the mountainous regions of eastern Africa).

Literature: Bunge, A. 1863. Über die Gattung *Echinops*. *Bull. Acad. Sci. Pétersb.*, **6**: 390–412.—Mulkidzhanian, Ya.I. 1950. Rod *Echinops* i ego Kavkazskie predstaviteli [The genus *Echinops* and its Caucasian representatives]. *Tr. Bot. Inst. Akad. Nauk ArmSSR*, **8**: 5–92.—Jeffrey, C. 1974. *Echinops ruthenicus*. *Curtis's Bot. Mag.*, **180**, 1: 77–81, tab. 677.

1. Outer and middle involucral bracts more or less densely (usually in upper half) covered with long stalked glandular hairs .... 2.
+ All involucral bracts dorsally without glandular hairs .... 3.
2. Involucre of 16–18 bracts, of which outer and middle densely covered with fluff of long-stalked, glandular hairs in dorsal upper half; leaves above densely covered with stalked, glandular hairs, usually not arachnoid-hairy; stem

---

[1]Treatment by O.V. Czerneva.

throughout covered with stalked, glandular hairs, also
with appressed white arachnoid-hairs in upper part .....
.... 3. **E. sphaerocephalus.**
+ Involucre of 22–26 bracts, of which outer and middle
dorsally with isolated, stalked, glandular hairs; leaves
arachnoid-hairy above, without or with isolated, glandular
hairs; stem throughout finely white-arachnoid-hairy, with
isolated, glandular hairs .... 4. **E. armatus.**
3. Leaves scatteredly short-setose, without glandular hairs,
cauline leaves amplexicaul, distinctly auriculate; glomerular
general inflorescence 4–6 cm in dia .... 5. **E. exaltatus.**
+ Leaves glabrous or sparsely arachnoid-hairy above,
sometimes with sessile glands; cauline leaves sessile,
without auricles, very shortly decurrent; glomerular general
inflorescence 2.5–4.0 cm in dia .... 4.
4. Stems in upper half densely white-tomentose ....
.... 1. **E. ruthenicus.**
+ Stems throughout short-glandular hairy, not tomentose
.... 2. **E. meyeri.**

Section 1. Ritro Endl. 1836, Gen. Pl.: 467, p. p.; Bunge, 1863,
Bull. Acad. Sci. Pétersb. **6**: 408.

Involucral bracts 18–25, in three rows; outer bracts dorsally
without glandular hairs, inner free.

Type: *E. ritro* L.

1. **E. ruthenicus** Bieb. 1819, Fl. Taur.-Cauc. **3**: 597; Wissjul.
1962, Fl. URSR, **11**: 415, Fig. 78; Jeffrey, 1974, Curtis's Bot. Mag.
180, 1: 77, tab. 677.—*E. ritro* L. 1753, Sp. Pl.: 815, p. p. (quoad
var. β.); Bobr. 1962, Fl. SSSR, **27**: 30; Kožuharov, 1976, Fl. Europ.
4: 214, p.p.

Type: Described from specimens probably collected from the
southeast of the European part of Russia, most probably from
Lower Volga Region.

Center (Volga-Kama; Volga-Don); *West* (Dnieper; Moldavia;
Black Sea); *East* (Lower Don; Trans-Volga; Lower Volga); *Crimea.*—
On rubbly and stony slopes, limestone, chalk and granite outcrops,
in dry meadows, steppes.—*General distribution*: Caucasus
(Ciscaucasia), Western Siberia, Russian Central Asia; Central
Europe, Mediterranean.—2n = 30, 32.

Note. The history of the complex taxonomy of *E. ritro* s. l.
group is described by Jeffrey (1974, loc. cit.). The six different
groups mentioned by K. Linnaeus in the description of this

species are now distributed in three species: *E. ritro* L., *E. ruthenicus* Bieb., and *E. latifolius* Tausch. *E. ruthenicus* is extremely variable in the form of leaves. The most commonly occurring form has bipinnatisect leaves with very narrow segments. No less common (and sometimes together with the first) is the form 210 with pinnately divided leaves having oblong-ovate or lanceolate lobes. In addition, a form has been reported in Bashkiria, which has sinuate-toothed leaves with shortly spinescent teeth.

2. **E. meyeri** (DC.) Iljin, 1932, Izv. Bot. Sada Akad. Nauk SSSR, **30**, 3–4: 343; Bobr. 1962, Fl. SSSR, **27**: 32.—*E. ritro* L. var. *meyeri* DC. 1838, Prodr. **6**: 524.—*E. ritro* L. var. *meyeri* DC. 1838, Prodr. **6**: 524.—*E. ritro* L. subsp. *meyeri* (DC.) Kožuharov, 1975, Bot. Journ. Linn. Soc. (London), **71**: 42; id. 1976, Fl. Europ. **4**: 214.

Type: Described without mentioning the locality; neotype (chosen by M.M. Iljin) from Lake Inder ("lacus Inder").

Center (Volga-Kama: south); *East* (Trans-Volga; Lower Volga).—On limestone and chalk outcrops, steppe slopes, sometimes in abandoned fields and by roadsides.—*General distribution*: Western Siberia (southwest), Russian Central Asia (north).

Section 2. Echinops.
Involucral bracts 16–18, in three or four rows; outer bracts dorsally glandular hairy, inner free.

Type: type species.

3. **E. sphaerocephalus** L. 1753, Sp. Pl.: 814; Bobr. 1962, Fl. SSSR, **27**: 40; Wissjul. 1962, Fl. URSR, **11**: 414; Kožuharov, 1976, Fl. Europ. **4**: 213; Tabaka and others, 1988, Fl. Sosud. Rast. LatvSSR: 134.—(Plate XXIII, 2).

Type: Italy ("in Italia").

Baltic (Latvia: ecdemic); *Center* (Ladoga-Ilmen: ecdemic; Upper Dnieper: northeast of Belorussia; Upper Volga: south and east; Volga-Kama: south and east; Volga-Don); *West*; *East*; *Crimea.*— In forest glades and edges, meadows, on steppe slopes, limestone and chalk outcrops, in scrublands and thinned-out forests, by roadsides.—*General distribution*: Caucasus, Western Siberia (south and southwest), Eastern Siberia (southwest), Russian Central Asia (northwest); Central Europe, Mediterranean, Dzhungaria-Kashgaria.—2n = 30, 32.

○ 4. **E. armatus** Stev. 1856, Bull. Soc. Nat. Moscou, **29**, 4: 382; Privalova and others, 1965, Novosti Sist. Vyssh. Rast. 1965:

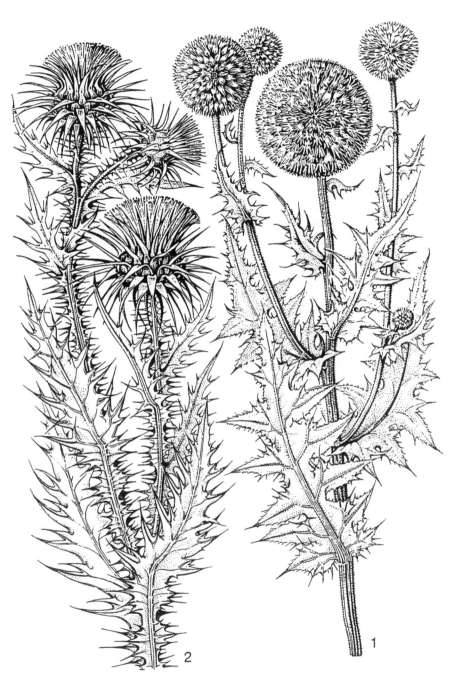

PlateXXIII.
1—*Onopordum tauricum* Willd.; 2—*Echinops sphaerocephalus* L.

302.—*E. bannaticus* auct. non Rochel ex Schrad. Wissjul. 1962, Fl. URSR, **11:** 417; Kožuharov, 1976, Fl. Europ. **4:** 214, p. p.—*E. tauricus* Willd. 1813, Enum. Pl. Horti Berol. Suppl.: 62, nom. nud.

Type: Crimea ("in Tauria meridionali").

Crimea.—In forest glades and edges, on stony and rubbly slopes, limestone rocks and rubble, sometimes by roadsides.—Endemic.—2n = 30.

Note. Closely related to *E. sphaerocephalus* L. as though being its xerophytic form. In the Steven's Herbarium in Helsinki, there are two sheets under *E. armatus,* which were kindly sent to me for study. One of them, bearing the label "*Echinops armatus* m. (an hic *E. binnaticus* DC.) in litt. merid. Tauriae, 7 Julii, 1847," should be considered the type of this species.

Section 3. Terma Endl. 1836, Gen. Pl.: 467; Bunge, 1863, Bull. Acad. Sci. Pétersb. **6:** 410.

Involucral bracts 16–18, in three or four rows, outer dorsally without glandular hairs, inner somewhat connate in lower part.

Lectotype: *E. exaltatus* Schrad.

5. E. exaltatus Schrad. 1809, Hort. Götting.: 15, tab. 9; Bobr. 1962, Fl. SSSR, **27:** 47; Kožuharov, 1976, Fl. Europ. **4:** 214.—*E. commutatus* Juratzka, 1858, Verh. Zool.-Bot. Ges. Wien, **8:** 17; Wissjul. 1962, Fl. URSR, **11:** 415.

Type: Described from plants raised in Göttingen Botanical Garden; as its origin erroneously mentioned Siberia ("in Sibiria").

West (Carpathians).—In forest glades and edges, thinned-out forests, scrublands.—*General distribution*: Central Europe, Atlantic Europe (ecdemic), Mediterranean.—2n = 30.

Note. The study of the specimens in the Herbarium of the Komarov Botanical Institute, raised in Göttingen Botanical Garden by Schrader, leaves no doubt that this species does not occur in Siberia but is widely distributed in Central Europe.

211   GENUS **93.** *CARLINA* L.[1]
1753, Sp. Pl.: 828; id. 1754, Gen. Pl., ed. 5: 360

Heads homogamous, many-flowered, with bisexual, tubular flowers, solitary at apices of stem and its branches or in center of rosette of basal leaves. Involucre (25)30–150(200) mm in dia, many-rowed; outer bracts leaf-like, spiny-toothed on margin; middle

---

[1]Treatment by O.V. Czerneva.

bracts lanceolate, cristate-spiny with branched spines in upper half; inner bracts linear, entire, exceeding flowers, coriaceous-membranous, lustrous, whitish or yellowish. Receptacle flat, fleshy, densely covered with scales. Corolla purple or yellow. Anthers with long, linear, basal appendages, ciliate on margin. Achenes oblong, almost terete, densely covered with short stiff hairs, with straight hilum; pappus one-rowed, persistent; its bristles plumose, more or less connate in lower part in fascicles. Annual, biennial or perennial herbaceous plants, sometimes stemless, occasionally shrubs (Canary islands), usually with more or less incised, less often undivided, spiny leaves.

Lectotype: *C. vulgaris* L.

About 28 species, distributed in Europe, northern Africa, and western Asia.

Literature: Meusel, H. and K. Werner. 1962. Über die Gliederung von *Carlina acaulis* L. und *Carlina vulgaris* L. *Wiss. Zeitschr. Univ. Halle*, **11**, 2: 279–291.—Meusel, H. and A. Kästner, 1972. Übersicht zur systematischen Gliederung der Gattung *Carlina. Feddes Repert.*, **83**, 4: 213–232.

1. Plants with well-developed stem, often branched in upper part, to 120 cm high; leaves undivided, uppermost subtending head, exceeding involucral bracts, as long, or shorter; heads 25–60 mm in dia.................................. 2.
+ Plants stemless or with simple stem, to 20(40) cm high; leaves pinnately lobed or pinnatisect, uppermost subtending head, considerably exceeding involucral bracts; heads 70–150(200) mm in dia.. 4.
2. Upper leaves subtending head, shorter than involucral bracts or almost as long .... 3.
+ Upper leaves subtending head, exceeding involucral bracts .... 2. **C. bibersteinii.**
3. Stem usually to 30 cm high; cauline leaves coarsely toothed-spiny, crimped, middle and upper oblong-ovate or ovate, short-acuminate .... 1. **C. vulgaris.**
+ Stem usually 30–65 cm high; cauline leaves flat, finely spiny-toothed or ciliate-serrate, middle and upper lanceolate, long-acuminate .... 3. **C. intermedia.**
4. Leaves pinnately lobed, with coarsely toothed lobes, arachnoid-hairy above, tomentose beneath; heads 150–200 mm in dia; pappus bristles connate at base in fascicles of 16–18................................. 6. **C. onopordifolia.**

302

+ Leaves pinnately divided or pinnatisect, glabrous above, arachnoid-hairy beneath or spiny-arachnoid-hairy on both sides; heads 70–150 mm in dia; pappus bristles connate at base in fascicles of 5–8 .... 5.

5. Plants stemless; leaves subglabrous above, finely arachnoid-hairy beneath .... 4. **C. acaulis.**

+ Plants with stem 20–40 cm high; leaves spiny-arachnoid-hairy on both sides, more densely on petiole and midrib .... 5. **C. cirsioides.**

212   1. **C. vulgaris** L. 1753, Sp. Pl.: 828; Lincz. 1962, Fl. SSSR, **27:** 79, p. p.; Klok. 1962, Fl. URSR, **11:** 426; Meusel and Werner, 1962, Wiss. Zeitschr. Univ. Halle, **11,** 2: 288.—*C. vulgaris* subsp. *vulgaris* Webb, 1976, Fl. Europ. **4:** 210.—*C. taurica* Klok. 1962, Fl. URSR, **11:** 558, 429, Fig. 80.—(Plate XXIV, 2).

Type: Italy ("in Italia").

Baltic; *West* (Carpathians; Dnieper: south; Moldavia; Black Sea); *Crimea.*—In forest glades and edges, more or less steppefied herbaceous slopes, limestone outcrops, scrublands, vineyards, some-times on dumps along roads.—*General distribution*: Caucasus; Central Europe, Mediterranean, Asia Minor, Iran (northwest).—2n = 20.

Note. Forms hybrids with *C. biebersteinii* and *C. intermedia.*

2. **C. biebersteinii** Bernh. ex Hornem. 1819, Hort. Hofn. Suppl.: 94; Lincz. 1962, Fl. SSSR, **27:** 82; Klok. 1962, Fl. URSR, **11:** 428, p. p.—*C. vulgaris* L. subsp. *longifolia* Nym. 1879, Consp. Fl. Eur.: 400; Webb. 1976, Fl. Europ. **4:** 210.—*C. vulgaris* L. f. *stricta* Rouy, 1903, Fl. Fr. **8:** 367.—*C. stricta* (Rouy) Fritsch, 1909, Excursionsfl. Österr., ed. 2: 635; Meusel and Werner, 1962, Wiss. Zeitschr. Univ. Halle, **11, 2:** 285.—(Plate XXIV, 3).

Type: Described from plants raised in Copenhagen Botanical Garden from seeds received from F.A. Bieberstein.

North (Dvina-Pechora: south); *Baltic*; *Center*; *West* (Carpathians; Dnieper; Moldavia); *East* (Trans-Volga).—In dry and more or less steppefied meadows, forest glades and edges, thinned-out forests, scrublands, on slopes of gullies, sometimes on chalk outcrops.—*General distribution*: Caucasus (western Transcaucasia), Western Siberia, Eastern Siberia (southwest); Scandinavia (south), Central and Atlantic Europe, Mediterranean (west), Asia Minor (north).—2n = 20.

Note. In the original description of the species, the source of its seeds was given as Crimea ("in Tauria"). However, apparently, the seeds came from the Volga Region or the Ukraine.

3. **C. intermedia** Schur, 1866, Enum. Fl. Transsilv.: 413; Meusel and Werner, 1962, Wiss. Zeitschr. Univ. Halle, **11,** 2: 289.—*C. vulgaris* L. subsp. *intermedia* (Schur) Hayek, 1931, Prodr. Fl. Penins. Balcan. **2:** 694; Webb, 1976, Fl. Europ. **4:** 210, p. p., excl. syn. *C. biebersteinii.*—*C. vulgaris* auct. non. L.: Lincz. 1962, Fl. SSSR, **27:** 79, p. p.—*C. biebersteinii* auct. non Bernh. ex Hornem.: Klok. 1962, Fl. URSR, **11:** 428, p. p.—(Plate XXIV, 1).

Type: Romania ("Oberhalb der Weinberge bei Hammersdorf, auf dem Schlossberge bei Kronstadt").

North (Karelia-Murmansk: south; Dvina-Pechora: southwest); *Baltic*; *Center*; *West*; *East* (Lower Don; Trans-Volga).—In forest glades and edges, usually with sandy or sandy-loam soil, on sands of terraces above floodplains, in thinned-out, predominantly pine forests, by roadsides.—*General distribution*: Caucasus, Russian Central Asia (southwest); Scandinavia, Central and Atlantic Europe, Mediterranean (west), Asia Minor (north), Iran (northwest)—2n = 20.

4. **C. acaulis** L. 1753, Sp. Pl.: 828; Lincz. 1962, Fl. SSSR, **27:** 86, Plate VI, Fig. 1; Klok. 1962, Fl. URSR, **11:** 425; Meusel and Werner, 1962, Wiss. Zeitschr. Univ. Halle, **11,** 2: 284; Webb, 1976, Fl. Europ. **4:** 210, p. p.

Type: Mountains of Europe ("in Italia, Germania montibus").

Center (Upper Dnieper: west); *West* (Carpathians).—On open stony slopes and rocks, in meadows, forest glades.—*General distribution*: Central Europe, Atlantic Europe (south), Mediterranean (northwest).—2n = 20.

Note. In the Carpathians, besides the typical stemless form there is also a form with short stems.

5. **C. cirsioides** Klok. 1954, Bot. Mat. (Leningrad), **16:** 355; Lincz. 1962, Fl. SSSR, **27:** 87; Klok. 1962, Fl. URSR, **11:** 422, Fig. 79.—*C. acaulis* L. subsp. *simplex* auct. non Nym.: Webb, 1976, Fl. Europ. **4:** 210, p. p.

Type: Kiev Region ("vicinity of the town of Belichi").

Center (Upper Dnieper); *West* (Carpathians; Dnieper; Moldavia).—In forest glades and edges, thinned-out forests.—*General distribution*: Central Europe (southeast).

214   6. **C. onopordifolia** Bess. ex Szaf., Kulcz. et pawl. 1924, Rosl. Polskie: 641; Lincz. 1962, Fl. SSSR, **27:** 84; Klok. 1962, Fl. URSR, **11:** 421.—*C. acanthifolia* auct. non All.: Webb, 1976, Fl. Europ. **4:** 210, p. p.—*C. utzka* auct. non Hacq.: Meusel and Kästner, 1972, Feddes Repert, **83,** 4: 231, p. p.

Plate XXIV
1—*Carlina intermedia* Schur; 2—*C. vulgaris* L.; 3—*C. biebersteinii* Bernh.
ex Hornem.

Type: Western Ukraine and Poland ("k. Rohatyna. k. Krzemienca, k. Chelmu, k. Pinczowa").

West (Carpathians; Dnieper: west; Moldavia).—On dry, often stony ˌslopes, in scrublands, rather rare.—*General distribution*: Central Europe (southeast).—2n = 20.

Note. Closer to the Mediterranean species *C. acanthifolia* All.; however, in *C. onopordifolia* the pappus bristles are basally connate into fascicle of 16–18, while in *C. acanthifolia* the fascicles comprise 9 or 10 bristles.

<h2 style="text-align:center">GENUS 94. XERANHEMUM L.[1]</h2>

1753, Sp. Pl.: 857; id. 1754, Gen. Pl., ed. 5: 369

Heads heterogamous, many-flowered, solitary at apices of stem and its branches, with tubular flowers, of which few peripheral pistillate, others bisexual. Involucre goblet-shaped or ovate, 5–15 mm in dia; involucral bracts membranous, imbricate, in many rows, inner longer, usually lingulately deflected sideways and brightly colored. Receptacle flat, densely covered with stiff, simple or split scales. Corolla of peripheral pistillate flowers arcuately bent, others regular with five very short teeth, whitish or pinkish. Anthers sagittate at base with short fimbriate appendages. Achenes of bisexual flowers oblong-cuneate, shortly appressed hairy, with straight hilum; pappus one-rowed, of 5–15 aristate, acuminate, scalelike bristles. Annual herbaceous plants, with undivided, narrow-lanceolate leaves without spines.

Type: *X. annuum* L.

Five or six species, distributed in southern Europe, north Africa, southwestern and Russian Central Asia.

Literature: Gay, J. 1827. Monographie des generes *Xeranthemum* et *Chardinia*. *Mém. Soc. Hist. Nat. Paris,* **3**: 325–371—Chrtek, J. 1961. The relation between genus *Xeroloma* Cass. and genus *Xeranthemum* L. *Novit. Bot. Prag.,* 1961: 3–5.

1. Outer involucral bracts dorsally glabrous, very short-acuminate, inner two times as long as outer, pink or pinkish-violet; pappus of five scalelike bristles... 1. **X. annuum.**
+ Outer involucral bracts dorsally, mainly on midrib, finely tomentose, subobtuse, inner somewhat exceeding outer, pale pink, later turning brown; pappus of (7)10–15 scalelike bristles .... 2. **X. cylindraceum.**

[1]Treatment by O.V. Czerneva.

1. **X. annuum** L. 1753, Sp. Pl.: 857; Lincz. 1962, Fl. SSSR, **27:** 62; Wissjul. 1962, Fl. URSR, **11:** 418; Webb, 1976, Fl. Europ. **4:** 212.

Type: Austria ("in Austria").

Center (Upper Dnieper: ecdemic; Volga-Kama: ecdemic; Volga-Don: south and east); *West* (Dnieper: south and east; Moldavia; Black Sea); *East*; *Crimea*.—In steppes, on dry herb slopes, chalk and limestone outcrops, sands, forest glades and edges, scrublands, sometimes by roadsides.—*General distribution*: Caucasus; Central Europe (south), Mediterranean, Asia Minor (west).—2n = 12.

Note. For southern Crimea, *X. inapertum* (L.) Mill. (Lincz. 1962, op. cit.: 64) has also been reported. It is distinguished from *X. annuum* by the erect (and not deflected sideways) reddish-brown-purple or pale violet (and not usually pink) inner involucral bracts. I could not lay my hands on the herbarium specimens from Crimea to confirm this.

215 2. **X. cylindraceum** Sibth. et Smith, 1831, Fl. Gràec. Prodr. **2:** 172; Lincz. 1962, Fl. SSSR, **27:** 66; Wissjul. 1962, Fl. URSR, **11:** 419; Webb, 1976, Fl. Europ. **4:** 212.

Type: Turkey ("in monte Olympo Bithyno").

West (Moldavia); *Crimea*.—On dry clayey and stony slopes, in pastures, forest glades and edges, steppes, by roadsides.—*General distribution*: Central and Atlantic Europe (south), Mediterranean, Asia Minor, Iran (northwest).—2n = 20.

GENUS 95. *ARCTIUM* L.[1]
1753, Sp. Pl.: 816; id. 1754, Gen. Pl., ed. 5: 357

Heads homogamous, many-flowered, with bisexual tubular flowers, clustered in corymbose, paniculate or subracemose inflorescences. Involucre more or less globose, 15–40 mm in dia (with acute conical involucral bracts); involucral bracts coriaceous, imbricate, many-rowed, linear, outer and middle merging into squarrose, subulate, inner bracts with uncinate cusp; inner bracts membranous, erect, acuminate or somewhat uncinate. Receptacle flat, setose, with somewhat flattened bristles. Corolla purple or purple-red, less often white. Anthers with filiform simple or much divided basal appendages; connective lanceolate, acuminate.

---

[1]Treatment by O.V. Czerneva.

Achenes oblong or oblong-obovate, usually longitudinally indistinctly ribbed, smooth or transversely rugose, with straight hilum; pappus many-rowed, with short, scabrous bristles detaching singly. Biennial herbaceous plants, with large undivided leaves without spines.

Lectotype: *A. lappa* L.

The genus includes 10 species, distributed in warm temperate and subtropical regions of Eurasia and Africa (particularly in countries of the Mediterranean), but many species are ecdemic in other extratropical countries and have become naturalized there. All species hybridize readily.

Literature: Arénes, J. 1950. Monographie du genere *Arctium* L. *Bull. Jard. Bot. Bruxelles,* **20**: 67–156.

1. Inner involucral bracts usually purple, somewhat broader in upper part, terminating in straight cusp; corolla limb densely glandular-hairy outside ....4. **A. tomentosum.**
+ Inner involucral bracts usually not colored, less often pale purple, not broader in upper part and terminating in uncinate cusp; corolla limb not glandular-hairy outside .... 2.
2. General inflorescence corymbose; heads on rather long peduncles, little differing in length .... 1. **A. lappa.**
+ General inflorescence more or less racemose; heads on short peduncles (to subsessile), terminal on branches and in leaf axils .... 3.
3. Heads (1.5)1.8–2.2(2.5) cm in dia; achenes 4–5 mm long .... 3. **A. minus.**
+ Heads 3–4 cm in dia; achenes 7–8 mm long ....
....2. **A. nemorosum.**

1. **A. lappa** L. 1753, Sp. Pl.: 816, p. p. (excl. var. β.); Juz. and E. Serg. 1962, Fl. SSSR, **27**: 97, Plate VII, figs. 10, 11; Klok. 1962, Fl. URSR, **11**: 436; Perring, 1976, Fl. Europ. **4**: 215.—*A. chaorum* Klok. 1962, Fl. URSR, **11**: 560, 437.

Type: Europe ("in Europae cultis ruderatis").

North (Karelia-Murmansk: south; Dvina-Pechora: south); *Baltic*; *Center*; *West*; *East*; *Crimea*.—In habitations, by roadsides, in weedy places, more or less trampled meadows and forest glades, on banks of water bodies.—*General distribution*: Caucasus, 216 Western and Eastern Siberia, Far East, Russian Central Asia; Central and Atlantic Europe, Mediterranean, Asia Minor, Iran, Mongolia, India—Himalayas, Japan-China; ecdemic in other extratropical countries.—2n = 32, 36.

Note. A. chaorum Klok. was described from the Crimea (vicinity of Alupka).

2. **A. nemorosum** Lej. 1833, Mag. Hort. (Liege), **1**: 289; Juz. and E. Serg. 1962, Fl. SSSR, **27**: 99, Plate VII, figs. 2, 3; Klok. 1962, Fl. URSR, **11**: 438, Fig. 82; Perring, 1976, Fl. Europ. **4**: 215.—*A. glabrescens* ("*flabrescens*") Klok. 1962, Fl. URSR, **11**: 560, 440, Fig. 83.

Type: Belgium.

Center (Upper Dnieper; Upper Volga; Volga-Kama; Volga-Don); *West*; *East* (Lower Don; Trans-Volga); *Crimea.*—In deciduous and mixed forests, forest glades and edges, on banks of water bodies, in scrublands, sometimes by roadsides.—*General distribution*: Caucasus (Ciscaucasia); Scandinavia, Central and Atlantic Europe.—2n = 36.

Note. A. glabrescens Klok. was described from the Crimea (the Alma River near the station of Alma).

3. **A. minus** (Hill) Bernh. 1800, Syst. Verz. Erfurt: 134; Juz. and E. Serg. 1962, Fl. SSSR, **27**: 100; Klok. 1962, Fl. URSR, **11**: 442; Perring, 1976, Fl. Europ. **4**: 215.—*Lappa minor* Hill, 1762, Veg. Syst. **4**: 28, tab. 25, fig. 3.

Type: Great Britain ("London, native of our dry waste grounds").

North (Karelia-Murmansk: south; Dvina-Pechora: south); *Baltic*; *Center*; *West*; *East*; *Crimea.*—In habitations, by roadsides, on banks of water bodies, in trampled forests, scrublands.—*General distribution*: Western Siberia (southwest), Far East (south); Scandinavia, Central and Atlantic Europe, Mediterranean, Asia Minor; ecdemic in other extratropical countries.—2n = 32, 36.

4. **A. tomentosum** Mill. 1768, Gard. Dict., ed. 8, No. 3; Juz. and E. Serg. 1962, Fl. SSSR, **27**: 104, Plate VII, figs. 8, 9; Klok. 1962, Fl. URSR, **11**: 433; Perring, 1976, Fl. Europ. **4**: 215.—*A. lappa* L. 1753, Sp. Pl.: 816, p. p. (quoad var. β.).—*A. leptophyllum* Klok. 1962, Fl. URSR, **11**: 559, 434, Fig. 81.

Type: Described from plants raised in botanical garden in England and originating from the Apennine. mountains (". . . grows naturally on the Apennine mountains").

Arctic (Arctic Europe: vicinity of Vorkuta); *North*; *Baltic*; *Center*; *West*; *East*; *Crimea.*—In habitations, by roadsides, in weedy forests, more or less weedy meadows and forest glades, scrublands.—*General distribution*: Caucasus (Ciscaucasia), Western and Eastern Siberia, Far East, Russian Central Asia;

Scandinavia, Central and Atlantic Europe, Mediterranean, Asia Minor, Iran, Dzhungaria-Kashgaria, Japan-China; ecdemic in other extratropical countries and naturalized there itself.—2n = 36.

Note. *A. leptophyllum* Klok. was described from the Crimea (village of Krasnoles'e in Simferopol District). Almost throughout its range, but less frequently, occurs a variety with glabrous or subglabrous involucre—var. *glabrum* (Koern.) Aren. All the species of the genus readily hybridize with each other producing hybrids with intermediate characters. These are: 1. *A. lappa* × *A. minus* = *A.* × *nothum* (Ruhm.) Weiss. (=*Lappa* × *notha* Ruthm.); 2. *A. lappa* × *A. nemorosum* = *A.* × *cimbricum* (Krause) Hayek (=*Lappa* × *cimbrica* Krause); 3. *A. lappa* × *A. tomentosum* = *A.* × *ambiguum* (Čelak.) Nym. (=*Lappa* × *ambigua* Čelak.); 4. *A. minus* × *A. nemorosum* = *A. maassii* (M. Schulze) Rouy (=*Lappa* × *maassii* M. Schulze); 5. *A. minus* × *A. tomentosum* = *A.* × *mixtum* (Simonk.) Nym. (=*Lappa* × *mixta* Simonk.); 6. *A. nemorosum* × *A. tomentosum* = *A.* × *neumannii* Rouy.

GENUS **96.** *COUSINIA* Cass.[1]
1827, Dict. Sci. Nat. **47**: 503

Heads homogamous, many-flowered, with bisexual, tubular flowers, solitary at apices of stem and its branches or clustered in various kinds of inflorescences. Involucre oblong-cylindrical to spherical, 5–45 mm in dia; involucral bracts imbricate, in many rows, coriaceous or membranous, entire or spiny-toothed, terminating in more or less long, straight or uncinate, often prickly cusp or expanded at tip into leaf-like appendage, inner often strongly differing from others. Receptacle flat, setose, with smooth or scabrous bristles. Corolla white to red or yellow. Anthers with fimbriately incised basal appendages; connectives of various forms. Achenes obovate to obpyramidal, glabrous and usually smooth, with straight hilum, pappus in one or two rows, its bristles scabrous, free and readily detaching singly, pappus sometimes absent. Annual, biennial, or perennial herbaceous plants, less often semishrublets, with undivided or more or less incised leaves.

Type: *C. carduiformis* Cass. (=*C. orientalis* (Adams) C. Koch).

More than 600 species, distributed predominantly in southwestern and Russian Central Asia but extending into the southeast of the European part of Russia, northern Kazakhstan, Dzhungaria and western Himalayas.

---

[1]Treatment by O.V. Czerneva.

310

Literature; Bunge, A. 1865. Übersichtliche Zusam menstellung der Arten der Gattung *Cousinia* Cass. Mém. Acad. Sci. Pétersb. (Sci. Phys.-Math.), ser. 7, **9**, 2: 1–56.—Winkler, C. 1893. Synopsis specierum generis *Cousinia* Cass. *Acta Horti Petropol.*, **12**, 2: 181–286.—Rechinger, K. 1972. Fl. Iran, **90**: 1–329; Rechinger, K. 1979. Fl. Iran, **139a**: 108–153.—Czerneva, O.V. 1988. Konspekt sistemy roda *Cousinia* Cass. (Asteraceae) [Conspectus of the system of the genus *Cousinia* Cass. (Asteraceae)]. *Bot. Zhurn.*, **73**, 6: 594–597.

1. **C. astracanica** (Spreng.) Tamamsch. 1954, Bot. Mat. (Leningrad), **16**: 468; Czern. 1962, Fl. SSSR, **27**: 152; Moore, 1976, Fl. Europ. **4**: 215.—*Carduus astracanica* Spreng. 1807, Mantissa: 49.—*Cousinia affinis* Schrenk, 1841, in Fisch. and C.A. Mey. Enum. Pl. Nov. **1**: 41; Czern. 1962, op. cit.: 153.

Heads solitary at apices of stem and its branches. Involucre 10–20 mm in dia; involucral bracts coriaceous, entire, terminating in arcuate cusp, straight outside, inner short-acuminate. Corolla whitish to pale yellow. Perennial herbaceous plants, with spiny-lobate or spiny-toothed leaves.

Type: Vicinity of Volgograd ("Sarepta").

East (Lower Volga: vicinity of Volgograd).—On clayey slopes.—*General distribution*: Russian Central Asia (north), Western Siberia (south); Dzhungaria-Kashgaria (Dzhungaria).

Note. Relates to section *Leiocaules* Bunge. For the vicinity of Riga, one more species, *C. tenella* Fisch. et C.A. Mey. (1834, *Index Sem. Hort. Petropol.* **1**: 25; Schultz, 1977, *Bot. Zhurn.*, **62**, 10: 1516) from section *Tenellae* Bunge, has been reported. This is an annual plant 8–40 cm high, with erect stem and rather soft, undivided (entire or finely toothed), sessile leaves, subglabrous above, grayish-tomentose beneath. The heads are subglobose, glabrous, solitary, 4–5 mm in dia.

GENUS 97. *SAUSSUREA* DC.[1]
1810, Ann. Mus. Hist. Nat. (Paris), **16**, 156, 196, nom. conser.

Heads homogamous, 5–25 mm in dia, with numerous pink, purple, or violet tubular flowers, solitary or clustered in corymbose inflorescence. Involucre many-rowed; involucral bracts imbricate, with or without membranous apical appendage. Receptacle with

[1]Treatment by G.Yu. Konechnaya.

membranous scales. Stamens with deltoid connectives and sagittate anthers having hairy apical appendages. Style branches thick, apically rounded, papillate outside. Achenes terete, glabrous and smooth, grayish or olive, with pappus of two or three rows of bristles, of which outer scabrous, one-third to half as long as inner plumose bristles. Perennial herbs, with undivided or pinnately incised alternate leaves.

Type: *S. alpina* (L.) DC.

About 400 species in arctic and temperate zones of Eurasia and North America, as also in Australia (one species).

Literature: Lipschitz, S.J. 1979. Rod *Saussurea* DC. [The Genus *Saussurea* DC.]. Leningrad: 1–282.

218

1. Involucral bracts apically expanded into orbicular, membranous, toothed appendage, pink in inner bracts, greenish in outer one or two rows; flowers pink-purple; leaves lanceolate to ovate, undivided, less often lower pinnatilobate or coarsely toothed .... 1. **S. amara.**

+ Involucral bracts acuminate, without appendage .... 2.

2. Lower leaves pinnatisect or lyrate; flowers pink .... 3.

+ All leaves undivided, lanceolate, ovate or cordate; flowers purple or lilac-violet .... 4.

3. Lower and middle leaves pinnatisect into coarsely toothed segments, upper lanceolate, undivided or coarsely toothed; heads 0.5–1.0 cm in dia .... 2. **S. turgaiensis.**

+ Lower leaves undivided, hastate or lyrate, with large, hastate, terminal lobe and two or three pairs of undivided lateral lobes; upper leaves lanceolate; heads 0.7–1.5 cm in dia .... 3. **S. salsa.**

4. Leaves glabrous or arachnoid-hairy beneath .... 5.

+ Leaves white-tomentose beneath .... 7.

5. Leaves lanceolate, lower petiolate, upper and middle sessile, with decurrent base .... 6.

+ Leaves narrow-ovate to linear, all petiolate, upper sometimes sessile but not decurrent on stem .... 4. **S. alpina.**

6. Stem narrow-winged, wings 1–2 mm wide. Plant of north and east of the European part of Russia .... 5. **S. parviflora.**

+ Stem broad-winged, wings 3–6 mm wide. Plant of the Carpathians .... 6. **S. porcii.**

7(4). Leaves cuneate or rounded at base ...... 9. **S. × uralensis.**

+ Leaves cordate .... 8.

8. Heads 2–10. Plant of the Carpathians ..... 7. **S. discolor.**
+ Heads 10–30. Plant of the Urals .... 8. **S. controversa.**

SUBGENUS **1. *THEODOREA*** (Cass.) Lipsch.
1961, Bot. Mat. (Leningrad), **21:** 379.—*Theodorea* Cass. 1828,
Dict. Sci. Nat. **53:** 463

Involucral bracts with rounded apical appendages.
Lectotype: *S. amara* (L.) DC.

1. **S. amara** (L.) DC. 1810, Ann. Mus. Hist. Nat. (Paris), **16:**
200; Lipsch. 1962, Fl. SSSR, **27:** 520; id. 1976, Fl. Europ. **4:** 216;
Tabaca and others, 1988, Fl. Sosud. Rast. LatvSSR: 139; Tretiakov,
1990, Bot. Zhurn. **75,** 2: 264.—*Serratula amara* L. 1753, Sp. Pl.:
819.—(Plate XXV, 2).

Type: Siberia ("in Sibiria").
North (Dvina-Pechora: south, ecdemic near the rail-road station
of Mikun); *Baltic* (ecdemic in vicinity of Daugavpils); *Center*
(Upper Dnieper: ecdemic in vicinity of Grodno and Slonim; Upper
Volga: ecdemic in vicinity of Moscow; Volga-Kama: east); *West*
(Dnieper: east); *East* (Trans-Volga; Lower Volga).—In saline
meadows and solonchaks, steppes.—*General distribution*: Western
and Eastern Siberia, Russian Central Asia; Mongolia, Japan-
China.—$2n = 26, 42$.

SUBGENUS **2. *SAUSSUREA***

Involucral bracts acute, without appendages.

Type: type species.
Section 1. Laguranthera Lipsch. 1959, Bot. Mat. (Leningrad),
15: 17.—*Saussurea* DC. a *Lagurostemon* DC. β. *Laguranthe* C.A.
Mey. ex Endl. 1838, Gen. Pl.: 468.

Lower leaves pinnatisect or lyrate.
Lectotype: *S. salicifolia* (L.) DC.

220 2. **S. turgaiensis** B. Fedtsch. 1910, Feddes Repert. **8:** 497;
Lipsch. 1962, Fl. SSSR, **27:** 508; id. 1976, Fl. Europ. **4:** 216.

Type: Kazakhstan ("in locis Turkestaniae septentrionalis: territ.
Turgai . . . in salsis prope lacum inter colles areneosos in
ditione Naursum secundo species haec detecta est; etiam in terr.
Akmolinsk in locis salsis ad fl. Saryusen prope cimeterium
Dshantai").

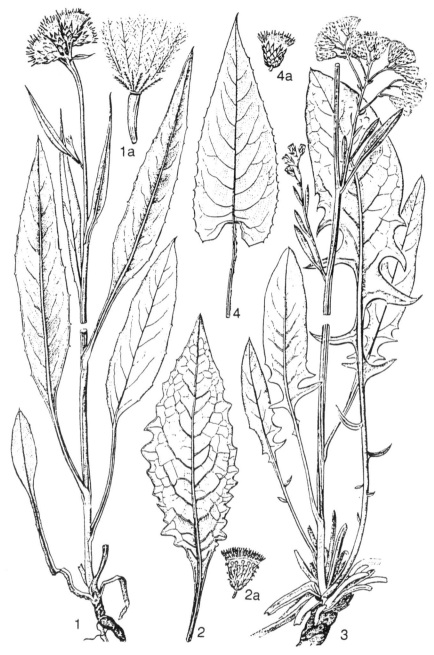

Plate XXV.

1—*Saussurea alpina* (L.) DC., la—achene with pappus; 2—*S. amara* (L.) DC., lower leaf, 2a—head; 3—*S. salsa* (Pall. ex Bieb.) Spreng.; 4—*S. controversa* DC., lower leaf, 4a—head.

East (Trans-Volga: southeast).—On chalk outcrops.—*General distribution*: Western Siberia (south), Russian Central Asia.

3. **S. salsa** (Pall. ex Bieb.) Spreng. 1826, Syst. Vet. **3**: 381; Lipsch. 1962, Fl. SSSR, **27**: 504; id. 1976, Fl. Europ. **4**: 216.— *Serratula salsa* Pall. ex Bieb. 1808, Fl. Taur.-Cauc. **3**: 266; Pall. 1776, Reise, **3**: 607, nom. nud.—(Plate XXV, 3).

Type: Lower reaches of the Terek River near Kizlyar and of Volga near Sarepta ("in graminosis subsalsis humidiusculis ad fluvium Terek, haud procul oppido Kisljar, nec non Volgam, circa coloniam Sareptanum locis salsuginosis reperitus").

West (Dnieper: east; Black Sea: east); *East*; *Crimea* (vicinity of Perekop).—In steppes, on solonetzes and solonchaks.—*General distribution*: Caucasus, Western and Eastern Siberia, Russian Central Asia; Iran, Mongolia.—2n = 28.

Section 2. Saussurea.

All leaves undivided, linear to ovate.

Type: type species.

4. **S. alpina** (L.) DC. 1810, Ann. Mus. Hist. Nat. (Paris), **16**: 198; Lipsch. 1962, Fl. SSSR, **27**: 489; id. 1976, Fl. Europ. **4**: 216.—*Serratula alpina* L. 1753, Sp. Pl.: 816.—*Saussura esthonica* Baer ex Rupr. 1845, Beitr. Pflanzenk. Russ. Reich. **4**: 21; Lipsch. 1962, op. cit.: 49.—*S. alpina* subsp. *esthonica* (Baer ex Rupr.) Kupff. 1902, Korrespondenzbl. Nautrf.-Vereins Riga, **45**: 94.— (Plate XXV, 1).

Type: Europe, Siberia ("in Alpibus Lapponiae, Austriae, Helvetiae, Arvoniae, Sibiriae").

Arctic (Arctic Europe); *North*; *Baltic* (Estonia and Latvia); *Center* (Ladoga-Ilmen: vicinity of Gatchina; Upper Volga: reported from Moscow Region; Volga-Kama: east); *West* (Carpathians: Chernogora Range).—In meadows, forest glades and edges, rocks and stony slopes, on banks of rivers and streams.—*General distribution*: Western and Eastern Siberia, Russian Central Asia; Central and Atlantic Europe, Mongolia.— 2n = 26, 48, 54.

Note. Smaller heads and narrower involucral bracts are reported for *S. esthonica*, described from the shores of Lake Chudskoe. However, these features are typical for all plants from the plains in the forest zone unlike tundra and alpine plants with larger heads and broader involucral bracts.

5. **S. parviflora** (Poir.) DC. 1810, Ann. Mus. Hist. Nat. (Paris), **16**: 200; Lipsch. 1962, Fl. SSSR, **27**: 472; id. 1976, Fl. Europ. **4**: 216.—*Serratula parviflora* Poir. 1805, Encycl. Méth. **6**: 554.

Type: Siberia ("a ete recueilie en Siberie").

North (Karelia-Murmansk: Kola Peninsula and vicinity of Segozero; Dvina-Pechora); *Center* (Volga-Kama: east); *East* (Trans-Volga).—In meadows and thinned-out forests, on banks of rivers and streams, edges of swamps.—*General distribution*: Western and Eastern Siberia; Mongolia, Japan China.—2n = 26, 52.

6. **S. porcii** Degen, 1904, Magyar Bot. Lapok, **12**: 311; Lipsch. 1962, Fl. SSSR, **27**: 474; id. 1976, Fl. Europ. **4**: 216.

Type: Eastern Carpathians ("In declivitate orientali montis Korongyis alpium rodnensium").

West (Carpathians).—In swamps in alpine zone.—*General distribution*: Central Europe (Romania).

7. **S. discolor** (Willd.) DC. 1810, Ann. Mus. Hist. Nat. (Paris), **16**: 199; Lipsch. 1962, Fl. SSSR, **27**: 439; id. 1976, Fl. Europ. **4**: 217.—*Serratula discolor* Willd. 1800, Sp. Pl., **3**, 3: 1641, p.p.

Type: Europe ("in alpibus Lapponiae, Scotiae, Helvetiae, Sibiriae").

221   *West* (Carpathians).—On rocks and taluses in alpine zone.— *General distribution*: Central Europe, Mediterranean,—2n = 26.

8. **S. controversa** DC. 1810, Ann. Mus. Hist. Nat. (Paris), **16**: 199; Lipsch. 1962, Fl. SSSR, **27**: 438; id. 1976, Fl. Europ. **4**: 217.— (Plate XXV, 4).

Type: Silberia ("in Sibiria").

North (Dvina-Pechora: Syktyvkar, ecdemic); *Center* (Volga-Kama: east); *East* (Trans-Volga).—In forest glades and edges, on rocks, coastal buffs, in pine forests.—*General distribution*: Western and Eastern Siberia; Mongolia.—2n = 26.

O 9. **S. × uralensis** Lipsch. 1954, Byull. Mosk. Obshch. Isp. Prir. Otd. Biol. **59**, 6: 75; id. 1962, Fl. SSSR, **27**: 442; id. 1976, Fl. Europ. **4**: 217 [=*S. alpina* (L.) DC. × *S. controversa* DC.].

Type: Urals ("Northern part of Middle Urals, Konzhakor Range, alpine zone, Iev mine").

North (Dvina-Pechora: east); *Center* (Volga-Kama: east).— On rocks and rubble in apline tundra belt.—Endemic.

GENUS **98.** *JURINEA* Cass.[1]
1821, Bull. Soc. Philom. Paris, 1821: 140

Heads usually many-flowered, with bisexual tubular flowers, solitary or less often in corymbose or corymbose-paniculate inflorescence. Involucre cylindrical to goblet-shaped, 3.5–4.0 mm in dia; involucral bracts coriaceous-herbaceous, imbricate, in three or four rows, inner as long as middle or often longer. Receptacle flat, setose. Corolla pale pink (sometimes peripheral) to dark purple, usually glandular-hairy. Anthers with setaceous basal appendages; connectives narrow-hastate. Achenes obpyramidal or oblong, more or less tetraquetrous or dorsally compressed, with scarcely oblique hilum; pappus in many rows, with bristles of variable length; inner row of two to five flat or broader bristles, as a rule, exceeding others. Perennial herbaceous plants, less often semishrublets, with undivided or more or less incised leaves.

Type: *J. alata* (Desf.) Cass.

Over 250 species, distributed in Central and Atlantic Europe, as also in Russian Central Asia.

Literature: Korshinsky, S. 1894. Notes sur quelques especes des *Jurinea*. *Izv. Akad. Nauk. Nov. Ser.* **5**, 1, 2: 113–129.—Iljin, M.M. 1925. Obzor turkestanskikh vidov roda *Jurinea* Cass. [Review of Turkestanian species of the genus *Jurinea* Cass.]. *Tr. Turkest. Nauch. Obshch.,* **2**: 1–29.—Klokov, M.D. 1951. Rid *Jurinea* Cass. ta iogo znacheniya v istorii razvitku flory Ukrainskoi RSR [The genus *Jurinea* Cass. and its importance in the evolutionary history of the flora of the Ukrainian SSR]. *Bot. Zhurn. Akad. Nauk URSR,* **7**, 4:39–53; **8**, 1: 47–70.—Czerneva, O.V. 1988. Chto takoe *Jurinea longifolia* DC. (Asteraceae) [What is *Jurinea longifolia* DC. (Asteraceae)]. *Novosti Sist. Vyssh. Rast.* **25**: 1–58–159.

1. Achenes glabrous and smooth, on edges parallel-striated or indistinctly transversely rugose.................................. 2.
+ Achenes on edges and ribs distinctly sculptured, often glandular-hairy .... 10.
2. Heads with cylindrical or goblet-shaped involucres, numerous, clustered in simple or compound corymbs; achenes usually with white-cartilaginous ribs .... 3.
+ Heads with cup-shaped involucres, usually solitary on long peduncles, less often 1–10, and then often clustered in lax corymbose-panicle; achenes usually unicolorous .... 4.

---

[1]Treatment by O.V. Czerneva.

3. Heads about 20 mm long and 6–7 mm in dia (at fruiting to 10 mm in dia), clustered in simple corymbs; achenes whitish-arachnoid-hairy, 7.0–7.5 mm long; leaves narrow-linear, sessile, not decurrent .... 9. **J. stoechadifolia.**

222 + Heads about 15 mm long and 3.5 mm in dia (at fruiting to 5 mm in dia), clustered in compound corymbs, glabrous or very weakly arachnoid-hairy, light green or often purple; achenes 3.5–5.0 mm long; leaves linear-lanceolate or narrowly oblong-ovate, amplexicaul, short-decurrent .... 8. **J. multiflora.**

4. Stems not winged; all cauline leaves sessile, but not decurrent .... 5.

 + Stems winged; middle and upper cauline leaves decurrent .... 7.

5. Outer and middle involucral bracts closely appressed, only their cups sharply decurved; leaves sharply scabrous above from dense cartilaginous papillae .... 6.

 + Outer and middle involucral bracts squarrose, long-acuminate, their cusps entangled and upward directed, sometimes slightly deflected outside; leaves smooth or slightly scabrous above from scattered subobtuse small tubercles .... 1. **J. cyanoides.**

6. Outer and middle involucral bracts linear, long-acuminate, their cusps, half as long as involucral bracts, sharply decurved .... 2. **J. ewersmannii.**

 + Outer and middle involucral bracts oblong-ovate, short-acuminate, their cusps, one-third as long as involucral bracts, sharply decurved .... 3. **J. salicifolia.**

7. Plants robust; stem at base 3–7 mm in dia; basal leaves 20–30 cm long; middle and upper cauline leaves broadly decurrent .... 5. **J. longifolia.**

 + Plants robust; stem at base 1–3 mm in dia; basal leaves small; middle and upper cauline leaves decurrent as narrow border or expanded auricles .... 8.

8. All involucral bracts closely appressed to each other, with short upward-directed cusps .... 7. **J. tenuiloba.**

 + Outer and middle involucral bracts somewhat patent, their cusps uncinate, deflected outward or entangled .... 9.

9. Stems branched in upper half; basal leaves pinnatisect or bipinnatisect into oblong-lanceolate or narrow-lanceolate, somewhat flat segments, less often leaves undivided, lanceolate, bright green, lustrous, glabrous

with scattered sessile glands above, white-tomentose beneath; cusps of outer and middle involucral bracts uncinate .... 6. **J. polyclonos.**

+ Stems simple, less often with solitary branches; basal leaves pinnatisect or bipinnatisect, their segments narrow-linear with margins involute to midrib, grayish-green, arachnoid-hairy, alveolate-rugose above, with glands in alveoli, white-tomentose beneath; cusps of outer and middle involucral bracts entangled and deflected .... 4. **J. creticola.**

10(1). Semishrublets; leaves unicolorous on both sides, grayish, densely tomentose .... 17. **J. kirghisorum.**

+ Perennial herbaceous plants; leaves unicolorous on both sides, green or grayish-geen, arachnoid-hairy or scatteredly hairy with thick multicellular hairs .... 11.

12. Outer and middle involucral bracts ovate to lanceolate, very short-acuminate .... 13.

+ Outer and middle involucral bracts oblong-ovate, gradually long-acuminate .... 14.

13. Heads tomentose; outer and middle involucral bracts closely appressed, their cusps glabrescent, upward directed .... 10. **J. arachnoidea.**

+ Heads arachnoid-hairy; outer and middle involucral bracts patent, their cusps arachnoid-hairy, obliquely divergent .... 13. **J. calcarea.**

14. Heads sparsely tomentose; cusps of outer and middle involucral bracts herbaceous, upward directed .... 11. **J. ledebourii.**

+ Heads arachnoid-hairy; cusps of outer and middle involucral bracts coriaceous, hard, arcuately bent .... 12. **J. michelsonii.**

15(11). Stems branched usually in upper half; leaves densely arachnoid-hairy above, not scabrous, with scattered small glands .... 14. **J. cretacea.**

+ Stems simple or weakly branched in lower half; leaves sparsely arachnoid-hairy above, scabrous from acute spines, with numerous deep-seated glands .... 16.

16. Heads densely arachnoid-hairy, grayish; outer and middle involucral bracts oblong-lanceolate, herbaceous in upper third, arachnoid-hairy, dark colored, deflected outside, inner bracts short-acuminate, densely glandular-hairy

dorsally; achenes black with ochraceous ribs ....
.... 15. **J. mollissima.**

+ Heads sparsely arachnoid-hairy, less often subglabrous, greenish; outer and middle involucral bracts narrow-lanceolate, sublinear, herbaceous in upper half, subglabrous, purple, erectopatent, inner bracts with long awn, scatteredly glandular-hairy dorsally; achenes brownish-olive with lighter ribs .... 16. **J. sordida.**

Section 1. Cyanoides (Korsh. ex Sosn.) Iljin, 1962, Fl. SSSR, **27**: 715, 541.—*Jurinea* subsect. *Cyanoides* Korsh. ex Sosn. 1926, Zhurn. Russk. Bot. Obshch. **11**, 1–2: 196.

Stem simple or more or less branched, with distant heads; many-flowered, solitary or two to five, less often more numerous and then clustered in corymbose-paniculate inflorescence; involucre goblet-shaped. Achenes obpyramidal, tetraquetrous, strongly narrowed toward base, glabrous and smooth, dentate at apex, without true crown, unicolorous, light-colored. Perennial herbaceous plants, with undivided or more or less incised leaves.

Type: *J. cyanoides* (L.) Reichenb.

1. **J. cyanoides** (L.) Reichenb. 1831, Fl. Germ. Excurs.: 290; Iljin, 1962, Fl. SSSR, **27**: 546; Kozuharov, 1976, Fl. Europ. **4**: 219, p. p.—*Carduus cyanoides* L. 1753, Sp. Pl.: 822, s. str., quoad var. *monoclonos.*—*Jurinea pseudocyanoides* Klok. 1951, Bot. Zhurn. Akad. Nauk URSR, **8**, 1: 51; Iljin, 1962, op. cit.: 547; Klok. 1962, Fl. URSR, **11**: 483, Fig. 94.—*J. charkoviensis* Klok. 1951, op. cit.: 49, Fig. 7; Iljin, 1962, op. cit.: 481; Klok. 1962, op. cit.: 481.—*J. centauroides* Klok. 1962, op. cit.: 564, 485, Fig. 95.—*J. ewersmanii* auct. non Bunge: Kožuharov, 1976, op. cit.: 219, p. p.

Type: Probably central part of the Don Basin ("Tataria").

Center (Upper Dnieper: southeast; Upper Volga: southeast; Volga-Kama: south; Volga-Don); *West* (Dnieper; Black Sea: vicinity of Nikolaev); *East.*—In pine forests and sandy steppes, forests, predominantly sandy glades and edges, on steppe slopes, chalk and limestone outcrops, sandy hills, sometimes by roadsides.— *General distribution*: Caucasus (Ciscaucasia?), Western Siberia (south); Central Europe.—2n = 30.

224 *Note.* The description of this species is based on the work of J.G. Gmelin (1749, *Fl. Sib.*, **2**: 42, Tab. 45), which in turn is based on T. Gerber's manuscript of *Fl. Tanaiensis* [Flora of the

Don River Area]. A critical study of the vast herbarium material from the entire range of the species, given its exceptional polymorphism in regard to the leaf cutting and the degree of roughness of the upper surface of the leaf blade, the degree of development of auricles, and deflection of the outer involucral bracts, it must be said that separation of *J. pseudocyanoides* Klok. (described from Kiev Region, Vassilkovs District), *J. charkoviensis* Klok. (described from Kharkov Region, vicinity of Merefa), and *J. centauroides* Klok. (described from Lugansk Region, Streletskaya steppe) is not valid. I have not seen materials of *J. cyanoides* from the Black Sea Region, though M.V. Klokov in *Flora URSR* refers it for the vicinity of Nikolaev.

2. **J. ewersmannii** Bunge, 1841, Flora (Regensb.), **24**, 1: 155; Iljin, 1962, Fl. SSSR, **27**: 549; Kožuharov, 1976, Fl. Europ. **4**: 219, p. p.—*J. granitica* Klok. 1951, Bot. Zhurn. Akad. Nauk URSR, **8**, 1: 48; Iljin, 1962, op. cit.: 552; Klok. 1962, Fl. URSR, **11**: 479, Fig. 93.

Type: Astrakhan Region, Mt. Bolshoi Bogdo ("ad Wolgam inferiorem ad montem Bogdo").

West (Black Sea: along the Kalmius, Berda and Kelchik rivers); *East.*—On dry hill slopes, ridges and low mountain granites, shell and sand rocks, chalk outcrops, solonchaks and sandy steppes.—*General distribution*: Western Siberia (southwest), Russian Central Asia (northwest).

Note. Extremely variable in the degree of leaf incision and roughness of the upper surface of leaf blades. Populations from granite outcrops of southeastern Ukraine were described as *J. granitica* Klok.

○ 3. **J. salicifolia** Grun. 1868, Bull. Soc. Nat. Moscou, **41**, 2: 431; Iljin, 1962, Fl. SSSR, **27**: 552; Klok. 1962, Fl. URSR, **11**: 476, Fig. 92.—*J. polyclonos* auct. non DC.: Kožuharov, 1976, Fl. Europ. **4**: 219, p. p.—(Plate XXVI, 2).

Type: Dnepropetrovsk Region, the Orel River (". . . ad. fl. Orel supra urbem Catherinoslaw").

West (Dnieper; Black Sea).—On steppe and stony slopes, limestone outcrops, sands of terraces above floodplain.—Endemic.

Note. Differs well from *J. cyanoides* by having smaller involucral bracts that are slightly deflected outside, with one-third part deflected. It also differs from *J. polyclonos* (L.) DC. by the cauline leaves that are not decurrent on stem.

○ 4. **J. creticola** Iljin, 1962, Fl. SSSR, **27**: 717, 555.—*J. cyanoides* (L.) Reichenb. subsp. *tenuiloba* auct. non Nyman: Kožuharov, 1976, Fl. Europ. **4**: 219, p. p.

Type: Saratov Region, vicinity of Khvalynsk ("in vicinitate Chvalynsk arenae cretarum "Pesczanaja Gora", prope "Sosnovaja Myza").

Center (Volga-Don: southeast).—On chalk outcrops, sands.—Endemic.

Note. A unique species collected only once. Very close to *J. albicaulis* Bunge, described from Altai. Additional collections are needed to establish reliable differences between these species.

○ 5. **J. longifolia** DC. 1838, Prodr. **6**: 674; Czern. 1988, Novosti Sist. Vyssh. Rast. **25**: 159.—*J. laxa* Fisch. ex Iljin, 1928, Izv. Glavn. Bot. Sada SSSR, **27**: 84; Iljin, 1962, Fl. SSSR, **27**: 556, Plate XLI; Klok. 1962, Fl. URSR, **11**: 489, Fig. 97.—*J. paczoskiana* Iljin, 1932, Izv. Bot. Sada Akad. Nauk SSSR, **30**, 3–4: 345; Iljin, 1962, op. cit.: 564; Klok. 1962, op. cit.: 487, Fig. 96.—*J. albicaulis* Bunge subsp. *laxa* (Fisch, ex Iljin) Kožuharov, 1975, Bot. Journ. Linn. Soc. (London), **71**: 42; id. 1976, Fl. Europ. **4**: 219.—*J. ambigua* auct. non DC.: Kožuharov, 1976, ibid.: 490.

Type: Lower reaches of the Dnieper River probably Aleshki sands ("in sabulosis Chersonensibus ad Borysthenem").

West (Dnieper: on the Dnieper River; Black Sea); *Crimea* (Kerch Peninsula).—On weakly turfaceous sands, in sandy steppes.—Endemic.

Note. In the Herbarium of De Candolle, under *J. longifolia* the lower part of the stem with basal and lower cauline leaves and upper part of the stem with one developed heads are preserved 226 as the type material. Apparently, the author of the species based his description of the leaves on the character of non-decurrent basal and lower cauline leaves, which led to the description of *J. laxa*. Analysis of the diagnosis and the photograph of the type of *J. longifolia* does not leave any doubt that this species is absolutely identical with *J. laxa*. The species varies in the form of leaves and the degree of deflection of outer involucral bracts, which is a feature typical of all species of section *Cyanoides*. For this reason, I think it is possible to combine with *J. longifolia* the species *J. paczoskiana* described from the sands of the South Bug River (vicinity of Nikolaev), based on the downward deflected outer and middle involucral bracts and partly

322

225

Plate XXVI.
1—*Jurinea polyclonos* (L.) DC.; 2—*J. salicifolia* Grun.; 3—*J. multiflora* (L.) B. Fedtsch.

pinnately divided leaves that differentiate it from *J. longifolia*. On the whole, the species is closer to *J. albicaulis* Bunge, described from Altai.

6. **J. polyclonos** (L.) DC. 1838, Prodr. **6**: 675; Iljin, 1936, Fl. Yugo-Vost. Evrop. Chasti SSSR, **6**: 392; id. 1963, Bot. Mat. (Leningrad), **22**: 270; Kožuharov, 1976, Fl. Europ. **4**: 219, p. p.— *Carduus cyanoides* L. var. (β.) *polyclonos* L. 1753, Sp. Pl.: 822.— ?*Centaurea amplexicaulis* S.G. Gmel. 1770, Reise Russland, **1**: 136, Tab. 24, nom. dub.—*Jurinea amplexicaulis* (S.G. Gmel.) Bobr. 1956, Bot. Zhurn. **43**, 11: 1544; Iljin, 1962, Fl. SSSR, **27**: 558.—*J. cyanoides* (L.) Reichenb. subsp. *tenuiloba* auct. non Nyman: Kožuharov, 1976, op. cit.: 219, p. p.—(Plate XXVI, 1).

Type: Don River area ("Tanaensibus legit").

Center (Volga-Don: southeast); *East* (Lower Don; Trans-Volga: south; Lower Volga).—On sands of terraces above floodplain, sandy glades of pine forests, in sandy steppes.—*General distribution*: Western Siberia (southwest), Russian Central Asia (Aralo-Caspian: north).

Note. The description of this species is based on the work of J.G. Gmelin (1749, *Fl. Sib.* **2**: 44, Tab. 16), which in turn is based on T. Gerber's manuscript of *Flora Tanaiensis* [Flora of the Don River Area]. The name *Centaurea amplexicaulis* S.G. Gmel. is based on very brief diagnosis and not very good illustration. Its affinity with *J. polyclonos* is doubtful.

7. **J. tenuiloba** Bunge, 1841, Flora (Regensb.), **24**, 1: 155; Iljin, 1962, Fl. SSSR, **27**: 565.—*J. cyanoides* (L.) Reichenb. subsp. *tenuiloba* (Bunge) Nyman, 1879, Consp. Fl. Eur. **2**: 415; Kožuharov, 1976, Fl. Europ. **4**: 219, p. p.—*J. thyrsiflora* Klok. 1951, Bot. Zhurn. Akad. Nauk URSR, **8**, 1: 53; Iljin, 1962, op. cit.: 567; Klok. XII, 1962, Fl. URSR, **11**: 491.—*J. tanaitica* Klok. VIII.1962, Fl. SSSR, **27**: 719, 566; Klok. XII, 1962, op. cit.: 565, 494; Kožuharov, 1976, op. cit.: 218.—*J. polyclonos* auct. non DC.: Kožuharov, op. cit.: 219, p. p.

Type: Specimens from the Herbarium of Eschscholtz of unknown origin ("Hebeo e herbario Eschscholtziano, nomine *Carduus cyanoides* e Tataria").

Center (Volga-Don: southeast); *West* (Dnieper: southeast; Black Sea: east); *East* (Lower Don).—On sands of terraces above floodplains, in sandy slopes.—*General distribution*: Russian Central Asia (Aralo-Caspian: northwest).

Note. I combine with this species the ones described later, viz. *J. thyrsiflora* (described from the vicinity of Dnepropetrovsk) and *J. tanaitica* (described from the Chir River near the village of Oblivskaya), which are differentiated by the structure of the involucre: in *J. thyrsiflora*, the outer bracts are more or less deflected but in *J. tanaitica* they are appressed. M.M. Iljin (1962, op. cit.) considered that *J. tenuiloba* is distributed only in the semidesert regions of the northeastern Caspian Region.

Section 2. Corymbosae Benth. 1873, in Benth. and Hook. fil., Gen. Pl. **2,** 1: 473.—*Jurinea* sect. *Stenocephalae* auct. non Benth.: Iljin, 1962, Fl. SSSR, **27:** 578.

Stem branched only in inflorescence; heads relatively few-flowered, numerous; clustered in corymbose inflorescence; involucre cylindrical or goblet-shaped; achenes oblong, tetraquetrous, gradually narrowed toward base, glabrous and smooth, apically with sharp-toothed cartilaginous crown, reddish-brown on edges, with whitish ribs, sometimes whitish both on edges and lower half. Perennial herbaceous plants, with mostly undivided and entire leaves.

Lectotype; *J. linearifolia* DC. (=*J. multiflora* (L.) B. Fedtsch.).

8. **J. multiflora** (L.) B. Fedtsch. 1911, in O. and B. Fedtsch. Perech. Rast. Turkest. **4:** 295; Iljin, 1962, Fl. SSSR, **27:** 579; Klok. 1962, Fl. URSR, **11:** 455.—*Serratula multiflora* L. 1753, Sp. Pl.: 817, s. str., excl. pl. J.G. Gmel.—*Jurinea linearifolia* DC. 1838, Prodr. **6:** 675; Kožuharov, 1976, Fl. Europ. **4:** 218.—(Plate XXVI, 3).

Lectotype: The Don River Area ("ad Tanaim major").

227   *Center* (Volga-Kama: south; Volga-Don: south and east); *West* (Dnieper; Moldavia; Black Sea); *East*; *Crimea*.—In steppes, steppefied meadows and forest glades, outcrops of chalk and limestone.—*General distribution*: Caucasus (Ciscaucasia), Western Siberia (south), Eastern Siberia (southwest), Russian Central Asia (north); Central Europe (southwest), Dzhungaria-Kashgaria.

Note. Of the two specimens in the Herbarium of Linnaeus, the plant collected by T. Gerber from the Don River area is selected as the lectotype. Another plant collected by J.G. Gmelin later, which apparently, Linnaeus himself wrote as "*Serratula salicifolia*," belongs to the genus *Saussurea*.

9. **J. stoechadifolia** (Bieb.) DC. 1838, Prodr. **6:** 674; Iljin, 1962, Fl. SSSR, **27:** 580; Klok. 1962, Fl. URSR, **11:** 456, Fig. 86, A; Kožuharov, 1976, Fl. Europ. **4:** 218.—*Serratula stoechadifolia* Bieb. 1808, Fl. Taur.-Cauc. **2:** 266.—*Jurinea brachycephala* Klok. 1961, Spisok Rosl. Gerb. Fl. URSR, **11,** 32, No. 197; id. 1962, op. cit.: 458, Fig. 86, B.

Lectotype: Crimea, vicinity of Belogorsk ("in Tauriae collibus lapidosis cretaceis, circa Karassubasar frequens").

Center (Volga-Don: south); *West* (Dnieper: southeast; Moldavia; Black Sea); *East* (Lower Don); *Crimea.*—On outcrops of limestone and chalk, in steppes.—*General distribution*: Caucasus (Ciscaucasia); Central Europe (southeast).

Section 3. Platycephalae Benth. 1873, in Benth. and Hook. fil. Gen. Pl. **2,** 1: 473, s. str.—*Jurinea* sect. *Molles* Iljin, 1962, Fl. SSSR, **27**: 657.

Stem simple, less often sparsely branched; heads many-flowered, solitary, less often few; involucre cup-shaped; achenes obpyramidal, tetraquetrous, ribs interruptedly cartilaginous, with distinct sculpture and scattered sessile glands, with cartilaginous toothed crown at apex, unicolorous, dark brown or black. Perennial herbaceous plants, with more or less dissected, less often undivided and entire leaves.

Lectotype: *J. mollis* (L.) Reichenb.

10. **J. arachnoidea** Bunge, 1841, Flora (Regensb.), **24,** 1: 157; Iljin, 1962, Fl. SSSR, **27**: 659, Plate XLV, 1; Klok. 1962, Fl. URSR, **11**: 460; Kozlovskaya, 1979, Fl. Beloruss., Zakonom. ee Formir.: 128.—*J. consanguinea* DC. subsp. *arachnoidea* (Bunge) Kožuharov, 1968, Izv. Bot. Inst. (Sofia), **18**: 69; id. 1976, Fl. Europ. **4**: 220, p. p.

Type: Volga River area, below Saratov ("ad Wolgam infra Saratow").

Center (Upper Dnieper: southeast; Volga-Don: south and east); *East* (Lower Don; Trans-Volga; Lower Volga: north); *West* (Dnieper: east; ?Black Sea).—In steppes, steppefied forest glades, on outcrops of chalk and limestone.—*General distribution*: Caucasus, Russian Central Asia (northeast).—2n = 34, 36.

Note. I have not seen the specimens of this species from the Black Sea Region of the *Flora.*

○ 11. **J. ledebourii** Bunge, 1841, Flora (Regensb.), **24,** 1: 157; Iljin, 1962, Fl. SSSR, **27**: 663; Kožuharov, 1976, Fl. Europ. **4**: 219, p. p.—*J. pseudomollis* Klok. 1951, Bot. Zhurn. Akad. Nauk URSR, **7,** 4: 46, nom. nud.; id. 1962, Fl. URSR, **11**: 462, Fig. 87, descr. ross.

Lectotype: Southern Russia ("Rossia meridionalis").

Center (Volga-Kama: south; Volga-Don: south and east); *West* (Dnieper); *East* (Trans-Volga); *Crimea* (Kerch Peninsula).—On chalk and limestone outcrops in steppefied forest glades and edges, in steppes.—Endemic.

326

Note. The problem about the type of this species so far remains unresolved. M.M. Iljin (op. cit.) selected as lectotype A. Lehmann's plants from Orenburg Region, that were identified by A. Bunge himself.

○ 12. **A. michelsonii** Iljin, 1962, Fl. SSSR, **27**: 726, 666.—*J. pachysperma* Klok. 1962, Fl. URSR, **11**: 562, 464, Fig. 88.—*J. tyraica* Klok. 1962, ibid.: 473, p. p.—*J. ledebourii* non Bunge: Kožuharov, 1976, Fl. Europ. **4**: 219, p. p.

Type: Ukraine, Ternopol Region ("statio Maximovka-Bogdanovka, prope pag. Romanovka").

228 *West* (Dnieper: southwest).—On stony steppe slopes.—Endemic.

13. **J. calcarea** Klok. 1951, Bot. Zhurn. Akad. Nauk URSR, **7**, 4: 47; Klok. 1962, Fl. URSR, **11**: 466, Fig. 89; Iljin, 1962, Fl. SSSR, **27**: 665.—*J. tyraica* Klok. 1962, op. cit.: **11**: 473, p. p.—*J. ledebourii* auct. non Bunge: Kožuharov, 1976, Fl. Europ. **4**: 219, p. p.

Type: Ukraine, Kherson Region ("Dit. Chersonensio, prope pag. Gavrilovka et Kaczkarovka").

West (Carpathians: L'vov and Chernovitsy regions; Dnieper: south and southwest; Moldavia; Black Sea).—On steppe slopes, limestone outcrops, steppefied forest glades and edges.—*General distribution*: Central Europe (southeast).

Note. J. arachnoidea, J. calcarea, J. ledebourii, and *J. michelsonii,* form a group of very close and difficult to separate species. *J. calcarea* is especially close to *J. arachnoidea* and replaces it in the southern and western regions of the European part of the former USSR, differing mainly by the arachnoid-hairy (and not tomentose) heads and somewhat obliquely directed (and not upward directed) short-acuminate tips of the outer and middle involucral bracts.

○ 14. **J. cretacea** Bunge, 1841, Flora (Regensb.) **24**, 1: 158; Iljin, 1962, Fl. SSSR, **27**: 661.—*J. kasakorum* Iljin, 1962, ibid.: 725, 662.—*J. talijevii* Klok. 1951, Bot. Zhurn. Akad. Nauk URSR, **7**, 4: 50; Iljin, 1962, op. cit.: 660; Klok. 1962, Fl. URSR, **11**: 474, Fig. 91.—*J. consanguinea* DC. subsp. *arachnoidea* (Bunge) Kožuharov, 1968, Izv. Bot. Inst. (Sofia), **18**: 69; id. 1976, Fl. Europ. **4**: 220, p. p.—*J. ledebourii* auct. non Bunge: Kožuharov, 1976, ibid.: 219, p. p.

Type: Volga River, between Kamyshin and Volgograd ("ad Wolgam in cretaceis ad Belaja Glinka, inter Kamyschin V Zarizin").

East (Lower Don; Lower Volga); *West* (Dnieper: southeast.—On chalk outcrops.—Endemic.

Note. J. kasakorum described from the chalks in the vicinity of the village of Kletskaya on the Don, is distinguished by M.M. Iljin by the achenes with four to six approximate ribs. Populations from the chalk of Derkul and Evsug (left tributaries of the North Donetz), described by M.M. Iljin by the name *J. talijevii,* have outer and middle involucral bracts with deflected part slightly shorter and usually obliquely upward directed, while in typical populations from the Don and Volga regions the herbaceous part of the outer and middle involucral bracts is usually deflexed. However, these differences are not clearly manifest and are linked with transitions.

15. **J. mollissima** Klok. 1951, Bot. Zhurn. Akad. Nauk URSR, **7**, 4: 49; id. 1962, Fl. URSR, **11**: 472; Iljin, 1962, Fl. SSSR, **27**: 664.—*J. tyraica* Klok. 1962, op. cit.: 563, 473, s. str., quoad typum.—*J. ledebourii* auct. non Bunge: Kožuharov, 1976, Fl. Europ. **4**: 219, p. p.

Type: Odessa Region ("prope pag. Budany").

West (Carpathians; Dnieper: southwest; Moldavia; Black Sea).—In steppes, often stony slopes.—*General distribution*: Central Europe (southeast).

Note. Very close to *J. sordida* Stev.

O 16. **J. sordida** Stev. 1856, Bull. Soc. Nat. Moscou, **29**, 4: 401; Iljin, 1962, Fl. SSSR, **27**: 667; Klok. 1962, Fl. URSR, **11**: 469, Fig. 90.—*J. ledebourii* auct. non Bunge: Kožuharov, 1976, Fl. Europ. **4**: 219, p. p.

Type: Crimea ("in montosis hinc inde").

Crimea.—On chalk and limestone outcrops.—Endemic.

Section 4. Chaetocarpae Iljin, 1962, Fl. SSSR, **27**: 677.

Stem simple or more or less branched; heads many-flowered, solitary or a few; involucre cup-shaped; achenes obpyramidal, tetraquetrous, sharply tuberculate on ribs, gradually narrowed toward base, with distinct sculpture, apically with distinct, toothed, cartilaginous crown, unicolorous, dark gray or ochraceous. Perennial herbaceous plants or semishrublets, with leaves undivided and entire or incised to varying degrees.

Type: *J. chaetocarpa* (Ledeb.) Ledeb.

229 17. **J. kirghisorum** Janisch. 1905, Tr. Obshch. Estestvoispit. Kazan. Univ. **40**, 1: 5; O. and B. Fedtsch. 1911, Perech. Rast.

Turkest. **4**: 298; Iljin, 1962, Fl. SSSR, **27**: 698; Kožuharov, 1976, Fl. Europ. **4**: 219.

Lectotype: Northern Kazakhstan ("Akchatau chalk mountains").

East (Lower Volga).—On chalk outcrops.—*General distribution*: Russian Central Asia (northwest).

GENUS **99.** *CARDUUS* L.[1]
1753, Sp. Pl.: 820; id. 1754, Gen. Pl., ed. 5: 358

Heads 5–80 mm in dia, homogamous, with bisexual flowers, solitary at apices of stem and its branches or clustered in paniculate or capitate inflorescences. Involucre 5–70 mm in dia and 7–40 mm long, spherical, hemispherical or campanulate to subcylindrical; involucral bracts imbricate, in many rows, highly unequal, lanceolate-ovate to linear, sometimes (section *Carduus*) narrowed in middle and geniculately bent, in upper part erect, deflected or uncinate, spinescent or cuspidate. Receptacle flat or somewhat concave, densely covered with long bristles. Flowers more or less similar, pink or purple, five-parted to more than one-third, with linear lobes. Achenes 3–6 mm long and 1.0–1.5 mm wide, oblong-ellipsoidal or subcylindrical, somewhat compressed on sides, glabrous and smooth, more or less lustrous, olive, grayish or light reddish-brown, with inconspicuous veins; pappus 8–15 mm long, readily detaching wholly, of several rows of basally connate scabrous bristles. Biennial or perennial, less often annual plants, with erect leafy, usually short-winged stems and entire to deeply pinnatipartite leaves with bristles or spines on margin.

Lectotype: *C. nutans* L.

About 100 species in temperate regions of Eurasia and North Africa, some species ecdemic in America and Australia.

Literature: Tamamschian, S.G. 1953. Novye i maloizvestnye vidy chertopolokha [New and less known species of bristle thistle]. *Bot. Mat. (Leningrad)*, **15**: 383–394.—Kazmi, S.M.A. 1964. Revision der Gattung *Carduus* (Compositae), Teil 2. *Mitt. Bot. Staatssamm. Münichen*, **5**: 279–550.—Moskalenko, Z.S. 1984. Izuchenie roda chetopolokha (*Carduus* L.) na yugo-vostoke Ukrainy [Study of the genus of bristle-thistle (*Carduus* L.) in southeast Ukraine]. In *Introduktsiya i Akklimatizatsiya Rastenii* [Introduction and Acclimatization of Plants], Kiev: 1: 34–35.—Kondratyuk, E.M. and Z.S. Gorlaczeva. 1985. Novi vidi rodu *Carduus* L. (Asteraceae)

---

[1]Treatment by D.V. Geltman.

[New species of the genus *Carduus* L. (Asteraceae)]. *Ukr. Bot. Zhurn.*, **42**, 5: 52–56.

1. Involucre spherical, hemispherical or campanulate, less often broadly cylindrical, shorter than wide, as long or less than 1.3 times longer than wide .... 2.

+ Involucre cylindrical or narrow-campanulate, usually 1.3–2.0 times as long as wide .... 13.

2. Heads nutant; involucral bracts narrowed in middle and geniculately bent at this point .... 1. **C. nutans.**

+ Heads erect; involucral bracts more or less erect, sometimes deflected or uncinate in upper part, but not geniculately bent in middle .... 3.

3. Stem, its branches, and peduncles spiny-winged or spiny throughout .... 4.

+ Stem, its branches, and peduncles without spines and wings at least in upper part .... 7.

4. Developed heads 1–3(6); stem and its main branches simple in upper part; heads solitary at tips of lateral branches .... 7. **C. hamulosus.**

+ Developed heads (3)5–15(20); stems and its main branches branched in upper part; heads on short stalks clustered in capitate inflorescence .... 5.

5. Plants glaucescent; leaves deeply pinnatisect, glabrous or with long segmented hairs beneath, not tomentose or arachnoid-hairy, largest spines at tips of leaf lobes (2.5)3.0–6.0 mm long .... 2. **C. acanthoides.**

+ Plants green; leaves pinnatisect to undivided, more or less densely white-tomentose and arachnoid-hairy beneath, longest spines at tips of leaf lobes 1.0–2.0(2.5) mm long .... 6.

6. Perennial mountain plants of the Carpathians; middle cauline leaves to 15 cm long, usually undivided, less often pinnatisect; largest of spines at tips of leaf lobes to 1 mm long; outer involucral bracts 1.5–2.0 mm wide in lower third, then abruptly narrowed, attenuate-deltoid, recurved, apically terminating in thin spines to 1 mm long .... 4. **C. bicolorifolius.**

+ Biennial plants of plains; middle cauline leaves pinnatisect or pinnatipartite, rarely undivided, usually to 10 cm long; largest spines at tips of leaf lobes, 1.0–1.5(2.0) mm long; outer involucral bracts 1.0–1.5 mm wide in lower part, then gradually narrowed, deltoid, erect or recurved,

330

terminating in more or less thick spines up to 1.5 mm long .... 3. **C. crispus.**
7(3). Stem not winged and not spiny; leaves undivided, elliptical, serrate, with thin spines .... 11. **C. glaucinus.**
  + Stem in lower and middle part spiny-winged; leaves pinnatisect, with rather thick spines on margin .... 8.
8. Involucre cylindrical or campanulate, glabrous or arachnoid-hairy, as long as wide or less than 1.3 times as long; involucral bracts (at least some deflected or uncinately bent; leaves white-tomentose beneath with mixture of long black hairs .... 9. **C. stenocephalus.**
  + Involucre hemispherical or broadly campanulate, much smaller than wide .... 9.
9. Involucre arachnoid-hairy; involucral bracts uncinate terminating in 1.0–1.5 mm long spines; leaves white-tomentose beneath, sometimes with mixture of long segmented hairs .... 8. **C. uncinatus.**
  + Involucre glabrous or weakly arachnoid-hairy; involucral bracts erect or deflexed terminating in about 0.5 mm-long spine; leaves with predominantly long-segmented hairs beneath, often with mixture of white tomentum .... 10.
10. Developed heads 1–3(6); stem and its main branches simple in upper part; heads solitary terminal on branches .... 11.
  + Developed heads (3)5–10(15); stem and its main branches branched in upper part; heads solitary or in groups of two or three at apices of branches, others considerably smaller and usually undeveloped .... 12.
11. Biennial plants of plains .... 7. **C. hamulosus.**
  + Perennial mountain plants of the Carpathians .... 10. **C. kerneri.**
12. Outer involucral bracts deflected in upper part; leaves densely white-tomentose beneath .... 5. **C. candicans.**
  + Outer involucral bracts erect, not deflected; leaves arachnoid-hairy beneath, sometimes with long, segmented hairs on veins .... 6. **C. collinus.**
232 13(1). Involucral bracts recurved in upper part .... 12. **C. pycnocephalus.**

---

231   Plate XXVII.
1—*Carduus crispus* L.; 2—*C. bicolorifolius* Klok., general inflorescence; 3—*C. cinereus* Bieb.; 4—*C. uncinatus* Bieb., head; 5—*C. hamulosus* Ehrh., head; 6—*C. nutans* L., head, 6a—involucral bract.

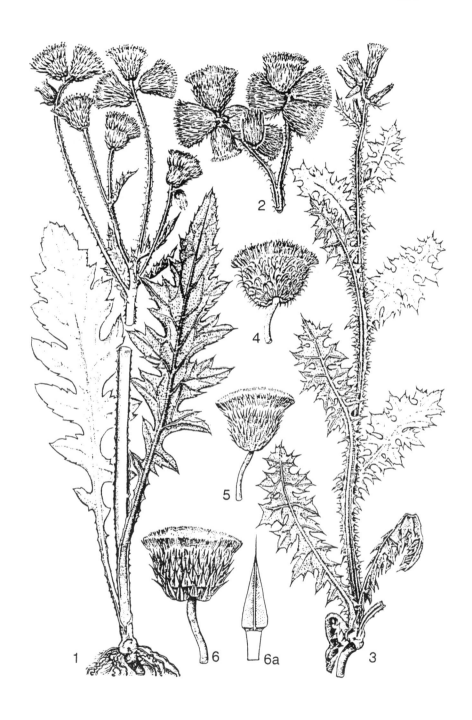

+ Involucral bracts appressed, sometimes only uppermost
  deflected .... 14.
14. Heads solitary or in pairs (second usually undeveloped),
  on short but distinct winged stalks ...... 14. **C. cinereus.**
+ Heads sessile, clustered in threes and fours (of these
  one usually larger) and borne in axils of leaves or at
  apices of winged or spiny-winged stalks ....
  .... 13. **C. arabicus.**

Section 1. Carduus.

Heads 4–7 cm in dia; involucre usually spherical or hemispherical, involucral bracts narrowed in middle and geniculately
bent.

Type: lectotype of genus.

1. **C. nutans** L. 1753, S. Pl.: 821; Tamamschian, 1963, Fl. SSSR,
28: 10; Franco and Afonso, 1976, Fl. Europ. **4**: 223.—*C. kondratjukii*
Gorlaczova, 1985, Ukr. Bot. Zhurn. **42**, 5: 52.—(Plate XXVII, 6).

Type: Europe ("in Europa ad pagos").

a. Subsp. *nutans.*—Leaves more or less pubescent beneath
with segmented hairs.

Baltic; *Center* (Ladoga-Ilmen; Upper Dnieper; Upper Volga);
West; *East* (Trans-Volga: probably ecdemic).—In steppes, on
abandoned lands, by roadsides, in habitations.—*General
distribution*: Western Siberia (ecdemic); Scandinavia (south),
Central and Atlantic Europe.—2n = 16.

b. Subsp. **leiophyllus** (Petrovič) Stojan. et Stef. 1948, Fl.
Blg.: 1183.—*C. leiophyllus* Petrovič, 1887, Add. Fl. Agry Nyss.:
105.—*C. thoermeri* Weinm. 1837, Bull. Soc. Nat. Moscou, **10,** 7:
69; Franco and Afonso, 1976, Fl. Europ. **4:** 223.—*C. attenuatus*
Klok. 1962, Fl. URSR, **11**: 569, 517.—Leaves glabrous beneath.

Type: Yugoslavia, vicinity of Nis.

North (ecdemic); *Baltic*; *Center*; *West* (Dnieper; Moldavia;
Black Sea); *East*; *Crimea.*—By roadsides, in habitations, steppes,
on abandoned lands.—*General distribution*: Caucasus, Western
Siberia, Russian Central Asia; Scandinavia (south), Central Europe,
Mediterranean, Asia Minor.

Section 2. Carduastrum Tamamsch. 1963, Fl. SSSR, **28**: 600, 20.

Heads medium-sized or large, 2–7 cm in dia; involucre spherical
to broadly campanulate; involucral bracts gradually narrowed
toward tip, often deflected in upper part, uncinate, or erect.

Type: *C. crispus* L.

2. **C. acanthoides** L. 1753, Sp. Pl.: 821; Tamamsch. 1963, Fl. SSSR, **28**: 24; Franco and Afonso, 1976, Fl. Europ. **4**: 224.—*C. fortior* Klok. 1962. Fl. URSR, **11**: 567, 504.

Type: Europe ("in Europae ruderatis").

Baltic; *Center*; *West*; *East*; *Crimea.*—By roadsides, in pastures, trampled meadows, on abandoned lands.—*General distribution*: Caucasus; Scandinavia (south), Central and Atlantic Europe, Mediterranean, Asia Minor.—2n = 22.

Note. Sometimes hybridizes with *C. nutans, C. crispus, C. hamulosus,* and *C. uncinatus.* M.V. Klokov considers that true *C. acanthoides* is found only in Western Europe; the Eastern European plant was described as *C. fortior* Klok. (loc. cit.).

3. **C. crispus** L. 1753, Sp. Pl.: 821; Tamamsch. 1963, Fl. SSR, **28**: 23; Franco and Afonso, 1976, Fl. Europ. **4**: 225.—*C. incanus* Klok. 1962, Fl. URSR, **11**: 567, 502.—(Plate XXVII, 1).

Type: Northern Europe ("in Europae septentrionalioris agris cultis").

North; *Baltic*; *Center*; *West*; *East*; *Crimea.*—In habitations, by roadsides, in crops, pastures, trampled meadows, forest glades 233 and edges.—*General distribution*: Caucasus, Western and Eastern Siberia, Far East, Russian Central Asia; Scandinavia, Atlantic and Central Europe, Mediterranean, Asia Minor, Iran, Mongolia; North America (ecdemic).—2n = 16, 18, 34.

Note. The plants with relatively narrow heads from Moldavia were described as *C. incanus* Klok. I do not think it purposeful to separate such plants in a separate species because they are linked through transitional forms and do not have a specific geographic range.

4. **C. bicolorifolius** Klok. 1962, Fl. URSR, **11**: 566, 499.—*C. personata* (L.) Jacq. subsp. *albidus* (Adamov.) Kazmi, 1964, Mitt. Bot. Staatssamm. München, **5**: 376; Franco and Afonso, 1976, Fl. Europ. **4**: 225.—*C. personata* auct. non (L.) Jacq.: Tamamsch. 1963, Fl. SSSR, **28**: 27.—(Plate XXVII, 2).

Type: Eastern Carpathians ("RSS Ucr. ditro Stanislavensis pag. Knjarb-dvir, in reservato Taki ad ripam dextram Prut").

West (Carpathians).—In beech and elm forests, meadows, less often by roadsides, in crops.—*General distribution*: Central Europe.

5. **C. candicans** Waldst. et Kit. 1801–1802, Descr. Icon. Pl. Rar. Hung. **1**:. 85, tab. 83; Nyárády, 1964, Fl. Rep. Pop. Romine, **9**: 652; Franco and Afonso, 1976, Fl. Europ. **4**: 228.

Type: Hungary ("in monte Domuglet, in sinistra ripa amnis Csernae ad thermas Herculis sito, versus limites Vallachiae").

West (Moldavia: Nyárády, loc. cit.).—By roadsides, in habitations.—*General distribution*: Central Europe, Mediterranean (east).—2n = 16–18.

6. **C. collinus** Waldst. et Kit. 1807, Descr. Icon. Fl. Rar. Hung. **3:** 257, tab. 232; Fodor, 1974, Fl. Zakarp.: 143; Franco and Afonso, 1976, Fl. Europ. **4:** 228.

Type: Hungary ("in collibus praesertim vitifers, Comitatus Hontensis. Neogradiensis, Hevesiensis, Borsodiensis, Zempliniensis").

West (Carpathians: reported for Chernogora).—In forests.— *General distribution*: Central Europe, Mediterranean (east).—2n = 16.

7. **C. hamulosus** Ehrh. 1792, Beitr. Naturk. **7:** 166; Tamamsch. 1963, Fl. SSSR, **28:** 22; Franco and Afonso, 1976, Fl. Europ. **4:** 226.—*C. pseudocollinus* (Schmalh.) Klok. 1953, Vizn. Rosl. URSR: 566; id. 1962, Fl. URSR, **11:** 510.—*C. tyraicus* Klok. 1962, ibid.: 568, 507.—*C. tauricus* Klok. 1962, ibid.: 569, 511.—*C. tortuosus* Gorlaczova et Kondratjuk, 1965, Ukr. Bot. Zhurn. **42,** 5: 54.—*C. collinus* auct. non Waldst. et Kit.: Ledeb. 1846, Fl. Ross. **2:** 721; Gorlaczova, 1987, Introd. Akklimatiz. Rastenii (Kiev), **7:** 55.— (Plate XXVII, 5).

Type: Apparently, Hungary ("Hungaria").

Baltic (Tallin: ecdemic); *Center* (Volga-Don); *West* (Dnieper; Moldavia; Black Sea); *East* (Lower Don); *Crimea*.—In steppes, on slopes of ravines and ridges, by roadsides, in habitations.— *General distribution*: Caucasus; Central Europe, Mediterranean, Asia Minor.—2n = 22.

Note. I support the broader sense (s. lat.) of this species. The plants with erect involucral bracts were described as *C. pseudocollinus*. *C. tortuosus* is a more branched plant, differing from *C. tyraicus* and *C. tauriana* mainly in the quantitative characteristics. All these features are highly variable and do not have any specific geographic affinity.

8. **C. uncinatus** Bieb. 1819, Fl. Taur.-Cauc. **3:** 553; Tamamsch. 1963, Fl. SSSR, **28:** 26; Franco and Afonso, 1976, Fl. Europ. **4:** 226; Kamyschev and Khmelev, 1976, Rast. Pokrov Voronezh. Obl.: 130.—(Plate XXVII, 4).

Type: South of the European part of the former USSR and northern Kazakhstan ("in Tauria meridionali, in ruderatis circa Kisljar, prope Astrachan, ad Hypanin circa Nikolaew").

North (Karelia-Murmansk: Khibiny Station, ecdemic); *Center* (Upper Volga: Moscow, ecdemic; Volga-Don: reported for south of Voronezh Region); *West* (Black Sea); *East*; *Crimea.*—In steppes, in saline meadows, on stony slopes, by roadsides, in habitations.— *General distribution*: Caucasus, Western Siberia (southwest), Russian Central Asia (northwest); Central Europe, Mediterranean (east).

234   *Note.* The plants from Crimea, with more incised leaves and densely pubescent involucre, were described as subsp. *davisii* Kazmi (1964, op. cit.: 404). They are found in other parts of the range of *C. uncinatus.*

9. **C. stenocephalus** Tamamsch. 1954, Bot. Mat. (Leningrad), **16**: 468; id. 1963, Fl. SSSR, **28**: 27.

Type: The Volga Delta ("Astrakhan, Volga Delta, near the former Tumak oil fields, at the foot of the Baer knoll*").

West (Black Sea); *East* (Trans-Volga; Lower Volga).—In sandy and clayey steppes, semideserts.—*General distribution*: Western Siberia (southwest), Russian Central Asia (northwest).

Note. Possibly of hybrid origin: *C. hamulosus* × *C. uncinatus*, although the first presumed ancestor is presently not found in the eastern part of the range of *C. stenocephalus.*

10. **C. kerneri** Simonk. 1886, Term. Füz. 10: 181; id. 1886, Enum. Fl. Transsilv.: 337; Tamamsch. 1963, Fl. SSSR, **28**: 31; Franco and Afonso, 1976, Fl. Europ. **4**: 228.

Type: Transylvania ("in alpibus Schuler, Bucsecs, Királyko").

West (Carpathians).—In subalpine meadows, forest glades and edges.—*General distribution*: Central Europe, Mediterranean (east).—2n = 16, 22.

11. **C. glaucinus** Holub, 1974, Folia Geobot. Phytotax. (Praha), **9**, 3: 272.—*C. glaucus* Baumg. 1816, Enum. Stirp. Transsilv. **3**: 58, non Cav. 1795; Tamamsch. 1963, Fl. SSSR, **28**: 28.—*C. defloratus* L. subsp. *glaucus* (Baumg.) Nym. 1879, Consp. Fl. Europ.: 412; Franco and Afons., 1976, Fl. Europ. **4**: 227.—*C. craifolius* Willd. subsp. *glaucus* (Baumg.) Kazmi, 1964, Mitt. Bot. Staatssamm München, **5**: 389.

Type: Transylvania ("in mont. calcareo rupestribus juxta pagum Torotzkó Szent György ibidem im M. Bedeliö, item in herbidis subalpinis silvisque montosis versus M. Surul et Aquam Mineralem Lövete; item in Sedibus Csik et Haromszek").

West (Carpathians).—On limestone rocks and stony slopes in forest zone.—*General distribution*: Central Europe.—2n = 24.

---

*A special kind of relief in the Lower Volga area—Sci. Editor.

Section 3. Leptocephali Reichenb. f. 1853, Icon. Fl. Germ. **15:**
86; Kazmi, 1964, Mitt. Bot. Staatssamm. München, **5:** 310.—*Carduus*
sect. *Stenocephalus* Rouy, 1905, Fl. Fr. **9:** 70; Tamamsch. 1963,
Fl. SSSR, **28:** 32.

Heads 1.0–1.5 cm in dia, narrow-cylindrical or narrow-
campanulate; involucral bracts gradually narrowed toward apex,
erect or deflected in upper part.

Type: *C. pycnocephalus* L.

12. **C. pycnocephalus** L. 1763, Sp. Pl., ed. 2: 1151; Tamamsch.
1963, Fl. SSSR, **28:** 32; Franco and Afonso, 1976, Fl. Europ. **4:**
224; Geideman, 1986, Opred. Vyssh. Rast. MoldSSR, ed. **3:** 555;
Katina, 1987, Opred. Vyssh. Rast. Ukrainy: 349.

Type: Southern Europe ("in Europa australi").
West (Carpathians: Katina, loc. cit.; Moldavia: Geideman, loc.
cit.); *Crimea* (vicinity of Yalta, possibly ecdemic).—By roadsides,
in habitations.—*General distribution*: Central Europe,
Mediterranean (east).—2n = 18, 32, 54, 64.

13. **C. arabicus** Jacq. 1784, in Murr., Syst. Veg., ed. 14: 724;
Jacq. 1786, Collect. Bot. **1:** 56; Tamamsch. 1963, Fl. SSSR, **28:** 34.

Type: Not mentioned, evidently northern Africa or southwestern
Asia.
Crimea.—By roadsides, in habitations, crops, on dry stony
slopes.—*General distribution*: Caucasus, Russian Central Asia;
Asia Minor, Iran.

14. **C. cinereus** Bieb. 1808, Fl. Taur.-Cauc. **2:** 270; Tamamsch.
1963, Fl. SSSR, **28:** 34.—(Plate XXVII, 3).

Type: Northern Caucasus, vicinity of Kizlyar ("in incultis
circa oppidum Kisljar").
Crimea.—In more or less turfaceous steppe areas, by
roadsides, in habitations.—*General distribution*: Caucasus, Russian
Central Asia; Mediterranean (east), Asia Minor, Iran.—2n = 22, 28.

235   GENUS **100.** *CIRSIUM* Mill.[1]
                1754, Gard. Dict. Abridg., ed. 4, 1, sine pag.

Heads 10–80 mm in dia, homogamous, usually bisexual, less
often (subgen. *Breea*) with unisexual flowers, solitary at apices
of stem and its branches or clustered in corymbose, capitate, or

---

[1]Treatment by N.N. Tzvelev.

corymbose-paniculate inflorescence. Involucre 5–60 mm in dia, 8–60 mm long, broadly ovate or subspherical to oblong-ovate; involucral bracts imbricate, in many rows, highly unequal, outermost lanceolate-ovate or lanceolate to innermost linear-lanceolate, herbaceous or subcoriaceous, usually with more or less distinct spine or cusp at apex. Receptacle flat or somewhat convex, densely covered with long bristles. Flowers more or less similar or peripheral somewhat enlarged; corolla pink, less often yellowish or whitish, one-third to half (rarely almost to base) divided in five lobes, with linear or linear-lanceolate lobes. Achenes (2.7)3.0–5.0(6.0) mm long and 1–2 mm wide, somewhat flattened, glabrous and smooth, often somewhat lustrous, light brownish or yellowish, often with dark brown longitudinal streaks, with inconspicuous veins; pappus 8–30 mm long, readily detaching wholly, of many rows of bristles plumose at base. Biennial or perennial plants, with erect, leafy stem (sometimes stemless); leaves undivided and subentire to deeply pinnately divided, with stiff bristles or spines on margin.

Lectotype: *C. heterophyllum* (L.) Hill.

About 300 species, distributed in extratropical regions of the Northern Hemisphere, and partly also in the montane regions of tropical Africa and Central America.

Literature: Petrak, F. 1912. Der Formenkreis des *Cirsium eriophorum* (L.) Scop. In *Europa. Biblioth. Bot. (Stuttgart)*, **78**: 1–92.—Charadze, A.L. 1963. Zametka o krymskom vide *Cirsium laniflorum* [Note on the Crimean species *Cirsium laniflorum*]. *Zam. Sist. Geogr. Rast. (Tbilisi)*, **23**: 104–109.—Gorlacheva, Z.S. 1986. Bodyaki resnitchatyi (*Cirsium ciliatum* (Murr.) Moench), melkozubchatyi (*C. serrulatum* (Bieb.) Bieb.) i ukrainskii (*C. ucranicum* Bess. ex DC.) na yugo-vostoke Ukrainy [Thistles: ciliate (*Cirsium ciliatum* (Murr.) Moench.), serrulate (*C. serrulatum* (Bieb.) Bieb.), and Ukrainian (*C. ucranicum* Bess. ex DC.) in southeastern Ukraine]. In *Introduktsiya i Akklimatizatsiya Rastenii* [Introduction and Acclimatization of Plants], (Kiev), **5**: 82–85.

1. Leaves with stiff squarrose bristles above. Strongly prickly biennial; heads 35–80 mm in dia, with pink flowers... 2.
+ Leaves glabrous above or more or less covered with soft crisped hairs or arachnoid-hairy tomentum. Usually less prickly perennials, less often biennials; heads 10–50 mm in dia... 10.

338

2. Heads 50–80 mm in dia; involucre grayish from dense arachnoid pubescence; middle and inner involucral bracts gradually narrowed above from base, later again more or less broadened and terminating in spine, more or less ciliate on margin; leaves pinnately divided, strongly prickly, with prickles to 1.5 mm long .... 3.

+ Heads 35–60 mm in dia; involucre glabrous or weakly arachnoid-hairy (or tomentose); all involucral bracts gradually narrowed above from base and spinescent .... 4.

3. Expanded part of involucral bracts below spine narrow and long, only slightly broader than basal part. Montane plant of the Carpathians .... 8. **C. decussatum.**

+ Expanded part of involucral bracts below spine, at least in some bracts, also short, considerably broader than basal part. Plant of the plains .... 9. **C. polonicum.**

4. Internodes spiny-winged; involucral bracts without cilia and bristles; leaves usually only weakly grayish-pubescent to subglabrous beneath, rarely white-tomentose ....
.... 1. **C. vulgare.**

236 + Internodes not winged; involucral bracts usually with stiff cilia or spines on margin, less often without them and then very narrow, linear-subulate; leaves whitish- or grayish-tomentose beneath .... 5.

5. Involucre subglabrous; outer and middle involucral bracts with spines and approximate stiff cilia on margin (more than four on each side) .... 6.

+ Involucre usually more or less grayish from arachnoid pubescence; outer and middle involucral bracts with spines and few distant stiff cilia on margin (one to four on each side), all or almost all lanceolate-subulate. Plants of Crimea
.... 8.

6. Outer involucral bracts lanceolate-subulate, usually sideways deflected, somewhat deflexed, gradually narrowed, terminating in spine more than 2 mm long; middle involucral bracts with still longer spines, at least some stiff cilia on involucral bracts more than 0.6 mm long .... 4. **C. serrulatum.**

+ Outer involucral bracts lanceolate, usually bent and deflexed, more abruptly narrowed, terminating in spine 0.7–1.5 mm long, rarely to 2 mm long; middle involucral bracts with spines usually to 2 mm long .... 7.

7. Outer and middle involucral bracts with stiff cilia more than 0.5 mm long; involucre usually about 30 cm in dia .... 2. **C. ciliatum.**

+ Outer and middle involucral bracts narrower in middle, with stiff cilia to 0.5 mm long, spinescent; involucre usually about 25 mm in dia .... 3. **C. ukranicum.**

8. Middle and upper leaves pinnately divided, rarely pinnatilobate, lobes and segments terminating in spine 5–8 mm long; outer and middle involucral bracts spiny, sideways deflected and slightly deflexed......................... .... 7. **C. laniflorum.**

+ Middle and upper leaves pinnatilobate, lobes terminating in spine 3–5 mm long... 9.

9. Outer and middle involucral bracts relatively short and with smaller (1–2 mm long) spine, usually deflected sideways and deflexed; involucre with rather abundant arachnoid pubescence .... 5. **C. tauricum.**

+ Outer and middle involucral bracts longer and with larger (usually exceeding 2 mm) spine, deflected sideways and slightly deflexed; involucre with less abundant arachnoid pubescence .... 6. **C. sublaniflorum.**

10(1). Stem well developed, spiny-winged to general inflorescence and partly also on peduncles; heads usually 5–25 on each stem, 12–30 mm in dia, with pink flowers .... 11.

+ Stems not always developed; internodes usually not winged, less often lower and partly middle with narrow weakly spiny wings, and then upper part of stem always not winged. Perennial plants, often with long rhizomes .... 12.

11. Cauline leaves undivided to pinnatilobate, with numerous spines longer than 2 mm; involucral bracts usually light green. Perennial plants of saline meadows, with short rhizomes and thick roots .... 21. **C. alatum.**

+ Cauline leaves usually all pinnately divided or pinnatilobate, with few thinner spines longer than 2 mm; involucral bracts pinkish in upper part. Biennial plants of swampy meadows, with think roots ...... 22. **C. palustre.**

12. Leaves whitish-, less often grayish-tomentose beneath; flowers pink .... 13.

+ Leaves green beneath, more or less covered with flexuous or crisped hairs to subglabrous .... 16.

237   13. Heads 10–25 mm in dia, usually numerous (5–30) on single stem; lower leaves usually withering by flowering, others numerous but usually small, undivided or pinnatilobate. Soboliferous dioecious plant ....
.... 24. **C. incanum.**
+ Heads 25–50 mm in dia, solitary or two to five(seven) on one stem; leaves predominantly in lower half of stems, often less numerous but rather large, basal usually persisting by flowering. Monoecious plants, with creeping rhizomes .... 14.

14. Heads usually three to seven, densely clustered at stem apex, more or less nutant; cauline leaves less numerous, ovate to broadly lanceolate, undivided, coarsely toothed or shallow pinnately lobed. Montane plant of the Carpathians .... 13. **C. waldsteinii.**
+ Heads solitary or two to five and then more or less distant, erect; cauline leaves more numerous, usually lanceolate, undivided and then serrulate or pinnatilobate. Meadow plants, absent in the Carpathians .... 15.

15. Rhizome thick, condensed; stems in middle and lower part covered with rather long, erect, flexuous hairs; leaves usually all undivided, with scattered flexuous hairs above
.... 14. **C. helenioides.**
+ Rhizome slender, long; stems in middle and lower part more or less covered with appressed tomentum to subglabrous; leaves undivided or pinnatilobate, glabrous above .... 15. **C. heterophyllum.**

16(12). Usually stemless, strongly prickly plants with short rhizome, rarely individual plants with thick stem to 30(40) cm high; heads 25–40 mm in dia, solitary or in groups of 4–10 and then densely clustered .... 17.
+ Plants with distinct stem, more than 30 cm high, weakly, less often strongly prickly and then with small heads not clustered .... 19.

17. All leaves in rosette pinnatipartite to narrow-winged rachis; heads often solitary, less often two to four; flowers pink. Plant of the west of the European part of Russia.........
.... 20. **C. acaule.**
+ Leaves from same rosette usually of different forms, undivided to pinnatilobate; heads usually 3–12, rarely solitary. Plants of the east and south of the European part of Russia .... 18.

18. Flowers light yellowish, often with pinkish tinge ....
.... 18. **C. roseolum.**

+ Flowers pink .... 19. **C. esculentum.**

19. Predominantly weedy soboliferous dioecious plants, with pink unisexual flowers; heads 10–25 mm in dia, usually numerous (more than six on a stem) and clustered in lax corymbose or corymbose-paniculate inflorescence; leaves relatively small but numerous (more than 15) on stem, lower ones withering by flowering .... 20.

+ Meadow, swamp, and forest rhizomatous monoecious plants, with bisexual flowers; heads one to five(six) on a stem, often densely clustered; leaves usually less numerous (2–15) on each stem, lower ones persisting at flowering, often large, upper usually strongly reduced ...
.... 21.

20. All leaves undivided or pinnatilobate with relatively broad and short lobes, usually elliptical or lanceolate-elliptical, weakly prickly, with several spines on them more than 2 mm long .... 23. **C. setosum.**

+ Only lower leaves withering by flowering, usually undivided, others pinnately divided or lobed, with narrow usually linear-lanceolate, strongly prickly lobes, prickles numerous, more than 2 mm long .... 25. **C. arvense.**

21. Heads 15–30 mm in dia, (one)two to five(six), densely clustered at stem tip, subtending leaves below heads ovate or lanceolate-ovate, dull green; flowers yellowish-whitish, often with weak pinkish tinge. Subglabrous plant, with fewer undivided or pinnately lobed cauline leaves
.... 10. **C. oleraceum.**

+ Heads solitary or two to five at stem apex and then may be densely clustered, but always more distant from narrow-lanceolate-linear subtending leaves .... 22.

22. All or almost all cauline leaves pinnately divided, green; heads 20–35 mm in dia, usually two to five on stem and then often clustered, less often solitary. Plants of the western regions of *Flora* .... 23.

+ Cauline leaves undivided, rarely partly pinnatilobate with distinct grayish tinge; heads 30–45 mm in dia, solitary, less often two to five and then more or less distant; flowers pink. Plants of the eastern and southern regions of *Flora* .... 24.

23. Peduncles and upper part of stem grayish-tomentose; involucre not sticky; flowers pink .... 11. **C. rivulare.**
 + Peduncles and upper part of stem densely short-crisped-hairy; involucre more or less sticky, with viscid glands on some involucral bracts; flower yellowish-white, sometimes with pinkish tinge .... 12. **C. erisithales.**
24. Rhizome rather long and slender, with thin distant roots; leaves always entire; heads almost always solitary; outer and middle involucral bracts apically with nondeflected or slightly deflected spines 0.3–0.7 mm long ....
.... 16. **C. pannonicum.**
 + Rhizome short and thick, with approximate and distinctly thickened roots; leaves undivided, entire or coarsely toothed, less often pinnatilobate; heads often two to five on a stem; outer and middle involucral bracts apically with deflected spine 0.7–2.0 mm long (at least some more than 1 mm long) .... 17. **C. canum.**

SUBGENUS **1.** *LOPHIOLEPIS* Cass.

1822, Dict. Sci. Nat. **25:** 225.—*Lophiolepis* (Cass.) Cass. 1823, Dict. Sci. Nat. **27:** 180.—*Eriolepis* Cass. 1825, Dict. Sci. Nat. **35:** 172.—*Cirsium* sect. *Epitrachys* DC. ex Duby, 1828, Bot. Gall., ed. 2, **1:** 286.—*Cirsium* sect. *Lophiolepis* (Cass.) DC. 1837, Prodr. **6:** 634.—*Cirsium* sect. *Eriolepis* (Cass.) DC. 1837, ibid. **6:** 635.

Heads 35–80 mm in dia, strongly prickly, with bisexual flowers. Leaves strongly prickly, with stiff squarrose bristles above. Biennials, with tap root.

Type: *C. ciliatum* (Murr.) Moench.

1. **C. vulgare** (Savi) Ten. 1836, Fl. Napol. **5:** 209; Charadze, 1963, Fl. SSSR, **28:** 135; K. Werner, 1976, Fl. Europ. **4:** 237.—*Carduus vulgare* Savi, 1798, Fl. Pis. **2:** 241.—*C. lanceolatus* L. 1753, Sp. Pl.: 821.—*Crisium lanceolatus* (L.) Scop. 1772, Fl. Carniol., ed. 2, **2:** 130, non Hill, 1769.—*C. sylvaticum* Tausch, 1829, Flora (Regensb.), **12,** 1: 38.—*C. nemorale* Reichenb. 1830, Fl. Germ. Excurs.: 286.—*C. lanceolatum* var. *nemorale* (Reichenb.) Naeg. ex Koch, 1845, Syn. Fl. Germ., ed. 2: 990.—*C. vulgare* subsp. *sylvaticum* (Tausch) Arén. 1954, Bull. Jard. Bruxell, **24,** 4: 130.—(Plate XXVIII, 1).

Type: Italy, vicinity of Pisa.

240 *North* (Karelia-Murmansk: ecdemic; Dvina-Pechora: south and west); *Baltic*; *Center*; *West*; *East*; *Crimea*.—In more or less saline meadows, forest glades and edges, on steppe slopes, by roadsides and forest paths, in habitations.—*General distribution*: Western and Eastern Siberia (south), Far East (south), Russian Central Asia; Scandinavia, Central and Atlantic Europe, Mediterranean, Asia Minor, Iran, Dzhungaria-Kashgaria; North America (ecdemic).—2n = 68.

Note. In var. *vulgare*, the leaves are pale green beneath because of relatively sparse pubescence but in var. *nemorale* (Reichenb.) Tzvel. comb. nova (=*C. nemorale* Reichenb. l. c.; =*C. sylvaticum* Tausch, l. c.) the leaves are grayish-tomentose on the lower side and it probably deserves separation as a subspecies—subsp. *sylvaticum* (Tausch) Arenes, l. c.

2. **C. ciliatum** (Murr.) Moench, 1802, Meth. Pl. Suppl.: 227; Charadze, 1963, Fl. SSSR, **28**: 130; K. Werner, 1976, Fl. Europ. **4**: 237.—*Carduus ciliatus* Murr. 1784, Comm. Goett.: 35.

Type: Described from the plants raised in Göttingen originating "from Siberia" ("in Sibiria"), but apparently from Volgograd or Saratov region.

Center (Upper Volga: ecdemic in vicinity of Kolomna; Volga-Don: south and east); *West* (Dnieper: east; Moldavia; Black Sea); *East* (Lower Don; Trans-Volga; Lower Volga: north).—In steppefied forest glades and edges, more or less trampled meadows, on steppe slopes, by roadsides.—*General distribution*: Caucasus, Western Siberia (southwest).

3. **C. ukranicum** Bess. 1838, in DC. Prodr. **6**: 635; Charadze, 1963, Fl. SSSR, **28**: 129; K. Werner, 1976, Fl. Europ. **4**: 237, sub hybr.

Type: Ukraine ("in Ucrania, Podolia australi").

Center (Volga-Don: vicinity of Boguchar, on chalk mountain in vicinity of the town of Petropavlovsk); *West* (Dnieper; Moldavia; Black Sea); *East* (Lower Don).—On steppe slopes, forest glades and edges, more or less weedy meadows, by roadsides.—*General distribution*: Central Europe (southeast).

4. **C. serrulatum** (Bieb.) Fisch. 1812, Cat. Jard. Gorenki, ed. 2: 35; Bieb. 1819, Fl. Taur.-Cauc. **3**: 557, p. p.; Charadze, 1963, Fl. SSSR, **28**: 128; K. Werner, 1976, Fl. Europ. **4**: 237, p. p.; Mikheev, 1984, Bot. Zhurn. **69**, 5: 693; Ignatov and others, 1990, Flor. Issl. Mosk. Obl.: 97.—*Cnicus serrulatus* Bieb. 1808, Fl. Taur.-Cauc. **2**: 275.

Type: Crimea, vicinity of Simferopol ("in Tauriae ruderatis, circa Simpheropolin").

239

Plate XXVIII.

1—*Cirsium vulgare* (Savi) Ten., 1a—achene; 2—*C. oleraceum* (L.) Scop., part of general inflorescence with some heads; 3—*C. heterophyllum* (L.) Hill, 3a—peripheral flower, 3b—inner flower without ovary.

Center (Upper Volga: ecdemic; Volga-Kama: southeast; Volga-Don: east); *West* (Moldavia); *East* (Trans-Volga; Lower Don); *Crimea.*—On steppe slopes, steppefied forest glades and edges, outcrops of chalk and limestones, by roadsides, in habitations.—*General distribution*: Western Siberia (southwest); Central Europe (eastern Romania).

○ 5. **C. tauricum** Soják, 1961, Novit. Bot. Prag. 1961: 35; Charadze, 1963, Fl. SSSR, **28:** 114.—*C. laniflorum* auct. non (Bieb.) Bieb.: K. Werner, 1976, Fl. Europ. **4:** 237, p. min. p.

Type: Crimea, yailas ("Taurus, prope jailam, ca 100–400 m").
Crimea (yailas and northern foothills).—On steppe slopes, outcrops of limestone, in forest glades and edges, by roadsides.—Endemic.

○ 6. **C. sublaniflorum** Soják, 1961, Nov. Bot. Prag. 1961: 36; Charadze, 1963, Fl. SSSR, **28:** 112.—*C. fimbriatum* Spreng. subsp. *bornmülleri* Petrak, 1910, Österr. Bot. Zeitschr. **10:** 395.—*C. lipskyi* Klok. 1962, Fl. URSR, **11:** 570, 533.—*C. laniflorum* auct. non (Bieb.) Bieb.: K. Werner, 1976, Fl. Europ. **4:** 237, p. min. p.

Type: Crimea, Karadag mountains ("Tauria, in silvis ad Karagatsch prope Sudak").
Crimea (predominantly southern foothills and Karadag).—On stony and clayey slopes, in forest glades and edges, thinned-out forests, by roadsides.—Endemic.
Note. C. lipskyi Klok. is described from the vicinity of Sudak.

241 ○ 7. **C. laniflorum** (Bieb.) Bieb. 1819, Fl. Taur.-Cauc. **3:** 557; Charadze, 1963, Fl. SSSR, **28:** 111; K. Werner, 1976, Fl. Europ. **4:** 237, p. p.—*Cnicus laniflorum* Bieb. 1808, Fl. Taur.-Cauc. **2:** 276.

Type: Crimea ("in Tauriae meridionalis collium sylvaticorum lapidosis").
Crimea (predominantly southern coast).—On stony and clayey slopes, rocks, taluses, by roadsides and pathways, in thinned-out forests, scrublands.—Endemic.

8. **C. decussatum** Janka, 1860, Linnaea, **30:** 582; K. Werner, 1976, Fl. Europ. **4:** 236, p. p.; Dubovic, 1977, Vizn. Rosl. Ukr. Karp.: 311; Tkaczik, 1981, Ukr. Bot. Zhurn. **38,** 5: 92.

Type: Romania, eastern Carpathians ("Alpes Rodnaensis").
West (Carpathians).—In forest glades and edges, on stony slopes, by roadsides, in thinned-out forests.—*General distribution*: Central Europe (southeast).

9. **C. polonicum** (Petrak) Iljin, 1932, Izv. Bot. Sada Akad. Nauk SSSR, **30,** 3–4: 350; Charadze, 1963, Fl. SSSR, **28:** 127.—*C. eriophorum* (L.) Scop. subsp. *decussatum* (Janka) Petrak var. *polonicum* Petrak, 1912, Biblioth. Bot. (Stuttgart), **78,** 35, Fig. 10.—*C. eriophorum* auct. non (L.) Scop.: Ledeb. 1846, Fl. Ross. **2:** 731.—*C. decussatum* auct. non Janka: Werner, 1976, Fl. Europ. **4:** 236, p. p.

Lectotype: Vinitsa Region of the Ukraine, vicinity of the town of Lipovets ("Fl. Polon. Exsicc. No. 957, Ucrainia, Zozól, distr. Lipowiec").

Center (Upper Volga: south; Volga-Don); *West*; *East*; (Lower Don).—In forest glades and edges, on steppe slopes, outcrops of chalk and limestone, in thinned-out forests, by roadsides.— *General distribution*: Central Europe (east).

Note. Occupies an intermediate position between the Carpathian mountain species *C. decussatum* and the western European *C. eriophorum* (L.) Scop. which has still larger appendages on the involucral bracts.

SUBGENUS **2.** *CIRSIUM*

Heads 12–45 mm in dia, weakly prickly, with bisexual flowers. Leaves usually weakly prickly, glabrous or more or less hairy above but without stiff squarrose bristles. Usually perennials, with more or less long rhizomes, rarely biennials.

Type: lectotype of genus.

Section 1. Cirsium
Perennials with more or less long rhizomes and usually weakly leafy stems; leaves weakly prickly, decurrent or weakly decurrent on stem, basal leaves usually persisting at flowering.

Type: type species of genus.

10. **C. oleraceum** (L.) Scop. 1769, Annus Hist.-Nat. **2:** 61; Charadze, 1963, Fl. SSSR, **28:** 161; K. Werner, 1976, Fl. Europ. **4:** 239.—*Cnicus oleraceus* L. 1753, Sp. Pl.: 826.—(Plate XXVIII, 2).

Type: Northern Europe ("in Europae septentrionalioris pratis subnemorosis").

North (Karelia-Murmansk: south; Dvina-Pechora: west and south); *Baltic*; *Center*; *West* (Carpathians; Dnieper; Moldavia); *East* (Trans-Volga).—In swampy forests, swampy meadows and forest glades, scrublands, near forest streams.—*General distribution*: Western Siberia (south); Scandinavia (south), Atlantic and Central Europe, Mediterranean.—2n = 34.

11. **C. rivulare** (Jacq.) All. 1789, Auct. Fl. Pedem.: 10; Charadze, 1963, Fl. SSSR, **28**: 162; K. Werner, 1976, Fl. Europ. **4**: 238; Haare, 1978, Novosti Sist. Vyssh. Rast. 15 (1978): 242.—*Carduus rivularis* Jacq. 1773, Fl. Austr. **1**: 57, tab. 91.—?*Cirsium montanum* auct. non (Waldst. et Kit. ex Willd.) Spreng.: Fodor, 1974, Fl. Zakarp.: 144.

Type: Austria and Hungary ("in Austriae, Pannoniae subalpinis").

Baltic (south); *Center* (Ladoga-Ilmen: vicinity of Gatchina and Lyuban; Upper Dnieper; Volga-Don: west); *West* (Carpathians; Dnieper).—In swampy meadows and forest glades, on banks of streams and rivers, in forest gullies.—*General distribution*: Scandinavia (south, ecdemic), Central and Atlantic Europe, Mediterranean.—2n = 34.

12. **C. erisithales** (Jacq.) Scop. 1769, Annus Hist. Nat. **2**: 62; id. 1772, Fl. Carn., ed. **2**, 2: 125; Charadze, 1963, Fl. SSSR, **28**: 163.—*Carduus erisithales* Jacq. 1762, Enum. Stirp. Vindób.: 279.—*Cnicus erisithales* (Jacq.) L. 1763, Sp. Pl., ed. 2: 1157.—*Cirsium ochroleucum* auct. non All.: Jundz. 1830, Opis. Rosl.: 330.

Type: Austria and France ("in Austriae, Galliae pratis subalpinis").

West (Carpathians; Dnieper: Ternopol Region; Moldavia: north).—In thinned-out forests, forest glades and edges, on stony slopes and rocks.—*General distribution*: Central Europe (southeast), Mediterranean (north of the Balkans).—2n = 34.

Note. Reports of this species from Latvia and Belorussia refer to the hybrid *C.* × *rigens* (Ait.) Wallr. (=*C. oleraceum* × *C. acaule*).

13. **C. waldstinii** Rouy, 1905, Fl. Fr., **9**: 84; Charadze, 1963, Fl. SSSR, **28**: 153; K. Werner, 1976, Fl. Europ. **4**: 239.—*Cnicus pauciflorus* Waldst. et Kit. 1803, in Willd. Sp. Pl. **3**, 3: 1677.—*Cirsium pauciflorum* (Waldst. et Kit.) Spreng. 1826, Syst. Veg. **3**: 375, non Lam. 1778.

Type: Carpathians ("in montibus Hungariae").

West (Carpathians).—In forest glades, edges and logged areas, thinned-out spruce forest.—*General distribution*: Central Europe (south), Mediterranean (north of the Balkans).—2n = 68.

14. **C. helenioides** (L.) Hill, 1768, Hort. Kew., 64; Charadze, 1963, Fl. SSSR, **28**: 166; Tzvel. 1970, Novosti Sist. Vyssh. Rast.

348

6 (1969): 300; K. Werner, 1976, Fl. Europ. **4**: 240, p. min. p.—
Carduus helenioides L. 1753, Sp. Pl.: 825, s. str. (quoad. pl.
sibir.).—*Cnicus helenioides* (L.) Willd. 1803, Sp. Pl., **3**, 3: 1674.—
Cirsium heterophylloides Pavl. 1938, Fl. Tsentr. Kazakhst. **3**: 313.

Type: England and Siberia; lectotype from Siberia ("in Anglia,
Sibiria").

Arctic (Arctic Europe: east); *North* (Dvina-Pechora: basin of
the Pechora River); *Center* (Volga-Kama: the Urals).—In swampy
meadows, forest glades and edges, thinned-out swampy forests.—
General distribution: Western Siberia (southeast), Eastern Siberia,
Far East (southwest), Russian Central Asia (melkosopochnik*);
Mongolia.—2n = 34.

Note. Although the original description of this species mentions
England, its diagnosis is clearly given based on the Siberian
plant, which is confirmed by the fact that Willdenow (loc. cit.)
cites this species for only Siberia ("in Sibiria").

15. **C. heterophyllum** (L.) Hill, 1768, Hort. Kew.: 64; Charadze,
1963, Fl. SSSR, **28**: 165.—*Carduus heterophyllus* L. 1753, Sp. Pl.:
824.—*Cirsium helenioides* auct. non (L.) Hill: K. Werner, 1976,
Fl. Europ. **4**: 240, p. max. p.—(Plate XXVIII, 3).

Type: Northern Europe ("in Europae frigidioris pratis
depressis").

Arctic (Arctic Europe); *North*; *Baltic*; *Center* (Dnieper: north
and east); *East* (Trans-Volga: southern Urals).—In meadows and
swamps, forest glades and edges, in thinned-out swampy forests.—
General distribution: Western Siberia; Scandinavia, Atlantic and
Central Europe, Mediterranean (mountains).—2n = 34.

16. **C. pannonicum** (L. fil.) Link, 1822, Enum. Pl. Hort. Berol.
alt. **2**: 299; Charadze, 1963, Fl. SSSR, **28**: 171; K. Werner, 1976,
Fl. Europ. **4**: 240; Mikheev, 1984, Bot. Zhurn. **69**, 5: 693.—*Carduus
pannonicus* L. fil. 1781, Suppl.: 348.—*Cirisium canum* (L.) All.
var. *pannonicum* (L. fil.) Schmalh. 1897, Fl. Sr. Yuzn. Ross. **2**: 104.

Type: Austria and Hungary ("in asperis Austriae, Pannoniae").

Center (Upper Dnieper: on the Dnieper River; Upper Volga:
on the Oka River; Volga-Don); *West* (Carpathians: vicinity of the
town of Chernovtsy; Dnieper; Moldavia; Black Sea: west, along
the Kuyalnik River).—In forest glades and edges, deciduous
forests.—*General distribution*: Central Europe, Mediterranean
(north).—2n = 34.

*Area of low, rounded, isolated hills—Translator.

17. **C. canum** (L.) All. 1785, Fl. Pedem. **1:** 151; Charadze, 1963, Fl. SSSR, **28:** 169; K. Werner, 1976, Fl. Europ. **4:** 241; Novikov and others, 1989, Nauchn. Dokl. Vyssh. Shkoly. Biol. Nauk, **4:** 61 —*Carduus canus* L. 1767, Mant. Pl.: 105.

243 Type: Austria ("in Austria").

Center (Volga-Kama: southeast; Volga-Don); *West*; *East.*—In meadows, forest glades and edges (particularly more or less saline and thinned-out forests, scrublands).—*General distribution*: Western Siberia; Central Europe, Mediterranean.—2n = 34.

Section 2. Acaulia (Petrak) Tzvel. 1991, Novosti Sist. Vyssh. Rast. **28:** 149.—*Cirsium* subsect. *Acaulia* Petrak, 1917, Bot. Centralbl. Beih. **35,** 2, 2–3: 258; Charadze, 1963, Fl. SSSR, **28:** 201.

Perennials with short thick rhizome and weakly leafy, strongly condensed stems, often acaulous; leaves rather strongly prickly, decurrent, basal persisting at fruiting.

Type: *C. acaule* (L.) Scop.

O 18. **C. roseolum** Gorlaczeva, 1989, Bot. Zhurn. **74,** 12: 1783.—*C. esculentum* auct. non (Siev.) C.A. Mey.: Charadze, 1963, Fl. SSSR, **28:** 201, p. p.; Oktryabreva and Tikhomir, 1987, Opred. Rast. Mentsery, **2:** 119.—*C. acaule* (L.) Scop. subsp. *esculentum* (Siev.) K. Werner, 1975, Bot. Journ. Linn. Soc. (London), **70:** 19, p. max. p.

Type: Donetzk Region ("Donetzk Region, vicinity of the town of Raigorodok, wet meadow").

Center (Upper Volga: southeast; Volga-Don); *West* (Dnieper and Black Sea: left bank of the Dnieper River); *East* (Lower Don: ecdemic).—In more or less saline meadows, forest glades and edges of swamps.—Endemic.

19. **C. esculentum** (Siev.) C.A. Mey. 1849, Mém. Acad. Sci. Pétersb., sér. 6, Sci. Nat. **6:** 42; Charadze, 1963, Fl. SSSR, **28:** 201, p. p.—*Cnicus esculentus* Siev. 1796, in Pall. Neueste Nord. Beitr. **3:** 362.—*C. gmelinii* Spreng. 1808, Hist. Rei. Herb. **2:** 270.— *Cirsium gmelinnii* (Spreng.) Tausch, 1828, Flora (Regensb.), **11,** 31: 482.—*C. acaule* (L.) Scop. var. *sibiricum* Ledeb. 1846, Fl. ross. **2:** 743.—*C. esculentum* var. *sibiricum* (Ledeb.) C.A. Mey. 1849, op. cit.: 42.—*C. acaule* (L.) Scop. subsp. *esculentum* (Stev.) K. Werner, 1975, Bot. Journ. Linn. Soc. (London), **70:** 19, p. min. p.

Type: Lake Balkhash Area ("Am Flusschen Ssun-Tass").

Center (Volga-Kama: southeast and along the Kolva River); *East* (Trans-Volga: east).—In saline meadows and solonchaks,

350

more or less saline swamps, sometimes on limestone outcrops.—
General distribution: Western and Eastern Siberia (south), Russian
Central Asia; Dzhungaria-Kashgaria, Mongolia.—2n = 34.

20. **C. acaule** (L.) Scop. 1769, Annus Hist.-Nat. **2**: 62; id.
1772, Fl. Carn. ed. 2, 2: 131; Charadze, 1963, Fl. SSSR, **28**: 205;
K. Werner, 1976, Fl. Europ. **4**: 240.—*Carduus acaulis* L. 1763, Sp.
Pl., ed. 2: 1156.—*Cnicus dubius* Willd. 1787, Fl. Berol. Prodr.: 260.

Type: Europe ("in Europae pascuis apricis depressis").
Baltic.—In meadows, forest glades and edges.—*General
distribution*: Scandinavia (south), Atlantic and Central Europe,
Mediterranean.—2n = 34.

Section 3. Orthocentron (Cass.) DC. 1837, Prodr. **6**: 641.—
Cnicus L. subgen. *Orthocentron* Cass. 1823, Dict. Sci. Nat. **35**:
173.—*Orthocentron* (Cass.) Cass. 1825, Dict. Sci. Nat. **35**: 173.

Perennials, with short rhizome and more or less thick roots;
stems rather densely leafy; leaves strongly prickly and prickly
decurrent on stem, basal often withering by flowering.

Type: *Orthocentron glomeratum* Cass. nom. illeg. (= *Cirsium
polyanthemum* (L.) Spreng.).

21. **C. alatum** (S.G. Gmel.) Bobr. 1958, Bot. Zhurn. **43**, 11:
1547; Chardze, 1963, Fl. SSSR, **28**: 192; K. Werner, 1976, Fl.
Europ. **4**: 241.—*Serratula alata* S.G. Gmel. 1770, Reise Russland,
1: 155, tab. 35, fig. 2.—*Cirsium desertorum* Fisch. ex Link, 1822,
Enum. Pl. Hort. Berol. alt. **2**: 300; Schischk. 1949, in Kryl. Fl. Zap.
Sib. **11**: 2888.—*C. setigerum* Ledeb. 1829, Icon. Pl. Fl. Ross. **1**:
9, tab. 35; Iljin, 1936, Fl. Yugo-Vost. Evrop. Chasti SSSR, **6**: 404;
Fig. 702.—*C. elodes* auct. non Bieb.: Klok. 1962, Fl. URSR, **11**:
534.

244   Type: Basin of the Lower Don, vicinity of the village of
Pyatiizbyanskaya ("Am Don . . . in Petis-benskaja").
West (Dnieper: southeast; Moldavia; Black Sea); *East* (Lower
Don; Trans-Volga: south and east; Lower Volga).—In saline
meadows and swamps, solonchaks.—*General distribution*: Western
Siberia (south), Russian Central Asia; Central Europe (southeast).

Section 4. Microcentron (Cass.) Tzvel. 1991, Novosti Sist.
Vyssh. Rast. **28**: 148.—*Onotrophe* Cass. sect. *Microcentron* Cass.
1825, Dict. Sci. Nat. **35**: 172.

Biennials, with rather densely leafy stem; leaves relatively
weakly prickly, decurrent on stem over entire internode; basal
leaves usually persistent.

Lectotype: *C. palustre* (L.) Scop.

22. **C. palustre** (L.) Scop. 1772, Fl. Carniol., ed. 2, 2: 128; Charadze, 1963, Fl. SSSR, **28**: 183; K. Werner, 1976, Fl. Europ. **4**: 241.—*Carduus palustris* L. 1753, Sp. Pl.: 822.—(Plate XXIX, 1).

Type: Europe ("in Europae pratis subpaludosis").
North; *Baltic*; *Center*; *West* (Carpathians; Dnieper; Moldavia); *East* (Trans-Volga).—In meadows, forest glades, swamps, swampy forests.—*General distribution*: Western Siberia, Eastern Siberia (southwest); Scandinavia, Central and Atlantic Europe, Mediterranean (rarely); North America (ecdemic).—2n = 34.

SUBGENUS **3.** *BREEA* (Less.) Tzvel.
1991, Novosti Sist. Vyssh. Rast. **28**: 148.—*Breea* Less, 1832, Syn. Gen. Compos.: 9.—*Cirsium* sect. *Cephalonoplos* Neck. ex DC. 1837, Prodr. **6**: 643.—*Cephalonoplos* (Neck. ex DC.) Fourr. 1869, Ann. Soc. Linn. Lyon, ser. 2, **17**: 25.—*Cnicus* L. subgen. *Breea* (Less.) Sch. Bip. ex Maxim. 1874, Bull. Acad. Sci. Petersb. **19**: 510.—*Cirsium* subgen. *Cephalonoplos* (Neck. ex DC.) Nakai, 1912, Bot. Mag. Tokyo, **26**: 355

Heads 10–25 mm in dia, somewhat prickly, with unisexual flowers. Leaves numerous, more or less prickly, glabrous or more or less hairy above but without stiff bristles. Soboliferous dioecious perennials.

Type: *C. arvense* (L.) Scop.

23. **L. setosum** (Willd.) Bess. 1816, Cat. Hort. Cremen: 39; Bieb. 1819, Fl. Taur.-Cauc. **3**: 560; Charadze, 1963, Fl. SSSR, **28**: 210.—*Serratula setosa* Willd. 1803, Sp. Pl., **3**, 3: 1654.—*C. arvense* (L.) Scop. var. *setosum* (Willd.) Ledeb. 1846, Fl. Ross. **2**: 735.— *Cephalonoplos setosum* (Willd.) Kitam. 1934, Acta Phytotax Geobot. (Kyoto), **3**, 1: 8.—*Cirsium arvense* subsp. *setosum* (Willd.) Iljin, 1936, Fl. Yugo-Vost. Evrop. Chast SSSR, **6**: 408.—*Breea setosa* (Willd.) Sojak, 1962, Novit. Bot. Prag. 1962: 42.

Type: Poland ("in Silesia").
Arctic (Arctic Europe: ecdemic); *North*; *Baltic*; *Center*; *West*; *East*; *Crimea*.—In fields and plantations of various crops, more or less trampled meadows and forest glades, by roadsides and pathways, in habitations, on banks of water bodies.—*General distribution*: Caucasus, Western and Eastern Siberia, Far East, Russian Central Asia; Scandinavia, Central and Atlantic Europe,

Mediterranean, Asia Minor, Iran, Dzhungaria-Kashgaria, Mongolia, Japan-China; North America (ecdemic).—2n = 34.

Note. Typical plants of this species (var. *setosum*) have all leaves undivided or the upper leaves are shallow pinnatilobate. In the predominantly more southern variety var. *mite* (Wimm. et Grab.) Tzvel. comb. nova (=*C. arvense* var. *mite* Wimm. et Grab. 1829, *Fl. Siles*, **2**, 2: 92), many leaves are pinnatilobate but their lobes are all the same broader and considerably less prickly than in *C. arvense* (L.) Scop.

24. **C. incanum** (S.G. Gmel.) Fisch. 1812, Cat. Gard. Pl. Gorenki: 35; Bieb. 1819, Fl. Taur.-Cauc. **3**: 561; Charadze, 1963, Fl. SSSR, **28**: 211; Efimova and others, 1981, Bot. Zhurn. **66**, 7: 1049; Tikhomir and Kharitontsev, 1984, Nauchn. Dokl. Vyssh. Shkoly, Biol. Nauk, **8**: 74.—*Serratula incana* S.G. Gmel. 1770, Reise Russland, **1**: 155, tab. 36, fig. 2.—*Cirsium arvense* (L.) Scop. var. *vestitum* Wimm. et Grab. 1829, Fl. Siles, **2**, 2: 92.—*C. arvense* var. *incanum* (S.G. Gmel.) Ledeb. 1846, Fl. Ross. **2**: 735.—*C. arvense* subsp. *incanum* (S.G. Gmel.) Petrak ex Iljin, 1936, Fl. Yugo-Vost. 246 Evrop. Chasti SSSR, **6**: 408.—*C. arvense* auct. non (L.) Scop.: K. Werner, 1976, Fl. Europ. **4**: 242, p. p.

Type: The Don Basin, near the village of Pyatizb yanskaya ("Am Don . . . in Petisbenskaja").

Arctic (Arctic Europe: along the Ness, Pechora, Usa, Shapkina, and Adzev rivers); *North* (Karelia-Murmansk: ecdemic in the village of Zasheek; Dvina-Pechora: Pechora Basin); *Center* (Volga-Kama: east; Volga-Don: east and vicinity of Trubchevsk in Bryansk Region); *West* (Carpathians: Uzhgorod District; Moldavia; Black Sea); *East*; *Crimea.*—In meadows (predominantly inundated) and forest glades, on banks of water bodies, in thinned-out forests, by roadsides.—*General distribution*: Caucasus, Russian Central Asia; Central Europe (east), Asia Minor, Iran.—2n = 28, 34.

Note. The type variety (var. *incanum*) has all leaves undivided. The variety var. *pinnatilobum* Sosn. ex Grossh. (1934, *Fl. Kavk.* **4**: 178), with almost all leaves pinnatilobate, is found predominantly in the more southern areas. The populations from the Dnieper sands of Kherson Region with narrower and less prickly leaves does not differ much from the typical plants.

25. **C. arvense** (L.) Scop. 1772, Fl. Carniol., ed. 2: 126; Charadze, 1963, Fl. SSSR, **28**: 213; K. Werner, 1976, Fl. Europ. **4**: 242, p. p.; Ignatov and others, 1990, Fler. Issled. Mosk. Obl.: 96.—*Serratula arvensis* L. 1753, Sp. Pl.: 820.—*Cirsium arvense* var. *horridum*

Wimm. et Grab. 1829, Fl. Siles, **2**, 2: 92.—*C. horridum* (Wimm. et Grab.) Stank. 1949, in Stank. and Taliev, Opred. Vyssh. Rast. Evrop. Chasti SSSR: 664, non (Bieb.) Bieb. 1819, nec (Adams) Petrak, 1912.—(Plate XXIX, 2).

Type: Europe ("in Europae cultis agris").

North (Karelia-Murmansk: ecdemic near the village of Yushkozero; Dvina-Pechora: ecdemic in the vicinity of Vologda and Arkhangelsk); *Baltic*; *Center* (Ladoga-Ilmen; Upper Dnieper: west; Upper Volga and Volga-Kama, ecdemic); *West* (Carpathians; Dnieper: west; Moldavia).—In weedy meadows, and forest glades, on banks of water bodies, by roadsides, in habitations, fields.—*General distribution*: Caucasus; Scandinavia, Central and Atlantic Europe, Mediterranean, Asia Minor; North America (ecdemic).—2n = 34.

HYBRIDS

1. *C.* × *affine* Tausch, 1829, Flora (Regensb.), **16**: 228 = *C. heterophyllum* × *C. oleraceum.*

Like *C. heterophyllum*, but the leaves are often large, lower leaves very weakly tomentose beneath, and usually more or less incised; heads are smaller, pale pinkish and often clustered in groups of a few heads. Known from *North* (Dvina-Pechora); *Baltic*; *Center* (Ladoga-Ilmen; Upper Dnieper; Upper Volga).

2. *C.* × *erucagineum* DC. 1805, in Lam. and DC. Fl. Fr. **4**: 115 = *C. oleraceum* × *C. rivulare.*

Intermediate between the parent species; the hybrid has pale pink flowers. Reliably known from *West* (Carpathians; Dnieper: northwest).

3. *C.* × *hybridum* Koch. 1815, in DC. Fl. Fr. Suppl.: 463 = *C. oleraceum* × *C. palustre.*

In comparison with *C. palustre*, the leaves are less numerous, larger, less prickly, and only somewhat decurrent; heads are larger, light pink, less numerous, known from *North* (Dvina-Pechora: Vologda Region); *Baltic*; *Center*; *West*, where it is a more common hybrid.

4. *C. ispolatovii* Iljin ex Tzvel. 1991, Novosti Sist. Vyssh. Rast. **28**: 151; Iljin, 1936, Fl. Yugo-Vost. Evrop. Chasti SSSR, **6**: 409, descr. ross. = *C. roseolum* × *C. vulgare.*

More like *C. vulgare,* but the leaves are less prickly, and with scattered stiff bristles above; heads are often clustered in twos to fours and the flowers are pale pink. Known from *West* (Trans-Volga: vicinity of Buguruslan).

245

Plate XXIX.
1—*Cirsium palustre* (L.) Scop., la—achene; 2—*C. arvense* (L.) Scop., 2a—achene; 3—*Picnomon acarna* (L.) Cass., 3a—achene.

5. *C.* × *rigens* (Ait.) Wallr. 1822, in Sched. Crit.: 466, non Spreng. 1822 = *C. acaule* × *C. oleraceum*.

Differs from *C. acaule* by having longer and more slender 247 stems, less prickly and larger leaves; heads are usually clustered in twos or threes at stem apex, with pale pinkish flowers. Known from *Baltic* where it is common.

6. *C.* × *reichardtii* Juratzka, 1859, Verh. Zool.-Bot. Ges. Wien, **9**: 317; Domin, 1929, Acta Bot. Bohem. **8**: 30 = *C. palustre* × *C. waldsteinii*.

A rare hybrid known only from *West* (Carpathians).

7. *C.* × *reichenbachianum* M. Loehr, 1852, Enum. Fl. Deutschl.: 364 = *C. arvense* × *C. oleraceum*.

Closer to *C. arvense*, but with lesser number of larger and less prickly leaves; heads are slightly larger but less numerous and often clustered, with light pinkish flowers. Known from *Baltic* where it is quite common.

8. *C.* × *silesiacum* Sch. Bip. 1889, in Nym. Consp. Fl. Eur.: 409; Marg. 1938, Bot. Közlem. 35: 63 = *C. canum* × *C. palustre*.

Known for *West* (Carpathians).

9. *C.* × *subalpinum* Gaud. 1829, Fl. Helv. **5**: 182 = *C. palustre* × *C. rivulare*.

Differs from *C. rivulare* by having more numerous and slightly decurrent cauline leaves. Known from *Center* (Ladoga-Ilmen: vicinity of Gatchina).

10. *C.* × *tataricum* (Jacq.) All. 1785, Fl. Pedem. **1**: 151 = *C. canum* × *C. oleraceum*.

Differs from *C. canum* by having lesser number of larger leaves, often pinnatilobate near base and heads three to five, often in clusters, with light pink flowers. Known from *Center* (Volga-Don: vicinity of Belgorod and Ulyanovsk) and *East* (Trans-Volga: vicinity of Buguruslan).

11. *C.* × *wankelii* Reichard, 1861, Verh. Zool.-Bot. Ges. Wien, **11**: 381.— *C. heterophyllum* × *C. palustre*.

Differs from *C. palustre* by having lesser number of less prickly leaves that are more or less tomentose beneath and only weakly decurrent; heads are somewhat larger and in lesser number. Known from *Center* (Ladoga-Ilmen).

356

GENUS **101.** *PICNOMON* Adans.[1]
1763, Fam. Pl. **2:** 116

Heads 7–15 mm in dia, homogamous, solitary or in groups of two to four at apices of stem and its branches, enclosed by spiny subtending leaves. Involucre 5–10 mm in dia, and 15–25 mm long, narrow-goblet-shaped, white-tomentose; involucral bracts imbricate, in many rows, highly unequal, outer lanceolate to innermost linear-lanceolate, more or less herbaceous, all except inner terminating in more or less deflexed spine or spine-like appendage, with one to four pairs of stiff cilia on sides. Receptacle almost flat, densely covered with long flattened bristles and with projecting margins of alveoli. Flowers all tubular, bisexual, with light pinkish five-lobed corolla. Achenes 4–5 mm long and 1.7–2.5 mm wide, somewhat flattened, glabrous, with inconspicuous veins, lustrous, yellowish-brown, apically with five-lobed tubercle; pappus 17–22 mm long, of many rows of plumose, almost equal bristles, connate only at base. Annual or biennial plants, more or less white-tomentose, with erect, usually strongly branched stem, spiny-winged throughout; leaves narrow-lanceolate, undivided or with one or two pairs of small lobes, subcoriaceous, with prickly bristles and 9–12 mm long spines on margin.

Type: *P. acarna* (L.) Cass.
A monotypic genus.

1. **P. acarna** (L.) Cass. 1826, Dict. Sci. Nat. **40:** 188; Tamamsch. 1963, Fl. SSSR, **28:** 218; K. Werner, 1976, Fl. Europ. **4:** 242.—
248 *Carduus acarna* L. 1753, Sp. Pl.: 820.—*Cirsium acarna* (L.) Moench, 1802, Meth. Pl. Suppl.: 226; Schmalh. 1897, Fl. Sr. Yuzhn. Ross. **2:** 104.—(Plate XXIX, 3).

Type: Spain ("in Hispania").
Crimea.—By roadsides, in habitations, on stony and clayey slopes, plantations of various crops, forest glades.—*General distribution*: Caucasus, Russian Central Asia (southwest); Mediterranean, Asia Minor, Iran.—2n = 34.

---

[1]Treatment by N.N. Tzvelev.

GENUS **102.** *LAMYRA* (Cass.) Cass.[1]
1822, Dict. Sci. Nat. **25:** 218, 222.—*Cirsium* Mill, subgen. *Lamyra* Cass. 1818, Bull. Soc. Philom. Paris, 1818: 168

Heads (1.5)2.0–4.0(5.0) mm in dia, homogamous, 1–6(10) clustered in terminal corymbose or racemose inflorescence. Involucre 15–28 mm in dia, 15–25 mm long, cup-shaped; involucral bracts imbricate, in many rows, highly unequal, outer lanceolate-ovate to linear-lanceolate innermost, lower and middle bracts with sideways inflected and somewhat deflected spines 5–12 mm long, innermost with long and narrow pinkish appendage. Receptacle somewhat convex, densely setose. Flowers all tubular, bisexual, with five-lobed pinkish corolla. Achenes 4–6 mm long and 3.2–4.0 mm wide, somewhat flattened, glabrous, with inconspicuous veins, not lustrous, brownish; pappus 18–25 mm long, comprising many rows of basally connate, plumose bristles, readily detaching entirely. Strongly prickly perennial, less often annual or biennial plant, often with many stems from base, white-tomentose, densely leafy, nonwinged stems; leaves subcoriaceous, oblong to linear, more or less pinnately divided, less often undivided, glabrous or scarcely pubescent above, white-tomentose beneath, with revolute margins and spines to 7 mm long.

Type: *L. stipulacea* Cass. non illeg. (=*L. stellata* (L.) Soják).

Literature: Tamamschan, S.G. 1954. Kriticheskie zametki po slozhnotsvetnym [Critical notes on the Compositae]. *Bot. Mat. (Leningrad)*, **16:** 468–478.—Soják, J. 1962. Bemerkungen zu einigen Compositen. II. *Novit. Bot. Prag.* 1962: 41–50.

Section 1. Platyraphium (Cass.) Tzvel. 1991, Novosti Sist. Vyssh. Rast. **28:** 152.—*Platyraphium* Cass. 1825, Dict. Sci. Nat. **35:** 173.

Perennial, strongly prickly plant; heads (1)2–8(10) clustered in more or less corymbose inflorescence.

Type: *L. diacantha* (Labill.) Cass.

1. **L. echinocephala** (Willd.) Tamamsch. 1954, Bot. Mat. (Leningrad), **16:** 470; id. 1963, Fl. SSSR, **28:** 220.—*Cnicus echinocephalus* Willd. 1803, Sp. Pl. **3,** 3: 1685; Bieb. 1808, Fl. Taur.-Cauc. **2:** 280.—*Cirsium echinocephalum* (Willd.) Bieb. 1819, Fl. Taur.-Cauc. **3:** 559.—*Chamaepeuce echinocephala* (Willd.) DC. 1837, Prodr. **6:** 660.—*Ptilostemon echinocephalus* (Willd.) Greuter, 1967, Boissiera, **13:** 146; K. Werner, 1976, Fl. Europ. **4:** 243.

---

[1]Treatment by N.N. Tzvelev.

358

Plant many-stemmed from base, 15–40(50) cm high; cauline leaves numerous, approximate, deeply pinnately divided into lanceolate-subulate lobes.

Type: Crimea ("in saxosis Tauriae").
Crimea.—On stony, predominantly limestone slopes, rocks and dumps.—*General distribution*: Caucasus.

249 GENUS **103.** *ONOPORDUM* L.[1]
1753, Sp. Pl.: 827; id. 1754, Gen. Pl., ed. 5: 359

Heads homogamous, many-flowered, with bisexual tubular flowers, solitary at apices of stem and its branches, less often some axillary. Involucre ovate or subglobose, 20–70 mm in dia, in many rows; involucral bracts imbricate, usually coriaceous, thickened or thin, narrow-lanceolate, terminating into prickly cusp. Receptacle flat, fleshy, without scales, alveolar, with alveoli bordered with membranous scaly processes. Corolla pink or purple, less often white or yellow. Anthers with short, subulate, basal appendage, connective cuneate or subulate. Achenes oblong or obovate, more or less tetragonous, longitudinally ribbed, transversely rugose, tuberculate on ribs, with straight or somewhat truncate hilum; pappus many-rowed, comprising numerous scabrous or plumose, basally connate bristles, readily detaching wholly. Biennial or perennial herbaceous plants, with prickly-winged cauline and large pinnatilobate or prickly toothed leaves.

Type: *O. acanthium* L.
About 40 species, distributed in Europe, northern Africa, and western Asia.
Literature: Rouy, M.G. 1896. Revision du Genre *Onopordon*. *Bull. Soc. Bot. Fr.*, **43**: 577, 599.—Eig, A. 1942. Revision of the *Onopordon* species of Palestine, Syria and adjacent countries. *Palest. Journ. Bot.*, ser. 2, **4**: 185–199.—Drees, W.J. 1966. Notes on the cultivated Compositae. 9: *Onopordum. Baileya.* **14:** 75–86.

1. Whitish arachnoid-hairy plant with broad (6–15 mm)-winged stems; involucral bracts with short-subulate cusp exceeding appressed lanceolate-ovate part of bract......
.... 1. **O. acanthium.**

[1]Treatment by O.V. Czerneva.

+ Plant green, more or less sticky from glandular pubescence, with narrow (2–5 mm wide)-winged stem; involucral bracts with prickly subulate cusp, shorter than appressed lanceolate part of bract............................ 2. **O. tauricum.**

1. **O. acanthium** L. 1753, Sp. Pl.: 827; Tamamsch. 1963, Fl. SSSR, **28**: 231; Amaral Franco, 1976, Fl. Europ. **4**: 245.

Type: Europe ("in Europae ruderatis, cultis").

Baltic; *Center*; *West*; *East*; *Crimea*.—In habitations, by roadsides, in weedy meadows and forest glades, edges of fields and plantations of various crops; in northern regions, only as a rare ecdemic plant.—*General distribution*: Caucasus, Western Siberia, Eastern Siberia (southwest), Russian Central Asia; Central and Atlantic Europe, Mediterranean, Asia Minor, Iran, ecdemic in other extratropical countries.—2n = 34.

2. **O. tauricum** Willd. 1803, Sp. Pl. **3**, 3: 1687; Tamamsch. 1963, Fl. SSSR, **28**: 233; Amaral Franco, 1976, Fl. Europ. **4**: 244.— (Plate XXIII, 1).

Type: Crimea ("in Tauria").

Crimea.—By roadsides, in habitations, trampled meadows, on gravel-beds and steppe slopes, edges of fields and plantations of various crops.—*General distribution*: Central Europe (ecdemic), Mediterranean, Asia Minor.—2n = 34.

GENUS 104. *CYNARA* L.[1]
1753, Sp. Pl.: 827; id. 1754, Gen. Pl., ed. 5: 359

250 Heads homogamous, many-flowered, with bisexual tubular flowers, at apices of stem and its branches. Involucre ovate or globose; involucral bracts imbricate, coriaceous, gradually attenuate into a cusp, less often obtuse, inner expanded in upper part into more or less rounded appendage. Receptacle flat, fleshy, setose. Corolla whitish to violet-azure. Anthers with short fimbriate basal appendage; connectives obtuse. Achenes obovate, indistinctly tetragonous, glabrous and smooth, with straight or somewhat truncate hilum; pappus many-rowed, of plumose bristles connate in ring at base, readily detached wholly. Perennial herbaceous plants, with large (to 1 m long) pinnately incised prickly leaves.

Type: *C. scolymus* L.

[1]Treatment by O.V. Czerneva.

The genus includes 10 or 11 species, distributed in the countries of the Mediterranean, including Canary Islands. One species is cultivated in many other extratropical countries.

1. **C. scolymus** L. 1753, Sp. Pl.: 827; Tamamsch. 1963, Fl. SSSR, **28**: 226; Amaral Franco, 1976, Fl. Europ. **4**: 249.

Involucre ovate-globose; involucral bracts glabrous, smooth, with fleshy base, outer with long cusp, middle and inner upward directed; pappus 20–30 mm long, longer than achenes; pappus bristles yellowish.

Type: Western Europe ("G. Narbonensis, Italiae, Siciliae agris").

West; *East* (Lower Don); *Crimea.*—Cultivated as ornamental plant.—*General distribution*: Central and Atlantic Europe, Mediterranean, Asia Minor; occasionally cultivated in other extratropical countries.—2n =34.

GENUS **105.** *SILYBUM* Adans[1]
1763, Fam. Pl. **2:** 116

Heads homogamous, many-flowered, with bisexual tubular flowers, solitary, drooping. Involucre subglobose, 30–60 mm in dia, many-rowed; involucre bracts imbricate, outer and middle expanded in upper half into foliate, prickly-toothed appendages, terminating in spine, inner with small appendage, innermost short-acuminate. Receptacle flat, densely setose. Corolla pink or purple, less often whitish. Anthers with ciliate basal appendage; connectives sagittate. Achenes obovate, laterally somewhat flattened, with inconspicuous longitudinal ribs and straight hilum; pappus of numerous scabrous bristles, basally connate in ring, readily detaching wholly. Annual or biennial herbaceous plants, with large pinnatilobate or pinnatisect usually white-spotted prickly leaves.

Type: *S. marianum* (L.) Gaertn.

Two species, one very widely distributed in Europe and central and southeastern Asia, another found only in northern Africa and Spain.

1. **S. marianum** (L.) Gaertn. 1791, Fruct. Sem. Pl. **2**: 378, tab. 168; Tamamsch. 1963, Fl. SSSR, **28**: 227; Amaral Franco, 1976, Fl. Europ. **4**: 249.—*Carduus marianus* L. 1753, Sp. Pl.: 823.

---

[1]Treatment by O.V. Czerneva.

Plant 50–150 cm high; leaves to 80 cm long and 30 cm wide, lustrous, with whitish spots; heads 40–60 mm in dia; pappus two to three times as long as achene.

Type: Europe ("Habitat in Angliae, Galliae, Italiae aggeribus ruderatis").

Baltic; *Center*; *West*; *East*; *Crimea.*—Cultivated as ornamental plant and occasionally found as escape or ecdemic plant in habitations, by roadsides.—*General distribution*: Caucasus, Western Siberia (southwest), Russian Central Asia; Central and Atlantic Europe, Mediterranean, Asia Minor, Iran.—2n = 34.

GENUS 106. *SERRATULA* L.[1]
1753, Sp. Pl.: 816; id. 1754, Gen. Pl., ed. 5: 357

Heads with numerous tubular flowers, usually bisexual or inner bisexual and peripheral pistillate, solitary or clustered in corymbose or paniculate-corymbose inflorescence. Involucre subglobose to ovate-cylindrical, of four to nine rows of imbricate, more or less coriaceous bracts; outer bracts usually ovate, cuspidate or spinescent, sometimes with large scaly appendage, inner lanceolate with elongate scaly appendage. Receptacle setose. Flowers exceeding involucre, pink or lilac-purple. Stamen filaments glabrous or with papillose hairs. Anthers sagittate with setaceous basal appendage. Achenes 4–6 mm long, oblong, terete, glabrous, smooth or finely sulcate, with oblique hilum; pappus many-rowed, comprising scabrous, less often plumose, persistent or gradually detaching unequal bristles. Perennial, hispid or subglabrous herbs, with erect, less often ascending, simple or more or less branched stem; leaves alternate usually pinnatilobate to pinnatisect, with entire, toothed or serrate lobes, less often undivided, lower petiolate, middle and upper usually sessile.

Lectotype: *S. tinctoria* L.

About 70 species, distributed in extratropical Eurasia and northern Africa.

1. Heads ovate-cylindrical, small, numerous, clustered in corymbose or corymbose-paniculate inflorescence; involucre 4–8 mm in dia .... 2.
+ Heads ovate or subglobose, medium or rather large, solitary or less numerous, and then clustered in more or less corymbose inflorescence; involucre 10–30 mm in dia .... 3.

[1]Treatment by E.V. Mordak.

2. Stems branched only in upper part; basal leaves usually undivided, cauline undivided to lyrate, all sharply ciliolate on margin; involucral bracts dark purple, sparsely arachnoid-hairy .... 1. **S. tinctoria.**

+ Stems spreadingly branched almost from base; all leaves pinnatilobate to pinnatisect; involucral bracts yellowish-green, subglabrous.................................... 3. **S. erucifolia.**

3. Involucre (10)20–30 mm in dia, outer involucral bracts ochraceous-tomentose, short-acuminate; leaves pinnati-partite or pinnatisect, lobes serrate and finely spiny-ciliate .... 2. **S. coronata.**

+ Involucre 10–20(25) mm in dia, outer involucral bracts not tomentose, long-acuminate or spinescent; leaves pinnatipartite, with entire or weakly toothed lobes, less often undivided or almost undivided .... 4.

4. Basal and lower cauline leaves usually undivided, less often few-lobed or only at base with few narrow lobes .... 5.

+ Basal and lower cauline leaves pinnatilobate to pinnatisect .... 7.

5. Basal and lower cauline leaves elliptical or oblong, obtuse or short-acuminate, entire or with occasional teeth, sometimes few-lobed, ciliolate on margin; heads less numerous, less often solitary; involucre (8)10–15(20) mm in dia, outer involucral bracts with brownish spot at tip, glabrous, more or less with deflected cusp or spine 0.5–1.0 mm long .... 10. **S. cardunculus.**

+ Basal and lower cauline leaves broadly ovate, at base often pinnatilobate, lyrate, acuminate, sinuate-coarsely toothed; heads solitary, less often two or three; involucre 12–23 mm in dia.. 6.

6. Outer involucral bracts greenish, with five to eight black veins at tip, short-cuspidate, with narrow reddish-brown membranous border.................................. 8. **S. lycopifolia.**

+ Outer involucral bracts greenish, apically brown including cusp, tip bordered with yellowish scaly appendage with undulate margin, as wide as almost half the length of involucral bract... 9. **S. bulgarica.**

7. Heads less numerous or solitary; involucre 10–15(20) mm in dia; outer involucral bracts with spine 1–2 mm long .... 4. **S. radiata.**

+ Heads solitary; involucre 15–20(25) mm in dia; outer involucral bracts short-cuspidate or with bent spine ... 8.

8. Basal and lower cauline leaves pinnatisect in lower part, pinnatipartite or pinnatilobate above, lobes 4–5 mm long and 5–12 mm wide; involucral bracts with short, readily detaching cusp .... 5. **S. gmelinii.**

+ Basal and lower cauline leaves pinnatipartite or pinnatisect, their lobes 2–4 cm long and 2–5(8) mm wide .... 9.

9. Leaves glabrous on both sides; outer involucral bracts subglabrous, yellowish, with short brown membranous cusp .... 6. **S. tanaitica.**

+ Leaves pubescent above, crisped-hairy or subglabrous beneath; outer involucral bracts glabrous, green, spinescent, spine curved, to 5 mm long ....

.... 7. **S. donetzica.**

Section 1. Serratula.

Heads homogamous, oblong-ovate or ovate-cylindrical, small, numerous, clustered in corymbose or corymbose-paniculate inflorescence; flowers unisexual or bisexual.

Type: lectotype of genus.

1. **S. tinctoria** L. 1753, Sp. Pl.: 816; Czer. 1973, Svod. Dop. Izm. "Fl. SSSR": 591; Cannon and Marshall, 1976, Fl. Europ. **4**: 250.—*S. inermis* Gilib. 1781, Fl. Lithuan. **3**: 183, nom. illeg.; Iljin, 1936, Fl. Yugo-Vost. Evrop. Chasti SSSR, **6**: 412; Boriss. 1963, Fl. SSSR, **28**: 265.

Type: Northern Europe ("in Europae borealioris pratis").

Baltic; *Center* (Ladoga-Ilmen: vicinity of Gdov and Velikie Luki; Upper Dnieper; Upper Volga; Volga-Don); *West* (Carpathians; Dnieper; Moldavia; Black Sea: Kherson Region); *East* (Trans-Volga: west).—In open broad-leaved and mixed forests, forest edges, glades and forest loggings, scrublands, in marshy meadows, sand ridges of river floodplain, outcrops of chalk and limestone.— *General distribution*: Scandinavia (Norway), Central and Atlantic Europe, Mediterranean, Asia Minor (northwest).—2n = 22.

Section 2. Mastrucium (Cass.) DC. 1837, Prodr. **4**: 667; Boriss. 1963, Fl. SSSR, **28**: 268.—*Mastrucium* Cass. 1825, Dict. Sci. Nat. **35**:173.

253 Heads heterogamous, ovate, large, clustered in corymbose inflorescence, less often solitary; peripheral flowers pistillate, with larger corolla, inner bisexual, numerous.

Type: *S. coronata* L.

2. **S. coronata** L. 1763, Sp. Pl., ed. 2: 1144; Iljin 1936, Fl. Yugo-Vost. Evrop. Chasti SSSR, **6**: 413; Boriss. 1963, Fl. SSSR, **28**: 268;

Klok. 1965, Fl. URSR, **12**: 11.—*S. wolffii* Andrae. 1855, Bot. Zeit. **13**: 321; Czer. 1973, Svod. Dop. Izm. "Fl. SSSR": 96; Cannon and Marshall, 1976, Fl. Europ. **4**: 250.

Type: Siberia ("in Sibiria, Italia").

Center (Ladoga-Ilmen: ecdemic, near the station of Mga; Upper Volga; Volga-Kama; Volga-Don); *West* (Dnieper; Moldavia: south; Black Sea: Kherson and Donetzk regions); *East.*—In broad-leaved forests, forest glades and edges, steppe scrubs, inundated meadows.—*General distribution*: Caucasus, Western and Eastern Siberia (south), Far East (south), Russian Central Asia (north); Central Europe (northeast), Mongolia, Japan-China.—2n = 22, 24, 30.

Section 3. Piptochaete Boiss. 1875, Fl. Or. **3**: 590; Boriss. 1963, Fl. SSSR, **28**: 269.

Heads homogamous, ovate-cylindrical, small, numerous, clustered in corymbose-paniculate inflorescence; all flowers bisexual. Plants branched almost from base.

Type: *S. erucifolia* (L.) Boriss.

3. **S. erucifolia** (L.) Boriss. 1963, Fl. SSSR, **28**: 270; Cannon and Marshall, 1976, Fl. Europ. **4**: 252.—*Xeranthemum erucifolium* L. 1753, Sp. Pl.: 858.—*Serratula xeranthemoides* Bieb. 1808, Fl. Taur.-Cauc. **2**: 265; Iljin, 1936, Fl. Yugo-Vost. Evrop. Chasti SSSR, **6**: 415; Klok. 1965, Fl. URSR, **12**: 13.

Type: Don ("ad Tanain prope oppidum Cavilnense").

Center (Volga-Don: south and southeast); *West* (Dnieper: west; Black Sea); *East* (Lower Don; Trans-Volga: south; Lower Volga); *Crimea.*—In steppes, on steppe slopes and solonetzes, sometimes in fields, by roadsides.—*General distribution*: Caucasus, Western Siberia (south), Russian Central Asia (north).

Section 4. Klasea (Cass.) DC. 1837, Prodr. **4**: 668; Boriss. 1963, Fl. SSSR, **28**: 272.—*Klasea* Cass. 1825, Dict. Sci. Nat. **35**: 173.

Heads homogamous, ovate or subglobose, medium or rather large, solitary or less numerous; all flowers bisexual; achenes with persistent pappus.

Lectotype: *S. centauroides* L.

4. **S. radiata** (Waldst. et Kit.) Bieb. 1819, Fl. Taur.-Cauc. **3**: 545; Boriss. 1963, Fl. SSSR, **28**: 274; Iljin, 1964, in Majevski, Fl. Sredn. Pol. Evrop. Chasti SSSR, ed. 9: 593; Klok. 1965, Fl. URSR, **12**: 15; Cannon and Marshall, 1976, Fl. Europ. **4**: 251.—*Carduus*

radiatus Waldst. et Kit. 1802, Descr. Icon. Pl. Rar. Hung. **1**: 9, tab. 11.—*Serratula bracteifolia* (Iljin ex Grossh.) Stank. 1949, in Stank. and Taliev, Opred. Vyssh. Rast. Evrop. Chasti SSSR: 670; Klok. 1965, op. cit.: 16.—*S. radiata* subsp. *bracteifolia* Iljin ex Grossh. 1934, Fl. Kavk. **4**: 194; Iljin, 1936, Fl. Yugo-Vost. Evrop. Chasti SSSR, **6**: 413.—*Klasea radiata* (Waldst. et Kit.) A. et D. Löve, 1961, Bot. Not. (Lund), **64**: 43.—*Serratula hungarica* Klok. et Dobrocz. 1965, Vizn. Ros. Ukr., ed. 2: 705, descr. ucrain; Zaverukha, 1987, Opred. Vyssh. Rast. Ukr.: 354, descr. ross.

Type: Hungary ("in montibus calcareis apricis ad Budam, Inotam, Palotam, etc.").

Center (Volga-Don: south and east); *West* (Dnieper: east; Moldavia; Black Sea); *East* (Lower Don; Trans-Volga); *Crimea.*— In thinned-out oak forests, forest glades and edges, on steppe slopes, outcrops of limestone.—*General distribution*: Caucasus; Central Europe, Mediterranean.—2n = 30, 60.

5. **S. gmelinii** Tausch, 1828, Flora (Regensb.), **11**: 485; Boriss. 1963, Fl. SSSR, **28**: 276; Cannon and Marshall, 1976, Fl. Europ. **4**: 252.—*S. isophylla* Claus, 1851, Beitr. Pflanzenk. Russ. Reich. **8**: 118, 301; Iljin, 1936, Fl. Yugo-Vost. Evrop. Chasti SSSR, **6**: 414.—*Klasea gmelinii* (Tausch) Holub, 1977, Folia Geobot. Phytotax. (Praha), **12**, 3: 305.

Type: Western Siberia, Ui River—a tributary of the Tobol River ("in Campis Uiensibus, Ucly caragaiensibus, Iaicensibus ut et in superiori regione Irtis fluvii").

Center (Volga-Kama: south; Volga-Don: east and southeast); *East* (Trans-Volga).—In steppefied meadows and solonetzes, forest glades and edges, outcrops of limestone, in scrublands, sometimes in abandoned fields.—*General distribution*: Western Siberia (south).

○ 6. **S. tanaitica** P. Smirn. 1940, Byull. Mosk. Obshch. Isp. Prir. Otd. Biol. **49**, 1: 92; Boriss. 1963, Fl. SSSR, **28**: 278; Klok. 1965, Fl. URSR, **12**: 21, Fig. 2; Cannon and Marshall, 1976, Fl. Europ. **4**: 252, in nota.—(Plate XXX, 1).

Type: Volgograd Region . . . upper reaches of the Golubaya River ("ad fontes Golubaya flum. in prov. Stalingradensi distr. Sairotinskij . . .").

Center (Volga-Don: southeast); *West* (Dnieper: east); *East* (Lower Don: north; Trans-Volga: southwest).—On chalk outcrops.—Endemic.

○ 7. **S. donetzica** Dubovik, 1965, Fl. URSR, **12**, 560: 19, Fig. 1; Cannon and Marshall, 1976, Fl. Evrop. **4**: 252, in nota.—(Plate XXX, 2).

Type: Severskyi Donetz ("ditio Rostoviensis, prope pag. Nizhnjaja Orechovka, in declivibus arenosis ad fl. Donetz . . .").

West (Dnieper: southeast); *East* (Lower Don: Donetz Ridge).— On outcrops of sandstone and chalk.—Endemic.

8. **S. lycopifolia** (Vill.) A. Kerner, 1872, Österr. Bot. Zeitschr. **22**: 13; Czer. 1973, Svod Dop. Izm. "Fl. SSSR": 591; Cannon and Marshall, 1976, Fl. Europ. **4**: 251.—*Carduus lycopifolius* Vill. 1779, Prosp. Hist. Pl. Dauph.: 30.—*Serratula heterophylla* auct. non (L.) Desf.; Iljin, 1936, Fl. Yugo-Vost. Evrop. Chasti SSSR, **6**: 415; Boriss. 1963, Fl. SSSR, **28**: 279; Klok. 1965, Fl. URSR, **12**: 21.—*Klasea lycopifolia* (Vill.) A. et D. Löve, 1961, Bot. Not. (Lund), **64**: 43.—(Plate XXX, 3).

Type: Hungary ("bei Inota auf dem Heigelzuge, welcher des Becken der Sárviz bei Stihlweissenburg nach Norden zu umrandet, Kalk, 150 m").

Center (Upper Volga: on the Oka River; Volga-Don); *West* (Dnieper; Moldavia; Black Sea); *East* (Lower Don; Trans-Volga).— In forest glades and edges, steppefied meadows, steppe slopes, in scrublands, sometimes on chalk outcrops.—*General distribution*: Central Europe, Mediterranean.—$2n = 30$.

Note. Basyonym of *S. heterophylla* (L.) Desf.—*Carduus heterophyllus* L. belongs to the genus *Cirsium* Mill.

9. **S. bulgarica** Act. et Stojan. 1932, Bull. Soc. Bot. Bulg. **5**: 111; Cannon and Marshall, 1976, Fl. Europ. **4**: 251.—*S. caputnajae* Zahar. 1946, Bull. Sect. Sci. Acad. Roum. **28**: 310, 318; Tory, 1979, Bot. Zhurn. **64**, 5: 727; Geidem. 1986, Opred. Vyssh. Rast. Mold. SSR, ed. 3: 566.—*Klasea bulgarica* (Acht. et Stojan.) Holub, 1977, Folia Geobot. Phytotax. (Praha), **12**, 3: 305.—(Plate XXX, 4).

Type: Northeastern Bulgaria ("prope urbem Eski-Džhumaja").

West (Moldavia: vicinity of the village of Batyr, Dumbreveni District).—In forest glades.—*General distribution*: Central Europe (southeast).

Note. Judged from the structure of the involucral bracts, *S. caput-najae,* described from Romania, does not differ from the earlier described *S. bulgarica.*

10. **S. cardunculus** (Pall.) Schischk. 1949, in Kryl. Fl. Zap. Sib. **11**: 2938; Boriss. 1963, Fl. SSSR, **28**: 280; Klok. 1965, Fl. URSR, **12**: 22; Cannon and Marshall, 1976, Fl. Europ. **4**: 252,— *Centaurea cardunculus* Pall. 1771, Reise, **1**: 500.—*Serratula nitida* Fisch. ex Spreng. 1826, Syst. Veg. **3**: 390; Iljin, 1936, Fl.

Yugo-Vost. Evrop. Chasti SSSR: 592.—*Klasea cardunculus* (Pall.) Holub, 1977, Folia Geobot. Phytotax. (Praha), **12**, 3: 305.

Type: Volga ("in ripis argillosis Volgae").

West (Dnieper: vicinity of Pavlograd; Black Sea: vicinity of Melitopol); *East.*—In saline meadows, solonetzes, on clayey-stony slopes, in steppes.—*General distribution*: Western Siberia (south), Russian Central Asia (north).

256 *Note.* In the vicinity of the Bolshoi [Greater] and Malyi [Lesser] Bogdo mountains, near Krasnoarmeisk (former Sarepta) and in some other regions of the Volga River area, there is var. *bogdensis* Iljin. It is characterized by the stems ascending at base and the basal leaves that are mostly divided and slightly undulate.

GENUS **107.** *RHAPONTICUM* Hill[1]
1762, Syst. Veg. **4**: 47

Heads homogamous, many-flowered, large. Involucre flattened-globose; involucral bracts many-rowed, coriaceous with scarious appendage; appendages of outer and middle involucral bracts large, lanceolate-ovate, mostly somewhat lacerate, convex, of inner shorter lanceolate or linear, always undivided. Flowers violet-purple, homogeneous, bisexual. Achenes weakly flattened from sides, ribbed, tetragonous, glabrous, lower areola of attachment (hilum) oblique; pappus two to two and one-half times as long as achene, readily detaching wholly, simple, biseriate, comprising short plumose bristles, connate at base into a ring. Perennial plants, with simple stems and undivided leaves (basal sometimes pinnatipartite in lower part), not decurrent on stem.

Type: *S. scariosum* Lam.

The genus includes 17 species, distributed in south of Europe, Siberia and Far East, in Russian Central Asia, Caucasus, northern Iran and farther in the east to the northern regions of China and to the Korean Peninsula, in Africa (Morocco) and in eastern Australia.

1. **S. serratuloides** (Georgi) Bobr. 1960, Bot. Mat. (Leningrad), **20**: 19; Sosk. 1963, Fl. SSSR, **28**: 313.—*Centaurea serratuloides* Georgi, 1775, Bemerk. Reise Russ. Reich, **1**: 231.—*C. altaica* Fisch. ex Spreng. 1813, Pl. Min. Cogn. Pugill. **1**: 59.—*Leuzea*

[1]Treatment by S.K. Czerepanov.

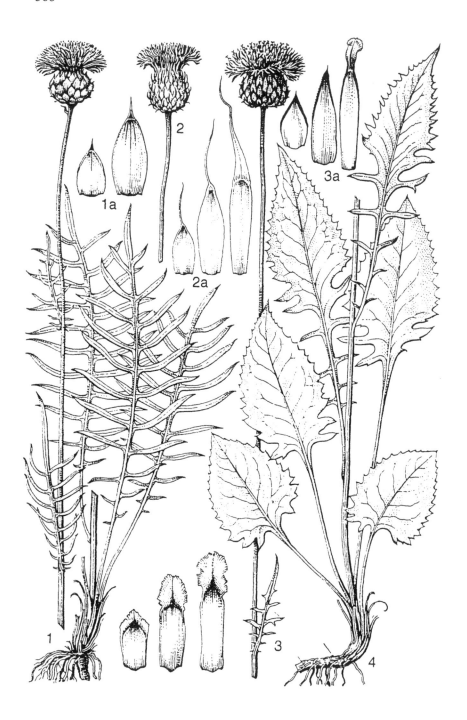

altaica (Fisch. ex Spreng.) Link, 1822, Enum. Pl. Horti Berol. Alt.
2: 356; Dostál, 1976, Fl. Europ. **4**: 243.—(Plate XXXI, 1).

Stem 40–100 cm high, sparsely arachnoid-hairy. Involucre 3–6 cm in dia. Achenes oblong, 6–8 mm long and 2–3 mm wide; pappus 15–16 mm long, creamish.

Type: Western Siberia (classical locality: steppes of western Baraba).

Center (Volga-Don: south); *West* (Dnieper: east; Moldavia; Black Sea); *East.*—On solonetzes, solonchaks, in saline meadows and river valleys.—*General distribution*: Western Siberia (south), Russian Central Asia; Central Europe (Romania: east).

Note. J.G. Gmelin's drawing in *Flora Sibiri* [Flora of Siberia] (J.G. Gmelin, 1749, *Flora Sibirica*, 2: tab. 37) is the type of this species.

GENUS 108. *ACROPTILON* Cass.[1]
1827, Dict. Sci. Nat. **50**: 464

Heads homogamous, many-flowered, rather small or medium-sized. Involucre oblong-ovate or ovate; involucral bracts many-rowed, thin-coriaceous with scaly appendage, appendages of outer and middle bracts as wide, often somewhat lacerate, border, of inner very narrow, linear-lanceolate. Flowers pink, all similar, bisexual. Achenes weakly flattened on sides, smooth, glabrous, lower areola (hilum) oblique; pappus two and one-half to three times as long as achene, very readily detaching, simple, many-rowed, of serrately setose flat bristles, not connate at base in ring; innermost bristles shortly plumose above. Perennial plants, with strongly branched stem and pinnatifid or toothed leaves not decurrent on stem.

Lectotype: *A. obtusifolium* Cass. (=*A. repens* (L.) DC.).

Two species of this genus are distributed in the south of the European part of Russia and Western Siberia, Russian Central Asia, Caucasus, Western and Central Asia.

1. **A. repens** (L.) DC. 1838, Prodr. **6**: 663; Czer. 1963, Fl. SSSR, **28**: 345; Dostál, 1976, Fl. Europ. **4**: 254.—*Centaurea repens* L.

Plate XXX.
1—*Serratula tanaitica* P. Smirn., 1a—involucral bracts; 2—*S. donetzica* Dubovik., head, 2a—involucral bracts; 3—*S. lycopifolia* (Vill.) A. Kerner, 3a—involucral bracts; 4—*S. bulgarica* Acht. et Stojan., involucral bracts.

---

[1]Treatment by S.K. Czerepanov.

258 1763, Sp. Pl., ed. 2: 1293.—*C. picris* Pall. ex Willd. 1803, Sp. Pl.
3, 3: 2302; Schmalh. 1897, Fl. Sredn. Yuzhn. Ross. **2**: 118.—
Acroptilon picris (Pall. ex Willd.) C.A. Mey. 1831, Verz. Pfl. Cauc.:
67; Stank. and Taliev, 1957, Opred. Vyssh. Rast. Evrop. Chasti
SSSR, ed. 2: 412.—(Plate XXXI, 2).

Plant with very long (to 6 m) root and long (to 1 m and more)
sobols. Stem 15–60(75) cm high, arachnoid-hairy, less often
subglabrous. Involucre 10–13 mm long and 5–7(8) mm in dia.
Achenes ellipsoidal-obovate, (2.5)3.0–4.0 mm long and (1.5)2.0
mm wide; pappus 8–11 mm long, whitish.

Type: East ("in Oriente").

North (Dvina-Pechora: Komi ASSR, near the Mikun Station,
ecdemic); *Baltic* (Latvia: city of Riga, ecdemic); *Center* (Ladoga-
Ilmen: Leningrad Region, St. Petersburg and Cholovo Station,
ecdemic; Upper Dnieper: Belorussia, vicinity of Borisovo and
Brest, ecdemic; Upper Volga: Tver and Moscow regions, ecdemic;
Volga-Kama: Kostroma Region, vicinity of Kostroma, Shar'ya
Station, ecdemic; Udmurt ASSR, Kamborka Station, Yuski Station,
ecdemic; Volga-Don: Tula Region, ecdemic, vicinity of Lipetsk,
Belgorod and Voronezh, ecdemic); *West* (Black Sea: east); *East*;
Crimea.—In steppes, saline meadows, and solonchaks, banks of
rivers and lakes, dry mountain slopes, as weed in crops, in old
fields, by roadsides, in habitations.—*General distribution*:
Caucasus, Western Siberia (south), Russian Central Asia; Asia
Minor, Iran, Dzhungaria-Kashgaria, Mongolia; North America
(ecdemic), Australia (ecdemic).—2n = 26.

GENUS 109. *PHALACRACHENA* Iljin[1]
1937, Bot. Mat. (Leningrad), **7**, 3: 51

Heads heterogamous, many-flowered, medium-sized. Involucre
broadly ovate or subglobose; involucral bracts many-rowed, more
or less coriaceous, with membranous appendage; appendages of
outer and middle bracts as wider broader, decurrent on bracts,
acuminate or short-spinescent, cristate-ciliate, inner larger, oblong-

257 Plate XXXI.
1—*Rhaponticum serratuloides* (Georgi) Bobr., 1a—involucral bracts;
2—*Chartolepis intermedia* Boiss., 2a—involucral bracts, 2b—achene; 3—
Acroptilon repens (L.) DC., 3a—head, 3b—achene.

[1]Treatment by S.K. Czerepanov.

372

ovate, almost not decurrent, finely toothed or ciliate. Flowers pink, heteromorphic, inner bisexual, peripheral sterile, large. Achenes very weakly compressed on sides, smooth, glabrous, with occasional hairs only near hilum, without pappus, lower areola (hilum) of achene lateral. Perennial plants, with simple, less often somewhat branched stems and entire leaves not decurrent on stem.

Type: *P. inuloides* (Fisch. ex Schmalh.) Iljin.

Two species of this genus are distributed in the European part of Russia and in the Kazakhstan.

○ 1. **P. inuloides** (Fisch. ex Schmalh.) Iljin, 1937, Bot. Mat. (Leningrad), **7,** 3: 51; Tzvel. 1963, Fl. SSSR, **28:** 580; Dostál, 1976, Fl. Europ. **4:** 254.—*Centaurea inuloides* Fisch. ex Schmalh. 1897, Fl. Sredn. Yuzhn. Ross. **2:** 126.—(Plate XXXII, 3).

Stem to 40(50) cm high, with very short papillate hairs and spinules, usually with inconsequential mixture of thin arachnoid tomentum. Involucre 15–20 mm long and 12–18 mm wide. Achenes oblong, 5–6 mm long.

Type: Rostov and Volgograd regions ["On clayey slopes (near Novocherkassk!), in Saratov steppes ("Tsarytsyn District!)"].

260 *West* (Black Sea); *East* (Lower Don; Lower Volga).—In alkaline and saline meadows, on clayey slopes and old fields.—Endemic

GENUS 110. *CENTAUREA* L.[1]
1753, Sp. Pl.: 909; id. 1754, Gen. Pl., ed. 5: 389, p. p.

Heads heterogamous, many-flowered, small to rather large. Involucre globose, ovate, oblong, sometimes cylindrical; involucral bracts many-rowed, coriaceous, with membranous, scarious, less often entirely thin-coriaceous or only in middle entire, irregularly toothed, fringed or cristate-ciliate appendage, not decurrent on bracts, often terminating in short cusp, spine, or small prickle, less often without appendage or outer and middle bracts with (3)5–7 palmately divergent spinules at tip, of which the central considerably longer (10–30 mm) and thicker than others. Flowers yellow, pink, pinkish-violet, purple, dark red, blue, azure, less often white or whitish, heteromorphic inner bisexual, peripheral sterile, not enlarged, slightly or highly enlarged, occasionally slightly reduced. Achenes very weakly compressed on sides, smooth, sometimes transversely rugose in upper part, scatteredly hairy, subglabrous (mature), less often glabrous, lower areola

[1]Treatment by S.K. Czerepanov.

Plate XXXII.
1—*Centaurea solstitialis* L., 1a—head; 2—*C. calcitrapa* L., 2a—head; 3—
Phalacrachena inuloides (Fisch. ex Schmalh.) Iljin, 3a—head.

259

374

(hilum) of achene lateral; pappus as long as achene, lower (not more than two times) or one-sixth to two-thirds as long, persistent, occasionally readily detaching (subgenus *Psephellus*), double, outer many-rowed, inner one-rowed, both of scabrous, serrate or serrate-ciliate bristles, less often setaceous scales, inner usually one-sixth to two-thirds as long as outer, very rarely pappus simple (subgenus *Psephellus*) or absent (achenes epappose).

Perennial, biennial or annual plants, with branched or simple stems and entire, remotely toothed, lobate or pinnately incised, nondecurrent or decurrent (cauline) leaves.

Lectotype: *C. centaurium* L.

Predominantly Mediterranean genus, with about 550 species.

Literature: Wagenitz, G. 1955. Pollen morphologie und Systematik in der Gattung *Centaurea* L. s. 1. *Flora*, **142**: 213–279.—Löve, A. and D. Löve, 1961. Some nomenclatural changes in the European flora. 1. *Bot. Not.* (Lund), **114**, 1: 33–47.—Wagenitz, G. 1963. Die Eingliederung der "Phacopappus"-Arten in des System von *Centaurea Bot. Jahrb.*, **82**, 2: 137–215.—Avetisiyan, E.M. 1964. K palinosistematike nekotorykh rodov podtriby Centaureinae semeistva Asteraceae [On the palinosystematics of some genera of the subtribe Centaureinae of the family Asteraceae]. *Tr. Bot. Inst. Akad. Nauk ArmSSR*, **14**: 31–47.—Dittrich, M. 1968. Karpologische Untersuchungen zur Systematik von *Centaurea* und verwandten Gattungen. Theil 1–2, *Bot. Jahrb.* **81**, 1: 70–122; **88**, 2: 123–162.—Gardou, Ch. 1972. Recherches biosystematiques sur la section *Jacea* Cass. et quelques sections voisines du genre *Centaurea* L. en France et dans less régions limitrophes. *Feddes Repert.* **83**, 5–6: 311–472.—Gochu, D.I. 1972. O vidovom sostave vasilkov Moldavskoi SSR [On the species composition of Moldavian centaurees]. *Izv. Akad. Nauk MoldSSR*, **2**: 15–18.—Dostál, J. 1973. Preliminary notes on the subtribe Centaureinae. *Acta Bot. Akad. Sci. Hung.*, **19**, 1–4: 73–79.—Wagenitz, G. 1974. Parallele Evolution von Merkmalen in der Gattung *Centaurea. Phyton* (*Austria*), **16**, 1–4: 301–312.—Plitmann, U. 1975. Taxonomic studies in *Centaurea* sect. *Calcitrapa* II. The section in Middle East. *Isreal Journ. Bot.*, **24**, 1: 10–25.—Dubovik, O.N. 1990. Rod *Centaurea* (Asteraceae) Krymsks-Novorossiiskoi provintsii i nekotorykh prilezhashchikh regionov [The genus *Centaurea* (Asteraceae) of Crimea-Novorossiisk Province and some adjoining regions]. *Bot. Zhurn.*, **75**, 11: 1573–1582.

1. Outer and middle involucral bracts obtuse or roundish, without appendage or appendages as membranous, fragile,

always entire but usually erose border; achenes entirely glabrous, in upper part weakly transversely rugose; flowers yellow. Perennial plants without stolons, with tap root ..... 2.

+ Outer and middle involucral bracts with membranous scarious, thin-coriaceous, or prickly mostly well developed apical appendage and rarely with few short cilia or terminating into spinule; achenes scatteredly short-hairy, sometimes mature achenes subglabrous; flowers usually colored .... 4.

2. Outer and middle involucral bracts without appendage, sometimes with narrow (to 1 mm) membranous border .. .. 1. **C. ruthenica.**

+ Outer and middle involucral bracts with appendages as more or less wide membranous border .... 3.

3. Plant 50–70 cm high, glabrous except at stem base, glaucescent-green; pappus 5–7 mm long, longest bristles of inner pappus half to two-thirds as long as longest outer pappus bristles .... 2. **C. kasakorum.**

+ Plant 80–100 cm high, throughout or only in lower part covered with sparse squarrose hairs, green; pappus 8–9 mm long, bristles of inner pappus one-sixth to one-fourth as long as longest outer pappus bristles .... 3. **C. taliewii.**

4(2). Annual, biennial or perennial plants, with tap root ... 5.

+ Perennial plant, with long or more or less short rhizome .... 54.

5. Outer and middle involucral bracts with almost completely membranous or in central part coriaceous and in peripheral part scarious, large, convex, orbicular, elliptical or broadly ovate, entire or irregularly toothed appendage not decurrent on bracts. Biennial, occasionally perennial plants; leaves (except uppermost) once or twice pinnately incised, with filiform, narrow-linear, linear-lanceolate, less often oblong segments .... 6.

+ Outer and middle involucral bracts with (three)five to seven palmately spreading spines or membranous, scarious or thin-coriaceous small or medium-sized, decurrent on more or less broad border on deltoid, lanceolate, ovate, elliptical, occasionally almost round, fringed, cristate-ciliate, irregularly toothed appendage, usually continued at tip as spinule, cusp or prickle, occasionally reduced to

261

short spine and some cilia; leaves entire and undivided to twice pinnately incised .... 22.

6. Appendage of involucral bracts often differentiated into thick central and scarious peripheral part; pappus almost as long as achene, less often slightly longer or distinctly shorter. Biennial psammophytes .... 7.

+ Appendage of involucral bracts occasionally differentiated into thick and coriaceous central part with five veins, and thin scarious peripheral part; pappus always considerably shorter (usually one-third to half as long) than achene. Perennial, occasionally biennial petrophyte .... 19.

7. Involucre 8–10 mm long, 7–8 mm wide; appendage of middle involucral bracts cuspidate; achenes 2.0–2.5 mm long; pappus 1.0–2.5 mm long. Plant of granite outcrops .... 70. **C. pseudoleucolepis.**

+ Involucre (11)12–22(27) mm long, 10–25(27) mm wide, occasionally 10–13 mm long, 6–10 mm wide, and then appendage of middle bracts not cuspidate; achenes (2.8)3.0–5.2 mm wide; pappus (2.5)3.0–7.0 mm long. Psammophytes .... 8.

8. Appendage of involucral bracts silky-lustrous, almost wholly opaque, translucent only on margin .... 9.

+ Appendage of involucral bracts with oily or micaceous coating, thin and almost wholly transparent, thickened only at base .... 11.

9. Corolla creamish, almost white; achenes biconvex, whitish, not turning black when mature ..... 59. **C. margaritacea.**

+ Corolla purple or pink; achenes almost tetragonous, black when mature .... 10.

10. Involucre globose, 15–20 mm in dia; appendage of involucral bracts brightly white; corolla purple, bright ... . 60. **C. margaritalba.**

+ Involucre broadly ellipsoidal or subglobose, 11–16 mm long, 11–15 mm in dia; appendage of involucral bracts yellowish to light orange; corolla pink or creamish .... 58. **C. protomargaritacea.**

11. Appendage of involucral bracts large, to 10 mm long and as much wide, very thin parchment-like, dorsally carinate, cuspidate, narrow strip of membranous tissue of appendage reaching on keel to very tip; on involucre all or some appendages connivent, forming hood.......................... 12.

+ Appendage of involucral bracts smaller, not so thin, without or with inconspicuous keel and apical spine membranous tissue of appendage forming triangle at base of appendage and not reaching apex; on involucre appendages more or less convex, not connivent in hood .... 13.

12. Corolla creamish, always white; involucre 22–27 mm in dia, with spreading appendages; stem sharply scabrous in upper part .... 62. **C. appendicata.**

+ Corolla pink; involucre 15–17 mm in dia, mostly with somewhat flat appendages; stem smooth in upper part .... 61. **C. konkae.**

13. Corolla pink or pale purple; appendage, at least in inner involucral bracts, bicolorous, with dark purple spot at base; stem more or less scabrous above on ridges ... 14.

+ Corolla creamish, almost white or light yellow; appendage of involucral bracts usually concolorous, yellowish or light yellow; stem smooth in upper part .... 17.

14. Appendage of involucral bracts yellowish, stiff, strongly convex, almost hemispherical, often with inconspicuous dark specks at base or entirely unicolorous; corolla pale pink or almost white .... 63. **C. protogerberi.**

+ Appendage of involucral bracts whitish or greenish, softer and weakly convex, almost all with distinct black specks at base; corolla pale pink or pink, brighter .... 15.

15. Involucre 10–13 mm long, 6–10 mm in dia; appendage of middle involucral bracts elliptical, greenish. Plant 50–100 cm and more high, with slender virgate branches and long lower leaves dissected into very narrow and long segments .... 66. **C. breviceps.**

+ Involucre 12–18 mm long and 10–16 mm in dia; appendage of middle involucral bracts more or less orbicular or elliptical, whitish. Plants 25–65 cm high, with thicker branches and shorter leaves with shorter and broader segments.... 16.

16. Involucre subglobose, 12–15 mm long, 10–16 mm in dia; achenes 4–5 mm long, lighter brown when mature, not turning black .... 64. **C. donetzica.**

+ Involucre ovate or oblong-ovate, 15–18 mm long, 10–14 mm in dia; achenes 3.5–4.7 mm long, turning black at maturity.. 65. **C. pineticola.**

17. Involucre 15–27 mm long, 15–25 mm in dia, often compressed-globose; appendage of involucral bracts weakly convex, completely covering bract, large, in

middle bracts 6–7 mm long, 8–9 mm wide ....

.... 69. **C. dubjanskyi.**

+ Involucre 11–18 mm long, 9–19 mm dia, not compressed from above; appendage of involucral bracts strongly convex, not covering bracts completely, smaller, with width not exceeding length .... 18.

18. Corolla creamish; appendage of involucral bracts yellowish, all unicolorous; mature achenes blackish, 3.5–4.5 mm long; pappus 3–4 mm long .... 68. **C. gerberi.**

+ Corolla light yellow; appendage of involucral bracts somewhat dark yellow, often with dark purple spot at base in middle and inner bracts; mature achenes whitish, 3.5–5.0 mm long; pappus 4.5–7.0 mm long ....

.... 67. **C. paczoskii.**

19(6). All appendages of involucral bracts with more or less distinct cusp; heads less numerous (to 10 on single stem) .... 20.

+ Appendages of middle and inner involucral bracts without cusp; heads more numerous .... 74. **C. sarandinakiae.**

20. Involucre ovate or oblong-ovate, 10–14 mm long, 6–10 mm in dia; stems to 100 cm high. Usually biennial plant of foothills and middle mountains .... 73. **C. sterilis.**

+ Involucre ovate, broadly ovate or globose, 14–22 mm long, 10–21 mm in dia; stems to 35 cm high. High-mountain perennial plant .... 21.

21. Involucre ovate or broadly ovate, 14–16 mm long, 10–14 mm in dia; appendage of middle involucral bracts 3–4 mm long, with large ovate-deltoid dark reddish-brown spot at base... 72. **C. vankovii.**

+ Involucre broadly ovate or subglobose, 16–22 mm long, 12–21 mm in dia; appendage of middle involucral bracts about 10 mm long, with oblong-lanceolate light reddish-brown spot dorsally .... 71. **C. semijusta.**

22(5). Flowers yellow .... 23.

+ Flowers pink of various shades, purple, dark red, blue, azure, occasionally whitish or white (albinos) .... 26.

23. Involucral bracts apically with five(seven) digitately spreading spines, of which the central usually considerably longer (to 30 mm) and thicker than others (2.5–4.0 mm long); achenes 2.5 mm long, outermost achenes without pappus; leaves (excluding lower) entire or subentire, sessile, long-decurrent on stem as narrow wings..................... 24.

+ Involucral bracts with large thin-coriaceous or membranous appendage at apex or as very narrow border, more or less decurrent on bract, cristate-ciliate, often terminating in spine or prickle; achenes 4–5 mm long, all with pappus; leaves (excluding upper) pinnately divided or lobed, sometimes some bipinnate, petiolate, not decurrent ..... 25.

24. Involucre 13–15(18) mm long, 7–12(15) mm in dia; outer and middle involucral bracts with five(seven) yellow spines, of which central considerably exceeding others .... 75. **C. solstitialis.**

+ Involucre 12–13(15) mm long, 7–8(10) mm in dia; outer and middle involucral bracts with five(seven) usually brownish spines, of which central thin, short, slightly exceeding others or almost as long .... 76. **C. adamii.**

25. Outer and middle involucral bracts with very large, yellowish or brownish-yellow appendage almost completely covering bract, cristate-ciliate (cilia shorter than width of appendage) usually terminating in spine to 2–3 mm long .... 34. **C. orientalis.**

264 + Outer and middle involucral bracts with appendage as very narrow short-ciliate (cilia to 1.0–1.5 mm long) border, often terminating in short spine or sideways deflected prickle to 0.8(1.0) cm long .... 32. **C. salonitana.**

26(22). Peripheral flowers blue or azure, strongly enlarged, obliquely tubular-infundibuliform, with corolla limb lobed to half or one-third into unequally long lobes; inner flowers lilac-pink or violet-lilac; stigma bifid to collar of hairs; style branches divergent; appendage of involucral bracts membranous, not prickly, very long-decurrent on bracts as border, together with border fringed or irregularly toothed. Annual plants .... 27.

+ All flowers pink of various shades, purple, less often dark red; peripheral flowers not enlarged or somewhat enlarged and then tubular flowers infundibuliform, with limb lobed almost to base into more or less equal lobes; style short bilobed; involucral bracts apically with long stiff spine or membranous or thin-coriaceous, cristate-ciliate, less often denticulate appendage, decurrent on bracts, terminating in spine, cusp or prickle .... 28.

27. Involucre 13(15–17) mm long, 9–13 mm in dia; achenes 5.0–5.5 mm long, 2.5–3.0 mm wide; pappus 6–8 mm long,

inner of scales; stem usually branched almost from base, coarse ....22. **C. depressa.**

+ Involucre 12–15 mm long, 5–9 mm in dia; achenes 3.0–4.0(4.5) mm long, 1.5–1.8 mm wide; pappus 3.0–3.5 mm long, inner setose; stem mostly branched from middle, stout .... 13. **C. cyanus.**

28. Outer and middle involucral bracts apically with (three) five to seven digitately spreading yellow spines, of which central considerably large (10–30 mm) and thicker than others (3–5 mm long) ....29.

+ Outer and middle involucral bracts apically with membranous or thin-coriaceous, cristate-ciliate, cristate-fringed, less often irregularly toothed or denticulate appendage, decurrent on bract, terminating in spine, cusp or prickle, occasionally reduced to short spine and some cilia .... 30.

29. Achenes with pappus; involucre ovate or subglobose, 13–16 mm long, 10–14 mm in dia; central spine of involucral bracts 10–30 mm long, others 3 mm long ....
.... 77. **C. iberica.**

+ Achenes without pappus; involucre oblong-ovate, 13–15 mm long, 6–8(10) mm in dia; central spine of involucral bracts 10–18 mm long, others 4–5 mm long ....
.... 78. **C. calcitrapa.**

30. Leaves with three prominent, almost parallel veins beneath, linear or linear-lanceolate, entire or undivided, sessile or subsessile. Perennial plant, with numerous stems, simple or branched near base, erect ....31. **C. trinervia.**

+ Leaves with one prominent vein, broader, pinnatilobate, pinnatipartite, once or twice pinnate occasionally all or some undivided or entire, and then lower and middle cauline leaves long-petiolate. Perennial or biennial plant, with one or few simple, weakly or strongly branched stems ....31.

31. Involucre (14)15–25(27) mm long, 9(11)–27(35) mm in dia; achenes 3.5–6.0 mm long; pappus (3.0)3.5–9.0 mm long. Perennial plant, with stems sparingly branched in upper part or simple, including leaves covered with short papillate hairs with mixture of crisped hairs and thin arachnoid tomentum and weakly scabrous, green; leaves (excluding upper) pinnatilobate, sometimes some twice pinnate with

linear oblong or broadly elliptical segments and lobes, occasionally some or all leaves undivided or entire, petiolate .... 32.

+ Involucre 8–12(15) mm long, (2.5)3.0–8.0 (occasionally 9–13) mm in dia; achenes 2.5–3.5(4.0) mm long; pappus 0.1–3.5(4.2) mm long, occasionally absent. Biennial plant, with stem branched from base or middle, mostly strongly, including leaves, arachnoid-hairy, arachnoid-tomentose and usually sharply scabrous; leaves (excluding upper) once or twice pinnate, with narrow-linear or linear segments; middle and upper cauline leaves sessile ..... 39.

32. Achenes 3.5–6.0 mm long; pappus (3.0)3.5–6.0 mm long, almost as long as achene or longer .... 33.

+ Achenes 3.5–5.0 mm long; pappus 1.5–2.5 mm long, one-third to half as long as achene; flowers pale pink, occasionally yellowish; appendage of involucral bracts deltoid-lanceolate .... 41. **C. stereophylla.**

33. Flowers dark red .... 34.

+ Flowers pink of various shades .... 35.

34. Involucral bracts with very large (almost completely overlapping bract) black-brown appendage; heads almost always solitary; pappus simple .... 35. **C. kotschyana.**

+ Involucral bracts with appendage as narrow brownish border; heads usually many on single stem; pappus double .... 33. **C. rubriflora.**

35. All leaves undivided and entire, occasionally few with one or two small lobes near base of leaf blade .... 37. **C. integrifolia.**

+ Leaves pinnatipartite or pinnatilobate, occasionally lower undivided .... 36.

36. Appendage of middle involucral bracts rather large, (1.8)2.0–6.0 mm long, excluding part decurrent on bract; peripheral flowers distinctly enlarged .... 37.

+ Appendage of middle involucral bracts very small, to 1.5(1.8) mm long, excluding part decurrent on bract and apical spinule; peripheral flowers not larger than inner .... 38.

37. Achenes 5–6 mm long; pappus 4.5–6.0 mm long .... 38. **C. ossethica.**

+ Achenes 3.5–4.5 mm long; pappus 4–5 mm long .... 36. **C. scabiosa.**

382

38. Involucral bracts without or with very small apical appendage to 0.8(1.0) mm long, with few very short cilia on margin ....40. **C. adpressa.**
  + Involucral bracts with apical appendage, 0.8–1.5(1.8) mm long, with rather long cilia on margin ...... 39. **C. apiculata.**
39(31). Involucre ovate, oblong-ovate, occasionally oblong, often bicolorous, with more or less colored-tipped bracts and appendages; involucral bracts with three to seven veins, their appendages, except thick middle part, scaly, with soft apical cusp and fimbriate ....40.
  + Involucre cylindrical or ovate-cylindrical, usually unicolorous, greenish, yellowish or stramineous-yellow; involucral bracts with one to three veins, their appendages coriaceous, with prickly apical cusp and stiff fimbriae ..
  ....53.
40. Involucral bracts with five to seven veins, their appendages cristate-fimbriate, without distinct scaly auricles at base, black or brown. Steppe plants ....41.
  + Involucral bracts with three to five veins, their appendages fimbriate, with more or less distinct scaly auricles at base, less often throughout irregularly toothed, weakly colored only in thick middle part. Plants of sands and stone outcrops ....43.
41. General inflorescence corymbose; appendage of outer and middle involucral bracts, including fimbria, black..
  ....42. **C. rhenana.**
  + General inflorescence paniculate; appendage of involucral bracts brown ....42.
42. Involucre oblong-ovate, 10–11 mm long, about 7 mm in dia; appendage of outer and middle involucral bracts light brown, with four to six fimbriae on each side; terminal leaf lobes narrow-linear, 1.0–1.5 mm wide.....................
  .... 44. **C. biebersteinii.**
  + Involucre ovate, 10–14 mm long, 10–13 mm in dia; appendage of outer and middle involucral bracts dark brown, with six to nine fimbriae on each side; terminal leaf lobes linear, 1–4 mm wide ...43. **C. pseudomaculosa.**
43. Involucre oblong-ovate, or ovate-conical, with soft fimbriate appendage, usually with distinct scaly auricles below appendage or irregularly toothed throughout; pappus always well-developed, as long as achene, slightly,

266

occasionally considerably shorter. Plants exclusively of sands .... 44.

+ Involucre subcylindrical, ovate-cylindrical, or oblong-ovate, with more or less somewhat stiff-fimbriate appendage and with inconspicuous auricles below appendage; pappus always shorter than achene, strongly reduced, occasionally absent. Petrophyllous plants (one species grows on shell-sands) .... 50.

44. Involucre 12–15 mm long; appendage of middle involucral bracts with six to eight fimbriae on each side, without distinct scaly auricles at base .... 51. **C. savranica.**

+ Involucre 9–12 mm long; appendage of middle involucral bracts with lesser number of fimbriae and with distinct scaly auricle at base or without distinct fimbriae, irregularly denticulate throughout .... 45.

45. Appendage of outer and middle involucral bracts fimbriate, thick middle part of appendage light reddish-brown or yellowish-brown with dark purple dots at base .... 46.

+ Appendage of outer and middle involucral bracts irregularly denticulate, often lacerate; thick middle part of appendage yellowish or brownish, without dots .... 47.

46. Stem, particularly in lower part, and leaves white-arachnoid-hairy-tomentose; pappus distinctly or considerably shorter than achene. Plant of marine sands .... 49. **C. odessana.**

+ Stem and leaves sparsely lanate-arachnoid-hairy; pappus as long as achene or shorter. Plant of river sands .... 50. **C. borysthenica.**

47. Involucre ovate-conical, 9–11 mm long, 6–8 mm in dia; stem in upper part sharply scabrous on ribs .... 45. **C. majorovii.**

+ Involucre oblong-ovate, 9–12 mm long, 3.5–6.0(6.5) mm in dia; stem entirely smooth, less often weakly scabrous on ribs in upper part .... 48.

48. Involucre 9–11 mm long, 3.5–5.0 mm in dia; stem sparsely arachnoid-hairy, glabrous in upper part, somewhat slender, with long branches; leaves smooth on margin .... 48. **C. wolgensis.**

+ Involucre 9–12 mm long (4.5)5.0–6.5 mm in dia .... 49.

49. Stem rather densely arachnoid-hairy, somewhat thick, with short branches; leaves on margin scabrous from conical tubercles .... 46. **C. sophiae.**

+ Stem sparsely arachnoid-hairy, glabrous in upper part, somewhat slender, with long branches; leaves smooth on margin .... 47. **C. arenaria.**

267 50(43). Middle involucral bracts with three veins; pappus 1.5–2.5 mm long, somewhat or distinctly shorter than achenes .... 51.

+ Middle involucral bracts with three to five veins; pappus strongly reduced, 0.1–0.5 mm long or absent .... 52.

51. Involucre 10–12 mm long, 5–6 mm in dia, ovate-cylindrical; stem and leaves weakly arachnoid-hairy ....
.... 52. **C. besseriana.**

+ Involucre 8–11 mm long, 3–5 mm in dia, subcylindrical, with conical base; stem and leaves rather densely arachnoid-hairy .... 53. **C. lavrenkoana.**

52. Involucre 10–12 mm long, 5–6 mm in dia; pappus about 0.5 mm long or absent .... 54. **C. steveniana.**

+ Involucre about 9 mm long, 3.5–4.0 mm in dia; pappus 0.1–0.2 mm long .... 55. **C. caprina.**

53(39). Stem and leaves scatteredly arachnoid-hairy; heads solitary at apices of very short or more or less elongated branches; involucre 8–10 mm long, 2.5–5.0 mm in dia, prickly cusp of appendages of involucral bracts weakly deflected outward; flowers pale pink or whitish, occasionally light purple .... 56. **C. diffusa.**

+ Stem and leaves rather densely arachnoid-hairy; heads usually more or less clustered at apices of second order branches; involucre 10–12 mm long, 3.0–3.5 mm in dia; prickly cusp of appendages of involucral bracts strongly squarrose and in lower bracts more or less deflexed; flowers purple .... 57. **C. aemulans.**

54(4). Achenes without pappus; appendages of outer and middle involucral bracts large, round, lacerate, irregularly toothed or fimbriate, less often entire, not decurrent on bracts; leaves entire, remotely toothed, often basal and lower cauline leaves sinuate-lobate .... 55.

+ Achenes with pappus .... 57.

55. Leaves elliptical-lanceolate or ovate-lanceolate to almost oblong, subacute or short-acuminate, green, rather thin, scabrous on both sides from flexuous, mostly scattered hairs; cauline leaves deflected, often subglabrous; appendage of involucral bracts usually brown or brownish
.... 15. **C. jacea.**

+ Leaves narrower, narrow-lanceolate to linear-lanceolate, acuminate, thick; cauline leaves obliquely upward directed; appendage of involucral bracts whitish, in middle part pale brownish .... 56.

56. Involucre ovate, (8)10–12 mm in dia. Plant green, subglabrous or sparsely arachnoid-hairy; stem in upper third less branched .... 16. **C. pannonica.**

+ Involucre globose-ovate or subglobose, 12–16(18) mm in dia. Plant grayish-green or grayish from more or less somewhat dense arachnoid pubescence, and also with bent stiff hairs; stem mostly rather strongly branched from middle or above .... 17. **C. substituta.**

57(54). Appendage of involucral bracts long decurrent on bract as border, together with border fimbriate or irregularly toothed .... 58.

+ Appendage of involucral bracts not decurrent on bract .... 66.

58. Fimbriae or teeth of appendages of involucral bracts black, occasionally blackish-reddish-brown .... 59.

+ Fimbriate of appendages of involucral bracts silver-white or whitish .... 61.

59. Appendage of involucral bracts short-toothed, teeth shorter than width of fimbria; rhizome long, creeping .... 60.

+ Appendage of involucral bracts short-fimbriate, fimbriae as long as wide or slightly longer; rhizome vertical or obliquely ascending .... 6. **C. stricta.**

60. Leaves rather thin, attenuate-acuminate, green on both sides, very weakly arachnoid-hairy-lanate; involucre 10–13(15) mm in dia .... 5. **C. marmarosiensis.**

+ Leaves more or less thick, gradually acuminate, greenish and weakly arachnoid-hairy above, grayish beneath from dense arachnoid-tomentum .... 4. **C. mollis.**

61. Rhizome with three to six oblong tuberously thickened roots .... 11. **C. thirkei.**

+ Rhizome without tuberously thickened roots .... 62.

62. Stem ribbed, not winged; involucre cylindrical, 7–10 mm in dia .... 9. **C. dominii.**

+ Stem narrow- or rather broad-winged; involucre ovate or oblong-ovate, 10–15 mm in dia .... 63.

63. Appendage of involucral bracts short-fimbriate, whitish fimbriae slightly or almost two times as long as wide; rhizome vertical or very weakly ascending .... 64.

268

+ Appendage of involucral bracts long-fimbriate, silvery-white fimbriae two to four times as long as wide; rhizome creeping .... 10. **C. fuscomarginata.**

64. Stem at base very slightly covered with relatively weakly expanded petiole bases of basal leaves; basal and lower cauline leaves withering early, strongly browning, more or less long-petiolate, others sessile, narrow and short- or somewhat long-decurrent on stem; pappus 1.0–1.5 mm long .... 6. **C. stricta** (cf. also couplet 59).

+ Stem at base entirely covered with strongly expanded petiole bases of basal leaves; basal and lower cauline leaves withering early, turning somewhat brown, short-petiolate, others sessile, somewhat broadly and long (from leaf to leaf)-decurrent on stem; pappus 1.5–3.0 mm long .... 65.

65. Stem somewhat thick, somewhat densely leafy, including leaves grayish from dense arachnoid-tomentum; involucre 16–23 mm long, (8)10–15 mm in dia .... 7. **C. tanaitica.**

+ Stem more slender, sparsely leafy, including leaves greenish-grayish from more sparse arachnoid pubescence; involucre 18–20 mm long, 9–11 mm in dia ....
.... 8. **C. angelescui.**

66(57). Pappus simple, of short, relatively readily detaching, scabrous, scaly bristles; peripheral flowers with staminodes; stigma shortly bilobate; appendage of involucral bracts fimbriate; leaves pinnately lobed .... 14. **C. declinata.**

+ Pappus distinctly double, persistent; peripheral flowers without staminodes; stigma biparted, with long branches
.... 67.

67. Outer pappus comprising finely setaceous bristles, inner of linear-lanceolate, smooth scales, converging by their apices into cone; appendages of involucral bracts fimbriate, denticulate or partly entire; shoots arising from axils of basal leaves, procumbent or ascending; basal leaves pinnately lobed, cauline pinnatipartite or undivided .... 68.

+ Pappus comprising serrate bristles of similar structure but bristles of inner pappus half to two-thirds as long as outer bristles; appendages of involucral bracts incised to filiform, mostly long cilia, attenuate into subulate or caudate non-prickly tip; shoots arising directly from rhizome .... 71.

68. Involucral bracts completely overlapped by large appendages; appendages of middle involucral bracts

orbicular or ovate, 5–9 mm long, 5.0–8.5 mm wide, with numerous (12–19 on each side) fimbriae; involucre subglobose .... 69.

+ Involucral bracts somewhat covered by small appendages or entirely uncovered; appendage of middle involucral bracts oblong-ovate, linear-lanceolate, less often broadly ovate, 2.0–3.5 mm long, 1–3 mm wide, with less numerous (three to five on each side) fimbriae or entire; involucre ovate ..... 70.

69. Stem usually ascending or almost erect, to 50 cm high; diameter of involucre exceeding its length; appendage of involucral bracts brownish-yellow .... 30. **C. sibirica.**

+ Stem procumbent, to 25 cm long; diameter of involucre not exceeding its length; appendage of involucral bracts darker, brownish-yellow .... 27. **C. carbonata.**

70. Involucre broadly ovate, about 15 mm long, 12–14 mm in dia; appendage of middle involucral bracts oblong-ovate or broadly ovate, toothed-fimbriate ....
.... 28. **C. marschalliana.**

+ Involucre ovate, 15–22 mm long, 8–14 mm in dia; appendage of middle involucral bracts linear-lanceolate, with three to five short fimbriae on each side or entire ....
.... 29. **C. sumensis.**

71(67). Involucre oblong-ovate, 13–15 mm long, 6–10 mm in dia; stem 20–40(70) cm high, rather strongly branched from middle or above, with more or less divergent stout branches
.... 26. **C. trichocephala.**

+ Involucre subglobose, ovate, occasionally ovate-cylindrical, (14)15–20(22) mm long, (8)10–20 mm in dia; stem to 100(120) cm high, sparingly branched in upper part, with obliquely upward directed thickish branches .... 72.

72. Appendages of inner involucral bracts projecting over middle bracts and not overlapped by appendages of the latter .... 73.

+ Appendages of inner involucral bracts not projecting over middle bracts and overlapped by appendages of the latter .... 76.

73. Appendages of outer and middle involucral bracts large, completely overlapping bracts below them .... 74.

+ Appendages of outer and middle involucral bracts rather small, partially overlapping bracts below them, blackish- or dark brown, attenuate into erect or scarcely deflected subulate tip; involucre 17–20 mm long, 12–15 mm in dia. Plant lanate or scabrous-lanate .... 21. **C. abbreviata.**

74. Stem lanate from scattered flexuous hairs; cauline leaves without auricles at base .... 75.

+ Stem glabrous; cauline leaves with large, toothed auricles at base enclosing stem; appendages of outer and middle involucral bracts black .... 20. **C. melanocalathia.**

75. Involucre 14–20 mm long, 12–17 mm in dia; appendages of outer and middle involucral bracts blackish- or dark brown, occasionally light brown; cauline leaves attenuate or rounded at base, not amplexicaul .... 18. **C. phrygia.**

+ Involucre 18–20 mm long, 17–18 mm in dia; appendages of outer and middle involucral bracts black; cauline leaves rounded at base and usually somewhat amplexicaul .....
.... 19. **C. carpatica.**

76(72). Involucre (18)20–22 mm long, (13)15–20 mm in dia. Plant of upper forest and subalpine mountain zones (Crimea)
.... 22. **C. alutacea.**

+ Involucre 10–15(20) mm long, 7–15 mm in dia. Plants of the plains .... 77.

77. Leaves green, scabrous from short stiff hairs, without arachnoid pubescence or with its inconsequential mixture; involucre 15–20 mm long, (8)10–15 mm in dia ....
.... 23. **C. pseudophrygia.**

+ Leaves gray or grayish-green from dense fine arachnoid pubescence, sometimes subtomentose .... 78.

78. Involucre 15–18 mm long, 9–14 mm in dia; stem cymosely branched; appendage of involucral bracts reddish-brown; middle involucral bracts with cusp 8–10 mm long ....
.... 25. **C. stenolepis.**

+ Involucre about 15 mm long, 8–10 mm in dia; stem weakly branched; appendages of involucral bracts dark reddish-brown, blackish at base; middle involucral bracts with cusp 6–7 mm long .... 27. **C. indurata.**

SUBGENUS 1. *CENTAUREA*

Involucral bracts without apical appendage or appendage as membranous border not decurrent on bracts, very narrow or rather wide, entire but usually more or less lacerate, as oblong or ovate appendage to 10 mm long in innermost involucral bracts. Flowers yellow, all tubular, peripheral not expanded. Stigma short-bilobate. Achenes 6–8 mm long, glabrous, transversely rugose in upper part; pappus 4.5–9.0 mm long, double, inner one sixth to two-thirds as long as outer, of scalelike bristles broader in lower part and gradually acuminate toward apex. Perennial plants with simple or weakly branched, erect stems and pinnatipartite or pinnatilobate, glabrous, less often scatteredly hairy, leaves but always with more or less profuse lanate tomentum in axils of basal leaves.

Type: lectotype of genus.

1. **C. ruthenica** Lam. 1785, Encycl. Méth. Bot. **1**: 663; Tzvel. 1963, Fl. SSSR, **28**: 380; Dostál, 1976, Fl. Europ. **4**: 263.

Type: European part of Russia ("Cette plante croit dans la Russie, la Moscovie").

Center (Upper Volga: Yaroslavl and Moscow regions, ecdemic; Volga-Kama: south; Volga-Don); *West* (Dnieper: east; Black Sea); *East.*—In steppes, on stony slopes, outcrops of chalk and limestone.—*General distribution*: Caucasus (Ciscaucasia: south; Transcaucasia: Daralgez); Western Siberia (south), Russian Central Asia; Central Asia; Central Europe (southeast), Mediterranean (Balkan Peninsula: northeast).—2n = 30.

2. **C. kasakorum** Iljin, 1937, Bot. Mat. (Leningrad), **7**, 3: 66; Tzvel. 1963, Fl. SSSR, **28**: 383; Dostál, 1976, Fl. Europ. **4**: 263.

Type: Western Kazakhstan ("distr. Temir, ad radicem montis cretacei Astau-Saldy").

East (Trans-Volga; Lower Volga: northeast).—On stony slopes, outcrops of chalk and limestone.—*General distribution*: Western Siberia, Russian Central Asia; Dzhungaria-Kashgaria.

○ 3. **C. taliewii** Kleop. 1927, Visn. Kiiv. Bot. Sadu, **5–6**: 87; Tzvel. 1963, Fl. SSSR, **28**: 386; Dostál, 1976, Fl. Europ. **4**: 263.— *C. ruthenica* Lam. var. *villosa* Taliev, 1900, Tr. Obshch. Isp. Prir. Khark. Univ. **34**: 247.

Type: Donetz Region ("Mariupil District....vicinity of the village of Lyapina").

West (Black Sea); *East*; *Crimea.*—In steppes, on stony slopes.—Endemic.

271   SUBGENUS **2.** *CYANUS* (Mill.) Spach
1841, Hist. Nat. Vég. (Phan.), **10**: 11, 68.—*Cyanus* Mill. 1754, Gard. Dict. Abridg. ed. 4, 1

Involucral bracts with membranous, nonprickly, quite long, fimbriate or irregularly toothed, apical appendage decurrent on bract as border. Inner flowers violet-lilac, lilac-pink, or violet-carmine, tubular, peripheral blue, azure, occasionally white (albinos), tubular-infundibuliform, strongly enlarged, with 4–8-lobed corolla. Stigma biparted to collar of hairs. Achenes 4.0–5.0(5.5) mm long, sparsely hairy, usually with tuft of hairs on edges of concave hilum, smooth; pappus 1.0–3.0(3.5) mm long (in *C. depressa* Bieb. 6–8 mm long), double, inner very slightly or one-third to half as long as outer, of narrower bristles, occasionally (*C. depressa*) of rather wide oblong-spatulate, apically setose or toothed scales. Perennials, annuals, occasionally biennial plants; stem simple or branched in upper part, very rarely almost from base, erect, often assurgent; leaves entirely remotely short-toothed, sometimes (predominantly basal) lobate or lyrate incised, weakly arachnoid-hairy or densely arachnoid-tomentose.

Type: *C. cyanus* L.

Section 1. Protocyanus Dobrocz. 1962, Ukr. Bot. Zhurn. **19,** 1: 43; id. 1949, Bot. Zhurn. Akad. Nauk Ukr. SSR, **6,** 2: 64, 68, nom. nud.

Heads one to four(seven); corolla limb of peripheral flowers more or less regularly tubular-infundibuliform, usually lobed at base, with linear, narrow-lanceolate-linear, occasionally narrow-oblong lobes mostly acuminate, almost of same length; pappus 1–3 mm long. Perennial rhizomatous plants; stems simple or sparingly branched in upper part; cauline leaves mostly decurrent on stem.

Type: *C. montana* L.

4. **C. mollis** Waldst. et Kit. 1806, Descr. Icon. Pl. Rar. Hung. **3:** 243, tab. 219; Czer. 1963, Fl. SSSR, **28:** 393; Dostál, 1976, Fl. Europ. **4:** 298.—*C. montana* L. subsp. *mollis* (Waldst. et Kit.) Gugl. 1907, Centaur. Ung. Nationalmus: 104, 110, p. p.—*Cyanus montanus* (L.) Hill subsp. *mollis* (Waldst. et Kit.) Soják, 1972, Čas. Nár. Muz. Odd. Přir. Praha, **140,** 3–4: 131.

Type: Hungary ("in subalpinis Zoliensis, Liptoviensis et Scepusiensis Comitatus").

West (Carpathians).—In meadows and on stony slopes in subalpine and alpine mountain zones.—*General distribution:* Central Europe (southeast).—2n = 44.

5. **C. marmarosiensis** (Jáv.) Czer. 1960, Bot. Mat. (Leningrad), **20:** 395; Czer. 1963, Fl. SSSR, **28:** 392; Dostál, 1976, Fl. Europ. **4:** 298.—*C. mollis* Waldst. et Kit. f. *marmarosiensis* Jáv. 1925, Magyar Fl. **3:** 1170.—*C. mollis* subsp. *marmarosiensis* (Jáv.) Soó, 1967, Acta Bot. Acad. Sci. Hung. **13,** 3–4: 309.—*Cyanus montanus* (L.)

Hill subsp. *marmarosiensis* (Jáv.) Soják, 1972, Čas. Nár. Muz. Odd. Přir. Praha, **140,** 3–4: 131.

Type: Eastern Carpathians ("Máramaros m.").

West (Carpathians: south).—Open areas in spruce-beech and spruce forests, occasionally in scrublands in subalpine mountain zone.— *General distribution*: Central Europe (Romania, Czechoslovakia).

6. **C. stricta** Waldst. et Kit. 1804, Descr. Icon. Pl. Rar. Hung. **2**: 194, tab. 178; Czer. 1963, Fl. SSSR, **28**: 402.—*C. triumfettii* All. subsp. *stricta* (Waldst. et Kit.) Dostál, 1931, Acta Bot. Bohem. **10**: 72; id. 1976, Fl. Europ. **4**: 299.—*C. ternopoliensis* Dobrocz. 1949, Bot. Zhurn. Akad. Nauk UkrSSR, **6**, 2: 71; Czer. 1963, op. cit.: 401; Dostál, 1976, op. cit.: 300, in nota.—*Cyanus strictus* (Waldst. et Kit.) Soják, 1972, Čas. Nár. Muz. Odd. Přir. Praha, **140,** 3–4: 131.

272 Type: Hungary ("in collibus et montibus vitiferis Comitatus Zempliniensis inde a Szerenes usque Sátor-allya-Ujhely").

West (Carpathians; Dnieper: southwest).—In forest edges, scrublands, dry meadows.—*General distribution*: Central Europe (southeast), Mediterranean (Balkan Peninsula: north).

O 7. **C. tanaitica** Klok. 1948, Nuk. Zap. Kiiv. Univ. **7,** 6: 81, 75; Czer. 1963, Fl. SSSR, **28**: 404, p. p. excl. pl. cauc.; Khmelev and Kunaeva, 1985, Bot. Zhurn. **70,** 10: 1414.—*Cyanus tanaiticus* (Klok.) Soják, 1972, Čas. Nár. Muz. Odd. Přir. Praha, **140,** 3–4: 132.—*Centaurea triumfettii* All. subsp. *tanaitica* (Klok.) Dostál, 1975, Bot. Journ. Linn. Soc. (London), **71,** 3: 208; id. 1976, Fl. Europ. **4:** 299.

Type:. Lugansk Region ("In steppis prope p. Novo-Olexandrivka, distr. Jevsug").

Center (Volga-Don: southeast); *West* (Black Sea: east); *East* (Lower Don).—In steppes, on chalk outcrops.—Endemic.

Note. Caucasian plants identified earlier as *C. tanaitica* should be referred to as *C. czerkessica* Dobrocz. et Kotov.

8. **C. angelscui** Grint. 1924, Bot. Pharmaceutica: 477, tab. 1, fig. 1; Czer. 1963, Fl. SSSR, **28**: 406.—*C. stricta* Waldst. et Kit. subsp. *angelescui* (Grint.), Prod. 1930, Centaur. Roman.: 66.—*C. triumfettii* All. subsp. *angelescui* (Grint.) Dostál, 1975, Bot. Journ. Linn. Soc. (London), **71,** 3: 209; id. 1976, Fl. Europ. **4:** 299.— *Cyanus angelscui* (Grint.) Holub, 1977, Folia Geobot. Phytotax. (Praha), **12:** 307.

Type: Romania.

West (Moldavia: south).—On glades in oak forests.—*General distribution*: Central Europe (Romania: southeast).

9. **C. dominii** (Dostál) Dubovik, 1990, Bot. Zhurn. **75**, 11:. 1579.—*C. triumfettii* All. subsp. *dominii* Dostál, 1931, Acta Bot. Bohem. **10**: 71.

Type: Czechoslovakia ("Slovakia centralis, in rupibus trachytis montis Br. nisko...").

West (Carpathians: south).—On stony slopes in scrublands, on andesite rocks.—*General distribution*: Central Europe (western Carpathians).

○ 10. **C. fuscomarginata** (C. Koch) Juz. 1951, Bot. Mat. (Leningrad), **14**: 41; Czer. 1963, Fl. SSSR, **28**: 408, p. p., excl. pl. cauc.—*C. axillaris* Willd. γ. *fuscomarginata* C. Koch, 1851, Linnaea, **24**: 426.—*C. triumfettii* All. subsp. *cana* (Sibth. et Smith) Dostál, 1975, Bot. Journ. Linn. Soc. (London), **71**, 3: 209, p. p.; id. 1976, Fl. Europ. **4**: 299, p. p.—*C. cana* auct. non Sibth. et Smith: Czer, 1981, Sosud. Rast. SSSR: 52, p. p.; Dobrocz. 1987, Opred. Vyssh. Rast. Ukr.: 359.

Type: Crimea ("Aus der Krym").

Crimea (south).—On stony slopes, taluses, forest edges and glades in subalpine and upper forest zone of mountains.—Endemic.

Note. Plants from Natukhaevka forestry division of Anapa District of Krasnodar Territory are referred to as *C. czerkessia* Dobrocz. et Kotov.

11. **C. thirkei** Sch. Bip. 1847, Linnaea, **19**: 314; Kononov and others, 1966, Bot. Zhurn. **51**, 9: 1309; Geideman, 1975, Opred. Vyssh. Rast. MoldSSR, ed. 2: 512.—*Cyanus thirkei* (Sch. Bip.) Holub, 1973, Preslia, **45**, 2: 145.—*Centaurea napulifera* Rochel subsp. *thirkei* (Sch. Bip.) Dostál, 1975, Bot. Journ. Linn. Soc. (London), **71**, 3: 210; id. 1976, Fl. Europ. **4**: 300.

Type: Asia Minor ("an der Nordküste Kleinasiens und am bithynischen Olymp").

West (Moldavia: south).—On dry slopes.—*General distribution*: Mediterranean (Balkan Peninsula: east), Asia Minor.

Section 2. Cyanus (Mill.) Dumort. 1827, Fl. Belg.: 72.

Heads clustered in paniculate or subcorymbose inflorescence; corolla limb of peripheral flowers obliquely tubular-infundibuliform, lobed one-third to half, lobes oblong-ovate, lanceolate-ovate or lanceolate, subobtuse, acute, less often acuminate, unequal; pappus (3.0)3.5–8.0 mm long. Annual, less often biennial, plants with tap

273 root; stems more or less strongly branched almost from base or middle; cauline leaves not decurrent on stem, sessile or scarcely auriculate.

Type: lectotype of subgenus.

12. **C. depressa** Bieb. 1808, Fl. Taur.-Cauc. **2**: 346; Czer. 1963, Fl. SSSR, **28**: 415; Dostál, 1976, Fl. Europ. **4**: 300.—*Cyanus depressus* (Bieb.) Soják, 1972, Čas. Nár. Muz. Odd. Přír. Praha, **140**, 3–4: 131.

Type: Georgia ("in collibus siccis Iberiae, circa Tiflin").
Crimea.—In fields, crops, weedy places in plains and foothills.—*General distribution*: Caucasus, Russian Central Asia (south); Mediterranean (Balkan Peninsula and Sicily, ecdemic), Asia Minor, Iran, Dzhungaria-Kashgaria (Kuldzha), Tibet (west), India (northwest).—2n = 16.

13. **C. cyanus** L. 1753, Sp. Pl.: 911; Czer. 1963, Fl. SSSR, **28**: 416; Dostál, 1976, Fl. Europ. **4**: 300.

Type: Europe ("inter Europae segetes biennes").
Arctic (Arctic Europe: Murmansk and westward); *North*; *Baltic*; *Center*; *West*; *East*; *Crimea.*—As weed in crop fields, on rubbish dumps, on railroad dumps.—*General distribution*: Caucasus, Western and Eastern Siberia (south), Far East (Kamchatka, Magadan Region and southward, Sakhalin), Russian Central Asia (south and east); Scandinavia, Central and Atlantic Europe, Mediterranean, Asia Minor (west), Iran, India (northwest); North America, Australia, Africa (north); grows in fields in Balkan Peninsula and Sicily.

SUBGENUS **3. *PSEPHELLUS*** (Cass.) Spach 1841, Hist. Nat. Vég. (Phan.), **10**: 11.—*Psephellus* Cass. 1826, Dict. Sci. Nat. **43**: 488

Involucral bracts with membranous, rather large, fimbriate apical appendage, not decurrent on bract; flowers pink, inner tubular, peripheral tubular-infundibuliform, not so large. Stigma short-bilobate. Achenes 3.0–3.5 mm long, scatteredly hairy, smooth; pappus 1.0–1.5 mm long, simple, of scabrous, readily detaching bristles. Perennial plants, with simple procumbent or assurgent stem; leaves pinnately lobed, white-tomentose beneath.

Type: *C. dealbata* Willd.

14. **C. declinata** Bieb. 1819, Fl. Taur.-Cauc. **3**: 590, p. p.; Sosn. 1963, Fl. SSSR, **28**: 430; Dostál, 1976, Fl. Europ. **4**: 297.—*Psephellus*

declinatus (Bieb.) C. Koch, 1851, Linnaea, **24**: 438, p. p.—*Centaurea cineraria* auct. non L. Bieb. 1808, Fl. Taur.-Cauc. **2**: 347.—*C. leucophylla* auct. non Bieb.: Sosn. 1963, op. cit.: 429, p. p., quoad pl. taur.: Dostál, 1976, op. cit.: 297.—(Plate XXXIII, 1).

Type: Crimea and Ciscaucasia ("in Tauriae montibus altioribus et promontorii caucasici collibus editis lapidosis").

Crimea (mountains).—In pine forests, on dry stony slopes.—*General distribution*: Caucasus (Ciscaucasia: west, western Transcaucasia: northwest).

SUBGENUS **4.** *JACEA* (Mill.) Spach

1841, Hist. Nat. Vég. (Phan.), **10**: 11, 67.—*Jacea* Mill. 1754, Gard. Dict. Abrig., ed. **4**: 1

Involucral bracts with membranous, nonprickly, apical appendage not decurrent on bracts and irregularly toothed, fimbriate or cristate-ciliate, obtuse or attenuate into subulate or caudate tip. Flowers pink, pinkish- or lilac-purple, occasionally white (albinos), inner tubular, peripheral tubular-infundibuliform, strongly enlarged, with five- or six-lobed corolla. Stigma bilobed to collar of hair. Achenes 2.7–3.5(4.0) mm long, scatteredly hairy, smooth; pappus 0.5–1.5(2.0) mm long, double, inner half to two-thirds as long as outer, of very slightly thinner bristles or pappus absent 274 (achenes without pappus). Perennial plants, with predominantly erect stems branched mostly in upper part; leaves remotely toothed, entire, occasionally some sinuate-lobate, stiff-hairy or lanate, sometimes mixed with arachnoid pubescence, occasionally glabrous, sharply scabrous only on margin.

Lectotype: *C. jacea* L.

Section 1. Jacea (Mill.) Dumort. 1827, Fl. Belg.: 73.

Appendage of outer and middle involucral bracts somewhat convex, orbicular, flabellately lacerate, irregularly toothed or fimbriate. Achenes without pappus.

Type: lectotype of subgenus.

15. **C. jacea** L. 1753, Sp. Pl.: 914; Czer. 1963, Fl. SSSR, **28**: 444; Dostál, 1976, Fl. Europ. **4**: 291.

Type: Northern Europe ("in Europa septentrionali").

Arctic (Arctic Europe: in the vicinity of Vorkuta and in the north of Murmansk Region, ecdemic); *North*; *Baltic*; *Center*; *West*; *East* (Lower Don: north; Trans-Volga: north); *Crimea* (mountains).—In meadows, forest glades and edges, forest clearings, scrublands, on dry slopes, as ecdemic plant on railroad dumps, near dwellings.—*General distribution*: Caucasus (vicinity of Sukhumi, ecdemic), Eastern Siberia (Krasnoyarsk Territory, ecdemic), Far East (Primorsky Territory, south of Khabarovsk Territory, Kunashir Island, ecdemic); Scandinavia, Central and Atlantic Europe, Mediterranean; North America (ecdemic).—2n = 27, 44.

Note. Often there is a hybrid *C. jacea* L. × *C. phrygia* L. (=*C.* × *livonica* Weinm. 1810, *Bot. Gart. Dorpat.*: 38—*Jacea* × *livonica* (Weinm.) Soják, 1972, *Čas. Nár. Muz. Odd. Přir. Praha*, **140**, 3–4: 132). Occasionally hybridizes with *C. diffusa* Lam. and *C. pseudophrygia* C.A. Mey.

16. **C. pannonica** (Heuff.) Simonk. 1891, Math. Term. Közl. **24**: 620; Czer. 1963, Fl. SSSR, **28**: 446; Dostál, 1976, Fl. Europ. **4**: 290, p. p.—*C. amara* L. β. *pannonica* Heuff. 1858, Verh. Zool.—Bot. Ges. Wien. **8**: 42.—*Jacea pannonica* (Heuff.) Soják, 1972, Čas. Nár. Muz. Odd. Přir. Praha, **140**, 3–4: 132.

Type: Hungary ("in dumetis collium arenosorum legionum Illyrico- et Teutonico-Banaticarum").

West (Carpathians; Dnieper: west; Moldavia).—In forest glades, clearings, scrublands, less often on open slopes.—*General distribution*: Central Europe (southeast), Mediterranean (Balkan Peninsula).—2n = 44.

17. **C. substituta** Czer. 1963, Fl. SSSR, **28**: 612, 448.—*Jacea substituta* (Czer.) Soják, 1972, Čas. Nár. Muz. Odd. Přir. Praha, **140**, 3–4: 133.—*Centaurea pannonica* (Heuff.) Simonk. subsp. *substituta* (Czer.) Dostál, 1975, Bot. Journ. Linn. Soc. (London), **71**, 3: 206; id. 1976, Fl. Europ. **4**: 290.

Type: Crimea ("Hortus Nikitensis, prope semitam").

Center (Volga-Don: Voronezh Region); *West* (Moldavia; Black Sea); *East* (Lower Don); *Crimea* (mountains).—In forest glades and edges, scrublands, meadows and on herbaceous slopes.—*General distribution*: Caucasus (Ciscaucasia: Northwest; western Transcaucasia: north).

Section 2. Lepteranthus (DC.) Dumort. 1827, Fl. Belg.: 73.—*Cyanus* Mill. sect. *Lepteranthus* DC. 1810, Ann. Mus. Hist. Nat. (Paris), **16**: 158.

Appendage of outer and middle involucral bracts flat, elongated, cristate-ciliate, apically attenuate into subulate or caudate tip; achenes with pappus.

Lectotype: *C. phrygia* L.

18. **C. phrygia** L. 1753, Sp. Pl.: 910; Czer. 1963, Fl. SSSR, **28:** 449.—*Jacea phrygia* (L.) Soják, 1972, Čas. Nár. Muz. Odd. Přír. Praha, **140,** 3–4: 132; Dostál, 1958, Klíč: 742, p. p. comb. invalid.—*Centaurea phrygia* subsp. *phrygia* Dostál, 1976, Fl. Europ. **4:** 294.

Lectotype: Finland ("in ... Finlandia").

276   *North* (in the north to 65° N. Lat.); *Baltic*; *Center*; *West* (?Carpathians; Dnieper).—In forest glades, edges and clearings, scrublands, meadows.—*General distribution*: Western Siberia (Sverdlovsk Region; Tyumen Region: west); Scandinavia, Central Europe.—2n = 22, 24.

Note. A hybrid *C. jacea* L. × *C. phrygia* L. is quite common. Occasionally hybridizes with *C. pseudophrygia* C.A. Mey.

19. **C. carpatica** (Porc.) Porc. 1885, Magyar Növ. Lapok, **9:** 128; Czer. 1963, Fl. SSSR, **28:** 451.—*C. plumosa* (Lam.) A. Kerner β. *carpatica* Porc. 1878, Enum. Pl. Phan. Distr. Quond Naszód: 34.—*Jacea carpatica* (Porc.) Soják, 1972, Čas. Nár. Muz. Odd. Přír. Praha, **140,** 3–4: 132.—*Centaurea phrygia* L. subsp. *carpatica* (Porc.) Dostál, 1975, Bot. Journ. Linn. Soc. (London), **71,** 3: 207; id. 1976, Fl. Europ. **4:** 294.

Type: Romania ("Districtus quondam Naszodiensis in Transsylvania").

West (Carpathians).—In forests, meadows in forest and subalpine mountain zones.—*General distribution*: Central Europe (Romania).—2n = 22.

20. **C. melanocalathia** Borb. 1889, Österr. Bot. Zeitschr. **39:** 235; Fodor, 1974, Fl. Zakarp.: 146; Dubovik, 1990, Bot. Zhurn. **75,** 11: 1579. —*C. nigriceps* Dobrocz. 1946, Bot. Zhurn. Akad. Nauk UkrSSR, **3,** 1–2: 31; Czer. 1963, Fl. SSSR, **28:** 452.—*C. phrygia* L. subsp. *nigriceps* (Dobrocz.) Dostál, 1925, Bot. Journ. Linn. Soc. (London), **71,** 3: 207; id. 1976, Fl. Europ. **4:** 294.

Type: Hungary ("... das Zipser Comitat").

West (Carpathians).—In forest glades and edges, meadows, alpine and forest zones of mountains.—*General distribution*: Central Europe.—2n = 24.

Note. Possibly, this species replaces *C. phrygia* L. in the eastern Carpathians.

21. **C. abbreviata** (C. Koch) Hand.-Mazz. 1909, Ann. Naturh. Mus. (Wien), **23:** 198; Czer. 1963, Fl. SSSR, **28:** 453.—*C. salicifolia*

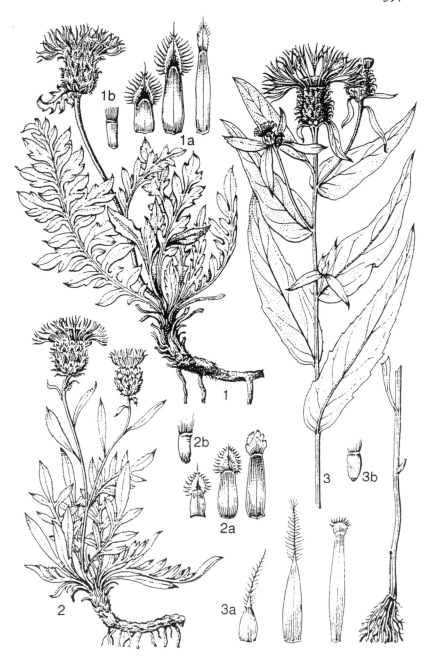

Plate XXXIII.

1—*Centaurea declinata* Bieb., 1a—involucral bracts, 1b—achene; 2—*C. carbonata* Klok., 2a—involucral bracts, 2b—achene; 3—*C. pseudophrygia* C.A. Mey., 3a—involucral bracts, 3b—achene.

Bieb. β. *abbreviata* C. Koch, 1843, Linnaea, **17**: 39.—*Jacea abbreviata* (C. Koch) Soják, 1972, Čas. Nár. Muz. Odd. Přir. Praha, **140**, 3–4: 132.—*Centaurea phrygia* L. subsp. *abbreviata* (C. Koch) Dostál, 1975, Bot. Journ. Linn. Soc. (London), **71**, 3: 287; id. 1976, Fl. Europ. **4**: 294.

Type: Armenia ("In tractu Daratschitschagk").

Crimea (Mt. Malyi Chuchen, terrain feature of Krasnyi Kamen).—In forest glades and edges in upper part of forest zone of mountains.—*General distribution*: Caucasus; Asia Minor (Pontic mountains), Iran (northwest).

22. **C. alutacea** Dobrocz. 1949, Bot. Zhurn. Akad. Nauk UkrSSR, **6**, 2: 74, p. p.; Czer. 1963, Fl. SSSR, **28**: 454; Dobrocz. 1987, Opred. Vyssh. Rast. Ukr.: 361; Dubovik, 1990, Bot. Zhurn. **75**, 11: 1579.

Type: Southern Ossetia ("Ciscaucasia. Prope Vaneli, faux fl. Liachva Major").

Crimea (Mt. Malyi Chuchel, Burulchi River).—In forest glades, scrublands in upper part of forest zone of mountains.—*General distribution*: Caucasus.

23. **C. pseudophrygia** C.A. Mey. 1845, Beitr. Pflanzenk. Russ. Reich. **4**: 82; Czer. 1963, Fl. SSSR, **28**: 456.—*C. phrygia* L. subsp. *pseudophrygia* (C.A. Mey.) Gugl. 1904, Mitt. Bayer. Bot. Ges. **1**: 408; Dostál, 1976, Fl. Europ. **4**: 294.—*Jacea pseudophrygia* (C.A. Mey.) Holub, 1965, Preslia, **37**, 1: 102.—(Plate XXXIII, 3).

Lectotype: Saratov Region ("versus prov. Saratow").

Center (Upper Volga; Volga-Kama; Volga-Don); *West* (Dnieper; Moldavia; Black Sea); *East* (Lower Don: north; Trans-Volga).— In meadows, forest glades and edges, scrublands, thinned-out forests.—*General distribution*: Central Europe (southeast).— 2n = 22.

Note. Hybridizes with *C. jacea* L., occasionally with *C. phrygia* L.

277   24. **C. indurata** Janka, 1858, Flora (Regensb.), **41**: 444; Dostál, 1976, Fl. Europ. **4**: 295; Dubovik, 1990, Bot. Zhurn. **75**, 11: 1579.— *C. phrygia* auct. non L.: Dubovik, 1977, Vizn. Rosl. Ukr. Karpat.: 315.

Type: Czechoslovakia ("in der Mezöség zwischen Szék und Szamos-Ujvar").

West (Carpathians).—In oak forests, on slopes in forest belt of mountains.—*General distribution*: Central Europe.

25. **C. stenolepis** A. Kerner, 1872, Österr. Bot. Zeitschr. **22:**
45; Czer. 1963, Fl. SSSR, **28:** 458; Dostál, 1976, Fl. Europ. **4:** 294,
p. p.—*Jacea stenolepis* (A. Kerner) Soják, 1972, Čas. Nár. Muz.
Odd. Přir. Praha, **140,** 3–4: 133.

Type: Hungary (large number of localities listed).
West (Dnieper: southwest; Moldavia).—On forest edges and
in forest fellings, scrublands, herbaceous transsylvan slopes.—
General distribution: Central Europe (south), Mediterranean (Balkan
Peninsula).—2n = 32.

26. **C. trichocephala** Bieb. 1803, in Willd. Sp. Pl. **3,** 3: 2286;
id. 1808, Fl. Taur.-Cauc. **2:** 344; Czer. 1963, Fl. SSSR, **28:** 462;
Dostál, 1976, Fl. Europ. **4:** 296, p. p.—*Jacea trichocephala* (Bieb.)
Soják, 1972, Čas. Nár. Muz. Odd. Př ir. Praha, **140,** 3–4: 133.

Type: European part of Russia ("ad Wolgam").
Center (Volga-Don: south); *West* (Dnieper: south; Moldavia;
Black Sea); *East.*—In steppes, on steppe slopes, in dry meadows,
scrublands, glades of dry forests.—*General distribution*: Caucasus
(Ciscaucasia: north), Western Siberia (Kuragan Region, Pritobolsky
district, vicinity of village of Pologovo, ecdemic).

SUBGENUS **5.** *HETEROLOPHUS* (Cass.) Spach
1841, Hist. Nat. Vég. (Phan.), **10:** 11.—*Heterolophus* Cass. 1827,
Dict. Sci. Nat. **50:** 250

Involucral bracts with membranous, nonprickly, fimbriate,
irregularly toothed or often entire, apical appendage not decurrent
on bracts. Flowers purple or pinkish-purple, inner tubular, peripheral
tubular-infundibuliform, enlarged. Stigma bilobed. Achenes 3.5–
5.4 mm long, scatteredly hairy, smooth; pappus 1.0–2.5 mm long,
double, inner shorter than outer, of linear-lanceolate smooth
connivent scales. Perennial plants, without distinct main stem,
but with creeping, procumbent or ascending, simple or weakly
branched, stems—fertile arising from axils of basal leaves; basal
leaves pinnately lobed or often undivided, densely pubescent or
tomentose.

Type: *C. sibirica* L.

O 27. **C. carbonata** Klok. 1948, Nauchn. Zap. Kiiv. Univ. **7,**
6: 82, 77; Klok. 1963, Fl. SSSR, **28:** 465; Dostál, 1976, Fl. Europ.
4: 297.—(Plate XXXIII, 2).

Type: Lugansk Region ("prope opp. Bjelovodsk").

Center (Volga-Don: south); *West* (Dnieper: east; Black Sea: east); *East* (Lower Don: north; Trans-Volga: south).—On chalk and limestone outcrops.—Endemic.

Note. Occasionally hybridizes with *C. sumensis* Kalenicz. and in the east of its range, apparently, with *C. sibirica* L.

28. **C. marschalliana** Spreng. 1826, Syst. Veg. **3**: 398; Klok. 1963, Fl. SSSR, **28**: 466; Dostál, 1976, Fl. Europ. **4**: 297.

Type: Ukraine, Caucasus ("Podol. Caucas").

West (Carpathians: south, on the Dniester River; Dnieper: on the Dniester River; Moldavia; Black Sea); *East* (Lower Don: south).—On limestone outcrops, glades in dry oak forests, steppe stony slopes.—*General distribution*: Caucasus (Stavropol District); Central Europe (Romania), Mediterranean (Bulgaria: northeast).— 2n = 18.

○ 29. **C. sumensis** Kalenicz. 1845, Bull. Soc. Nat. Moscou, **18**, 1: 238; Klok. 1963, Fl. SSSR, **28**: 467; Dostál, 1976, Fl. Europ. **4**: 297.

Type: Sumy Region ("... circa oppidum Sumy, quoque prope pagos sumensis Schelez-niak, Popovka et Bezdryk ejusdem districtus").

Center (Upper Volga; Volga-Kama; Volga-Don); *West* (Dnieper; Black Sea: northeast); *East* (Lower Don: north).—In pine forests, steppe and stony slopes.—Endemic.

Note. A hybrid *C. carbonata* Klok. × *C. sumensis* Kalenicz. is found; some of its forms resemble *C. marschalliana* Spreng. in habit.

30. **C. sibirica** L. 1753, Sp. Pl.: 913; Klok. 1963, Fl. SSSR, **28**: 469; Dostál, 1976, Fl. Europ. **4**: 297.

Type: Siberia ("in Sibiria").

Center (Volga-Kama); *East* (Trans-Volga).—On stony and steppe slopes.—*General distribution*: Western Siberia (south).— 2n = 30.

SUBGENUS **6. *ODONTOLOPHUS*** (Cass.) Spach 1841, Hist. Nat. Vég. (Phan.), **10**: 11.—*Odontolophus* Cass. 1827, Dict. Sci. Nat. **50**: 252

Involucral bracts with membranous acuminate apical appendage, decurrent on bracts, occasionally separate, irregularly toothed or cristate-ciliate. Flowers pink, inner tubular, peripheral tubular-infundibuliform, somewhat enlarged. Stigma very shortly bilobate. Achenes 5–6 mm long, predominantly in upper part and near hilum scatteredly short-hairy, smooth; pappus 2.0–2.5 mm long, double inner one-third to half as long as outer, like outer scabrous bristles. Perennial plant, many-stemmed from base, simple or branched near base, with erect branches and entire, narrow (oblong-elliptical to narrow-linear), very weakly pubescent or grayish-tomentose leaves.

Type: *C. trinervia* Steph.

31. **C. trinervia** Steph. 1803, in Willd. Sp. Pl. **3,** 3: 2301; Tzvel. 1963, Fl. SSSR, **28:** 483; Dostál, 1976, Fl. Europ. **4:** 297.

Type: European part of Russia ("in Sibiria").

West (Moldavia; Black Sea); *East* (Lower Don); *Crimea.*—In steppes, scrublands, glades of dry forests, on stony and clayey slopes.—*General distribution*: Caucasus (Ciscaucasia); Central Europe (Romania).—2n = 32.

SUBGENUS 7. LOPHOLOMA (Cass.) Spach
1841, Hist. Nat. Vég. (Phan.), 10: 11, 70.—Lopholoma Cass. 1826,
Dict. Sci. Nat. 44: 37

Involucral bracts with thin-coriaceous or membranous, apical appendage decurrent on bracts (but not more than half or one-third their length), usually as more or less broad, cristate-ciliate border often terminating in spine, cusp or prickle, less often rather large, broadly lanceolate, ovate or orbicular, cristate-ciliate, sometimes appendage almost absent and then involucral bracts terminating in large spine and few cilia. Flowers slightly enlarged, less often distinctly enlarged and then tubular-infundibuliform. Stigma short-bilobate. Achenes 3.5–6.0 mm long, hairy; pappus 3.5–9.0 mm long, as long as achene or longer, occasionally (in *C. stereophylla* Bess.), 1.5–2.5 mm long, one-third to half as long as achene, double, inner one-seventh to two-fifths as long as outer, of scalelike bristles. Perennial plants, with stems branched in upper part, less often simple and erect; leaves pinnatipartite or pinnatilobate, occasionally (in *C. integrifolia* Tausch) undivided or subentire, hairy, sometimes subglabrous.

Type: *C. scabiosa* L.

Section 1. Acrocentron (Cass.) DC. 1838, Prodr. **6:** 586, p. p.—*Acrocentron* Cass. 1826, Dict. Sci. Nat. **44:** 37.

402

Flowers yellow, less often dark red; involucral bracts with apical appendage as very narrow, more or less pectinate-ciliate border, often terminating in cusp or prickle; pappus slightly or almost two times as long as achene.

Type: *C. collina* L.

32. **C. salonitana** Vis. 1829, Flora (Regensb.), **12,** 1, Erg.: 23; Tzvel. 1963, Fl. SSSR, **28:** 498; Dostál, 1976, Fl. Europ. **4:** 264.—*Colymbada salonitana* (Vis.) Holub, 1972, Folia Geobot. Phytotax. (Praha), **7,** 3: 316.

Type: Yugoslavia, Dalmatia ("prope Salonas ...").
West (Moldavia; Black Sea); *East* (Lower Don); *Crimea.*—In steppes, on dry stony slopes, sandstone and limestone outcrops, as weed in abandoned lands, by roadsides.—*General distribution*: Caucasus (Ciscaucasia: west; western Transcaucasia: neighborhood of Novorossiisk); Central Europe (southeast), Mediterranean (Balkan Peninsula).—2n = 20.

33. **C. rubriflora** Illar. 1957, Bot. Mat. (Leningrad), **18:** 309; Tzvel. 1963, Fl. SSSR, **28:** 498.

Type: Crimea ("Karadag, along road from Feodosia to Planerskoe (Koktebel)").
Crimea (south).—On dry stony slopes, by roadsides.—*General distribution*: Mediterranean (Balkan Peninsula).
Section 2. Orientales (Hayek) Tzvel. 1963, Fl. SSSR, **28:** 500.—*Centaurea* sect. *Acrocentron* D. *Orientales* Hayek, 1901, Centaurea-Art. Österr.-Ung.: 53.

Flowers yellow, less often dark red; involucral bracts with large, lanceolate, ovate or orbicular, apical appendage weakly decurrent on bracts and more or less ciliate or fimbriate, often terminating into short cusp or spine; pappus scarcely shorter than achene or almost as long.

Type: *C. orientalis* L.

34. **C. orientalis** L. 1753, Sp. Pl.: 913; Tzvel. 1963, Fl. SSSR, **28:** 507; Dostál, 1976, Fl. Europ. **4:** 265.—*Colymbada orientalis* (L.) Holub, 1972, Folia Geobot. Phytotax. (Praha), **7,** 3: 316.

Type: European part of Russia ("in Sibiria").
Center (Volga-Don: south); *West* (Dnieper; Moldavia; Black Sea); *East* (Lower Don); *Crimea.*—In steppes, on dry stony and sandy slopes.—*General distribution*: Caucasus (Ciscaucasia); Central Europe (southeast), Mediterranean (Balkan Peninsula: north).—2n = 20.

35. **C. kotschyana** Heuff. 1835, Flora (Regensb.), **18,** 1: 245; Tzvel. 1963, Fl. SSSR, **28:** 502; Dostál, 1976, Fl. Europ. **4:** 266.— *Colymbada kotschyana* (Heuff.) Holub, 1972, Folia Geobot. Phytotax. (Praha), **7,** 3: 315.

Type: Romania ("in graminosis alpinis supra ipsas cataractas Bisztrae in alpe Szorko Banatus (ego) et in vicina Transylvaniae alpe Retyezt").

West (Carpathians).—In meadows in upper mountain zone.— *General distribution*: Central Europe (southeast), Mediterranean (Balkan Peninsula: north).—2n = 20, 22.

Section 3. Lopholoma (Cass.) DC. 1838, Prodr. **6:** 580, p. p.

Flowers pink of various shades; involucral bracts with small, (0.5)1.8–4.0(6.0) mm long, obtusely deltoid, apical appendage always strongly decurrent on bracts, more or less cristate-ciliate and usually terminating into short spine or cusp, sometimes without appendages and cilia, only with short spine; pappus as long or almost as long as achene.

Type: *C. scabiosa* L.

280   36. **C. scabiosa** L. 1753, Sp. Pl.: 913; Tzvel. 1963, Fl. SSSR, **28:** 503; Dostál, 1976, Fl. Europ. **4:** 267.—*Colymbada scabiosa* (L.) Holub, 1972, Folia Geobot. Phytotax. (Praha), **7,** 3: 316.

Type: Northern Europe ("in Europae septentrionalis pratis").

North (Karelia-Murmansk: south; Dvina-Pechora); *Baltic*; *Center*; *West* (Carpathians; Dnieper: north; Moldavia); *East* (Trans-Volga).—In meadows, forest glades, scrublands, by roadsides.— *General distribution*: Western and Eastern Siberia, Far East (Primorye: ecdemic); Scandinavia, Central and Atlantic Europe, Mediterranean (Balkan Peninsula: north).—2n = 20.

37. **C. integrifolia** Tausch, 1828, Flora (Regensb.), **11:** 485; Tzvel. 1963, Fl. SSSR, **28:** 504.

Type: Siberia ("in Sibiria").

Center (Volga-Kama: southeast); *East* (Trans-Volga: east).— In meadows, forest glades, birch forests.—*General distribution*: Western Siberia.

38. **C. ossethica** Sosn. ex Tzvel. 1963, Fl. SSSR, **28:** 614, 505; Sosn. 1952, Fl. Gruzii, **8:** 568, nom. invalid.: Dubovik, 1990, Bot. Zhurn. **75,** 11: 1580.—*C. alpestris* auct. non Hegetsch.: Didukh, 1976, Ukr. Bot. Zhurn. **33,** 4: 399; Dobrocz, 1987, Opred. Vyssh. Rast. Ukr.: 364.

Type: Southern Ossetia ("Ossetia ausralis, in fauce Ediss").
Crimea (Ai-Petri yaila).—In meadows of upper mountain zone.—*General distribution*: Caucasus.

39. **C. apiculata** Ledeb. 1824, Index Sem. Hort. Dorpat.: 3; Tzvel. 1963, Fl. SSSR, **28**: 506.—*C. pseudocoriacea* Dobrocz. 1948, Bot. Zhurn. Akad. Nauk UkrSSR, **4**, 3–4: 78; id. 1987, Opred. Vyssh. Rast. Ukr.: 362.—*Colymbada apiculata* (Ledeb.) Holub, 1972, Folia Geobot. Phytotax. (Praha), **7**, 3: 314.—*Centaurea apiculata* subsp. *apiculata* Dostál, 1976, Fl. Europ. **4**: 268.

Type: Crimea ("inter Sympheropolin et Karassu-Basar").
Center (Upper Dnieper: south; Upper Volga: south; Volga-Kama: east; Volga-Don); *West*; *East*; *Crimea*.—In steppes, meadows, on dry slopes, scrublands and abandoned fields.—*General distribution*: Caucasus (Ciscaucasia, Dagestan), Western Siberia; Central Europe, Mediterranean.—2n = 20.

40. **C. adpressa** Ledeb. 1824, Index Sem. Hort. Dorpat.: 3; Tzvel. 1963, Fl. SSSR, **28**: 507.—*Colymbada adpressa* (Ledeb.) Holub, 1974, Preslia, **46**, 3: 228.—*Centaurea apiculata* Ledeb. subsp. *adpressa* (Ledeb.) Dostál, 1975, Bot. Journ. Linn. Soc. (London), **71**, 3: 196; id. 1976, Fl. Europ. **4**: 268.

Type: Moldavia ("in deserto Bessarabiae").
Center (Volga-Don: south and east); *West* (Dnieper: southeast; Moldavia; Black Sea); *East*; *Crimea*.—In steppes, on dry, stony slopes, meadows, sands and old fields.—*General distribution*: Caucasus (Ciscaucasia, Dagestan, eastern Transcaucasia: northeast), Western Siberia (south), Russian Central Asia (north and east); Central Europe (Romania: east), Mediterranean (Balkan Peninsula), Dzhungaria-Kashgaria.—2n = 16.

Section 4. Stereophyllae (Tzvel.) Tzvel. 1963, Fl. SSSR, **28**: 511.—*Centaurea* sect. *Acrocentron* subsect. *Stereophyllae* Tzvel. 1959, Bot. Mat. (Leningrad), **19**: 437.

Flowers pale pink, less often pale yellow; involucral bracts with broadly lanceolate apical appendage weakly decurrent on bracts and short-ciliate, usually terminating in short spine; pappus one-third to half as long as achene.

Type: *C. stereophylla* Bess.

41. **C. stereophylla** Bess. 1822, Enum. Pl. Volhyn.: 35; Tzvel. 1963, Fl. SSSR, **28**: 511; Dostál, 1976, Fl. Europ. **4**: 268.—*Colymbada stereophylla* (Bess.) Holub, 1973, Preslia, **45**, 2: 144.

Type: Ukraine ("in Podolia australi").

West (Dnieper: south; Moldavia; Black Sea).—In steppes, old fields, occasionally as weed in crops.—*General distribution*: Central Europe (southeast), Mediterranean (Balkan Peninsula).

281   SUBGENUS **8. *ACROLOPHUS*** (Cass.) Spach
1841, Hist. Nat. Vég. (Phan.), **10**: 11.—*Acrolophus* Cass. 1827, Dict. Sci. Nat. **50**: 253

Outer and middle involucral bracts with scaly or semicoriaceous apical appendage decurrent on bracts, cristate-fimbriate, less often finely toothed, deltoid, oblong-ovate, oval or orbicular, 1.2–3.5(6.0) mm long, terminating in soft or prickly cusp 0.1–1.5 (occasionally to 4) mm long, with or without scaly auricles; inner involucral bracts with short-fimbriate, irregularly toothed or subentire, shorter and narrower appendage. Flowers pink, less often purple. Achenes 2.5–3.5(4.0) mm long, hairy or glabrous; pappus 0.1–3.0(4.2) mm long, as long as achene, often considerably shorter, double, sometimes absent. Biennial plants, usually with strongly branched, erect or basally assurgent, non-winged stem; basal leaves simple or twice pinnate, arachnoid-hairy, arachnoid-tomentose, arachnoid-lanate or scabrous, petiolate; lower cauline leaves decurrent, other cauline leaves sessile.

Type: *C. maculosa* Lam.

42. **C. rhenana** Boreau, 1857, Fl. Centr. Fr., ed. 3, 1: 355; Klok. 1963, Fl. SSSR, **28**: 517.—*Acosta rhenana* (Boreau) Soják, 1972, Čas. Nár. Muz. Odd. Přir. Praha, **140**, 3–4: 134.—*Centaurea rhenana* subsp. *rhenana*; Dostál, 1976, Fl. Europ. **4**: 278.

Type: France ("La plante de la vallae du Rhin").

Baltic (Estonia: ecdemic; Latvia: ecdemic; Lithuania: predominantly southern half; Kaliningrad Region); *Center* (Upper Dnieper: Belorussian Polesie, western districts of Bryansk Region); *West* (Carpathians: south; Dnieper: west; Moldavia: northern half).—In steppes, on stony, chalk and limestone outcrops, forest edges, glades, scrublands, as weed on railroad dumps, on berms of roads, sand-gravel pits.—*General distribution*: Central Europe (up to the Rhine River in the west and the Dunai in the south).—$2n = 18$.

43. **C. pseudomaculosa** Dobrocz. 1949, Bot. Zhurn. Akad. Nauk UkrSSR, **6**, 2: 73; Klok. 1963, Fl. SSSR, **28**: 518.—*Acosta pseudomaculosa* (Dobrocz.) Soják, 1972, Čas. Nár. Muz. Odd.

Přir. Praha, **140**, 3–4: 134.—*Centaurea rhenana* Boreau subsp. *pseudomaculosa* (Dobrocz.) Dostál, 1975, Bot. Journ. Linn. Soc. (London), **7,** 3: 200; id. 1976, Fl. Europ. **4:** 278.

Type: Donetzk Region ("distr. Mekejevskiensis. Sovchoz Shachtar").

North (Dvina-Pechora: Syktyvkar Station, ecdemic); *Center* (Ladoga-Ilmen: St. Petersburg, Pskov Station, near the village of Pushkinskie Gory, ecdemic; Upper Dnieper: southern half of Belorussia, Bryansk Region; Upper Volga: Moscow and Kaluga regions; Volga-Kama: south; Volga-Don); *West* (Dnieper: east; Moldavia; Black Sea: north); *East* (Lower Don: north; Trans-Volga).—In steppes, on steppe slopes, stony limestone outcrops, in forest glades and edges, sometimes as weed in fields, pastures, by roadsides.—*General distribution*: Western Siberia (west).

Note. This species was mistakenly reported for Crimea. A hybrid has been described: *C. pseudomaculosa* Dobrocz. × *C. substituta* Czer. (=*C.* × *klokovii* Tzvel. 1985, *Novosti Sist. Vyssh. Rast.* **22:** 275).

44. C. biebersteinii DC. 1838, Prodr. **6:** 583.—*C. micranthos* S.G. Gmel. ex Hayek, 1901, Centaurea-Art. Österr.-Ung.: 92, non *C. micrantha* Duf. 1831; S.G. Gmel. 1770, Reise, **1:** 135, nom. invalid.; Klok. 1963, Fl. SSSR, **28:** 519.—*Acosta micanthos* (S.G. Gmel. ex Hayek) Soják, 1972, Čas. Nár. Muz. Odd. Přir. Praha, **140,** 3–4: 134.—*Centaurea biebersteinii* subsp. *bibersteinii*; Dostál, 1976, Fl. Europ. **4:** 279.—*Acosta bibersteinii* (DC.) Dostál, 1984, Folia Muz. Rer. Nat. Bohem. occid., Bot. **21:** 13.

Type: Caucasus, Moldavia ("in campestribus ad Caucasum et in Bessarabia").

Center (Upper Dnieper: Bryansk Region, ecdemic; Upper Volga: Moscow Region, Serpukhov, ecdemic; Volga-Don); *West* (Dnieper: south; Moldavia; Black Sea); *East* (Lower Don; Trans-Volga); *Crimea*.—In dry steppes, on outcrops of chalk and clay, rubbly slopes, forest glades and edges.—*General distribution*: Caucasus (Ciscaucasia), Western Siberia (south); Central Europe (southeast), Mediterranean (Balkan Peninsula: north).—2n = 18.

45. C. majorovii Dumb. 1946, Dokl. Akad. Nauk ArmSSR, **5,** 2: 48; Klok. 1963, Fl. SSSR, **28:** 520.—*C. arenaria* Bieb. subsp. *majorovii* (Dumb.) Dostál, 1975, Bot. Journ. Linn. Soc. (London), **71,** 3: 197; id. 1976, Fl. Europ. **4:** 273.

Type: Dagestan ("Kumtorkali").

Center (Upper Dnieper: southwest, Bryansk Region, ecdemic; upper Volga: vicinity of Moscow, Bologoe station, ecdemic; Volga-Kama: ecdemic; Volga-Don); *West* (Dnieper: south; Black Sea: northeast); *East* (Lower Don; Trans-Volga; Lower Volga).—On river, less often marine dry sands.—*General distribution*: Caucasus (Dagestan, eastern Transcaucasia.—2n = 36.

O 46. **C. sophiae** Klok. 1963, Fl. SSSR, **28**: 615, 521.—*C. arenaria* Bieb. subsp. *sophiae* (Klok.) Dostál, 1975, Bot. Journ. Linn. Soc. (London), **71**, 3: 197; id. 1976, Fl. Europ. **4**: 273.

Type: Rostov Region ("regio Salakiensis, 3–4 km SW versus a st. Tzymljamskaja").

Center (Volga-Don: east of Voronezh Region); *East* (Lower Don).—On river sands.—Endemic.

47. **C. arenaria** Bieb. 1803, in Willd. Sp. Pl., **3**, 3: 2278, p. p.; id. 1808, Fl. Taur.-Cauc. **2**: 347; Klok. 1963, Fl. SSSR, **28**: 522, p. p.; Dubovik, 1990, Bot. Zhurn. **75**, 11: 1575.—*Acosta arenaria* (Bieb.) Soják, 1972, Čas. Nár. Muz. Odd. Přir. Praha, **140**, 3–4: 133.

Type: Lower reaches of the Kuma River ("Ex deserto Cumono").

Baltic (Latvia: city of Riga, ecdemic); *East* (Lower Don: area from Temryuk to Primorsko-Akhtarsk); *Crimea* (south).—On marine and river sands.—*General distribution*: Caucasus (Ciscaucasia: Taman Peninsula, vicinity of Anapa, basin of the Kuma River).

48. **C. wolgensis** DC. 1838, Prodr. **6**: 581; Dubovik, 1990, Bot. Zhurn. **75**, 11: 1576.—*C. arenaria* Bieb. subsp. *arenaria*: Dostál, 1976, Fl. Europ. **4**: 273.—*C. arenaria* auct. non Bieb.: Klok. 1963, Fl. SSSR, **28**: 522, p. p.

Type: Lower reaches of the Volga River ("ad Wolgam").

East (Lower Volga).—On marine and river sands, gravel-beds facing offshore dunes.—*General distribution*: Caucasus (Dagestan), Russian Central Asia (Aralo-Caspian district).

O 49. **C. odessana** Prod. 1928, Contr. Bot. Cluj, **1**, 17: 10; Klok. 1963, Fl. SSSR, **28**: 523.—*Acosta odessana* (Prod.) Soják, 1972, Čas. Nár. Muz. Odd. Přir. Praha, **140**, 3–4: 134.—*Centaurea arenaria* Bieb. subsp. *odessana* (Prod.) Dostál, 1975, Bot. Journ. Linn. Soc. (London), **71**, 3: 197; id. 1976, Fl. Europ. **4**: 273.

Type: Odessa Region ("in campestribus circa Odessam frequens").

West (Black Sea); *Crimea* (Arbat Spit).—On marine sands.—Endemic.

○ 50. **C. borysthenica** Grun. 1868, Bull. Soc. Nat. Moscou, **41**, 2, 4: 426; Klok. 1963, Fl. SSSR, **28**: 523.—*Acosta borysthenica* (Grun.) Soják, 1972, Čas. Nár. Muz. Odd. Přír. Praha, **140**, 3–4: 133.—*Centaurea arenaria* Bieb. subsp. *borysthenica* (Grun.) Dostál, 1975, Bot. Journ. Linn. Soc. (London), **71**, 3: 197; id. 1976, Fl. Europ. **4**: 273; O. Baran. and others, 1992, Konsp. Fl. Udmurt.: 98.

Type: Zaporozhye Region ("Inter frutices, locis arenosis ad Borysthenem fluvium frequens").

Center (Volga-Kama: ecdemic in vicinity of Izevsk); *West* (Dnieper; Black Sea: west).—On marine sands.—Endemic.

○ 51. **C. savranica** Klok. 1948, Nauk. Zap. Kiiv. Univ. **7**, 6: 80, 67; Klok. 1963, Fl. SSSR, **28**: 524.—*Acosta savranica* (Klok.) Holub, 1977, Folia Geobot. Phytotax. (Praha), **12**, 3: 304.—*Centaurea rhenana* Boreau subsp. *savranica* (Klok.) Dostál, 1975, Bot. Journ. Linn. Soc. (London), **71**, 3: 200; id. 1976, Fl. Europ. **4**: 278.—*C. arenaria* auct. non Bieb.: Geideman, 1975, Opred. Vyssh. Rast. MoldSSR, ed. 2: 514.

283   Type: Odessa Region ("In arenosis ad fl. Sarvanj, prope pag. Puzhajkove").

West (Moldavia; Black Sea: Odessa Region).—On river sands and sandy steppe slopes.—Endemic.

○ 52. **C. besseriana** DC. 1838, Prodr. **6**: 585; Klok. 1963, Fl. SSSR, **28**: 526.—*Acosta besseriana* (DC.) Soják, 1972, Čas. Nár. Muz. Odd. Přír. Praha, **140**, 3–4: 133.—*Centaurea ovina* Pall. ex Willd. subsp. *besseriana* (DC.) Dostál, 1975, Bot. Journ. Linn. Soc. (London), **73**, 3: 197; id. 1976, Fl. Europ. **4**: 273.—*C. steveniana* auct. non Klok.: Geideman, 1975, Opred. Vyssh. Rast. MoldSSR, ed. 2: 514.

Type: Moldavia, Podolia ("in Bessarabia et Podolia australiori").

West (Dnieper: southwest; Moldavia; Black Sea: west).—On limestone and rubbly slopes, outcrops of granite, sometimes in forest glades.—Endemic.

○ 53. **C. lavrenkoana** Klok. 1948, Nauk. Zap. Kiiv. Univ. **7**, 6: 81, 73; id. 1963, Fl. SSSR, **28**: 526.—?*C. pseudoovina* Illar. 1957, Bot. Mat. (Leningrad), **18**: 308; Illar. 1969, in E. Wulf, Fl. Kryma, **3**, 3: 278; Dubovik, 1990. Bot. Zhurn. **75**, 11: 1575.—*C. ovina* Pall. ex Willd. subsp. *lavrenkoana* (Klok.) Dostál, 1975, Bot. Journ. Linn. Soc. (London), **71**, 3: 197; id. 1976, Fl. Europ. **4**: 273.

Type: Ukraine ("Swjatogorsk ('Montes Sanctae' ad fl. Donetz). In cretaceis").

West (Black Sea: east); *East* (Lower Don: west); *Crimea* (predominantly Tarkhankut Peninsula).—On chalk, limestone and shale outcrops.—Endemic.

Note. Differentiation of *C. pseudoovina* Illar. from *C. lavrenkoana* Klok. is more quantitative, which does not allow the former to be considered as a separate subspecies.

○ 54. **C. stevenIana** Klok. 1963, Fl. SSSR, **28:** 616, 527.— *Acosta steveniana* (Klok.) Holub, 1974, Preslia, **46,** 3: 227.— *Centaurea ovina* Pall. ex Willd. subsp. *steveniana* (Klok.) Dostál, 1975, Bot. Journ. Linn. Soc. (London), **71,** 3: 197, p. p.; id. 1976, Fl. Europ. **4:** 273, p. p.

Type: Crimea ("Eupatoria, in arenis litoralibus ad Pontum Euxinum").

Crimea (west, vicinity of the city of Evpatoria).—On coastal shell-sands.—Endemic.

55. **C. caprina** Stev. 1856, Bull. Soc. Nat. Moscou, **29,** 2, 4: 394; Illar. 1969, in E. Wulf, Fl. Kryma, **3,** 3: 279; Dubovik, 1990, Bot. Zhurn. **75,** 11: 1574.—*C. koktebelica* Klok. 1963, Fl. SSSR, **28:** 617, 528.—*C. ovina* Pall. ex Willd. subsp. *koktebelica* (Klok.) Dostál, 1975, Bot. Journ. Linn. Soc. (London), **75,** 3: 198; id. 1976, Fl. Europ. **4:** 273.

Type: Crimea ("... in valle Sudak").

Crimea (mountains).—On stony and rubbly slopes, taluses, in warmwood steppes.—Endemic?

Note. Hybridizes with *C. sterilis* Stev. (=*C.* × *iljiniana* Illar. 1957, *Bot. Mat. (Leningrad),* **18:** 307), *C. substituta* Czer. (=*C. comperiana* Stev. 1856, *Bull. Soc. Nat. Moscou,* **29,** 1, 2: 219), and *C. vankovii* Klok. (=*C.* × *ninae* Juz. 1951, *Bot. Mat. (Leningrad),* **14:** 42).

56. **C. diffusa** Lam. 1783, Encycl. Méth. Bot. **1:** 675; Klok. 1963, Fl. SSSR, **28:** 532; Dostál, 1976, Fl. Europ. **4:** 282.—*Acosta diffusa* (Lam.) Soják, 1972, Čas. Nár. Muz. Odd. Přir. Praha, **140,** 3–4: 133.

Type: Western Europe ("Cette plante croit le Levant").

North (Karelia-Murmansk: Belomorsk; Dvina-Pechora: Izuya southwest of Kosyu station, ecdemic throughout; *Baltic* (Estonia, vicinity of Pius; Latvia: city of Riga; Lithuania: Kaliningrad Region, vicinity of Kaliningrad, ecdemic throughout); *Center* (Ladoga-Ilmen: Leningrad Region—Ornienbaum station, city of Gatchina, Sortirovochnaya station in the vicinity of St. Petersburg, Pskov

Region—Nevel station, city of Velikie Luki, Novgorod Region—
vicinity of Okulovka station, ecdemic throughout; Upper Dnieper:
city of Minsk, Radiatornyi station, Gomel and Bryansk regions,
throughout ecdemic; Upper Volga: Tver Region—city of Torzhok,
city of Bezhetsk, Lazurnaya, Redkino and Doronikha stations,
Moscow Region—predominantly in vicinity of Moscow, throughout
ecdemic; Volga-Kama: Kostroma Region—vicinity of Kostroma:
Kirov Region—city of Vyatskie Polyany, Perm Region—Kungur
station, Udmurt ASSR—often throughout ecdemic; Volga-Don:
south); *West*; *East* (Lower Don); *Crimea*.—On dry stony and
284 clayey slopes, in steppes, forest glades and edges, weedy places,
pastures, on edges of fields and by roadsides, railroad dumps.—
General distribution: Caucasus, Eastern Siberia (Buryat ASSR—
Vydrino station, ecdemic); Mediterranean (Balkan Peninsula), Asia
Minor. As ecdemic plant in Scandinavia and many regions of
central and southern Europe.—2n = 18.

Note. Forms hybrids with *C. margaritalba* Klok. (=*C.* ×
hypanica Pacz.), *C. pseudomaculosa* Dobrocz. (=*C.* × *dobroczaevae*
Tzvel. 1985, *Novosti Sist. Vyssh. Rast.* **22**: 275), and *C. sterilis* Stev.
(=*C.* × *longiaristata* Illar. 1957, *Bot. Mat. (Leningrad)*, **18**: 306).

57. **C. aemulans** Klok. 1963, Fl. SSSR, **28**: 617, 533; Dostál,
1976, Fl. Europ. **4**: 282.—*Acosta aemulans* (Klok.) Holub, 1974,
Preslia, **46**, 3: 226.

Type: Eastern Crimea ("... regio Sudakiensis, prope pag.
Planerskoje").

West (Black Sea: Melitopol, Kuyuktuk Island); *Crimea* (east).—
On dry stony slopes, in wormwood steppes, in damp places.—
General distribution: Caucasus (Ciscaucasia: Taman Peninsula,
city of Nevinnomyssk).

SUBGENUS **9. *PHALOLEPIS*** (Cass.) Spach
1841, Hist. Nat. Vég. (Phan.), **10**, 11.—*Phalolepis* Cass. 1827,
Dict. Sci. Nat. **50**: 248

Involucral bracts with almost entirely membranous or coriaceous
in central and scarious in peripheral part, large, convex apical
appendage not decurrent on bracts, orbicular, elliptical, or broadly
ovate, irregularly toothed, scarcely toothed or undivided, dorsally
carinate or not, with or without short, soft cusp 0.4–3.0 mm long.
Flowers pink, purple, light yellow, less often white, all tubular,
often peripheral tubular-infundibuliform, somewhat enlarged.

Achenes 2–5 mm long, scatteredly hairy, sometimes mature achenes subglabrous; pappus 0.4–5.5(7.0) mm long, almost as long as achene, slightly longer, or considerably shorter, double. Biennial, less often perennial plants, with mostly strongly branched, erect, occasionally assurgent stems and once or twice pinnately incised (excluding upper cauline) arachnoid-hairy, decurrent (cauline) leaves.

Type: *C. alba* L.

Section 1. Pseudophalolepis Klok. 1963, Fl. SSSR, **28**: 618, 541.

Appendage of involucral bracts almost entirely membranous or scarious, with small deltoid thick tissue at base, less often with narrow strip in middle, without distinct dark spots, less often with spots; thicker part of appendages not clearly demarcated from scarious part; pappus almost as long as achene, less often slightly longer or distinctly shorter. Exclusively biennial plants; stems more or less strongly light yellow.

Type: *C. gerberi* Stev.

O 58. **C. protomargaritacea** Klok. 1936, Tr. Nauk. Issl. Inst. Bot. Khirkiv. Derzh. Univ. **1**: 102, 8̕1; id. 1963, Fl. SSSR, **28**: 543.—*C. margaritacea* Ten. subsp. *protomargaritacea* (Klok.) Dostál, 1975, Bot. Journ. Linn. Soc. (London), **73**, 3: 204; id. 1976, Fl. Europ. **4**: 286.

Type: Nikolaev Region ("In arenosis inferioribus fl. Bug. meridiem ab opp. Nikolajev versus").

West (Black Sea: left bank of Bug estuary, between Ozharskaya and Krivaya spits, below Nikolaev").—On sands.—Endemic.

O 59. **C. margaritacea** Ten. 1830, Syll. Neap. (App. 3): 628; Klok. 1963, Fl. SSSR, **28**: 543.—*Jacea margaritacea* (Ten.) Soják, 1972, Čas. Nár. Muz. Odd. Přir. Praha, **140**, 3–4: 132.— *Centaurea margaritacea* subsp. *margaritacea* Dostál, 1976, Fl. Europ. **4**: 286.

285   Type: Nikolaev Region (evidently, described from issued specimens of Lang and Szowitz from vicinity of Nikolaev).

West (Black Sea: right bank of the South Bug River above Nikolaev).—On sands.—Endemic.

O 60. **C. margaritalba** Klok. 1936, Tr. Nauchn.-Ills. Inst. Bot. Kharkiv. Derzh. Univ. **1**: 100, 81; id. 1963, Fl. SSSR, **28**: 544.—*C. margaritacea* Ten. subsp. *margaritalba* (Klok.) Dostál, 1975, Bot. Journ. Linn. Soc. (London), **71**, 3: 204; id. 1976, Fl. Europ. **4**: 286.

Type: Nikolaev Region ("Nikolajev. In arenosis prope arboretum urbicum").

West (Black Sea: left bank of the Bug estuary in vicinity of Nikolaev).—On sands.—Endemic.

○ 61. **C. konkae** Klok. 1936, Tr. Nauk.-Issl. Inst. Bot. Kharkiv. Derzh. Univ. 1: 100, 93; id. 1963, Fl. SSSR, 28: 545.—*C. margaritacea* Ten. subsp. *konkae* (Klok.) Dostál, 1975, Bot. Journ. Linn. Soc. (London), 71, 3: 204; id. 1976, Fl. Europ. 4: 286.

Type: Zaporozhe Region ("Zaporozhje. In arena "Velyki Kuczuhury" ad fl. Konka, confl. Dnjepr").

West (Black Sea: left bank of the Dnieper River, Konka River).—On sands.—Endemic.

○ 62. **C. appendicata** Klok. 1936, Tr. Nauk.-Issl. Inst. Bot. Kharkiv. Derzh. Univ. 1: 101, 94; id. 1963, Fl. SSSR, 28: 545.—*C. margaritacea* Ten. subsp. *appendicata* (Klok.) Dostál, 1975, Bot. Journ. Linn. Soc. (London), 71, 3: 204; id. 1976, Fl. Europ. 4: 286.

Type: Zaporozhye Region ("Distr. Zaporozhje. "Lysa Hora" (meridiem a p. Bilenjke versus").

West (Black Sea: right bank of the Dnieper River below Zaporozhye).—On sands.—Endemic.

○ 63. **C. protogerberi** Klok. 1936, Tr. Nauk.-Issl. Inst. Bot. Kharkiv. Derzh. Univ. 1: 102, 89; id. 1963, Fl. SSSR, 28: 546.—*C. margaritacea* Ten. subsp. *protogerberi* (Klok.) Dostál, 1975, Bot. Journ. Linn. Soc. (London), 71, 3: 204; id. 1976, Fl. Europ. 4: 286.

Type: Lugansk Region ("Prope p. Petropavlovka in distr. Lugansk. In arenosis ad fl. Donetz").

West (Dnieper: Lugansk Region, Krasnodons District, left bank of the Seversky Donetz River, below the town of Krasnyi Yar); *East* (Lower Don: left bank of the Seversky Donetz River in lower reaches).—On open sands.—Endemic.

64. **C. donetzica** Klok. 1936, Tr. Nauk.-Issl. Inst. Bot. Kharkiv. Derzh. Univ. 1: 101, 86; id. 1963, Fl. SSSR, 28: 547.—*C. margaritacea* Ten. subsp. *donetzica* (Klok.) Dostál, 1975, Bot. Journ. Linn. Soc. (London), 71, 3: 204; id. 1976, Fl. Europ. 4: 286.

Type: Lugansk Region ("In arenosis ad pinetum magnum prope p. Kremennaja").

West (Dnieper: Seversky Donetz River from the mouth of the Oskol River to Lisichansk District; Black Sea: Donetz Region, Seversky Donetz River).—In pine forests.—Endemic.

○ 65. **C. pineticola** Iljin, 1927, Izv. Glavn. Bot. Sada Akad. Nauk SSSR, **26**, 1: 34; Klok. 1963, Fl. SSSR, **28**: 548. —*C. margaritacea* Ten. subsp. *pineticola* (Iljin) Dostál, 1975, Bot. Journ. Linn. Soc. (London), **71**, 3: 204; id. 1976, Fl. Europ. **4**: 286.

Type: Voronezh Region ("Khrenovo pine forest").
Center (Volga-Don: Voronezh Region, Khrenovo pine forest).— On pine forest sands.—Endemic.

○ 66. **C. breviceps** Iljin, 1927, Izv. Glavn. Bot. Sada Akad. Nauk SSSR, **26**, 1: 35; Klok. 1963, Fl. SSSR, **28**: 548.—*C. margaritacea* Ten. subsp. *breviceps* (Iljin) Dostál, 1975, Bot. Journ. Linn. Soc. (London), **71**, 3: 204; id. 1976, Fl. Europ. **4**: 286.

Type: Kherson Region ("on sands in lower reaches of the Dnieper River in the vicinity of Aleshky (now Tsyurupinsk)").
286   *West* (Black Sea: left bank of the Dnieper River in lower reaches and Dnieper estuary).—On open sands, in sandy steppes.— Endemic.

○ 67. **C. paczoskii** Kotov ex Klok. 1936, Tr. Nauk.-Issl. Inst. Bot. Kharkiv. Derzh. Univ. **1**: 101, 92; Klok. 1963, Fl. SSSR, **28**: 549.—*C. margaritacea* Ten. subsp. *paczoskii* (Kotov ex Klok.) Dostál, 1975, Bot. Journ. Linn. Soc. (London), **71**, 3: 204; id. 1976, Fl. Europ. **4**: 286.

Type: Kherson Region ("In arenosis ripae sinistrae fl. inhuletz prope p. Novogrednivka").
West (Black Sea: Inguletz River—right tributary of the Dnieper River).—On open river sands.—Endemic.

○ 68. **C. gerberi** Stev. 1856, Bull. Soc. Nat. Moscou, **29**, 2, 4: 391; Klok. 1963, Fl. SSSR, **28**: 550.—*C. margaritacea* Ten. subsp. *gerberi* (Stev.) Dostál, 1975, Bot. Journ. Linn. Soc. (London), **71**, 3: 204; id. 1976, Fl. Europ. **4**: 286.

Type: Volga River ("Saratov").
Center (Volga-Kama: south); *East.*—On open hummocky sands, in sandy steppes.—Endemic.

○ 69. **C. dubjanskyi** Iljin, 1927, Izv. Glavn. Bot. Sada Akad. Nauk SSSR, **26**, 1: 36; Klok. 1963, Fl. SSSR, **28**: 550.—*C. margaritacea* Ten. subsp. *dubjanskyi* (Iljin) Dostál, 1975, Bot. Journ. Linn. Soc. (London), **71**, 3: 204; id. 1976, Fl. Europ. **4**: 286.

Type: Voronezh Region ("Boguchavsk District, the town of Bereznyagi").

Center (Volga-Don: Voronezh Region).—On hummocky sands.—Endemic.

Section 2. Phalolepis (Cass.) DC. 1838, Prodr. **6:** 568, p. p.

Appendage of involucral bracts sharply differentiated into dense, absolutely opaque, coriaceous central part with five veins passing through it and thin-scarious, transparent, peripheral part; pappus always considerably shorter (usually one-third to half as long) than achene. Perennial, less often biennial, plant; stems more or less branched; corolla purple, light purple, occasionally dull pink.

Type: *C. alba* L.

70. **C. pseudoleucolepis** Kleop. 1926, Izv. Kiev. Bot. Sada, **4:** 20; Klok. 1963, Fl. SSSR, **28:** 542.—*C. margaritacea* Ten. subsp. *pseudoleucolepis* (Kleop.) Dostál, 1975, Bot. Journ. Linn. Soc. (London), **71,** 3: 204; id. 1976, Fl. Europ. **4:** 286.

Type: Donetzk Region ("District Mariupol, in rupibus graniticis "Kamjani Mogyly" ").

West (Black Sea: Donetz Region, Volodarskoe District, "Kamennye Mogily" Preserve).—On granite outcrops.—Endemic.

O 71. **C. semjusta** Juz. 1951, Bot. Mat. (Leningrad), **14:** 43; Klok. 1963, Fl. SSSR, **28:** 551.—*C. sterilis* Stev. subsp. *semijusta* (Juz.) Dostál, 1975, Bot. Journ. Linn. Soc. (London), **71,** 3: 204; id. 1976, Fl. Europ. **4:** 287.

Type: Crimea ("Mt. Chatyrdag, rocky slope and Eklizburun peak").

Crimea (Mt. Chatyrdag, yailas of Babugan, Nikitskaya, Demerdzhi).—On rocky slopes and talus in upper mountain zone.—Endemic.

72. **C. vankovii** Klok. 1963, Fl. SSSR, **28:** 618, 552.—*C. nikitensis* Illar. 1957, Bot. Mat. (Leningrad), **18:** 304, p. p. excl. typo.—*C. sterillis* Stev. subsp. *vankovii* (Klok.) Dostál, 1975, Bot. Journ. Linn. Soc. (London), **71,** 3: 204; id. 1976, Fl. Europ. **4:** 287.

Type: Crimea ("Jajla ajpetrica").

Crimea (yaila of Ai-Petri, Mt. Kopka below Simeiz).—On stony slopes, in upper mountain zone.—Endemic.

O 73. **C. sterilis** Stev. 1856, Bull. Soc. Nat. Moscou, **29,** 2, 4: 390; Klok. 1963, Fl. SSSR, **28:** 553.—*C. stankovii* Illar. 1957, Bot. Mat. (Leningrad), **18:** 305.—*C. nikitensis* Illar. 1957, ibid.: 304, p. max. p. incl. typo.—*C. sterilis* subsp. *sterilis* Dostál, 1976, Fl. Europ. **4:** 287.

287   Type: Crimea ("in cretaceis circa Karassubasar").

Crimea (mountains).—On dry stony slopes, in pine forests from seacoast to middle mountain zone.—Endemic.

○ 74. **C. sarandinakiae** Illar. 1957, Bot. Mat. (Leningrad), **18:** 303; Klok. 1963, Fl. SSSR, **28:** 555.—*C. transcaucasica* auct. non Grossh.: Dostál, 1976, Fl. Europ. **4:** 286.

Type: Crimea ("Karadag, northwestern slope and peak of Mt. Karadag").

Crimea (southern coast from the village of Privetnoe to the village of Planerskoe and Mt. Acharmysh in the vicinity of Staryi Krym).—On stony slopes and taluses near the coast.—*General distribution*: Caucasus (Ciscaucasia, Novorossiisk District).

SUBGENUS **10. *SOLSTITIARIA*** (Hill) Dobrocz. 1949, Bot. Zhurn. Akad. Nauk UkrSSR, **6,** 2: 64, 69.—*Solstitiaria* Hill, 1762, Syst. Veg. **4:** 21

Outer and middle involucral bracts apically with five(seven) palmately spreading spines, of which central one usually considerably longer (to 30 mm) and thicker than others (2.5–4.0 mm long); inner involucral bracts with small membranous, irregularly toothed, apical appendage. Flowers yellow, all tubular, peripheral very slightly shorter than inner, with three- to five-lobed corolla. Stigma short-bilobate or undivided (its branches fused to tip). Achenes 2.5 mm long, scatteredly hairy, mature ones glabrous; pappus 5 mm long, twice as long as achene, double, inner several times shorter than outer, comprising one row of bristles. Biennial plants, with branched, narrow-winged stems and entire, undivided or (lower) lyrately incised, arachnoid-hairy-tomentose, decurrent (cauline) leaves.

Type: *C. solstitialis* L.

75. **C. solstitialis** L. 1753, Sp. Pl.: 917; Czer. 1963, Fl. SSSR, **28:** 571.—*C. solstitialis* subsp. *solstitialis* Dostál, 1976, Fl. Europ. **4:** 284.—(Plate XXXII, 1).

Type: Western Europe ("in Gallia, Anglia, Italia").

Baltic (Estonia: ecdemic); *Center* (Upper Dnieper: vicinity of Minsk; Volga-Kama: Kirov Region; Volga-Don: Kaluga Region, throughout ecdemic); *West* (Dnieper: vicinity of the city of Kharkov, ecdemic; Moldavia; Black Sea: western half); *Crimea*.—On dry rubbly and clayey slopes, in forest glades in foothills and lower mountains, as weed in pastures, fields, weedy places.—*General distribution*: Caucasus, Far East (Vladivostok, ecdemic), Russian Central Asia; Scandinavia (Finland, ecdemic), Central and Atlantic Europe (ecdemic), Mediterranean, Asia Minor, Iran; North America (ecdemic).—2n = 16.

76. **C. adamii** Willd. 1803, Sp. Pl. **3**, 3: 2310; Czer. 1963, Fl. SSSR, **28**: 573.—*C. solstitialis* L. subsp. *adamii* (Willd.) Nym. 1879, Consp.: 430; Dostál, 1976, Fl. Europ. **4**: 284; O. Baran. and others, 1992, Konsp. Fl. Udmurt.: 98.

Type: Georgia ("in Iberia").

Center (Volga-Kama: Izhevsk. ecdemic); *West* (Moldavia; Black Sea: western Odessa Region); *Crimea*.—On dry slopes, in steppes, weedy places.—*General distribution*: Caucasus; Central Europe (extreme southeast), Mediterranean (east).

SUBGENUS **11. *CALCITRAPA*** (Heist. ex Fabr.) Spach 1841, Hist. Nat. Vég. (Phan.), **10**: 11, 72.—*Calcitrapa* Heist. ex Fabr. 1759, Enum. Meth. Pl.: 94

Outer and middle involucral bracts apically with (three)five to seven palmately spreading spines, of which central considerably longer (10–30 mm) and thicker than others (3–5 mm long); inner involucral bracts with small membranous, slightly unevenly toothed appendage. Flowers pinkish-violet, all tubular, inner scarcely enlarged, with five- or six-lobed corolla. Stigma somewhat bilobate. Achenes 3–4 mm long, scatteredly hairy, mature subglabrous; pappus absent or half to two-thirds as long as achene, double, 288 inner one-fourth to one-third as long as outer, comprising one row of bristles. Biennial plants, with strongly divaricately branched, nonwinged stems; leaves pinnately incised, lanate, scabrous-lanate, more or less scabrous, occasionally subglabrous, decurrent (cauline).

Type: *C. calcitrapa* L.

77. **C. iberica** Trev. 1826, in Spreng. Syst. Veg. **3**: 406; Czer. 1963, Fl. SSSR, **28**: 574; Dostál, 1976, Fl. Europ. **4**: 282.

Type: Armenia ("Armenia, Caucas.").

Baltic (ecdemic); *Center* (Upper Dnieper; Upper Volga: vicinity of Serpukhov; Volga-Kama: vicinity of Izhevsk: throughout ecdemic); *West* (Moldavia: southwest); *Crimea* (south).—In weedy places, pastures, abandoned lands, gardens, by roadsides, in foothills and lower mountains.—*General distribution*: Caucasus, Russian Central Asia; Central Europe (Romania), Mediterranean (Balkan Peninsula), Asia Minor, Iran, Dzhungaria-Kashgaria, Himalayas.—2n = 16.

78. **C. calcitrapa** L. 1753, Sp. Pl.: 917; Czer. 1963, Fl. SSSR, **28**: 575; Dostál, 1976, Fl. Europ. **4**: 282.—(Plate XXXII, 2).

Type: Western Europe ("in Helvetia, Anglia, et Europa australiori secus vias").

Baltic (Latvia: cities of Riga, Daugavpils, ecdemic); *Crimea* (south).—On dry slopes, in weedy places, by roadsides, predominantly in foothills.—*General distribution*: Scandinavia (Finland: ecdemic), Central Europe (south), Atlantic Europe (ecdemic), Mediterranean, Asia Minor; North America (ecdemic).—2n = 20.

GENUS **111.** *HYALEA* (DC.) Jaub. et Spach[1]
1847, Ill. Pl. Or. **3**: 19, p. p.—*Centaurea* L. sect. *Hyalea* DC. 1838, Prodr. **6**: 565

Heads heterogamous, few-flowered, rather small. Involucre oblong; involucral bracts many-rowed, thin-coriaceous, glabrous, with almost entire (except thicker and usually darker central part) membranous appendage, strongly convex on outer side, entire (often lacerate) decurrent on bract; appendages of outer and middle bracts broadly semilunar to oval, innermost as wide membranous border. Flowers pinkish or whitish with weakly pinkish tinge, heteromorphic: inner bisexual, peripheral sterile, not or very slightly enlarged. Achenes compressed laterally, narrowed toward base, without distinct longitudinal riblets, uniformly covered with sparse and very short hairs, later subglabrous, lower areola of attachment (hilum) lateral; pappus as long as achene or slightly longer, persistent, double: outer three to three and one-half times as long as inner, many-rowed, of weakly scabrous bristles; inner of eight broadly linear, upward directed, dorsally sparsely hairy scales, terminating in tuft of also upward directed, rather long hairs. Annual plants, with branched stem and denticulate (basal often somewhat lobate) leaves not decurrent on stem.

Type: *H. pulchella* (Ledeb.) C. Koch.

418

Two species of this genus are distributed in the Altai Region, western areas of China up to western Asia. One ecdemic species is found in the territory of our "Flora".

1. **H. pulchella** (Ledeb.) C. Koch, 1851, Linnaea, **24:** 418; Tzvel. 1963, Fl. SSSR, **28:** 369; Safonov, 1974, Bot. Zhurn. **59,** 7: 1019; id. 1982, Byull. Glavn. Bot. Sada Akad. Nauk SSSR, **124:** 49.—*Centaurea pulchella* Ledeb. 1829, Ic. Pl. Fl. Ross. **1:** 22; id. 1833, Fl. Alt. **4:** 47.

Stems to 70–80 cm high, more or less arachnoid-tomentose in lower part, subglabrous above, branched in upper part, less often almost from base, with slender, whitish, somewhat lustrous branches. Involucre 10–12 mm long and 4–6 mm in dia. Achenes 3 mm long; pappus 3.0–3.6 mm long, whitish.

Type: Russian Central Asia ("in subsalsis arenosis deserti soongoro-Kirghisici").

East (Lower Volga: Cis-Volga sands, ecdemic).—On turfaceous sands.—*General distribution*: Caucasus, Russian Central Asia; Asia Minor, Iran, Dzhungaria-Kashgaria.—2n = 18.

GENUS 112. *CRUPINA* (Pers.) Cass.[1]
1818, Dict. Sci. Nat. **12:** 67

Heads heterogamous, few-flowered, medium. Involucre cylindrical; involucral bracts few-rowed, herbaceous, punctate-glandular outside, without appendage. Flowers purple, heteromorphic, inner bisexual, peripheral sterile, not enlarged. Achenes not compressed from sides, smooth, appressed-sericeous, hilum basal, straight; pappus one and one-half times as long as achene, persistent, double, outer many-rowed, comprising serrate scabrous bristles, inner one-rowed, considerably shorter than outer, of 5–10 small scales. Annual and biennial plants, with branched stems and pinnatipartite (basal undivided) leaves not decurrent on stem.

Type: *C. vulgaris* Cass.

Three or four species of this genus are distributed in southern Europe, in south of Russian Central Asia, North Africa and West Asia (in the east to the Caucasus and Iran).

1. **C. vulgaris** Cass. 1818, Dict. Sci. Nat. **12:** 68; Czer. 1963, Fl. SSSR, **28:** 254; Franco, 1976, Fl. Europ. **4:** 301.—*Centaurea crupina* L. 1753, Sp. Pl.: 909.

---

[1]Treatment by S.K. Czerepanov.

Stem 20–60(80) cm high, pubescent in lower part from whitish plumose thin short hairs. Involucre (12)14–18(20) mm long and 4–6(8) mm in dia. Achenes short-cylindrical, 3.5–6.0 mm long and 2–3 mm wide; pappus 5–8 mm long, blackish-brown.

Type: Southern Europe ("in Hetruria").

West (Moldavia; Black Sea); *Crimea.*—On dry herbaceous and stony slopes, rocks, taluses, in scrublands, sometimes on sands and in abandoned lands.—*General distribution*: Caucasus, Russian Central Asia (south); Central and Atlantic Europe (south), Mediterranean, Asia Minor, Iran.—2n = 30.

GENUS 113. *AMBERBOA* (Pers.) Less.[1]

1832, Syn. Compos.: 8, nom. conserv.—*Centaurea* L. *Amberboa Pers. 1821, Sp. Pl. **5**: 316, p. p.

Heads heterogamous, many-flowered, medium-signed. Involucre elongated-ovate or ovate; involucral bracts many-rowed, subcoriaceous, outer and middle without appendage, obtuse, inner with thin-membranous, small, lanceolate, entire appendage. Flowers light yellow or yellowish white, heteromorphic, inner bisexual, peripheral sterile, somewhat elongated (slightly exceeding inner). Achenes weakly compressed from sides, smooth, rather densely appressed-hairy, hilum lateral; pappus scarcely shorter than achene, persistent, simple, many-rowed, comprising serrate-dentate bristles. Annual plants, with simple or weakly branched stems and pinnatilobate or pinnatipartite (basal, usually undivided) leaves not decurrent on stem.

Type: *A. moschata* (L.) DC. (typ. cons.).

Seven species of this genus are distributed in the Caucasus, south of Western Siberia and Asia Minor, Russian Central Asia and Central Asia (in the east to Dzhungaria).

290   1. **A. turanica** Iljin, 1932, Izv. Bot. Sada Akad. Nauk SSSR, **30**, 1–2: 110; Tzvel. 1963, Fl. SSSR, **28**: 330; Dostál, 1976, Fl. Europ. **4**: 253.

Stem 2–50 cm high, more or less crisped-hairy. Involucre 10–16 mm long and 5–12 mm in dia; achenes oblong, 4.0–5.5 mm long.

Type: Western Kazakhstan ("inter fl. Emba et Ustj-Urt. mons Ak-tau").

---

[1]Treatment by S.K. Czerepanov.

420

East (Lower Volga).—On clayey and stony slopes, gravel-beds, sands and solonetzes.—*General distribution*: Caucasus (eastern Transcaucasia), Western Siberia (south), Russian Central Asia; Iran, Dzhungaria-Kashgaria.

GENUS 114. *CHARTOLEPIS* Cass.[1]
1826, Dict. Sci. Nat. **44**: 36

Heads heterogamous, many-flowered, medium-sized. Involucre oblong-ovate; involucral bracts many-rowed, coriaceous, with thin membranous, rather large, suborbicular to ovate, erose-toothed, in most part lacerate appendage. Flowers yellow, heteromorphic, inner bisexual, peripheral sterile, not enlarged. Achenes weakly compressed from sides, smooth, scatteredly hairy, hilum lateral; pappus almost one and one-half times as long as achene, persistent, double, outer many-rowed, comprising plumose bristles, connate at base into ring, inner one-rowed, considerably shorter than outer, of scalelike bristles unevenly ciliate above. Perennial plants, with branched, broad-winged stems and entire (lower sometimes indistinctly toothed) leaves long decurrent on stem.

Type: *C. glastifolia* (L.) Cass.
Seven species of this genus are distributed in the south of the European part of Russia and Western Siberia, Caucasus, Russian Central Asia, Asia Minor, and Central Asia (in the east to Mongolian Altai).

1. **C. intermedia** Boiss. 1856, Diagn. Pl. Or., ser. 2, **3**: 64; Czer. 1963, Fl. SSSR, **28**: 337.—*C. glastifolia* auct. non Cass.: Dostál, 1976, Fl. Europ. **4**: 301.—(Plate XXXI, 2).
Stem 30–80(100) cm high, weakly arachnoid-hairy, less often subglabrous, sparsely covered with sessile golden glands. Involucre 22–25 mm long and 10–15 mm in dia. Achenes oblong-elliptical or oblong, 5.5–6.0 mm long and about 2.3 mm wide, brownish-cream; pappus 8–10 mm long, somewhat ochreous.

Type: European part of Russia, Siberia, Dzhungaria ("in Rossia meridionali! Siberia! Songaria!).
Center (Volga-Kama: extreme southeast; Volga-Don: south); *West* (Dnieper: southeast; Moldavia; Black Sea); *East*.—In solonetzic wet meadows and meadow solonchaks, in river valleys and on lake shores, on herbaceous slopes.—*General distribution*:

[1]Treatment by S.K. Czerepanov.

Western Siberia (south), Russian Central Asia (north and east); Dzhungaria-Kashgaria (Kuldzha), Mongolia (Mongolian Altai).— $2n = 36$.

GENUS 115. *CARTHAMUS* L.[1]
1753, Sp. Pl.: 830; id. 1754, Gen. Pl., ed. 5: 361

291 Heads 15–30 mm in dia, homogamous, with bisexual flowers, one to several at apices of stem and its branches. Involucre 12–30 mm in dia and 20–30 mm long, ovate to goblet-shaped; involucral bracts imbricate, in several rows, unequal, outer foliaceous, usually spiny-toothed, often exceeding flowers, others lanceolate-ovate to lanceolate, spinescent or some with spiny toothed appendage. Receptacle flat, densely covered with flattened bristles. Flowers more or less homomorphic or peripheral somewhat enlarged, their corollas yellow, orange, or pink, five-lobed. Achenes 3–6 mm long and 3–4 mm wide, broadly conical, obtusely-tetragonous, glabrous, smooth or more or less rugose, whitish, with lateral hilum; pappus 4–9 mm long, absent in peripheral and sometimes in all flowers, persistent, of several rows of strongly flattened, more or less scabrous bristles, of which outer shorter, innermost also often shorter. Annual or biennial plants, with erect, more or less branched stems; leaves alternate, undivided to pinnatipartite, subcoriaceous, decurrent, spiny on margin or without spines.

Lectotype: *C. tinctorius* L.

About 15 species in southern Europe, northern Africa, Russian Central Asia, and southwest Asia. One species, *C. tinctorius* L., is cultivated in some other countries.

1. Cauline leaves undivided or with very short and remote spiny teeth, oblong-ovate to lanceolate, entire or scarcely toothed; flowers orange-yellow; pappus absent or of fewer strongly flattened bristles, shorter than achenes .... 3. **C. tinctorius.**
+ Cauline leaves pinnatipartite to coarsely toothed, occasionally, in part, undivided, lanceolate, more or less hairy, grayish- or glaucous-green; outer involucral bracts lanceolate or broadly lanceolate, spiny-toothed; pappus always present, one and one-half to three times as long as achene .... 2.

[1]Treatment by N.N. Tzvelev.

2. Flowers pink; heads terminal on stem and branches, usually in one to three, less often solitary. Plant glaucous-green, puberulent .... 1. **C. glaucus.**

  + Flowers yellow; heads solitary. Plant grayish-green, more or less arachnoid-hairy .... 2. **C. lanatus.**

Section 1. Lepidopappus Hanelt, 1963, Feddes Repert. **67,** 1–3: 99.

Flowers pink; heads clustered; achenes with pappus.

Type: *C. glaucus* Bieb.

1. **C. glaucus** Bieb., 1798, Tabl. Prov. Casp.: 58; Shostakovsky, 1963, Fl. SSSR, **28:** 584; Hanelt, 1976, Fl. Europ. **4:** 302.

Type: Eastern Carpathians.

Crimea (vicinity of Sevastopol, Balaklava, Alupka, Gurzuf).— On stony and clayey slopes, in habitations, by roadsides, in plantations of various crops.—*General distribution*: Caucasus; Asia Minor, Iran, Africa (northeast).

Section 2. Kentrophyllum (DC.) Baill. 1882, Hist. Pl.: 86, p. p.— *Kentrophyllum* DC. 1810, Ann. Mus. Hist. Nat. (Paris), **16:** 158.

Flowers yellow; heads solitary, with spiny-toothed outer involucral bracts; achenes with pappus.

Type: *C. lanatus* L.

2. **C. lanatus** L. 1753, Sp. Pl.: 830; Shostakovsky, 1963, Fl. SSSR, **28:** 583; Hanelt, 1976, Fl. Europ. **4:** 303; Schultz, 1977, Bot. Zhurn. **62,** 10: 1517; Tretyakov, 1988, Bot. Zhurn. **73,** 6: 904.

Type: Europe ("in Gallia, Italia, Creta").

Baltic (ecdemic in Riga); *Center* (Upper Dnieper: ecdemic in Gomal Region); *West* (Moldavia; Black Sea); *East* (Lower Don: south and Ergeni); *Crimea.*—On stony and clayey slopes, in plantations of various crops, by roadsides, in habitations.—*General distribution*: Caucasus, Russian Central Asia; Atlantic and Central Europe (south), Mediterranean, Asia Minor, Iran.—2n = 44.

292 **Section 3. Carthamus.**

Flowers orange-yellow; heads solitary, with entire or subentire involucral bracts; achenes usually without pappus.

Type: lectotype of genus.

3. **C. tinctorius** L. 1753, Sp. Pl.: 830; Shostakovsky, 1963, Fl. SSSR, **28:** 586; Hanelt, 1976, Fl. Europ. **4:** 302.

Type: Egypt ("in Aegypto").

West (Moldavia; Black Sea); *East* (Lower Don: south; Lower Volga); *Crimea*.—Occasionally cultivated as oil and dye plant, sometimes as an escape, found by roadsides, in habitations.— *General distribution*: Caucasus, Russian Central Asia; Central Europe (south), Mediterranean, Asia Minor, Iran, Japan-China, southern Asia; not known in wild.—2n = 24.

GENUS 116. *CNICUS* L.[1]
1753, Sp. Pl.: 826, nom. conserv.; id. 1754, Gen. Pl., ed. 5: 358

Heads 10–30 mm in dia, with sterile peripheral and bisexual inner flowers, solitary at apices of stem and its branches, surrounded by subtending leaves. Involucre goblet-shaped, 8–25 mm in dia and 15–30 mm long; involucral bracts imbricate, in several rows, unequal, lanceolate-ovate to lanceolate, outer with apical spine, middle and inner with sideways deflected, stiff-ciliate, apical appendage, spiny on margin. Receptacle weakly convex, densely covered with bristles. Peripheral flowers sterile, filiform, with two- or three-lobed yellow corolla, others bisexual, with yellow bilabiate corolla, its limb consisting of entire, narrow or four-toothed, broader lobes. Achenes 6–9 mm long, 2.1–3.0 mm wide, almost not compressed, glabrous, usually more or less brownish, with distinct longitudinal ribs, apically with toothed crown 0.5–0.8 mm long; pappus double, comprising flattened bristles, outer with (9)10(12) bristles 10–12 mm long, inner with 10 bristles 2–3 mm long. Annual plants, with erect, usually strongly branched (often from base) stem; leaves pinnately divided or undivided, more or less toothed, with stiff cilia on margin, to 1.2 mm long, somewhat prickly, more or less arachnoid-hairy and with scattered crisped hairs, lower with narrowly winged petioles, middle and upper sessile, uppermost subtending heads, ovate to broadly lanceolate, usually undivided, spiny-ciliate on margin, with spines to 2 mm long.

Type: *C. benedictus* L.
A monotypic genus.

1. **C. benedictus** L. 1753, Sp. Pl.: 826; Iljin, 1963, Fl. SSSR, **28:** 587; Franco, 1976, Fl. Europ. **4:** 301.

Type: Southern Europe ("in Chio, Lemno, Hispania").
West (Black Sea); *East* (Lower Don); *Crimea*.—Cultivated as ornamental and oleiferous plant, sometimes as escape.—*General distribution*: Caucasus, Russian Central Asia; Mediterranean, Asia Minor, Iran, Dzhungaria-Kashgaria, Himalayas.—2n = 22.

---
[1]Treatment by N.N. Tzvelev.

Addendum[1]

GENUS *SPHAERANTHUS* L.
1753, Sp. Pl.: 927; id. 1754, Gen. Pl., ed. 5: 399 (Inuleae)

Small and few-flowered heads with peripheral pistillate and inner bisexual flowers, clustered in very dense general inflorescence—glomerules, at apices of tem and its branches. Involucre comprising four or more numerous herbaceous bracts in one or two rows. Corolla small, often almost not exserted from involucre, whitish or yellowish, in peripheral flowers often undeveloped, in bisexual usually four-toothed. Ovary glabrous or short-hairy; achenes without or with pappus of few bristles. Annual or perennial herbaceous plants, with erect stem and alternate, undivided, less often pinnatilobate leaves.

Type: *S. indicus* L.

About 50 species in the tropics and partly subtropics of the Old World, and in the Volga Delta.

Literature: Barmin, A.N., V.N. Pilipenko, V.V. Sinyakina and N.N. Tzvelev. 1991. *Sphaeranthus* (Asteraceae)—novyi rod flory SSSR [*Sphaeranthus* (Asteraceae)—a new genus in the flora of the USSR]. *Bot. Zhurn.*, **76**, 12: 1768–1771.

SUBGENUS 1. *PSEUDOSPHAERANTHUS* Robyns
1924, Kew Bull. **5**: 179

Heads with four to six flowers and four or five involucral bracts of various structure, of which outer larger and often with cusp.

Lectotype: *S. amaranthoides* Burm.

Section 1. Cuspidella DC. 1836, Prodr. **5**: 370. Heads clustered in very dense ovoid glomerules. Lectotype; *S. angustifolius* DC.

1. **S. volgensis** Tzvel. 1991, Bot. Zhurn. **76**, 12: 1770, with fig.

---

[1]Treatment by N.N. Tzvelev.

Leaves linear-lanceolate or laneolate, more or less toothed or entire, decurrent on stem and forming rather broad wings. Capitula numerous, ovate, very dense, purple, very short and scatteredly hairy. Outer of four involucral bracts, obovate, cuspidate, cusp 0.5–1.2 mm long, two lateral carinate, inner obovate. Of the six flowers in a head two inner bisexual, with four-toothed whitish corolla and glabrous ovary with sessile glands, four outer pistillate with almost filiform corolla and short-hairy ovary; achenes without pappus.

Type: "Astrakhan Region, Volga Delta, Volodarsk District, wet meadow, 2–3 km northeast of the town of Yamnoe, VIII.1990, A. Barmin, V. Pilipenko, V. Sinyakina" (LE).

East (Lower Volga: Volga Delta).—In wet inundated meadows, on banks of water bodies and swamps.—Endemic.

Note. A more complete description of the genus and the new species is available in the above-cited work. The species is very close to *S. strobiliferus* Boiss. et Noë (1836, in Boiss. *Diagn. Pl. Or.*, ser. 2, **3**: 6), described from Iraq, but differs by having smaller and weakly hairy capitula.

ON THE GENERA *GNAPHALIUM* L. AND *FILAGINELLA* OPIZ.

For the genus *Gnaphalium* L., Holub (1976, *Fl. Europ.* **4**: 128), according to rules of priority, considered *G. luteo-album* L. as the lectotype and for species of the *G. uliginosum* L. group proposed the generic name *Filaginella* Opiz. However, Jeffrey (1979, Taxon, **28**, 4: 350) proposed *G. uliginosum* as lectotype of the genus *Gnaphalium*, and with this lectotype the genus name was proposed to be conserved. Therefore, the name of the genus and species used earlier in this volume of *Flora* must be changed as following:

294   GENUS **48.** *LAPHANGIUM* (Hill. et Burtt) Tzvel. 1993, Byull. Mosk. Obshch. Ispyt. Přir., Otd. Biol. **98**, 6: 105.— *Gnaphalium* subgen. *Laphangium* Hill. et Burtt, 1981, Bot. Journ. Linn. Soc. (London), **82**, 3: 205

1. **L. luteo-album** (L.) Tzvel. 1993, loc. cit.—*Gnaphalium luteo-album* L. 1753, Sp. Pl.: 851 (Typus generis).

GENUS **50.** *GNAPHALIUM* L.

1. **G. rossicum** Kirp.

2. **G. kasachstanicum** Kirp.

3. **G. pilulare** Wahlenb.

4. **G. uliginosum** L.

NEW SPECIES FOR "FLORA OF RUSSIA...."

1. **Rudbeckia bicolor** Nutt. 1834, Journ. Acad. Philad. 7: 81; O. Baran. and others, 1992, Konsp. Fl. Udmurt.: 105.

The North American species of *R. hirta* L. group differs from *R. hirta* s. str. by having narrower and entire or scarcely serrate (and not coarsely toothed) leaves. Reported as ecdemic in Izhevsk. However, a large part of reports of *R. hirta* s. 1. for eastern Europe [European part of Russia] relates to it. Many later authors combine it with *R. hirta* s. str.

2. **Coreopsis atkinsiana** Dougl. ex Lindl. 1830, Bot. Beg. **16:** tab. 1376; O. Baran. and others, 1992, op. cit.: 100.

The North American species, reported as ecdemic in Izhevsk. Closer to *C. tinctoria* but taller and with smooth (not tuberculate) achenes narrow-winged on sides.

3. **Bidens bipinnata** L. 1753, Sp. Pl.: 832; Vass. 1959, Fl. SSSR, **25:** 561; O. Baran. and others, 1992, op. cit.: 97; Papchenkov, 1993, Bot. Zhurn. **78, 9:** 74.

The North American species, found as ecdemic plant in Udmurtia (Izhevsk, Mozhga) and in Tatarstan (railroad station of Sviyazhsk).

4. **Tagetes minuta** L. 1753, Sp. Pl.: 887; Gorschk. 1959, Fl. SSSR, **25:** 561; O. Baran. and others, 1992, op. cit.: 108.

American species ecdemic in Udmurtia (Izhevsk, Glazev, Sarapul).

5. **Helichrysum nogaicum** Tzvel. 1993, Byull. Mosk. Obshch. Ispyt. Prir., Otd. Biol. **98, 6:** 103.

This species replaces *H. arenarium* on sands of eastern Ciscaucasia and northern Cis-Caspian Region (including the Lower Volga Region of our *Flora*). It differs by having a more lax general inflorescence and pale yellowish (and not yellow) involucres.

6. **Pyrethrum coccineum** (Willd.) Worosch. 1954, Spisok. Sem. Glavn. Bot. Sada Akad. Nauk SSSR, **9:** 21; Tzvel. 1961, Fl. SSSR, **26:** 218; O. Baran. and others, 1992, op. cit.: 105.

Together with the closely related species *P. roseum* (Adam) Bieb., it is sometimes grown in gardens and experimental fields as ornamental and insecticidal (Florist's pyrethrum) plant. Reported from Izhevsk as an escape of grassy patches.

7. **Artemisia canadensis** Michx. 1803, Fl. Bor.-Amer. **2**: 129; Kondratyuk and others, 1985, Konsp. Fl. Yugo-Vost. Ukr.: 174; Mosyak. 1991, Ukr. Bot. Zhurn. **48**, 4: 32.

Psamophilic species, ecdemic in vicinity of Mariopol. It differs from the relatively closer *A. scoparia* Waldst. et Kit. by having hemispherical heads in more dense panicles and entirely membranous inner involucral bracts. From *A. campestris* L., it is differentiated by smaller heads.

295   8. **A. feddei** Levl. et Vaniot. 1910, Feddes Repert. **8**: 138; Korobkov, 1992, Sosud. Rast. Sov. Daln. Vost. **6**: 134; O. Baran. and others, 1992, op. cit.: 95.—*A. lavandulifolia* auct. non DC.: Poljak. 1961, Fl. SSSR, **26**: 453.

An eastern Asiatic species ecdemic in Izhevsk.

9. **A. oelandica** (Bess.) Krasch. 1946, Mat. Ist. Fl. i Rastit. SSSR, **2**: 126.—*A. punctata* Bess. var. *oelandica* Bess. 1832, Nouv. Mem. Soc. Nat. Moscou, **3**: 44.

This species, earlier considered endemic in Åland in Sweden, was found in 1993 by N.N. Tzvelev in the vicinity of Luga in Leningrad [St. Petersburg] Region, on the left bank of the Ordezh River near its mouth, in thinned-out pine forest with glades, near the foothill of a large hillock. Found in abundance and decidedly in wild, but flowers very rarely. It occupies an intermediate position between *A. laciniata* Willd. and *A. tanacetifolia*, differing from the former by having broader leaf lobes (especially of basal leaves) and from the latter by smaller heads.

10. **A. opulenta** Pamp. 1930, Nouv. Giorn. Bot. Ital. N.S. 36: 464; Poljak. 1961, Fl. SSSR, **26**: 439; O. Baran and others, 1992, op. cit.: 96.

Eastern asiatic species, ecdemic in Udmurtia (railroad station of Tukhtyan).

11. **A. serotina** Bunge, 1854, Mem. Pres. Acad. Sci. Petersb. Div. Sav. **7**: 341; Poljak. 1961, op. cit.: 608; O. Baran. and others, 1992, op. cit.: 96.

Central Asian species, ecdemic in Udmurtia near railroad tracks in the vicinity of Izhevsk, Votkinsk, Uva).

12. **Erigeron macrophyllus** Herbich, 1853, Strip. Rar. Bucov.: 57, s. str.—*E. acris* L. subsp. *macrophyllus* (Herbich) Guterm. 1973, Phyton (Austria), **15**: 268; Halliday, 1976, Fl. Europ. **4**: 118, p. p.

Carpathian species, also penetrating the Ukrainian Carpathians. Very close to *E. podolicus* Bess., differing by subglabrous stem and leaves and from *E. droebachiensis* C.F. Muell. by having cauline leaves that somewhat but gradually (and not abruptly) reducing upward. Also found in the Caucasus mountains.

13. **Centaurea montana** L. 1753, Sp. Pl.: 911; Kurtto, 1984, Retkeilykasvio: 374.

Perennial soboliferous species of the section *Protocyanus* Dobrocz., with large violet-blue heads and decurrent leaves. Distributed in Central Europe, but cultivated considerably widely; moreover, in Finland, it has a tendency for naturalization (Kurtto, loc. cit.). Found by N.N. Tzvelev as escape, in large numbers, in the forest on the coast of Vyborg Gulf in the vicinity of Vyborg (Leningrad Region).

14. **C. nigra** L. 1753, Sp. Pl.: 911; Kurtto, 1984, loc. cit.: 374.

Closer to *C. phrygia* L., but almost black from very large involucral bracts with their noncaudate appendages completely covering the green part of involucral bracts. Heads often densely clustered. Known as ecdemic or introduced plant in many places in Finland, including on border with Vyborg District of Leningrad Region (Kurtto, loc. cit.); it was found by N.N. Tzvelev in the meadow in Zapadnyi Berezovyi island in vicinity of Vyborg. Distributed predominantly in Central Europe.

Alphabetic Index of Latin Names of Plants[1],*

---

[1]Names of families and genera are printed in boldface, synonyms in italics. An asterisk sign with the name of a species or subspecies indicates illustration.

*Reproduced from the Russian original. Page numbers of Russian original appear in the left-hand margin in the text—General Editor.

436

444

Erigeron *acer* subsp. *macrophyllus* auct. 200
— — subsp. *politus* (Fries) Lindb. fil. 202
— — var. *elongatus* (Ledeb.) Mela et Cajand. 202
— acris* L. 198, 199, 200
— — subsp. *macrophyllus* (Herbich) Guterm.
— alpinus L. 197, 202
— — var. *eriocalyx* Ledeb. 202
— angulosus Gaud. 198, 202
— *annuus* (L.) Pers. 204
— *annuus* auct. 204
— — subsp. *septentrionalis* (Fern. et Wieg.) Wagenitz 204
— — subsp. *strigosus* (Muehl. ex Willd.) Wagenitz 204
— *asteroides* Andrz. ex Bess. 200
— borealis* (Vierh.) Simm. 197, 202
— *brachycephalus* Lindb.fil. 200
— *canadensis* L. 205
— cannabinum L.
— *ciliatus* Ledeb. 195
— *consanguineus* (Ledeb.) Novopokr. 199
— *decoloratus* Lindb.fil. 200
— × decoloratus Lindb.fil. 200
— droebachiensis O.F. Mull. 198, 200
— *elongatiformis* Serg. 200
— Erigeron *elongatus* Ledeb. 202
— eriocalyx (Ledeb.) Vierh. 198, 202
— eriocephalus Vahl. 198, 203
— *glaucus* Ker-Gawl. 199
— komarovii Botsch. 197, 199
— macrophyllus Herbich 295
— muirii A. Gray 199
— *muirii* auct. 199
— orientalis Boiss. 198, 199
— podolicus Bess. 198, 200
— politus Fries 198, 200, 202
— *politus* auct. 200
— *pseudoannuus* Makino 204
— *ramosus* (Walt.) Britt., Sterns et Pogg. var. *septentrionalis* (Fern. et Wieg.) Wagenitz 204
— silenifolius (Turcz.) Botsch. 197, 199

Erigeron speciosus (Lindl.) DC. 197, 199
— *strigosus* Muehl. ex Willd. 204
— uniflorus L. 197, 198, 202
— *uniflorus* auct. 202
— — subsp. *eriocalyx* (Ledeb.) A. et D. Löve 202
— — subsp. *eriocephalus* (Vahl) Cronq. 203
— uralensis Less. 198, 200
Eriolepis Cass. 238
Eupatorieae Cass. 206
Eupatorium L. 12, 206
— cannabinum L. 206
Evax Gaertn. sect. *Filaginoides* Smoljian. 104
— *filaginoides* Kar. et Kir. 105

Filaginella Opiz 21, 97, 100
— kasachstanica (Kirp.) Tzvel. 101
— pilularis (Wahlenb.) Tzvel. 101, 102
— rossica (Kirp.) Tzvel. 101
— uliginosa* (L.) Opiz 101, 102
— — subsp. *kasachstanica* (Kirp.) Holub 101
— — subsp. *rossica* (Kirp.) Holub 101
— — subsp. *sibiricum* (Kirp.) Holub 102
Filago 15, 21, 103
— sect. Filaginoides (Smoljian.) Wagenitz 104
— sect. Filago 104
— *arvensis* L. 103
— *filaginoides* (Kar. et Kir.) Wagenitz 104, 105
— *gallica* L. 102
— *germanica* Huds. 104
— *germanica* L. 104
— *minima* (Smith) Pers. 103
— *montana* L. 103
— *montana* auct. 103
— pyramidata* L. 104
— *spathulata* C. Presl 104
— vulgaris* Lam. 104

Gaillardia Foug. 14, 43
— aristata Pursh 43
— × hybrida Lort. 43

Senecio jacquinianus Reichenb.*
55, 61
— *jailicola* Juz. 67
— *kirghisicus* DC. 62
— nemorensis* L. 55, 61
— *nemorensis* L. 61
— — subsp. *fuchsii* (C.C. Gmel.)
Čelak. 62
— — subsp. *nemorensis* 61
— — var. *polyglossus* Rupr. 61
— noeanus Rupr.* 53, 57
— *octoglossus* DC. 61
— ovatus (Gaertn., Mey. et
Scherb.) Willd.* 55, 61
— paludosus L.* 55, 63
— *papposus* (Reichenb.) Less. 67
— *papposus* auct. 68
— paucifolius S.G. Gmel.* 54, 62
— *praealtus* Bertol. B.
borysthenicus D.C. 57
— *resedifolius* Less. 65
— *rivularis* (Waldst. et Kit.) DC.
67
— *sarracenicus* L. 62
— *sarracenicus* auct. 62
— schvetzovii Korsch. 54, 62
— *scopolii* Hoppe et Hornsch. ex
Bluff et Fingerh. 63
— subalpinus Koch* 54, 59
— sylvaticus* L. 53, 55
— tataricus Less.* 55, 63
— tauricus Konechn.* 54, 59
— *tenuifolius* Jacq. 61
— *tundricola* Tolm. 67
— umbrosus Waldst. et Kit. 54,
62
— vernalis Waldst. et Kit. 53, 55
— viscosus L.* 53, 55
— vulgaris L.* 53, 54
— *vulgaris* auct. 55
Senecioneae Cass. 52
Seriphidium (Bess.) Poljak. 169
— *compactum* (Fisch. ex DC.)
Poljak. 173
— *fragrans* (Willd.) Poljak. 169
— *gracilescens* (Krasch. et Iljin)
Poljak. 173
— *lerchianum* (Web. ex Stechm.)
Poljak. 169
— *lessinglanum* (Bess.) Poljak.
173
— *maritimum* (L.) Poljak. 169

Seriphidium *monogynum* (Waldst.
et Kit.) Poljak. 171
— *nitrosum* (Web. ex Stechm.)
Poljak. 173
— *pauciflorum* (Web. ex
Stechm.) Poljak. 173
— *tauricum* (Willd.) Poljak. 171
— *terrae-albae* (Krasch.) Poljak.
173
Serratula L. 21, 251
— sect. Klasea (Cass.) DC. 253
— sect. Mastrucium (Cass.) DC.
252
— sect. Piptochaete Boiss. 253
— sect. Serratula 252
— *alata* S.G. Gmel. 243
— *alpina* L. 220
— *amara* L. 218
— *arvensis* L. 246
— *bracteifolia* (Iljin ex Grossh.)
Stank. 253
— bulgarica* Acht. et Stojan.
252, 254
— *caput-najae* Zahar. 254
— cardunculus (Pall.) Schischk.
252, 254
— — var. bogdensis Iljin 256
— *caspia* Pall. 90
— *centauroides* L. 253
— *coronata* L. 251, 253
— *discolor* Willd. 220
— donetzica* Dubovik 252, 254
— *erucifolia* (L.) Boriss. 251,
253
— *gmelinii* Tausch 252, 253
— *heterophylla* (L.) Desf. 254
— *heterophylla* auct. 254
— *hungarica* Klok. ex Dobrocz.
253
— *incana* S.G. Gmel. 244
— *inermis* Gilib. 252
— *isophylla* Claus 253
— lycopifolia* (Vill.) A. Kerner
252, 254
— *multiflora* L. 226
— *nitida* Fisch. ex Spreng. 254
— *parviflora* Poir. 220
— radiata (Waldst. et Kit.) Bieb.
252, 253
— — subsp. *bracteifolia* Iljin ex
Grossh. 253